Stresses in Shells

By

AF172946

Wilhelm Flügge

Dr.-Ing., Professor of Engineering Mechanics
Stanford University

Third Printing

With 244 Figures

Springer-Verlag
Berlin Heidelberg GmbH
1966

© Springer-Verlag Berlin Heidelberg 1960
Ursprünglich erschienen bei Springer-Verlag Berlin/Heidelberg 1960
Softcover reprint of the hardcover 3rd edtion 1960
Library of Congress Catalog Card Number A 61—608

ISBN 978-3-662-28216-8   ISBN 978-3-662-29730-8 (eBook)
DOI 10.1007/978-3-662-29730-8

Titel-Nr. 0248

PREFACE

There are many ways to write a book on shells. The author might, for example, devote his attention exclusively to a special type, such as shell roofs or pressure vessels, and consider all the minor details of stress calculations and even the design. On the other hand, he might stress the mathematical side of the subject to such an extent that he virtually writes a book on differential equations under the guise of the mechanical subject. The present book has been kept away from these extremes. At first sight it may look to many people like a mathematics book, but it is hoped that the serious reader will soon see that it has been written by an engineer and for engineers.

In a theoretical subject such as this one, it is, of course, not possible to get very far with the multiplication table and elementary trigonometry alone. The mathematical prerequisites vary widely in different parts of the book, depending on the subject. In some parts ordinary differential equations with constant coefficients are all that is needed. In other sections ordinary equations with variable coefficients, product solutions of partial differential equations, the theory of complex variables, or numerical analysis will be encountered. However, the author wishes to assure his readers that nowhere in this book has an advanced mathematical tool been used just for the sake of displaying it. No matter which mathematical tool has been used, it had to be used to solve the problem at hand.

The book may be divided into four parts. Chapter 1 contains preliminary matter, and a reader sufficiently familiar with the basic definitions may omit this chapter until he finds that a real need for studying it arises.

Chapters 2 to 4 contain the membrane theory, i. e. the theory of shells whose bending rigidity may be neglected. The spectacular simplification thus obtained makes it possible to examine a wide variety of shapes and support conditions. In particular, the stress problems of tanks and shell roofs have benefited from this fact, and many examples of these applications have been included. There is, of course, a heavy penalty to be paid for the simplification, and the shortcomings of the membrane theory are pointed out at many places in these chapters. It has been considered important to show that the inadequacies of the membrane theory can be discovered by a critical inspection of the membrane solutions, without any need for first solving the bending problem—a task which often enough is out of reach of the practical engineer and even of the research worker.

Chapters 5 and 6 are devoted to the bending theory of the two most important types of shells, the circular cylinder and the general shell of revolution. It is in this field that most of the development of the last two decades has taken place. Since the solution of most problems of this category requires a rather elaborate preparation, a careful choice of subject matter had to be made; otherwise the proper balance between the simple and the complicated would have been lost. In these two chapters an attempt has been made to cover a wide variety of questions and to carry every theory to a definite end, viz. to a set of formulas giving all the stress resultants and the displacements in terms of the constants of integration and the coordinates. In many cases it has been possible to present these results in the form of a table. It has, however, mostly been left to the reader to adapt a solution to his particular case of boundary conditions.

Chapter 7 is concerned with the stability of shells. From a research man's point of view this is a rather unrewarding subject. A long struggle through the mechanics and mathematics of a problem and a tedious numerical evaluation ultimately yield a curve or only a single numerical factor in a simple formula. And, after all, there is only a rather loose correlation between the actual collapse of a shell and the buckling load obtained from a linear theory. While in some cases a large-deformation analysis has thrown light into a dark corner of our understanding, the numerical labor involved is so prohibitive that the designer cannot expect too much help from this side. In this book a choice of stability problems has been made which is considered representative of the present state of knowledge.

Some material contained in this book has been used in courses on shell theory and on shell design, which the author has been giving for many years at Stanford University. However, much of what is found in this book goes beyond the possibilities even of an elaborate university course. It has been written essentially for graduate engineers. Among these, the author has been thinking principally of two groups, namely, design engineers and stress analysts who need shell theory for their work, and research workers entering the field or working in it. For the first group, the book offers a body of well-established knowledge that will help them in most cases or may show them what can be expected of the services of a special consultant. The second group may use the book as a reference work and as a starting point for their own endeavors. The annotated bibliography should be particularly helpful in locating additional information on a specific subject.

The author owes many thanks to colleagues and former students. Professor S. TIMOSHENKO gave the first encouragement to undertake the big task, and he has followed its progress through the years with steady interest. Professor J. N. GOODIER read substantial parts of Chapters 1 and 2 at a very early stage and gave the author much good advice during his first steps of writing a book in English. After completion of the manuscript, Dr. H. L. ENGEL read it painstakingly and made many helpful suggestions. Parts of the manuscript have been read at different stages by Drs. R. E. PAULSEN, F. T. GEYLING, R. H. STIVERS, H. V. HAHNE, D. A. CONRAD, P. M. RIFLOG, F. A. LECKIE, and R. A. EISENTRAUT, and smaller parts by many more of the author's students. All of them deserve the author's sincere thanks for constructive suggestions and for checking formulas.

Los Altos, Calif., February 1960

W. Flügge

CONTENTS

Chapter 3

DIRECT STRESSES IN CYLINDRICAL SHELLS 107

Appendix
FORCES AND DEFORMATIONS IN CIRCULAR RINGS 478

TABLES

Chapter 1

GENERAL PROPERTIES OF STRESS SYSTEMS IN SHELLS

1.1 Definitions

1.1.1 Definition of a Shell

Every part of a structure, of a machine or of any other object is a three-dimensional body, however small its dimensions may be. Nevertheless, the three-dimensional theory of elasticity is not often applied when stresses in such a body are calculated. There is a simple reason for this: Every structural element is created for a certain purpose, one of the most frequent being the transmission of a force from one point to another. Cables, shafts and columns are typical examples of such elements which receive a force or a couple at one end and transmit it to the other, whereas beams and arches usually transmit loads to supports at both ends. The stress analyst does not envisage these elements as three-dimensional but rather as lines having some thickness, a kind of "physical lines" as opposed to the mathematical meaning of the word. When he wants to describe the stresses in them, he first defines a cross section and then calculates the resultant of the stresses acting in it. Instead of describing this resultant by its magnitude, direction and location in space, he usually gives its three components and its moments about three axes. These quantities, commonly known as the normal force, two (transverse) shearing forces, two bending moments, and the torque, are called the "stress resultants" in the cross section.

Not all structural elements are of the kind just described. A second large group consists of all those which are made to bound or enclose some space: walls, in the widest sense of the word, e. g., the wall of a tank, the metal hull of an airplane, or the cloth-and-rubber hull of a balloon. All these objects cannot be described by a line, but can be described by a plane or curved surface, and consequently, their stress analysis must be built on the concept of a "physical surface", a surface made of some more or less solid material, capable of transmitting loads from one part to another and of undergoing consequent deformations.

In the development of the mathematical theory of such structural elements, it has become necessary to distinguish between two types:

Plane walls are called plates, while all walls shaped to curved surfaces are called shells.

Summarizing these considerations, we may define a shell as an object which, for the purpose of stress analysis, may be considered as the materialization of a curved surface. This definition implies that the thickness of a shell is small compared with its other dimensions, but it does not require that the smallness be extreme. It also does not require that the shell be made of elastic material. The occurrence of plastic flow in a steel shell would not prevent its being a shell; a soap bubble is also a shell, although made of liquid. Even the surface of a liquid, because of the surface tension acting in it, has all the properties of a true shell and may be treated by the methods of shell theory (see p. 39).

Most shells, of course, are made of solid material, and generally in this book we shall assume that this material is elastic according to HOOKE's law.

In most cases, a shell is bounded by two curved surfaces, the faces. The thickness t of the shell may be the same everywhere, or it may vary from point to point. We define the middle surface of such a shell as the surface which passes midway between the two faces. If we know the shape of the middle surface and the thickness of the shell for every one of its points, then the shell is geometrically fully described. Mechanically, the middle surface and the thickness represent the shell in the same way as a bar is represented by its axis and the cross section.

However, not every shell fits this description. A parachute, for instance, is made of cloth, i. e., of threads crossing each other and leaving holes in between. Nevertheless, it is a shell, and the "middle surface" which represents it is fairly well defined, although not by the definition just given. However, the thickness t is not easily defined in such a case. Another example of this kind is culvert pipe used in highway work. For most purposes it may be treated as a shell in the shape of a circular cylinder, and its middle surface may easily be defined. The real pipe, however, is corrugated, and in alternate regions all of the material lies either on one side of the "middle surface" or on the other. For some special purposes one may, of course, consider the corrugated surface which really bisects the thickness, as the middle surface of this pipe, but in many cases this is not done, and shell theory may still be applied.

1.1.2 Stress Resultants

Before we can define stresses in a shell, we need a coordinate system. Since the middle surface extends in two dimensions, we need two coordinates to describe the position of a point on it. Let us assume that some system of coordinates x, y has been defined on the middle surface so that the lines $x = $ const. meet the lines $y = $ const. at right angles (GAUSSIAN coordinates). We may then cut an element from the shell by cutting along two pairs of adjacent coordinate lines as shown in fig. 1. The cuts are made so that the four sides of the element are normal to the middle surface of the shell.

Since it is not always possible for the distance ds_x or ds_y between two adjacent coordinate lines to be the same everywhere, opposite sides of the element will differ slightly in length. However, for the present purpose this difference is of no importance.

The front side of the element is part of a cross section $x = $ const through the shell and has the area $ds_y \cdot t$. The stresses acting on this

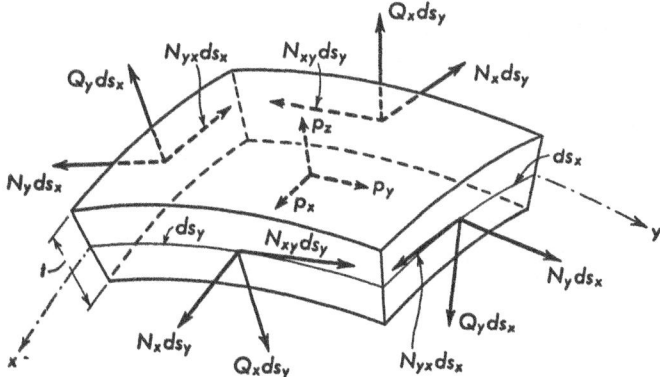

Fig. 1. Stress resultants and loads acting on a shell element

area have a certain resultant which, of course, depends on the length ds_y. When ds_y approaches zero, the resultant decreases proportionately, and the quotient "force divided by length of section" has a finite limit. It is therefore reasonable to call this quotient the "stress resultant". It is a force per unit length of section and may be measured in lb/ft or kg/m, for example.

For all analytical work we must resolve the stress resultant into components. We choose as a reference frame the tangent to the line element ds_y, another tangent to the middle surface at right angles to ds_y (i. e., normal to the cross section), and a normal to the shell. For the force components in these directions we give the following definitions:

In a section $x = $ const, the force in direction x, transmitted by a unit length of section (measured on the middle surface) is called the normal force N_x. It is considered positive if tensile and negative if compressive. The normal force N_y in a section $y = $ const is defined correspondingly.

In a section $x = $ const., the force transmitted by a unit length of this section and directed tangent to ds_y is called the shearing force N_{xy}. It is considered positive if it points in the direction of increasing y on the same side of the shell element where a tensile force N_x would point in the direction of increasing x. Correspondingly, in the section $y = $ const. the shearing force N_{yx} is defined with a similar rule for its positive sign (fig. 1). Evidently the sign of both shearing forces depends on the choice of the coordinates. It changes when the positive direction of one of them is reversed.

In a section $x = $ const., the force normal to the middle surface transmitted by a unit length is called the transverse force Q_x. The positive sign of this force will be defined later [eq.(1c)].

Each of the three components thus defined for a section is the resultant of a certain kind of stresses (fig. 2), normal stresses (σ_x, σ_y), shear stresses parallel to the middle surface ($\tau_{xy} = \tau_{yx}$), and shear stresses normal to it (τ_{xz}, τ_{yz}). Consequently they also deserve the name of "stress resultant", and so they will be called in this book.

The foregoing definitions apply to every shell, including shells in which the faces and thickness are not defined. In the common case of a shell consisting of solid material included between its faces, it is possible to express the stress resultants as integrals of the stresses acting on a section. Then one may consider these integral expressions which are derived from the foregoing definitions, as the definitions themselves. We shall now derive these integrals.

In the section $x = $ const. (fig. 2), the total force normal to this section is by definition $N_x\,ds_y$. It is the resultant of the stresses σ_x which act on this area. Since the width ds_y is of differential magnitude, we may disregard a possible variation in this direction, but we have to consider a variability of all stresses across the thickness of the shell. It is therefore necessary to consider first an element in the cross section which has differential magnitude in all directions. Such an element has been shaded in fig. 2. Because of the curvature of the shell, its width is not simply ds_y, but $ds_y(r_y + z)/r_y$, and the force transmitted through it is

$$\sigma_x\,ds_y\,\frac{r_y + z}{r_y}\,dz.$$

The total normal force for the element $ds_y \cdot t$ is found when this expression is integrated between the limits $-t/2$ and $+t/2$:

$$N_x\,ds_y = \int_{-t/2}^{+t/2} \sigma_x\,ds_y\,\frac{r_y + z}{r_y}\,dz.$$

When the factor ds_y on both sides is dropped, this is the equation which relates the normal force and the normal stress. In the same way the shearing stresses τ_{xy} and τ_{xz} must be integrated to obtain the forces N_{xy} and Q_x. Altogether, we have

$$N_x = \int\limits_{-t/2}^{+t/2} \sigma_x \frac{r_y + z}{r_y}\, dz, \qquad N_{xy} = \int\limits_{-t/2}^{+t/2} \tau_{xy} \frac{r_y + z}{r_y}\, dz,$$

$$Q_x = -\int\limits_{-t/2}^{+t/2} \tau_{xz} \frac{r_y + z}{r_y}\, dz. \tag{1 a--c}$$

The minus sign which has been added to the equation for Q_x, stipulates that a positive transverse force shall have the direction shown in fig. 1, which is opposite to the direction of τ_{xz} in fig. 2.

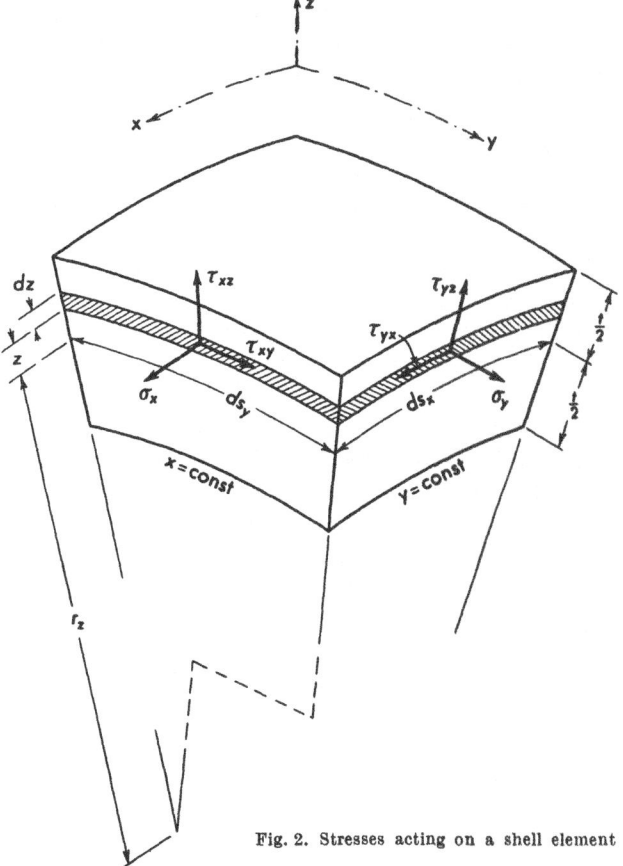

Fig. 2. Stresses acting on a shell element

We may apply the same reasoning to a section $y = $ const and write three more equations for the other three stress resultants; we must, of course, keep in mind that the line element ds_x has a different radius of curvature, say r_x. We have then

$$N_y = \int_{-t/2}^{+t/2} \sigma_y \frac{r_x + z}{r_x}\, dz, \qquad N_{yx} = \int_{-t/2}^{+t/2} \tau_{yx} \frac{r_x + z}{r_x}\, dz,$$

$$Q_y = -\int_{-t/2}^{+t/2} \tau_{yz} \frac{r_x + z}{r_x}\, dz. \tag{1 d--f}$$

When we compare eqs. (1 b) and (1 e), we see that the equality of the shearing stresses, $\tau_{xy} = \tau_{yx}$, does not imply the equality of the shearing forces. The difference between N_{xy} and N_{yx} vanishes only if $r_x = r_y$ (e. g., for a sphere), or if τ_{xy} does not depend on z. In a thin shell t and z are small compared with the radii r_x, r_y; then the difference between the two shearing forces is not large and may often be neglected.

When the stresses are not distributed uniformly across the thickness t, some of them have moments with respect to the center of the section. Since these moments influence the equilibrium of the shell element, we must consider them. The moment of the stresses σ_x in a section $x = $ const. is referred to a tangent to the line element ds_y of the middle surface. The moment is of differential magnitude and proportional to ds_y. If it is designated by $M_x\, ds_y$, the quantity M_x is finite and represents a moment per unit length of section. Consequently, it may be measured in such units as ft.lb/ft or in.lb/ft or others of the same kind. M_x is called the bending moment of the section.

When the stresses τ_{xy} are distributed non-uniformly across the thickness t, their resultant may lie anywhere in the plane of the cross section and has a moment with respect to an axis which is normal to the section and passes through the center of the line element ds_y. This moment is also proportional to ds_y and is denoted by $M_{xy}\, ds_y$. The finite quantity M_{xy} is called the twisting moment.

One may easily read from fig. 2 the relations

$$M_x = -\int_{-t/2}^{+t/2} \sigma_x \frac{r_y + z}{r_y}\, z\, dz, \qquad M_{xy} = -\int_{-t/2}^{+t/2} \tau_{xy} \frac{r_y + z}{r_y}\, z\, dz \tag{1 g, h}$$

which may be considered as the definitions of the bending and twisting moments. The minus signs are arbitrary and fix the sign convention used in this book. (See also figs. V–1 b and VI–1 b).

When the same ideas are applied to a section $y = $ const., another bending moment and another twisting moment are obtained:

$$M_y = -\int_{-t/2}^{+t/2} \sigma_y \frac{r_x + z}{r_x}\, z\, dz, \qquad M_{yx} = -\int_{-t/2}^{+t/2} \tau_{yx} \frac{r_x + z}{r_x}\, z\, dz. \tag{1 i, j}$$

Again, as in the case of the shearing forces, the shear stresses in eqs. (1h, j) are equal, but the resultant moments are different. And again the difference is not large and may often be neglected (see p. 216), but may sometimes be the key to the exact formulation of a problem (see p. 421). It will be noticed that, because of the factors $(r_x + z)/r_x$ and $(r_y + z)/r_y$, the moments are not zero when the stresses are independent of z, i. e., uniformly distributed across the thickness. These factors are required because of the curvature of the shell and represent the fact that the sides of an element are not rectangles, but trapezoids, and that their centroids do not lie exactly on the middle surface.

It should be noted that eqs. (1g–j) do not imply any particular law of distribution of the stresses across the thickness. Whether or not the distribution is linear, these equations are always valid as definitions of the moments.

The transverse shearing stresses τ_{xz} and τ_{yz} do not lead to moments. The ten quantities

$$N_x, \quad N_y, \quad N_{xy}, \quad N_{yx}, \quad Q_x, \quad Q_y, \quad M_x, \quad M_y, \quad M_{xy}, \quad M_{yx}$$

describe the forces and moments acting on the sides of a rectangular shell element. A common name for the whole group is needed, and we shall call them the "stress resultants". It is the main purpose of Chapters 2–6 of this book to explain the methods which allow their computation in shells of different shapes.

Once the stress resultants are known, the stresses may be found by elementary methods. In thin shells of homogeneous material the stress distribution is generally not far from linear, and we may obtain the stresses from the simple relations derived for beams of rectangular cross section, subjected to a normal force and a bending moment:

$$\sigma_x = \frac{N_x}{t} - \frac{12 M_x z}{t^3}, \qquad \sigma_y = \frac{N_y}{t} - \frac{12 M_y z}{t^3}. \qquad \text{(2a, b)}$$

The N-term in these formulas is called the direct stress, and the M-term is called the bending stress. If the shell thickness is not very small compared with the radii of curvature, it may be worth while to take the trapezoidal shape of the cross section into account; but then one should also make use of the basic ideas of bars of great curvature and consider the corresponding non-linearity in the stress distribution.

The tangential shearing stresses follow the same pattern as the bending stresses and must be handled in the same way. However, the two formulas

$$\tau_{xy} = \frac{N_{xy}}{t} - \frac{12 M_{xy} z}{t^3}, \qquad \tau_{yx} = \frac{N_{yx}}{t} - \frac{12 M_{yx} z}{t^3} \qquad \text{(2c, d)}$$

will not necessarily yield identical results. This indicates that there is a logical objection to the assumption of linear stress distribution. Since the discrepancy is not large in thin shells, we may usually disregard it at this stage of the stress analysis.

If the bending and twisting stresses are distributed linearly, the transverse shearing stresses will have the parabolic distribution of the shearing stresses in a beam of rectangular cross section:

$$\tau_{xz} = -\frac{3Q_x}{2t}\left(1 - \frac{4z^2}{t^2}\right), \quad \tau_{yz} = -\frac{3Q_y}{2t}\left(1 - \frac{4z^2}{t^2}\right). \qquad (2\,e,\,f)$$

If the shell is not made of homogeneous material, or if there is a system of ribs or stiffeners incorporated into the shell, other formulas must be set up. In the case of reinforced concrete shells some particular problems arise.

1.1.3 Membrane Forces

Let us consider two examples of shells which behave very differently. First, roll a sheet of paper to cylindrical form and paste the edges together. This is a cylindrical shell. Very feeble lateral forces will suffice to produce in it a considerable deformation. The resistance of this shell to loads is contingent upon the bending moments, and in more complicated cases of this kind the whole group of bending and twisting moments may come into action.

Second, take the shell of an egg or an electric light bulb. Both are very thin and are made of rather fragile materials, but they can withstand remarkable forces without breaking and without undergoing a visible deformation. In these shells a quite different mechanism of load carrying is at work. It consists essentially of normal and shearing forces N_x, N_y, N_{xy}, N_{yx}. Since there is not much deformation, we would expect the bending and twisting moments to be small, at least in thin shells. A detailed study shows that this is true.

While the first kind of shell is not very attractive for design purposes, the second one is, and whenever it is possible, engineers attempt to shape and to support a shell so that it carries its load essentially by normal and shearing forces. If this is done, it seems reasonable to neglect the moments altogether in the stress analysis. The simple version of shell theory which is obtained in this way is called the Membrane Theory of Shells. We shall study it in the following Chapters, and we shall see its merits and its limitations.

If the bending and twisting moments are zero, only the forces shown in fig. 1 act on the sides of the shell element. In addition there may be a load, proportional to the area $ds_x \cdot ds_y$ of the element, applied at its centroid in an arbitrary direction. We shall now consider the moment equilibrium of this force system. First, we choose as a reference axis a normal to the middle surface, passing through the center of the element (marked "p_z" in fig. 1). The only moments with respect to this axis are those of the shearing forces N_{xy} and N_{yx}. The two forces $N_{xy}\,ds_y$ form a couple with the arm ds_x, turning counterclockwise if we look on the upper face of the shell. The other two shearing forces, $N_{xy}\,ds_x$, form a clockwise couple, and there is equilibrium if

$$N_{xy}\,ds_y \cdot ds_x = N_{yx}\,ds_x \cdot ds_y,$$

that is, if the two shearing forces are equal:

$$N_{xy} = N_{yx}. \tag{3}$$

Next, we choose the line marked "p_y" as a reference axis. It is a tangent to a line $x = $ const. on the middle surface. With respect to this axis, there is the moment of the transverse forces $Q_x\,ds_y$ which form a couple with the arm ds_x, but all other forces either are parallel to the axis or intersect it, or they pass it so closely that their moments are infinitely small compared with $Q_x\,ds_y \cdot ds_x$. It follows that $Q_x = 0$. From the moment equilibrium for the axis "p_x" we find in the same way that $Q_y = 0$.

Thus we arrive at a remarkable simplification of shell theory: Of the ten unknown stress resultants, only three are lef , N_x, N_y, and $N_{xy} = N_{yx}$. The three equations of force equilibrium, which have not yet been used, are available and sufficient in number for calculating these forces (see pp. 19, 109, 167). When the normal and shearing forces have been found, the corresponding deformations may be calculated, and we may check whether or not they lead to bending stresses. In many cases it is found that the bending stresses are negligibly small, and this justifies the basic assumption of the membrane theory. In other cases it is found that the deformations derived from the membrane theory contain a discrepancy or a contradiction, and that, therefore, bending and twisting moments must be an important part of the stress system.

When speaking of membrane theory, membrane stresses, or membrane forces (i. e., N_x, N_y, N_{xy}), we do not imply that the normal forces are necessarily tensile forces. In many shells they are compressive; nevertheless the theory is exactly the same and is also called membrane theory.

1.2 Membrane Forces in Arbitrary Directions

1.2.1 Rectangular Coordinates

The membrane forces at a point of a shell represent a plane stress system in a tangential plane to the middle surface. When these stresses or the stress resultants N_x, N_y, N_{xy} have been calculated for the sections $x = $ const and $y = $ const passing through that point, the question may be raised as to what forces would be found if the shell were cut in another direction, making an arbitrary angle α with the x direction.

For a plane stress system σ_x, σ_y, τ_{xy} the answer is well known and may be found in textbooks on elementary strength of materials. We need only repeat the essential facts in the notation used for the stress resultants of shells.

We consider a certain point of the shell (i. e., of its middle surface) and define there two rectangular reference frames x, y and ξ, η (fig. 3 a). The directions x and y may be those of the Gaussian coordinates used on the preceding pages for defining the normal and shearing forces N_x, N_y, N_{xy}, N_{yx}, and we assume now that these forces are known. We wish to find the forces in sections $\xi = \text{const}$ and $\eta = \text{const}$ as defined

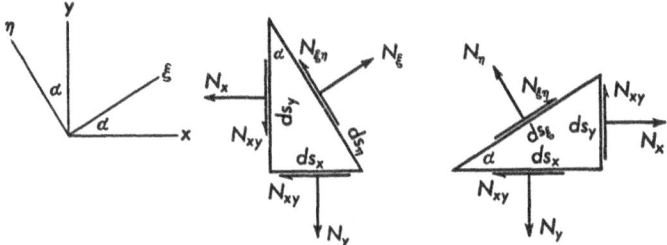

Fig. 3. Equilibrium of triangular shell elements

by the second reference frame (which need only be defined locally). We obtain them by cutting from the shell one of the triangular elements shown in fig. 3 b, c.

The first of these elements has two sides ds_x and ds_y in which the forces are known, and one side ds where two of the desired forces, N_ξ and $N_{\xi\eta}$, appear. The equilibrium of the six forces yields the following equations:

$$N_\xi\, ds_\eta = \quad N_x\, ds_y \cos\alpha + N_{xy}\, ds_y \sin\alpha + N_y\, ds_x \sin\alpha + N_{xy}\, ds_x \cos\alpha,$$

$$N_{\xi\eta}\, ds_\eta = -N_x\, ds_y \sin\alpha + N_{xy}\, ds_y \cos\alpha + N_y\, ds_x \cos\alpha - N_{xy}\, ds_x \sin\alpha.$$

With the angle α as shown in fig. 3 a, we have

$$\frac{ds_x}{ds_\eta} = \sin\alpha, \qquad \frac{ds_y}{ds_\eta} = \cos\alpha,$$

and so we obtain the first and third of the following formulas:

$$N_\xi = N_x \cos^2\alpha + N_y \sin^2\alpha + 2 N_{xy} \cos\alpha \sin\alpha,$$

$$N_\eta = N_x \sin^2\alpha + N_y \cos^2\alpha - 2 N_{xy} \cos\alpha \sin\alpha, \qquad (4\,\text{a–c})$$

$$N_{\xi\eta} = (N_y - N_x) \cos\alpha \sin\alpha + N_{xy}(\cos^2\alpha - \sin^2\alpha).$$

Eq. (4b) is obtained in the same way from the shell element shown in fig. 3c. The equations may also be written in the following form:

$$N_\xi = \frac{1}{2}(N_x + N_y) + \frac{1}{2}(N_x - N_y)\cos 2\alpha + N_{xy}\sin 2\alpha,$$

$$N_\eta = \frac{1}{2}(N_x + N_y) - \frac{1}{2}(N_x - N_y)\cos 2\alpha - N_{xy}\sin 2\alpha, \quad (5\,\text{a–c})$$

$$N_{\xi\eta} = -\frac{1}{2}(N_x - N_y)\sin 2\alpha + N_{xy}\cos 2\alpha.$$

Eq. (5a) gives the normal force as a function of the direction of the section. When α varies through $180°$, N_ξ must have at least one maximum and one minimum. We find the angles $\alpha = \alpha_0$ for which these extrema occur, from the condition $dN_\xi/d\alpha = 0$. It yields

$$\tan 2\alpha_0 = \frac{2N_{xy}}{N_x - N_y} \qquad (6)$$

and thus determines two directions at right angles to each other which are called the principal directions of the membrane forces at this point of the shell. From eqs. (6) and (5c) it may easily be seen that the shear is zero for $\alpha = \alpha_0$. The extreme normal forces are called the principal forces and are denoted by N_a, N_b. One of them is the maximum and the other one the minimum that the normal force N_ξ or N_η can assume for any direction at this point. From eqs. (5) and (6) one may obtain the following formulas for these forces:

$$N_a = \frac{1}{2}(N_x + N_y) + \frac{1}{2}\sqrt{(N_x - N_y)^2 + 4N_{xy}^2},$$

$$N_b = \frac{1}{2}(N_x + N_y) - \frac{1}{2}\sqrt{(N_x - N_y)^2 + 4N_{xy}^2}. \qquad (7)$$

One of the principal forces makes an angle α_0 with the x axis, the other one with the y axis, but eqs. (7) do not indicate which of them is N_a and which N_b. To find this out, one must use either eqs. (4) or MOHR's circle (see p. 12).

When the principal directions are known at every point of the shell, one may draw a net of curves which have these directions as tangents. They are called the trajectories of the normal forces. They indicate the paths along which the loads are carried to the supported edges by a system of tensile and compressive forces in the shell. These trajectories may give a very suggestive picture of the stresses in a shell (figs. II–26, II–31), but they are laborious to obtain and not easy to represent on paper. Therefore they are not often used in practical stress analysis work. However, they indicate in which direction a thin shell may best be reinforced by ribs, and in which directions the steel rods in reinforced concrete shells should preferably be placed.

1.2.2 Mohr's Circle

Equations (4) indicate that the membrane forces at a point of a shell represent a two-dimensional, symmetric tensor, just as do two-dimensional stresses $(\sigma_x, \sigma_y, \tau_{xy})$ or strains $(\epsilon_x, \epsilon_y, \frac{1}{2}\gamma_{xy})$, and the moments and products of inertia of a cross section (I_x, I_y, I_{xy}). In all these cases there exists a set of formulas identical with eqs. (6) and (7), and there are several graphical methods available which do the same service as these equations (e. g., the different ellipses of inertia, LAND's circle, MOHR's circle). Among all these devices, MOHR's circle appears to be the most useful one, and although graphical methods have lost

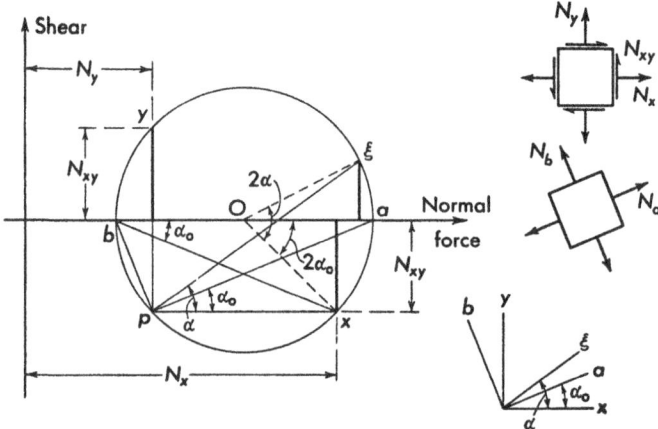

Fig. 4. MOHR's circle for normal and shearing forces

much of their former importance, we shall describe it here in some detail because of its usefulness for the qualitative understanding of stress patterns.

We consider a certain point of the shell and the normal and shearing forces which may be found from eqs. (4) for various sections passing through this point. In a rectangular coordinate system we mark the points x and y with the coordinates N_x, $-N_{xy}$ and N_y, N_{xy}, respectively, and then we draw a circle which has the line $x\,y$ as a diameter (fig. 4). The center of this circle has the coordinates $\frac{1}{2}(N_x + N_y)$, 0, and its radius is

$$\sqrt{\left(\frac{N_x - N_y}{2}\right)^2 + N_{xy}^2}.$$

It follows that the points a and b have the abscissas N_a and N_b as given by eqs. (7), their ordinates being zero. Consequently, the points x, y, a, b represent the forces transmitted through sections which pass through the shell point under consideration in four different directions.

Since the circle is unequivocally determined by the principal forces N_a, N_b, we should necessarily have found this same circle, if we had started from the forces N_ξ, N_η, $N_{\xi\eta}$ for an arbitrary pair of orthogonal sections passing through the same point of the shell. Hence, this circle is the locus for all points whose coordinates are the normal and shearing forces in sections of arbitrary direction and is a graphical representation of the stress resultants at the particular point of the shell. It is called MOHR's circle.

From eq. (6) we see that $\measuredangle\, x\, o\, a = 2\alpha_0$, and from a well known theorem of elementary geometry it follows that $\measuredangle\, x\, b\, a = \alpha_0$.

In the lower right-hand corner of fig. 4 are shown the reference frames x, y and a, b which define the directions of the sections in which the different forces N_x, N_{xy}, etc. are transmitted. The force N_x, for example, has the direction x and is transmitted in a section at right angles to the x axis.

We may define a pole p on MOHR's circle by drawing through one of the points, x, y, a, b, a straight line parallel to the corresponding line of the reference frame. All such lines lead to the same point p, and the angle α_0 is found again there. When we now choose an arbitrary ξ direction and draw parallel to it the line $p\xi$ through the pole p, we may read from the figure the following relations for the coordinates of the point ξ: Its abscissa is

$$\frac{1}{2}(N_x + N_y) + \overline{o\,x}\cos 2(\alpha - \alpha_0) =$$

$$= \frac{1}{2}(N_x + N_y) + \overline{o\,x}\cos 2\alpha_0 \cos 2\alpha + \overline{o\,x}\sin 2\alpha_0 \sin 2\alpha$$

$$= \frac{1}{2}(N_x + N_y) + \frac{1}{2}(N_x - N_y)\cos 2\alpha + N_{xy}\sin 2\alpha,$$

i. e., exactly the normal force N_ξ as given by eq. (5). The ordinate of the point ξ is

$$\overline{o\,x}\sin 2(\alpha - \alpha_0) = -\overline{o\,x}\sin 2\alpha_0 \cos 2\alpha + \overline{o\,x}\cos 2\alpha_0 \sin 2\alpha$$

$$= -N_{xy}\cos 2\alpha + \frac{1}{2}(N_x - N_y)\sin 2\alpha,$$

and this is equal to $-N_{\xi\eta}$.

Evidently, every point of MOHR's circle corresponds to one possible section through the shell, and the direction of the normal force is parallel to the line ξp in the MOHR diagram. When this direction is rotated through 180°, the corresponding point runs just once around the whole circle.

It should be observed that MOHR's circle requires a sign convention of its own. While positive normal forces are always plotted to the right and negative ones to the left, the same positive shearing force N_{xy} had to be plotted downward when it was associated with N_x and upward when associated with N_y. We may easily verify the rule that the right angle between the normal and shearing forces in a section and the right angle between the directions in which they are plotted must always be of opposite sense, one of them clockwise and the other one counterclockwise. As an example, we may look at the forces N_x and N_{xy} in fig. 4. At the shell element they point right and up, in the MOHR diagram they point right and down.

1.2.3 Oblique Coordinates and Skew Forces

On the curved middle surface of a shell the coordinates cannot be simple cartesian coordinates but must be some kind of orthogonal curvilinear coordinates. In many cases it is advisable to use, instead,

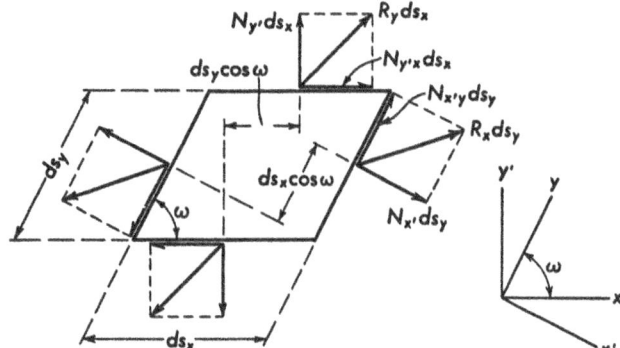

Fig. 5. Orthogonal force components at an oblique shell element

a non-orthogonal system which is better adapted to the general shape of the middle surface or to the boundaries of the shell (see Chapter IV). In such cases the lines $x = $ const and $y = $ const meet each other at an angle ω which may be constant or even vary from point to point. The shell element is then in the first approximation a parallelogram (fig. 5).

The membrane force $R_x\, ds_y$ which is transmitted in the side ds_y of the element, is certainly situated in the tangential plane to the middle surface. There are different ways of resolving it into two components. One might think of using rectangular components $N_{x'}\, ds_y$ and $N_{x'y}\, ds_y$. These correspond to the definitions of normal and shearing forces given on p. 4, if we use a rectangular reference frame x', y. The force $R_y\, ds_x$ on the adjacent side of the element should then be resolved into the rectangular components $N_{y'}\, ds_x$ and $N_{y'x}\, ds_x$ shown in fig. 5, and these forces require the use of another reference frame x, y'.

The two shearing forces $N_{x'y}$ and $N_{y'x}$ are, of course, not equal since equality can be expected only for sections at right angles to each other. Therefore, the tensor of the membrane forces is now described by four quantities insted of three. These four quantities, however, are not

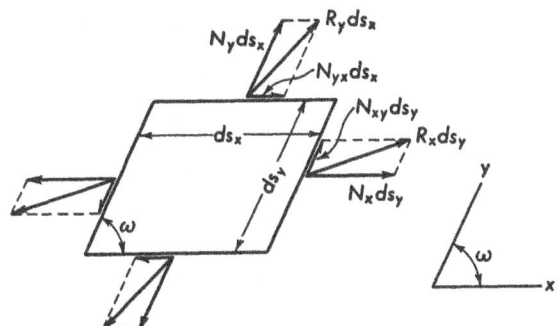

Fig. 6. Skew force components

independent of each other but are connected by the condition of moment equilibrium with respect to a normal to the shell:

$$N_{x'} \, ds_y \cdot ds_x \cos\omega - N_{x'y} \, ds_y \cdot ds_x \sin\omega - N_{y'} \, ds_x \cdot ds_y \cos\omega +$$
$$+ \, N_{y'x} \, ds_x \cdot ds_y \sin\omega = 0$$

which yields the relation

$$N_{x'y} - N_{y'x} = (N_{x'} - N_{y'}) \cot\omega. \tag{8}$$

We may avoid complications and arrive at a more natural description of the state of stress at a point (i. e., of the membrane force tensor) if we resolve the forces $R_x \, ds_y$ and $R_y \, ds_x$ in oblique components following the directions of the lines $x = \text{const}$ and $y = \text{const}$ (fig. 6). On the sides ds_y of the element we have then per unit length the "skew fiber force" N_x and the "skew shearing force" N_{xy} which has the same direction as the orthogonal shear $N_{x'y}$ but not the same magnitude. From fig. 7 we easily read the relations between the orthogonal and the skew forces:

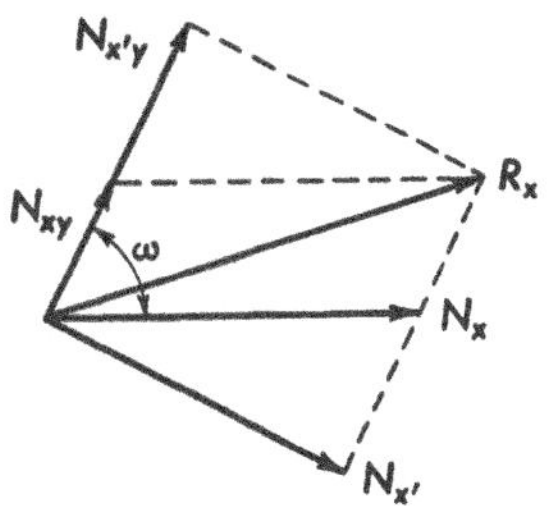

Fig. 7
Relation between orthogonal and skew force components

$$N_x = \frac{N_{x'}}{\sin\omega}, \qquad N_{xy} = N_{x'y} - N_{x'} \cot\omega.$$

Applying the same ideas to R_y, we obtain the skew forces N_y and N_{yx} in the section $y = \text{const}$:

$$N_y = \frac{N_{y'}}{\sin\omega}, \qquad N_{yx} = N_{y'x} - N_{y'} \cot\omega.$$

Like the normal forces on a rectangular shell element, the skew forces N_x or N_y on opposite sides of the oblique element fall on the same line and do not yield a couple. Thus the shearing forces are again alone in the equation of moment equilibrium:

$$N_{xy}\,ds_y \cdot ds_x \sin\omega - N_{yx}\,ds_x \cdot ds_y \sin\omega = 0,$$

and hence they are again equal to each other:

$$N_{xy} = N_{yx}.$$

Having solved a shell problem in oblique coordinates x, y, we may desire to find from the skew forces N_x, N_y, N_{xy} the components N_ξ, N_η, $N_{\xi\eta}$ for an orthogonal pair ξ, η of sections or the principal forces N_a, N_b. The set of transformation formulas needed may be found by the

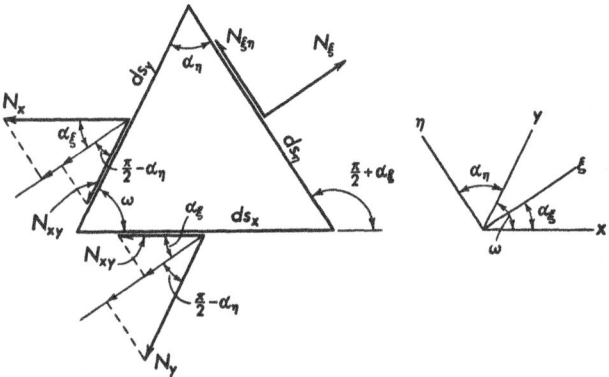

Fig. 8. Triangular shell element in oblique coordinates x, y

method which led to eqs. (4). We cut from the shell a triangular element having one side parallel to one of the new rectangular axes, and the other two sides parallel to the directions x and y (fig. 8). The equilibrium of all forces in the direction ξ yields the equation

$$N_\xi\,ds_\eta = N_x\,ds_y\cos\alpha_\xi + N_{xy}\,ds_y\sin\alpha_\eta + N_y\,ds_x\sin\alpha_\eta + N_{xy}\,ds_x\cos\alpha_\xi,$$

and a similar equation will be found for the η-components:

$$N_{\xi\eta}\,ds_\eta = -N_x\,ds_y\sin\alpha_\xi + N_{xy}\,ds_y\cos\alpha_\eta + N_y\,ds_x\cos\alpha_\eta - N_{xy}\,ds_x\sin\alpha_\xi.$$

Between the three sides of the element we have the geometric relation

$$\frac{\sin\omega}{ds_\eta} = \frac{\sin\alpha_\eta}{ds_x} = \frac{\cos\alpha_\xi}{ds_y}.$$

We multiply each term in the preceding equations by one of the three identical factors and thus obtain two of the three following equations, the third of which can be derived from another triangular element:

$$N_\xi \ \sin \omega = N_x \cos^2 \alpha_\xi + N_y \sin^2 \alpha_\eta + 2 N_{xy} \cos \alpha_\xi \sin \alpha_\eta,$$
$$N_\eta \ \sin \omega = N_x \sin^2 \alpha_\xi + N_y \cos^2 \alpha_\eta - 2 N_{xy} \sin \alpha_\xi \cos \alpha_\eta,$$
$$N_{\xi \eta} \sin \omega = N_y \cos \alpha_\eta \sin \alpha_\eta - N_x \cos \alpha_\xi \sin \alpha_\xi +$$
$$+ N_{xy} (\cos \alpha_\xi \cos \alpha_\eta - \sin \alpha_\xi \sin \alpha_\eta). \tag{9}$$

To find the principal forces N_a, N_b we must put $N_{\xi \eta} = N_{ab} = 0$. This is an equation for the unknown angles α_ξ and α_η, which we now call α_a and α_b. Using well-known trigonometric formulas, we may bring this equation into the form

$$N_y \sin 2\alpha_b - N_x \sin 2\alpha_a + 2 N_{xy} \cos(\alpha_a + \alpha_b) = 0.$$

From fig. 8 we find

$$\alpha_b = \frac{\pi}{2} + \alpha_a - \omega,$$

which enables us to eliminate α_b. Subsequent trigonometric transformation leads to an equation in which only the functions $\cos 2\alpha_a$ and $\sin 2\alpha_a$ occur. It has the solution

$$\tan 2\alpha_a = \frac{N_y \sin 2\omega + 2 N_{xy} \sin \omega}{N_x + N_y \cos 2\omega + 2 N_{xy} \cos \omega}. \tag{10a}$$

By a similar calculation we find also

$$\tan 2\alpha_b = - \frac{N_x \sin 2\omega + 2 N_{xy} \sin \omega}{N_y + N_x \cos 2\omega + 2 N_{xy} \cos \omega}. \tag{10b}$$

If we put $\omega = \pi/2$, both formulas coincide with our formula (6) for rectangular coordinates.

1.3 Transformation of Moments

All the questions we have asked and answered on the preceding pages for the normal and shearing forces may also be formulated for the bending and twisting moments. The answers may be found easily by reducing each moment problem to the corresponding force problem. We simply replace each moment by a couple of forces parallel to the middle surface. The arms of all these couples must be equal, but are otherwise arbitrary. We choose them equal to the thickness of the shell. We have then in its upper surface a system of normal and shearing forces and in the lower surface a system which is identical except that the direction of each force is reversed. We may now cut triangular and other elements from the shell and write for each one of the two force systems the equations of equilibrium as we did in the preceding sections. The resultant forces may then be recombined to yield bending and twisting moments.

It follows that in all the equations and in all the diagrams of Section 1.2 we may simply replace everywhere the letter N by M to obtain valid results for the transformation of the bending and twisting moments to a new set of axes.

Chapter 2

DIRECT STRESSES IN SHELLS OF REVOLUTION

2.1 General Differential Equations

2.1.1 Geometrical Relations

The particular type of shell which we are going to treat in this chapter appears in many technical applications, especially in the construction of tanks, pressure vessels and domes.

Before we enter into the investigation of the stress resultants in these shells, we must examine the geometry of their middle surfaces.

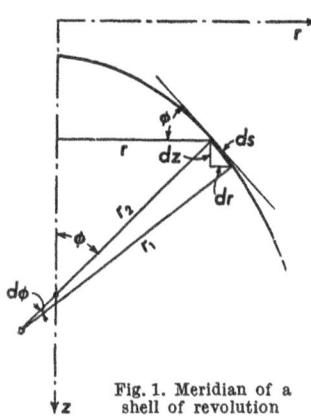

Fig. 1. Meridian of a shell of revolution

A surface of revolution is generated by the rotation of a plane curve about an axis in its plane. This generating curve is called a meridian, and an arbitrary point on the middle surface of the shell is described by specifying the particular meridian on which it is found and by giving the value of a second coordinate which varies along the meridian and is constant on a circle around the axis of the shell. Since all these circles for different values of the second coordinate are parallel to each other, they are called the "parallel circles".

We shall identify a meridian by the angular distance θ of its plane from that of a datum meridian and choose as second coordinate the angle ϕ between a normal to the shell and its axis of revolution. If the middle surface of our shell is a sphere, these coordinates are the spherical coordinates used in geography: θ is the longitude and ϕ is the complement to the latitude: hence the nomenclature of the meridians and the parallel circles.

Fig. 1 shows a meridian of the shell. Let r be the distance of one of its points from the axis of rotation and r_1 its radius of curvature. In our equations we also need the length r_2, measured on a normal to the meridian between its intersection with the axis of rotation and the middle surface. It is the second radius of curvature of the shell, and we read from fig. 1 the relation

$$r = r_2 \sin\phi. \tag{1}$$

For the line element ds of the meridian we have

$$ds = r_1 \, d\phi, \tag{2}$$

and since

$$dr = ds \cos\phi, \quad dz = ds \sin\phi \tag{3a, b}$$

we have the relations

$$\frac{dr}{d\phi} = r_1 \cos\phi, \quad \frac{dz}{d\phi} = r_1 \sin\phi. \tag{4a, b}$$

Finally we obtain from (1) and (4a)

$$\frac{1}{r} \frac{dr}{d\phi} = \frac{r_1}{r_2} \cot\phi. \tag{5}$$

2.1.2 Equilibrium of the Shell Element

The shell element (fig. 2) is cut out by two meridians and two parallel circles, each pair indefinitely close together. The conditions of its equilibrium will furnish three equations, just enough to determine the three unknown stress resultants: the meridional force N_ϕ, the hoop force N_θ, and the shear $N_{\phi\theta}$.

To find these equations, let us begin with the forces parallel to the tangent to the meridian. The shear transmitted by one of the meridional edges of the element is $N_{\theta\phi} \, r_1 d\phi$, on the opposite side edge it is

$$\left(N_{\theta\phi} + \frac{\partial N_{\theta\phi}}{\partial \theta} d\theta\right) r_1 d\phi.$$

These two forces are of opposite direction and

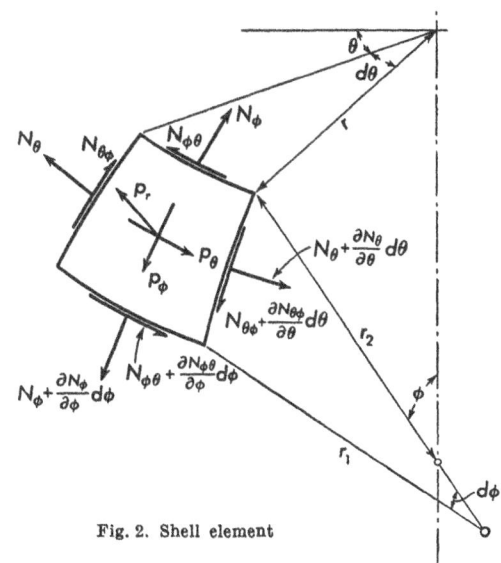

Fig. 2. Shell element

therefore almost cancel each other. Only their difference

$$\frac{\partial N_{\theta\phi}}{\partial \theta} r_1 d\theta \, d\phi$$

enters the equilibrium condition. In the same way we have the difference of the two meridional forces, but in computing it, we must bear in mind that both the force per unit length of section, N_ϕ, and the length of section $r \, d\theta$ vary with ϕ. Therefore we have to introduce the increment

$$\frac{\partial}{\partial \phi} (r N_\phi) \, d\phi \, d\theta$$

into the condition of equilibrium. But that is not all. The hoop forces N_θ also contribute. The two forces $N_\theta r_1 d\phi$ on either side of the element lie in the plane of a parallel circle where they include an angle $d\theta$. They therefore have a resultant force $N_\theta r_1 d\phi \cdot d\theta$, situated in the same plane and pointing towards the axis of the shell. We resolve this force into two rectangular components normal to the shell and in the direction of the tangent to the meridian. The latter one,

$$N_\theta r_1 d\phi \, d\theta \cdot \cos\phi,$$

enters our condition of equilibrium, and since its direction is opposite to that of the increments of $N_{\theta\phi}$ and N_ϕ, it requires a negative sign. Finally we have to introduce a component of the external force, which is the product of the load component per unit area of shell surface, p_ϕ, and the area of the element, $r \, d\theta \cdot r_1 d\phi$. The equilibrium condition thus reads:

$$\frac{\partial N_{\theta\phi}}{\partial \theta} r_1 d\theta \, d\phi + \frac{\partial}{\partial \phi} (r N_\phi) \, d\phi \, d\theta - N_\theta r_1 d\phi \, d\theta \cos\phi +$$
$$+ p_\phi r r_1 d\theta \, d\phi = 0.$$

All its terms contain the product of the two differentials $d\theta \, d\phi$. Dividing by this, we get the partial differential equation

$$\frac{\partial}{\partial \phi} (r N_\phi) + r_1 \frac{\partial N_{\theta\phi}}{\partial \theta} - r_1 N_\theta \cos\phi + p_\phi r r_1 = 0. \qquad (6a)$$

By quite similar reasoning we obtain an equation for the forces in the direction of a parallel circle. For the difference of the two shearing forces which are transmitted in the horizontal edges of the shell element, we must take into account the variability of the length of the line element:

$$\frac{\partial}{\partial \phi} (r N_{\phi\theta}) \, d\phi \, d\theta.$$

Then we have a term representing the difference of the two forces $N_\theta \cdot r_1 d\phi$ and another one with the load component p_θ. Furthermore, we have a contribution from the shear acting on the meridional edges. The two forces $N_{\theta\phi} \cdot r_1\,d\phi$ are not exactly parallel. Their horizontal components make an angle $d\theta$ and therefore have a resultant

$$N_{\theta\phi} \cdot r_1\,d\phi \cdot \cos\phi \cdot d\theta$$

which has the direction of the tangent to a parallel circle and thus enters our equation. If we drop the factor $d\theta\,d\phi$, common to all terms, we have:

$$\frac{\partial}{\partial\phi}\,(r\,N_{\phi\theta}) + r_1\frac{\partial N_\theta}{\partial\theta} + r_1 N_{\theta\phi}\cos\phi + p_\theta\,r\,r_1 = 0. \qquad (6\,\mathrm{b})$$

The third equation refers to the forces which are perpendicular to the middle surface of the shell. It contains contributions from both normal forces N_ϕ and N_θ and the third load component, p_r.

In formulating eq. (6a), we have already seen that the two forces $N_\theta\,r_1\,d\phi$ have a horizontal resultant $N_\theta\,r_1\,d\phi\,d\theta$. It has a component

$$N_\theta\,r_1\,d\phi\,d\theta\,\sin\phi,$$

directed normally to the shell and pointing toward its inner side. Similarly, the two forces $N_\phi\,r\,d\theta$, including the angle $d\phi$, have the resultant

$$N_\phi\,r\,d\theta\,d\phi$$

in the same direction. These two forces and the component

$$p_r\,r\,r_1\,d\theta\,d\phi$$

of the load must be in equilibrium. This yields the equation

$$N_\theta\,r_1\sin\phi + N_\phi\,r - p_r\,r\,r_1 = 0.$$

We divide by $r\,r_1$, use the geometric relation (1), and thus get the third of our equations:

$$\frac{N_\phi}{r_1} + \frac{N_\theta}{r_2} = p_r. \qquad (6\,\mathrm{c})$$

This equation not only is valid for shells in the form of a surface of revolution, but may be applied to all shells when the coordinate lines $\phi = \mathrm{const}$ and $\theta = \mathrm{const}$ are the lines of curvature of the surface. Therefore, we shall meet it again in the next chapter, and we shall see in Chapter 4 what becomes of it when the coordinates no longer follow the lines of curvature of the shell.

It is notable that eq. (6c) does not contain any derivatives of the unknowns. It may therefore always be used to eliminate one of the normal forces and to reduce our problem to two differential equations, with the shear and one of the normal forces as unknowns.

Till now, we have used two angular coordinates θ and ϕ. This is adequate for many shells with meridians of simple shape and has been done quite generally in the theory of shells of revolution. However the angle ϕ is very inconvenient if the meridian has a point of inflection. At such a point, ϕ passes a maximum and afterwards begins to decrease. The stress resultants must therefore be double-valued functions of ϕ, the two branches belonging to the two parts of the meridian above and below the point of inflection. Even worse is the fact that the sign of the shear $N_{\phi\theta}$ depends on the direction in which ϕ increases. Since this is reversed beyond the inflection point, the shear must suddenly have the opposite sign, without passing through zero. It is evident that an analytical solution fulfilling all these requirements cannot be very simple and that numerical methods for the solution of the differential equations will also meet with difficulties. For such cases it is useful to replace ϕ by a coordinate which avoids all these difficulties, and that is the length s of the meridian, measured from any datum point, say from the vertex of the shell if such a point exists, or otherwise from its edge. Consequently, we then replace the subscript ϕ by s.

Between s and ϕ we have the relation (2) and introducing this into the eqs. (6a–c), we get

$$\frac{\partial}{\partial s}(r\,N_s) + \frac{\partial N_{\theta s}}{\partial \theta} - N_\theta \cos\phi + p_s r = 0,$$

$$\frac{\partial}{\partial s}(r\,N_{\theta s}) + \frac{\partial N_\theta}{\partial \theta} + N_{\theta s}\cos\phi + p_\theta r = 0, \qquad (7\,\text{a–c})$$

$$\frac{N_s}{r_1} + \frac{N_\theta}{r_2} = p_r.$$

There is still a third way of formulating the fundamental equations, using rectangular coordinates r, z in the plane of the meridian (fig. 1). From (4b) we find that

$$\frac{\partial}{\partial \phi} = r_1 \sin\phi \cdot \frac{\partial}{\partial z},$$

and when we introduce that into (6a, b), we find

$$\frac{\partial}{\partial z}(r\,N_\phi)\sin\phi + \frac{\partial N_{\theta\phi}}{\partial \theta} - N_\theta \cos\phi + p_\phi r = 0,$$

$$\frac{\partial}{\partial z}(r\,N_{\phi\theta})\sin\phi + \frac{\partial N_\theta}{\partial \theta} + N_{\theta\phi}\cos\phi + p_\theta r = 0. \qquad (8\,\text{a, b})$$

There is some advantage in using this form of the equations, if the shape of the meridian is given by its equation in rectangular coordinates r, z. However, there is no particular reason to prefer for structures shells whose meridians have a simple cartesian equation to those which yield simple relations between ϕ and the radii.

2.2 Loads Having Axial Symmetry

2.2.1 Differential Equations

In many practical problems the external forces have the same symmetry as the shell itself. Then the stresses are independent of θ, and all derivatives with respect to this coordinate disappear from eqs. (6). The equations (6a, c) then read

$$\frac{d}{d\phi} (r N_\phi) - r_1 N_\theta \cos\phi = -p_\phi r r_1,$$

$$\frac{N_\phi}{r_1} + \frac{N_\theta}{r_2} = p_r. \qquad (9\,\text{a, b})$$

Eq. (6b) becomes independent of these equations and contains only the shear:

$$\frac{d}{d\phi} (r N_{\phi\theta}) + r_1 N_{\theta\phi} \cos\phi = -p_\theta r r_1.$$

It describes a kind of torsion of the shell about its axis, a very simple state of stress which may be treated separately. We eliminate it from our further considerations by putting $p_\theta \equiv 0$ and $N_{\phi\theta} \equiv 0$. When we solve eq. (9b) for N_θ and substitute the result into eq. (9a), we obtain a first order differential equation for N_ϕ. After multiplication by $\sin\phi$ it reads

$$\frac{d(r N_\phi)}{d\phi} \sin\phi + r N_\phi \cos\phi = r_1 r_2 p_r \cos\phi \sin\phi - r_1 r_2 p_\phi \sin^2\phi.$$

The two terms at the left may be combined to form a total derivative,

$$\frac{d}{d\phi} (r N_\phi \sin\phi) = \frac{d}{d\phi} (r_2 N_\phi \sin^2\phi),$$

and N_ϕ may be found by an integration:

$$N_\phi = \frac{1}{r_2 \sin^2\phi} \left[\int r_1 r_2 (p_r \cos\phi - p_\phi \sin\phi) \sin\phi \, d\phi + C \right]. \qquad (10)$$

N_θ may then be found from (9b).

Eq. (10) may be interpreted as a condition of equilibrium for the part of the shell above a parallel circle $\phi = $ const. Indeed, if we cut the shell along this circle, $2\pi r_2 \sin\phi$ is its circumference, and $2\pi N_\phi r_2 \sin^2\phi$ is the vertical resultant of all internal forces transmitted in this section. The integral times 2π represents the distributed load applied above this circle, if we write it as a definite integral between appropriate limits. The upper limit, of course, will be the value ϕ for the circle in question, and the lower limit will be the value $\phi = \phi_0$ with which the meridian begins (see figs. 4, 9). When the shell has a flat top (figs. 6, 14) we have $\phi_0 = 0$. The constant C represents the effect of loads which may be applied above the circle $\phi = \phi_0$ (see fig. 4), $2\pi C$ being the resultant of these forces.

If the shell is closed at the vertex, such an additional load can only be a concentrated force P applied at this point. If no other load is present, we have in eq. (10)

$$p_\phi \equiv p_r \equiv 0, \quad 2\pi C = -P,$$

and hence the meridional force

$$N_\phi = -\frac{P}{2\pi r_2 \sin^2\phi} \tag{11a}$$

and from eq. (9b) the hoop force

$$N_\theta = +\frac{P}{2\pi r_1 \sin^2\phi}. \tag{11b}$$

At the top both forces have a singularity of the second order, i. e. they tend toward infinity as ϕ^{-2}. We shall see later (p. 350) that the immediate vicinity of this point where the concentrated load is applied will be subjected to severe bending stresses but that at some distance the membrane forces as given by eqs. (11) still represent the real state of stress.

2.2.2 Solution for some Typical Cases

2.2.2.1 Spherical Dome

As a first example we consider a spherical dome as shown in fig. 3. We ask for the stress resultants produced by a dead load p (weight per unit area of the middle surface). To apply our formula (10), we must first resolve this load into its components tangential and normal to the shell. These are

$$p_\phi = p \sin\phi, \quad p_r = -p \cos\phi. \tag{12}$$

Introducing this into (10), we find with $r_1 = r_2 = a$:

$$N_\phi = -\frac{1}{a \sin^2\phi} \int_0^\phi a^2 p \sin\phi \, d\phi = -p \, a \, \frac{1-\cos\phi}{\sin^2\phi}.$$

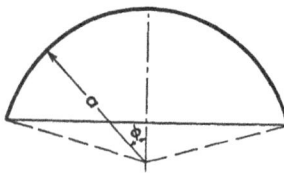

Simplifying the trigonometric expression and then using (9b) we find for the stress resultants the formulas

$$N_\phi = -\frac{p \, a}{1 + \cos\phi},$$

$$N_\theta = p \, a \left(\frac{1}{1 + \cos\phi} - \cos\phi \right). \tag{13}$$

Fig. 3. Spherical dome

It is interesting to discuss these forces in some detail. When we put $\phi = 0$, we find $N_\phi = N_\theta = -pa/2$. The meridional force N_ϕ is negative throughout the hemisphere, but N_θ decreases in absolute value with increasing ϕ and changes sign at a value $\phi = 51.82°$. which follows from the equation $\cos^2\phi + \cos\phi - 1 = 0.$

If the shell is so flat that ϕ does not exceed this limit, no tensile stress appears, assuming that the dead load is the only load and that a proper abutment is provided. This abutment has to resist the thrust N_ϕ which has the direction of the tangent to the meridian. Such an abutment usually consists of a continuous vertical support and a ring, which resists the horizontal component of N_ϕ and from it receives a tensile force

$$\mathsf{N} = -N_\phi\, a\, \sin\phi \cos\phi.$$

This ring is the source of a perturbation of the membrane stresses given by our formulas. In flat domes its stress is of opposite sign to the hoop stress in the shell, and in high domes, where the hoop stress at the springing line is positive, it is usually much smaller than the stress in the ring. Therefore, after the elastic deformations, the ring and the shell do not fit together. The continuity of deformation is re-established by an additional bending of the shell, which will be treated in Chapter 6. It may be mentioned here that the bending stresses are confined to a border zone of limited width and that the major part of the shell has, in fact, the simple stresses given by the membrane theory.

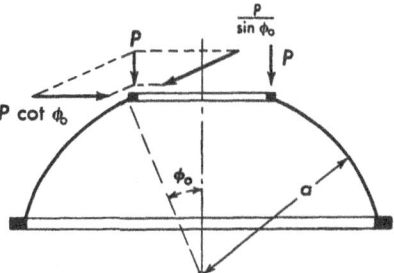

Fig. 4. Shell dome with skylight

Most domes are not closed at the vertex but have a sky-light, or a ventilation opening, covered by a superstructure, the lantern. Its weight, say $2\pi \cdot P \cdot a \sin\phi_0$, acts on the upper edge of the shell as a vertical line load. Since the shell can resist only tangential forces, this edge also needs a stiffening ring, which takes the other component (fig. 4) and gets a compressive force from it. We find the stress resultants in such a shell with its own dead load p and the lantern load P by returning to the integral (10) and determining C so that for $\phi = \phi_0$ we have $N_\phi = -P/\sin\phi_0$. The simple computation leads to the following formulas:

$$N_\phi = -p\,a\,\frac{\cos\phi_0 - \cos\phi}{\sin^2\phi} - P\,\frac{\sin\phi_0}{\sin^2\phi},$$

$$N_\theta = p\,a\left(\frac{\cos\phi_0 - \cos\phi}{\sin^2\phi} - \cos\phi\right) + P\,\frac{\sin\phi_0}{\sin^2\phi}.$$

The difference of the two cosines is disadvantageous for numerica work, in particular for small angles ϕ, and it is better to write the formulas in the following form:

$$N_\phi = -\frac{2\,p\,a}{\sin^2\phi}\sin\frac{\phi+\phi_0}{2}\sin\frac{\phi-\phi_0}{2} - P\frac{\sin\phi_0}{\sin^2\phi},$$

$$N_\theta = -N_\phi - p\,a\cos\phi.$$

(14)

Some figures may interpret this result. The roof represented in fig. 5 carries a uniformly distributed load $p = 45$ lb/ft², and the lantern ring has a line load of 460 lb/ft, applied along its center line, i. e. on a circle of 13' 5'' radius. The edge of the shell has a slightly greater radius,

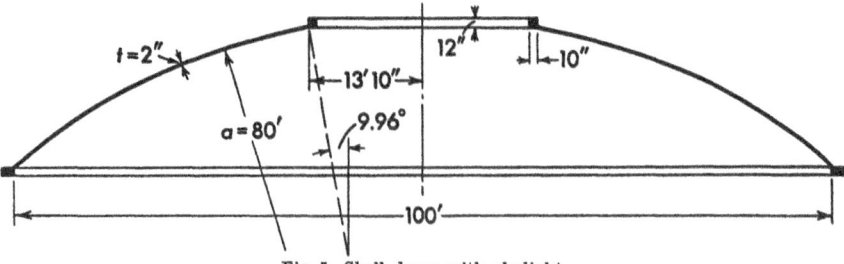

Fig. 5. Shell dome with skylight

$r = 13'\ 10''$, and the vertical line load P transmitted at this edge is correspondingly smaller, $P = 446$ lb/ft. When we introduce these values in eqs. (14), we obtain

$$N_\phi = -\frac{7200\ \text{lb/ft}}{\sin^2\phi}\sin\frac{\phi+9.96°}{2}\sin\frac{\phi-9.96°}{2} - \frac{77.1\ \text{lb/ft}}{\sin^2\phi},$$

$$N_\theta = -N_\phi - 3600\ [\text{lb/ft}]\cos\phi.$$

At the upper edge ($\phi = 9.96°$) these formulas yield $N_\phi = -2580$ lb/ft, $N_\theta = -966$ lb/ft, and at the springing line ($\phi = 38.7°$): $N_\phi = -2087$ lb/ft, $N_\theta = -723$ lb/ft. The highest stress occurs at the upper edge and is $\sigma_\phi = N_\phi/t = -107.5$ lb/in². With an admissible stress of 500 or 600 lb/in² there is sufficient margin for additional bending stresses.

2.2.2.2 Boiler End

Pressure vessels of all kinds are built as shells of revolution, consisting of a cylindrical drum and two ends which may be shaped as hemispheres, half ellipsoid or in any other suitable form. They have to resist an internal pressure p, constant and perpendicular to the wall.

When we put $p_\phi = 0$, $p_r = p$, the integral (10) may be simplified considerably. Making use of (3a), we find

$$N_\phi = \frac{1}{r_2\sin^2\phi}\int_0^\phi r_1 r_2\,p\cos\phi\sin\phi\,d\phi = \frac{p}{r_2\sin^2\phi}\int_0^r r\,dr,$$

and this integral may be evaluated independently of the shape of the meridian. Eq. (9b) then yields the hoop force N_θ. Thus we get the following simple expressions for the stress resultants in pressure vessels:

$$N_\phi = \frac{1}{2}\, p\, r_2, \qquad N_\theta = p\, r_2 \frac{2r_1 - r_2}{2r_1}. \tag{15}$$

We shall use these formulas to study some typical forms of boiler ends.

Boiler ends are often shaped as flat ellipsoids of revolution (fig. 6). As we find easily by well-known methods of analytical geometry, the

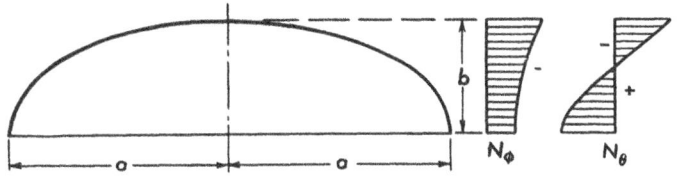

Fig. 6. Ellipsoid as boiler end

elliptic meridian has the radius of curvature

$$r_1 = \frac{a^2 b^2}{(a^2 \sin^2\phi + b^2 \cos^2\phi)^{3/2}},$$

and the radius of transversal curvature of the ellipsoid is

$$r_2 = \frac{a^2}{(a^2 \sin^2\phi + b^2 \cos^2\phi)^{1/2}}.$$

Introducing these expressions and the given load into the eq. (15), we find

$$N_\phi = \frac{p\, a^2}{2} \frac{1}{(a^2 \sin^2\phi + b^2 \cos^2\phi)^{1/2}},$$

$$N_\theta = \frac{p\, a^2}{2b^2} \frac{b^2 - (a^2 - b^2)\sin^2\phi}{(a^2 \sin^2\phi + b^2 \cos^2\phi)^{1/2}}.$$

At the vertex $\phi = 0$ we have $N_\phi = N_\theta$. This is no peculiarity of the ellipsoid but is true for any surface of revolution. At the vertex all meridians meet, and any direction is parallel to one of them and at right angles to another. Since in a surface of continuous curvature we have at the vertex $r_1 = r_2$, the common magnitude of both longitudinal forces may be found immediately from (9b):

$$N_\phi = N_\theta = \frac{p_r\, r_1}{2}$$

and this may be used as a boundary condition to determine C in (10).

Fig. 6 shows the distribution of the stress resultants in the shell. The hoop force changes sign and becomes negative near the equator. The zero is found where

$$\sin\phi = \frac{b}{\sqrt{a^2 - b^2}}.$$

This formula yields a real angle only if $a/b \geqq \sqrt{2}$. If the ellipsoid is flatter than indicated by this ratio of its axes, an equatorial zone exists where the hoop stress is a compression. The elastic deformation of such a shell must be such that the diameter of its border decreases. On the other hand, the cylindrical part of the boiler has a positive hoop force $N_\theta = p\,a$ everywhere as we see from eq. (9b) by putting $r_1 = \infty$, $r_2 = a$. On the parallel circle where the two parts meet, they have quite different deformations and will not fit together without an additional deformation. This is furnished by bending stresses, which

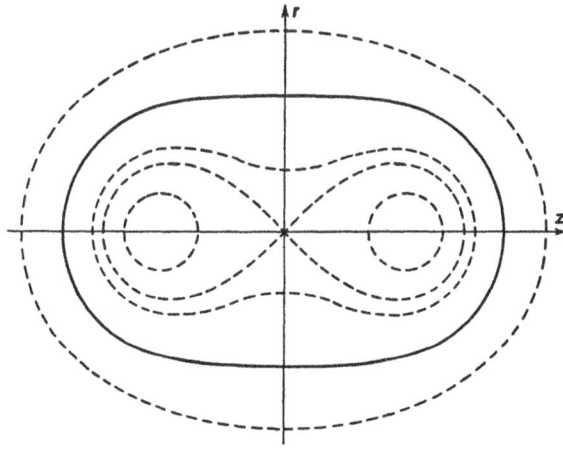

Fig. 7. CASSINI's curves

bend the cylinder inward and the ellipsoid outward. We shall study them in detail in Chapters 5 and 6.

The discrepancy of the hoop forces of the boiler end and the boiler drum may be avoided by choosing another shape of the meridian. The only requirement is that the radius $r_1 = \infty$ for $\phi = 90°$. There are, of course, many curves which fulfill this condition. One of them may be found among the Cassinian curves (fig. 7). Its equation is

$$(r^2 + z^2)^2 + 2a^2(r^2 - z^2) = 3a^4.$$

This curve is rather lengthy and therefore not particularly fit for the end of a pressure vessel, but its property of zero curvature at $z = 0$ is preserved when we subject it to an affine transformation, substituting $n\,z$ for z with $n > 1$:

$$(r^2 + n^2 z^2)^2 + 2a^2(r^2 - n^2 z^2) = 3a^4.$$

To find the stress resultants in a boiler end having this curve as a meridian, we need the radii r_1 and r_2. A simple but somewhat lengthy computation yields the following formulas:

$$r_1 = 2 \frac{[r^2(a^2 + n^2 z^2) + n^4 z^2 (a^2 - r^2)]^{3/2}}{3 n^2 a^3 (a^2 - r^2 + n^2 z^2)},$$

$$r_2 = 2 a \frac{[r^2(a^2 + n^2 z^2) + n^4 z^2 (a^2 - r^2)]^{1/2}}{a^2 + r^2 + n^2 z^2}.$$

Introducing this into (15), we find the stress resultants

$$N_\phi = p a \frac{[r^2(a^2 + n^2 z^2) + n^4 z^2(a^2 - r^2)]^{1/2}}{a^2 + r^2 + n^2 z^2},$$

$$N_\theta = N_\phi \left[2 - \frac{3 n^2 a^4 (a^2 - r^2 + n^2 z^2)}{(a^2 + r^2 + n^2 z^2) [r^2(a^2 + n^2 z^2) + n^4 z^2(a^2 - r^2)]} \right].$$

Fig. 8 gives an example of the distribution of the stress resultants in such a boiler end. It shows the continuity of the hoop force. There is

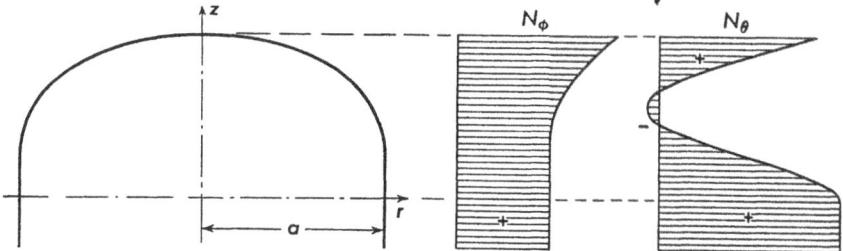

Fig. 8. Boiler end without discontinuity in the hoop forces

a small zone in which N_θ is negative. This may be avoided by choosing $n < 1.9$. If n is chosen much greater than 2, the compressive zone is wider and the maximum compressive stress higher.

2.2.2.3 Pointed Shells

It is not necessary that the meridian meet the axis of the shell at a right angle. If it does not, a shell with a pointed apex results. Such shells have some particularities which we shall now study in a typical example. The meridian of the dome, fig. 9, is a circle whose center does not lie on the axis of revolution. Although the radius of curvature $r_1 = a$ of the meridian is a constant, the radius of transversal curvature is variable:

$$r_2 = \frac{r}{\sin\phi} = a \left(1 - \frac{\sin\phi_0}{\sin\phi} \right).$$

We ask for the stress resultants produced by the weight of the structure, assuming a constant wall thickness. The load is then given by (12). We find N_ϕ from eq. (10) and avoid the determination of the constant C from a boundary condition by using the mechanical interpretation ·f this formula, writing the integral between the limits ϕ_0 and ϕ and dropping C:

$$N_\phi = -\frac{p\,a}{(\sin\phi - \sin\phi_0)\sin\phi} \int_{\phi_0}^{\phi} (\sin\phi - \sin\phi_0)\,d\phi$$

$$= -p\,a\,\frac{(\cos\phi_0 - \cos\phi) - (\phi - \phi_0)\sin\phi_0}{(\sin\phi - \sin\phi_0)\sin\phi}.$$

The hoop force then follows from (9b):

$$N_\theta = -\frac{p\,a}{\sin^2\phi}\,[(\phi - \phi_0)\sin\phi_0 - (\cos\phi_0 - \cos\phi) +$$

$$+ (\sin\phi - \sin\phi_0)\cos\phi\sin\phi].$$

At the vertex $\phi = \phi_0$ these formulas yield $N_\theta = 0$, but N_ϕ becomes indefinite. We find in the usual way by differentiating the numerator

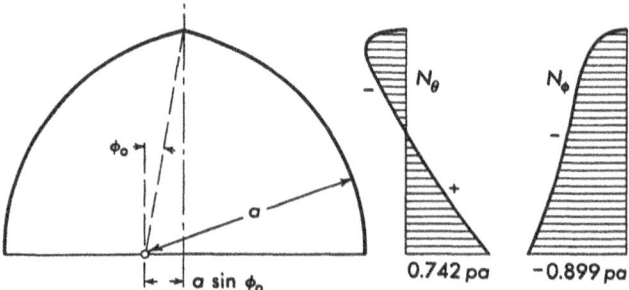

Fig. 9. Ogival shell. Force diagrams for $\phi_0 = 10°$

and denominator that N_ϕ also becomes zero. The stress distribution is shown in fig. 9.

In the limiting case $\phi_0 = 0$ the ogival dome becomes a sphere, and the preceding formulas give the stress resultants of a spherical dome. In this limiting case N_ϕ and N_θ are no longer zero at the top. One may easily see from fig. 9, how the limiting case is approached when $\phi_0 \to 0$: For very small values of ϕ_0, the normal forces rise rather suddenly from zero to approximately $-p\,a/2$. Such a sudden local change of the stress resultants sometimes occurs in membrane theory formulas, but it does not represent a physical reality. It would lead to almost discontinuous deformations, and the shell avoids such states of stress by additional bending stresses, as will be discussed in Chapter 6.

We now consider a modification of the ogival dome, in which the meridian begins at the axis with a negative value of ϕ, say $\phi = -\phi_0$. This results in a cupola of the type of fig. 10, having a downward point at its center. Let us compute the stresses for a snow load, distributed uniformly over the projected area. Its components are

$$p_\phi = p \cos\phi \sin\phi, \quad p_r = -p \cos^2\phi.$$

From eq. (10) we find

$$N_\phi = -\frac{p\,a}{(\sin\phi + \sin\phi_0)\sin\phi} \int_0^\phi (\sin\phi + \sin\phi_0)\cos\phi\,d\phi + \frac{C}{a(\sin\phi + \sin\phi_0)\sin\phi}$$

and after evaluation of the integral

$$N_\phi = -\frac{p\,a}{2}\frac{\sin\phi + 2\sin\phi_0}{\sin\phi + \sin\phi_0} + \frac{C}{a(\sin\phi + \sin\phi_0)\sin\phi}.$$

The denominator is zero for $\phi = -\phi_0$, and if we put $C = 0$, N_ϕ will become infinite at this point. It is possible to give C such a value that the numerator vanishes too, leading to $N_\phi = 0$, as we had in the ogival shell. But then N_ϕ would be infinite on the whole top circle $\phi = 0$, and that would be much worse. We choose tentatively $C = 0$ and we shall see at once what the sin-.gularity at the center means. Our formula now reads

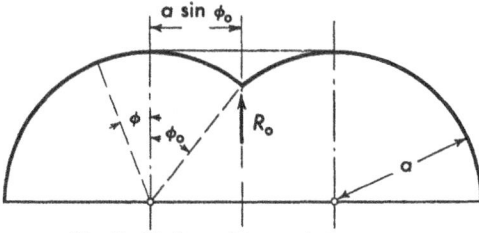

Fig. 10. Shell requiring central support

$$N_\phi = -\frac{p\,a}{2}\frac{\sin\phi + 2\sin\phi_0}{\sin\phi + \sin\phi_0},$$

and the hoop force follows from (9b):

$$N_\theta = \frac{p\,a}{2}(2\sin\phi_0\sin\phi - \cos2\phi).$$

To study the singularity, we cut the shell in a parallel circle having a negative ϕ, say $\phi = -\phi' < 0$ and compute the resultant of the forces N_ϕ which act on it. It is a vertical force of magnitude

$$R = \frac{p\,a}{2}\frac{-\sin\phi' + 2\sin\phi_0}{-\sin\phi' + \sin\phi_0} \cdot \sin\phi' \cdot 2\pi a(\sin\phi_0 - \sin\phi')$$

$$= p\,a^2\,\pi(2\sin\phi_0 - \sin\phi')\sin\phi'.$$

For $\phi' = 0$, in the top-circle, R is zero. This means that the meridional forces there which are horizontal, cannot carry any load from the inner part of the shell to the outer part. The inner part must therefore find its equilibrium by a special support, and that is possible only in the center. Indeed, for $\phi' = \phi_0$, the resultant R is

$$R_0 = p\,a^2\,\pi\sin^2\phi_0,$$

and this indicates what the singularity of the stress resultants means: that forces of infinite intensity, acting on a circle of radius zero, carry the total load applied on the part of the shell within the top circle. A support, say a column, which can exert a vertical force R_0 is needed there. Then the infinity disappears if the thin shell extends only to the circumference of this column.

The stress system, which we now have found, shows nothing special on the top circle $\phi = 0$ and seems to be quite harmless. But on p. 99, when discussing the deformations of toroidal shells, we shall see that this stress system cannot be realized because it would lead to an impossible deformation. We therefore have to expect additional bending stresses in a certain zone near the top circle, but since they are needed only to remedy an impossible deformation, they will be much smaller than those which would be needed in the absence of a central support, and which would have to transmit an important part of the total load. It is this argument which finally justifies our choice for the constant C.

2.2.2.4 Toroidal Shell

A toroid is generated by the rotation of a closed curve about an axis passing outside. A toroidal shell encloses an annular volume and may be considered as a pressure vessel. Figs. 11 and 12 show meridional sections of two typical cases.

The shell, fig. 11, may be cut in two parts as indicated by the broken line. The meridian of each part begins and ends with a horizontal

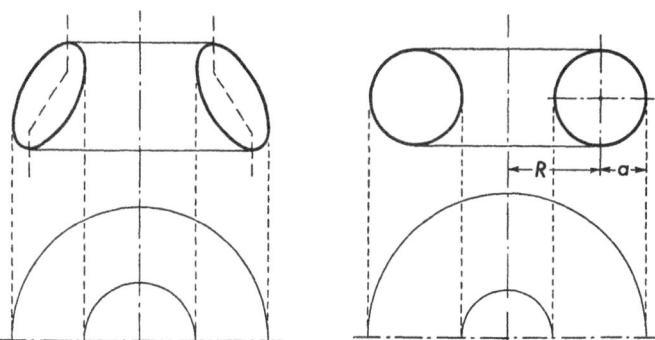

Figs. 11 and 12. Toroidal shells

tangent. Therefore, the meridional forces acting at each edge do not have a vertical component and cannot transmit any vertical force from one half of the shell to the other. Now, when the shell is filled with gas of pressure p, this pressure has a downward resultant on the inner half and an upward resultant of the same magnitude on the outer half, and neither part can be in equilibrium under the action of the pressure p and the forces on its edges. It follows that a membrane stress system with finite values N_ϕ, N_θ is not possible in this shell under this load.

This difficulty disappears when the two top circles have the same radius, e. g. when the meridian of the shell is a circle (fig. 12). Then eq. (10) gives with $p_\phi = 0$, $p_r = p$:

$$N_\phi = \frac{p\,a}{(a\sin\phi + R)\sin\phi}\left[\int (a\sin\phi + R)\cos\phi\,d\phi + C\right]$$

$$= \frac{p\,a}{(a\sin\phi + R)\sin\phi}\left[-\frac{a}{4}(\cos^2\phi - \sin^2\phi) + R\sin\phi + C\right],$$

and here we can determine C so that the singularities at $\phi = 0$ and at $\phi = \pi$ disappear simultaneously. This yields

$$N_\phi = \frac{p\,a}{a\sin\phi + R}\left(\frac{a}{2}\sin\phi + R\right), \qquad N_\theta = \frac{p\,a}{2}.$$

However, this solution also cannot be realized in the vicinity of the top and bottom circles without additional bending, because it again leads to an incompatibility of deformations which we shall discuss on p. 99.

2.2.2.5 Tanks

Our next example we choose in the domain of steel tanks. Fig. 13 shows a spherical tank, as used for storing water or gas. It is a complete sphere, supported along one of its parallel circles, $A\,A$. The essential

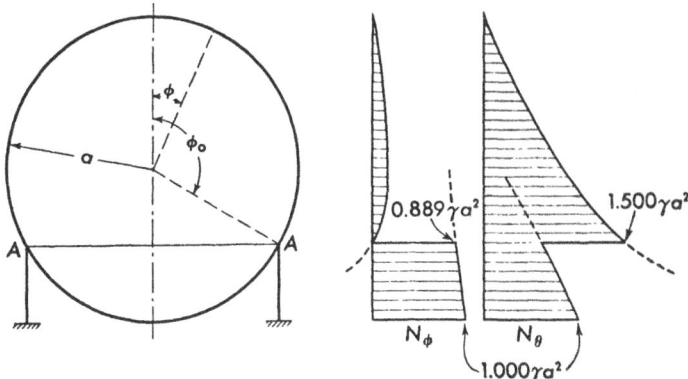

Fig. 13. Spherical water tank; support at $\phi_0 = 120°$

load for a water tank is the pressure of the water (specific weight γ). It is normal to the shell ($p_\phi = 0$) and proportional to the depth. If the tank is completely filled, we have

$$p_r = \gamma\,a(1 - \cos\phi).$$

By a simple integration, we find from eq. (10) the meridional force

$$N_\phi = \frac{\gamma\, a^2}{\sin^2\phi} \left[\int (1 - \cos\phi)\cos\phi \sin\phi\, d\phi + C \right]$$

$$= \frac{\gamma\, a^2}{6\sin^2\phi} [(2\cos\phi - 3)\cos^2\phi + 6C].$$

At the vertex $\phi = 0$ the denominator vanishes. To obtain a finite value of N_ϕ the factor in brackets must also become zero. This leads to $C = 1/6$, and after some simple transformation we find

$$N_\phi = \frac{\gamma\, a^2}{6} \frac{1 - \cos\phi}{1 + \cos\phi}\, (1 + 2\cos\phi)$$

and, from eq. (9b),

$$N_\theta = \frac{\gamma\, a^2}{6} \frac{1 - \cos\phi}{1 + \cos\phi}\, (5 + 4\cos\phi).$$

These formulas are valid above the supporting circle $\phi = \phi_0$. In the lower part of the shell we have to apply another value of C, which makes N_ϕ finite at $\phi = \pi$. It is $C = 5/6$ and hence we have

$$N_\phi = \frac{\gamma\, a^2}{6} \frac{5 - 5\cos\phi + 2\cos^2\phi}{1 - \cos\phi},$$

$$N_\theta = \frac{\gamma\, a^2}{6} \frac{1 - 7\cos\phi + 4\cos^2\phi}{1 - \cos\phi}.$$

The distribution of these forces is shown in fig. 13.

The location of the supporting circle does not influence the two values of C. If we give it a higher or lower position, only the domains of validity of the two pairs of formulas are changed. The corresponding changes in the stress resultants are indicated by dotted lines in fig. 13. They show that a position of the support below $\phi = 120°$ leads to compressive forces in the meridian, which in a thin-walled structure like this one should be avoided, and that a higher position cuts off the peak value of N_θ which determines the wall thickness, but of course it leads to a larger and more expensive support.

At the supporting ring both stress resultants change their values discontinuously. The difference of the meridional forces is a load applied to the ring. We resolve it into a vertical component

$$\frac{2\gamma\, a^2}{3\sin\phi_0},$$

directed downward, which the ring must pass to its numerous supports by bending and torsion, and into a horinzontal component

$$\frac{2\gamma\, a^2}{3} \frac{\cot\phi_0}{\sin\phi_0},$$

which is a radial load applied to the ring, producing in it a compressive hoop stress.

Here we have again a case in which the direct stresses lead to a deformation which is incompatible with the continuity of the structure. A discontinuity in the hoop force means a discontinuity of the elastic extension of the parallel circles. A membrane-stress system which avoids this discrepancy cannot exist, since we have already used all available constants to fulfill other, more important conditions. The continuity of deformations can be reestablished only by an additional bending of the border zones of both halves of the shell, and again we have to refer to the treatment of this problem in Chapter 6.

A similar disturbance, but of greater intensity, is caused by the connection of the shell to the supporting ring, if this is supported by vertical forces, as shown in fig. 13. Then the ring is subject to compressive stresses which fit the positive hoop stresses in both parts of the shell even more poorly than these fit each o her. For this reason it is preferable to support the ring by inclined bars, tangential to the meridians of the shell, or even by a conical steel plate. Then the ring is relieved of its hoop stress and causes less disturbance of the membrane forces of the shell.

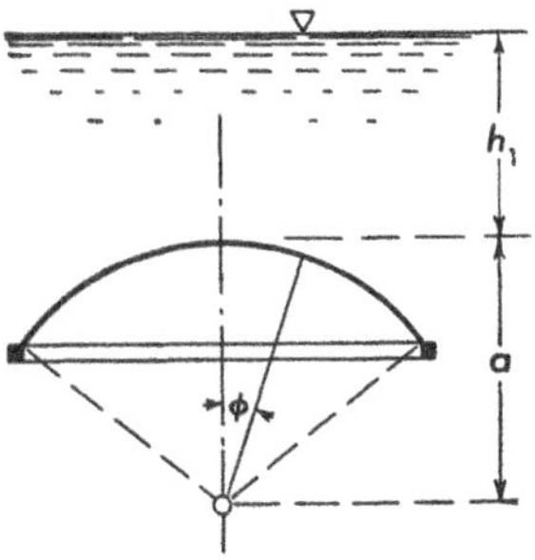

If we change the formula for p_r slightly, writing

$$p_r = -\gamma(h_1 + a - a\cos\phi),$$

we may obtain the membrane forces in a spherical tank bottom such as that shown in fig. 14. The evaluation of the integral (10) and subsequent application of (9 b) yield

Fig. 14. Spherical tank bottom

$$N_\phi = -\frac{\gamma a}{6}\left[3h_1 + a\frac{1-\cos\phi}{1+\cos\phi}(1+2\cos\phi)\right],$$

$$N_\theta = +\frac{\gamma a}{6}\left[3h_1 + a\frac{1-\cos\phi}{1+\cos\phi}(5+4\cos\phi)\right].$$

These are both compressive forces, and at the edge of the shell there must be a ring to take care of the horizontal component of the meridional force N_ϕ.

Another kind of tank bottom which is of practical interest is shown in fig. 15a. It is the lower half of an ellipsoid of revolution. Some formulas concerning its geometry have already been given on p. 27. We add here the relation

$$z = \frac{-b^2\cos\phi}{(a^2\sin^2\phi + b^2\cos^2\phi)^{1/2}}.$$

The load on the shell is $p_r = \gamma(h + z)$. When this is introduced into the integral (10), a somewhat lengthy integration must be performed.

It remains, however, within the domain of elementary functions and yields finally

$$N_\phi = \frac{\gamma\, h\, a^2}{2} \frac{1}{(a^2 \sin^2\phi + b^2 \cos^2\phi)^{1/2}} + \frac{\gamma\, b}{3} \frac{b^3 \cos^3\phi + (a^2 \sin^2\phi + b^2 \cos^2\phi)^{3/2}}{(a^2 \sin^2\phi + b^2 \cos^2\phi) \sin^2\phi}.$$

This may be introduced into eq. (9b). It is not of much use to do this in general terms, since a rather clumsy formula would result. We prefer to write simply

$$N_\theta = -\frac{\gamma\, a^2 (h + z)}{(a^2 \sin^2\phi + b^2 \cos^2\phi)^{1/2}} - \frac{a^2 \sin^2\phi + b^2 \cos^2\phi}{b^2}\, N_\phi$$

and to use this formula for numerical work.

At the bottom of the tank, $\phi = 180°$, we obtain

$$N_\phi = N_\theta = \frac{\gamma\, a^2}{2\, b} (h + b).$$

At the edge, $\phi = 90°$, the meridional force is

$$N_\phi = \frac{\gamma\, a}{6} (3h + 2b).$$

This force transmits the whole water weight to the cylindrical wall. We shall see on p. 195 how it may be transferred from there to a support. The hoop force at $\phi = 90°$ is

$$N_\theta = \frac{\gamma\, a}{6\, b^2} [3h(2b^2 - a^2) - 2b\, a^2].$$

If $b > a/\sqrt{2}$, this may be positive when h is large enough, but it always becomes negative when the water level in the tank is lowered. If $b < a/\sqrt{2}$, the hoop force at the edge of the bottom is always a compression, independent of h.

For a tank bottom with $h = 1.5a$, $b = 0.6a$ the stress resultants are plotted in fig. 15b over the horizontal projection of the meridian.

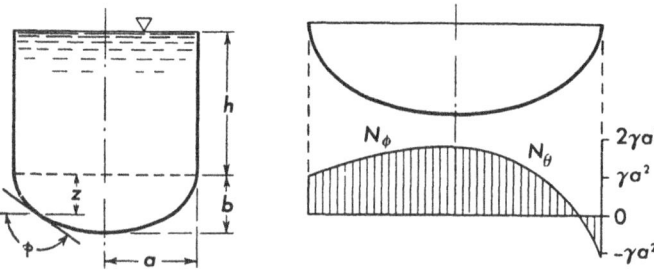

Fig. 15. Ellipsoid as tank bottom, (a) Tank, (b) Tank bottom and stress resultants

The figure illustrates that $N_\phi = N_\theta$ at the center and that the hoop force changes sign near the edge of the shell. The greatest compressive force is slightly more than one half of the greatest tension.

2.2.2.6 Conical Shell

In conical shells, the slope angle ϕ is a constant and can no longer serve as a coordinate on the meridian. We replace it by the arc length s, measured from the top of the cone (fig. 16). Accordingly, we have to use eqs. (7 a–c). Simplifying them for axial symmetry and putting $\phi = \alpha$. $r = s \cos\alpha$, $r_1 = \infty$, $r_2 = s \cot\alpha$, we find from them the following set:

$$\frac{d}{ds}(N_s s) - N_\theta = -p_s s, \qquad (16)$$
$$N_\theta = p_r s \cot\alpha.$$

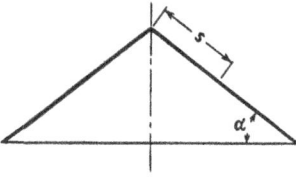

Fig. 16. Conical shell

The second of these equations yields N_θ immediately, as a function of the local intensity of the normal load, there being no chance of adapting it to a boundary condition. A similar situation exists in cylindrical shells, and we shall discuss its consequences in Chapter 3 in complete detail.

By simple addition of both our equations we obtain a first-order differential equation for N_s:

$$\frac{d}{ds}(N_s s) = -(p_s - p_r \cot\alpha)\,s,$$

from which we find the meridional force by simple integration:

$$N_s = -\frac{1}{s}\int (p_s - p_r \cot\alpha)\,s\,ds. \qquad (17)$$

As an example we consider a mushroomlike shelter as shown in fig. 17. The weight p of the shell has the components

$$p_s = p \sin\alpha, \qquad p_r = -p \cos\alpha.$$

We find

$$N_\theta = -p\,s\cos\alpha \cot\alpha$$

and

$$N_s = -\frac{1}{s}\frac{p}{\sin\alpha}\frac{s^2}{2} + \frac{C}{s}.$$

At the free edge $s = l$ of the roof, this must give zero, whence $C = p\,l^2/(2\sin\alpha)$ which gives

$$N_s = \frac{p}{2}\frac{l^2 - s^2}{s\sin\alpha}.$$

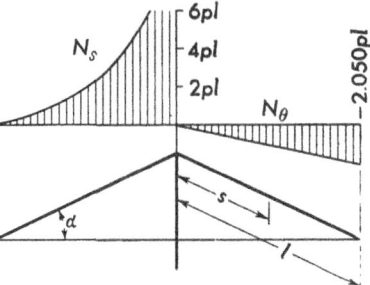

Fig. 17. Conical shell roof

Fig. 17 illustrates this result. At the top $s = 0$ the meridional force becomes infinite of the first order, as we must expect at a point support. It may easily be checked that the vertical resultant of the forces N_s transmitted in a parallel circle approaches the total load of the shell, when that circle is contracted into the point $s = 0$.

If the shell is not supported at the center but along the edge, the hoop force N_θ will be the same, but in the general expression for N_s the value of the constant must be chosen $C = 0$ to make the meridional force finite at $s = 0$. This yields

$$N_s = -\frac{ps}{2\sin\alpha}.$$

Of course, the support must be adequate to resist the thrust of the shell. If it can resist only vertical forces, a pure direct stress system is not possible in the shell, and the additional bending will be of such magnitude that it upsets the stress system thoroughly in the border zone.

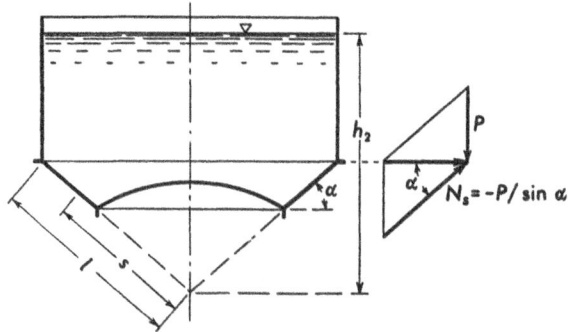

Fig. 18. INTZE tank bottom

When we assume all distributed loads to be zero, the hoop force will also be zero, and there remains the homogeneous solution

$$N_s = \frac{C}{s}.$$

It corresponds to a concentrated force applied at the top of the cone in the direction of its axis. Its magnitude is

$$P = C\pi\sin 2\alpha,$$

as may be found by integrating the vertical component of N_s along an arbitrary parallel circle.

For the bottom of elevated water tanks, a conical and a spherical shell are often combined as shown in fig. 18 (Intze tank). The solution for the spherical part is given on p. 35. For the cone we have with

$$p_s = 0, \qquad p_r = \gamma(h_2 - s\sin\alpha)$$

the stress resultants

$$N_\theta = \gamma s\left(\frac{h_2}{\sin\alpha} - s\right)\cos\alpha,$$

$$N_s = \frac{\gamma\cot\alpha}{s}\left[\int(h_2 - s\sin\alpha)s\,ds + C\right].$$

The constant follows from a condition at the outer edge $s = l$ of the cone. From the weight of the cylindrical wall of the tank and from loads which may be applied to it, there will be a vertical load, say P, per unit of circumference. Since the cone can absorb forces only in the direction of its generators, a ring must be provided. It receives a radial load $P \cot\alpha$, producing the hoop force

$$F = P\, l \cos\alpha \cot\alpha.$$

The cone receives the component $P/\sin\alpha$, and the boundary condition is therefore $N_s = -P/\sin\alpha$. This determines C and then

$$N_s = \frac{\gamma \cot\alpha}{6s}\,[2\,(l^3 - s^3)\sin\alpha - 3h_2\,(l^2 - s^2)] - \frac{P\, l}{s\sin\alpha}.$$

Where the conical and spherical parts of the bottom meet, a ring must be provided which resists the difference of the horizontal components of the meridional forces N_s in the cone and N_ϕ in the sphere. This ring may be omitted or reduced to what is needed for structural purposes, if the dimensions of the shells are so chosen that the thrusts of cone and sphere balance each other. This condition can, of course, be fulfilled only for a certain load, e. g. that one belonging to the highest water level in the tank.

2.3 Shells of Constant Strength

2.3.1 Drop-shaped Tank

In the upper part of a spherical water tank, fig. 13, the meridional stress $\sigma_\phi = N_\phi/t$ is rather small compared with the hoop stress $\sigma_\theta = N_\theta/t$. In a cylindrical tank there is no meridional stress at all and the hoop stress alone carries the load. From eq. (6c) one might expect to save some steel, if one could shape the tank so that at every point the two stresses are equal to each other and to a certain design stress σ. For a tank of constant wall thickness this amounts to having everywhere $N_\phi = N_\theta = \sigma\, t$.

The problem of finding the shape of this tank is identical to that of finding the shape of a drop of liquid resting on a plane surface. $N_\phi = N_\theta$ is then the capillary force which, exactly as the stress resultants in the tank wall, must be in equilibrium with the hydraulic pressure. Therefore we speak of a drop-shaped tank.

To establish the differential equation for the meridian, we assume that the tank is completely filled with a liquid of specific weight γ and that at the highest point there is still a pressure, which we call $\gamma\, h$. It might be maintained by a safety valve or a standpipe, connected anywhere with the interior of the tank and having a free liquid level at the altitude h above the top of the tank. The origin of the coordinates r, z is chosen on this level (fig. 19).

In this notation the load acting on the shell is

$$p_\phi = 0, \qquad p_r = \gamma z.$$

Since we want constant forces $N_\phi = N_\theta = \sigma t$, eq. (9b) becomes

$$\sigma t \left(\frac{1}{r_1} + \frac{1}{r_2} \right) = \gamma z, \tag{18}$$

while (9a) degenerates into the simple geometric relation (4a).

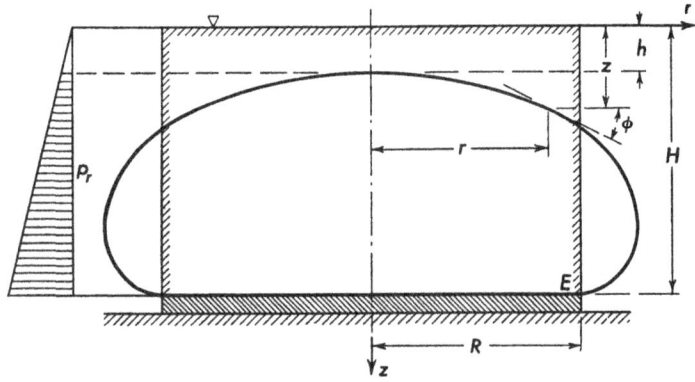

Fig. 19. Drop-shaped tank

We now use eqs. (4a) and (1) to express the radii r_1 and r_2 in eq. (18) in terms of the coordinates r and z and obtain

$$\frac{d \sin\phi}{dr} + \frac{\sin\phi}{r} = \frac{\gamma}{\sigma t} z. \tag{19a}$$

This is a differential equation in which r appears as the independent variable and where ϕ and z are the unknowns. The geometrical relation

$$\tan\phi = \frac{dz}{dr} \tag{19b}$$

is another equation of the same kind, and the solution of this pair determines the shape of the meridian we seek.

Eqs. (19) can be solved only by numerical integration. In order to make the results of such a computation as useful as possible, it is advisable to introduce dimensionless notations. We do so by putting

$$a^2 = \frac{\sigma t}{\gamma}, \qquad \varrho = \frac{r}{a}, \qquad \zeta = \frac{z}{a}, \qquad \eta = \sin\phi$$

and bring eqs. (19) into the following form:

$$\frac{d\eta}{d\varrho} = \zeta - \frac{\eta}{\varrho}, \qquad \frac{d\zeta}{d\varrho} = \frac{\eta}{\sqrt{1 - \eta^2}}. \tag{20a, b}$$

If for some value of ϱ the values of η and ζ are known, one may introduce them on the right-hand side of these equations and calculate the derivatives and, from them, the increments $\varDelta\eta$, $\varDelta\zeta$ for a small increment of the independent variable. This yields η and ζ for a new value of ϱ, and now the procedure may be repeated, advancing step by step to higher values of ϱ. This method is rather crude, but in many cases good enough. It may be refined in many ways which are explained in books on numerical integration[1].

All these methods progress easily once the computation has been started, but they require some outside help in the beginning, the more refined methods requiring more help.

We know that for $\varrho = 0$ the tangent to the meridian must be horizontal, that is, $\eta = 0$, and that $z = h$, that is, $\zeta = h/a$. When we try to introduce this into eqs. (20a, b), we find that η/ϱ assumes the form $0/0$. We may overcome this deadlock remembering that at the apex of a surface of revolution $r_1 = r_2$. It follows from eq. (18) that

$$r_1 = \frac{2\sigma t}{\gamma h} = \frac{2a^2}{h}$$

and hence for sufficiently small values of r that

$$r = r_1 \sin\phi = \frac{2a^2}{h}\sin\phi$$

and

$$\varrho = \frac{2a}{h}\,\eta.$$

Eq. (20a) must therefore be started as

$$\frac{d\eta}{d\varrho} = \frac{h}{a} - \frac{h}{2a} = \frac{h}{2a},$$

and then there is no difficulty in continuing.

For the more elaborate methods of numerical integration it is necessary to have values of η and ζ for several equidistant values of ϱ, beginning with $\varrho = 0$. Such starting values may be procured in the following way.

Near the top of the shell $\sin\phi$ is small, and may be neglected, compared with 1 in eq. (20b). The equations then become linear in η and ζ, and η may easily be eliminated. When this is done, a second-order equation results:

$$\frac{d^2\zeta}{d\varrho^2} + \frac{1}{\varrho}\frac{d\zeta}{d\varrho} - \zeta = 0. \tag{21}$$

---

[1] SCARBOROUGH, J. B.: Numerical Mathematical Analysis. 3rd ed. Baltimore 1955, Chapter XI. – MILNE, W. E.: Numerical Solution of Differential Equations, Chapter V. New York 1952. – COLLATZ, L.: Numerische Behandlung von Differentialgleichungen. 2nd ed., Berlin 1955, Chapter II.

Except for the minus sign this is BESSEL's equation, and (21) is solved by the modified BESSEL functions of order zero:

$$\zeta = A\,I_0(\varrho) + B\,K_0(\varrho).\tag{22}$$

Since $I_0(0) = 1$, $K_0(0) = \infty$, one must choose $A = h/a$, $B = 0$ to meet the condition that $z = h$ for $r = 0$, whence

$$\zeta = \frac{h}{a}\,I_0(\varrho)\tag{23a}$$

and consequently

$$\eta = \frac{d\zeta}{d\varrho} = \frac{h}{a}\,I_1(\varrho).\tag{23b}$$

From these formulas η and ζ may be computed for small values of ϱ, and with them the numerical integration may be started.

Whatever method may be used for the integration, it will be found that it does no work well beyond $\phi \approx 50°$, corresponding to $\eta \approx 0.75$. This is easy to understand since ϱ is not suitable as an independent variable when ϕ approaches $90°$. It is therefore necessary to change the procedure and to write everything in terms of ζ. From eq. (4b) we derive that

$$\frac{1}{r_1} = -\frac{d\cos\phi}{dz},$$

and from this and eq. (1) we see that eq. (18) may be brought into the form

$$-\frac{d\cos\phi}{dz} + \frac{\sin\phi}{r} = \frac{z}{a^2}.\tag{19c}$$

This and eq. (19b) are again a pair of differential equations for ϕ and z. As before, we use ϱ and ζ as dimensionless variables, but instead of η it is better to introduce now

$$\xi = \cos\phi.$$

With these notations eqs. (19b, c) assume the form

$$\frac{d\xi}{d\zeta} = \frac{\sqrt{1-\xi^2}}{\varrho} - \zeta, \qquad \frac{d\varrho}{d\zeta} = \frac{\xi}{\sqrt{1-\xi^2}}.\tag{24a, b}$$

These equations may be used to continue the computation to $\phi \approx 140°$, and the rest of the meridian is determined by returning to the set (20).

Fig. 19 shows a typical shape of meridian obtained by the procedure described. The meridian ends at the point E with a horizontal tangent, $\phi = 180°$. The tank may be closed by a flat bottom into which the force N_ϕ from the edge of the shell is introduced as a radial load and which, therefore, is in the same state of uniform biaxial stress as the shell. However, the flat bottom cannot resist the vertical pressure γH of the liquid and must be put on a smooth support, say a concrete slab, which can exert on it an upward pressure of the same magnitude. The total reaction of the support is the product of this pressure and of the area of the plane bottom, $\gamma\,H \cdot \pi\,R^2$. Since the tank is in equilibrium, this force must be equal to the weight of the liquid, $\gamma\,V$, where V is the capacity of the tank, whence

$$V = \pi\,R^2\,H.$$

Comparing this value with the volume obtained from direct integration of the meridian offers a valuable check for the numerical integration.

The computation depends on only one parameter h/a. When it is repeated for a series of different values of this parameter, a series of

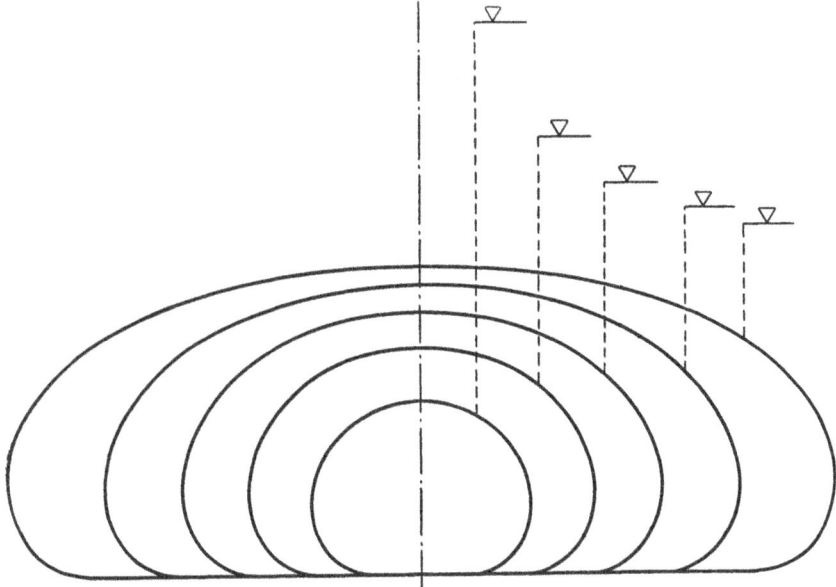

Fig. 20. Series of drop-shaped tanks

curves is obtained from which, by mere change of scale, all possible solutions of the problem may be derived. Fig. 20 represents such a series in which the scale for each curve has been chosen so that all have the same value of a. The higher the pressure head h at the top of the tank, the greater the curvature of the meridian and the smaller the tank. The small tanks are almost spherical, while the larger ones are rather flat. This corresponds to the fact that a small drop of water on a plane plate looks almost like a sphere and that a large one forms a shallow puddle of almost constant depth.

For $h = 0$ eqs. (23a, b) yield $\eta \equiv \zeta \equiv 0$ as far as they are applicable. Since they are good as long as $\eta^2 \ll 1$ (see eq. (20b)), they never

cease to be so. The shell degenerates into a plane plate, and no tank of
reasonable shape can be obtained. Therefore, drop tanks are not particularly fit for storing water, but they have been built for the storage
of gasoline. When the tank contains a volatile liquid, the pressure $\gamma\,h$
at the top is welcome to prevent evaporation losses in hot weather.

When a drop tank is to be constructed there will be given the
specific weight γ of the contents, the desirable pressure head h, the
working stress σ in the steel plates, and the capacity V. The first three
of these data enter into the parameter

$$\frac{h}{a} = h\sqrt{\frac{\gamma}{\sigma\,t}},$$

but V does not. Instead of this the wall thickness t appears. One has to
start the computation with an assumed thickness and at the end check

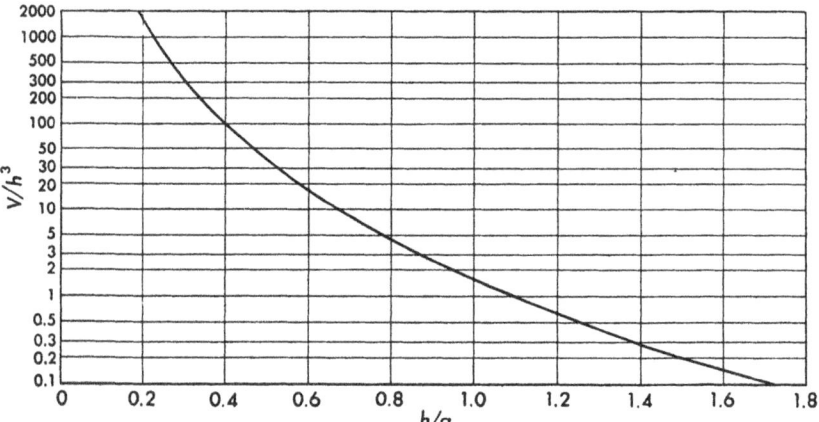

Fig. 21. Capacity of drop-shaped tanks vs. h/a

the volume of the resulting tank. Then the computation must be repeated with a better fitting wall thickness until agreement between the
resulting and the required capacity is reached. This will be facilitated
by fig. 21, where V/h^3 is plotted against h/a. With the help of this
diagram the computation may be started at once with the right value
of h/a.

The drop-shaped tank cannot be expected to have uniform stress
if the actual load is different from the design load, whether it be that
the top pressure is not exactly as assumed or that the tank is only
partially filled, with or without some gas pressure on the liquid level.
In all these cases N_ϕ must be found from eq. (10) by numerical integration and N_θ from eq. (9b). One essential result of such computations
may be predicted without going through the details of the numerical

work: When we separate the tank from its foundation, we find two external forces acting on it, the weight of the contents and the upward force exerted by the foundation on the flat bottom. The latter force equals the pressure γH at this level multiplied by the area of the bottom plate. Under arbitrary loading conditions one cannot expect that this reaction and the weight will be equal. The force N_ϕ at the edge E, fig. 19, cannot take care of the difference because it is horizontal, and therefore a transverse shear Q_ϕ is needed at the edge. Since the membrane theory denies the existence of transverse shears, it will yield $N_\phi = \pm\infty$, and N_θ will then become infinite, with the opposite sign. The practical application of the drop shape should therefore be limited to the upper part of the tank, say to $\phi < 150°$ or $160°$, and the rest should be completely cut away or replaced by an arc which leads

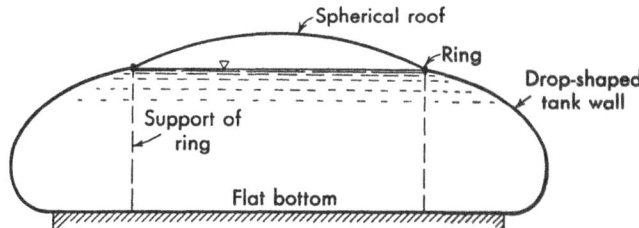

Fig. 22. Partially drop-shaped tank

without discontinuity of the curvature to a ring before its tangent becomes horizontal.

The height h depends on the liquid to be stored. Tanks of great capacity will always become rather flat and may not be able to support their own weight when empty. Such tanks may be built in an open form and closed by a roof which is not touched by the liquid (fig. 22). The calculation of the shape is very similar to that described here. The meridian starts with a set of finite values ϱ, η, ζ. If η is small enough, the approximate formula (22) may be used to start the integration, but now both terms must be employed, and the constants A, B must be chosen so as to meet the initial values of η and ζ.

2.3.2 Dome of Constant Strength

A shell dome looks almost like a three-dimensional arch structure. This raises the question whether or not for a given load there also exists a best shape, analogous to the funicular curve for the arch.

This question shows plainly the fundamental difference between the shell and the arch. Only an arch shaped like the funicular curve is free from bending moments; any other one needs them for its equilibrium. Exactly the contrary is true for a shell dome. We have seen how we can have equilibrium without bending in almost any shell for almost any load, and the additional bending which occurs in boundary zones is of somewhat the same importance as the bending moments in a statically indeterminate funicular arch.

From this situation it follows that we can ask for more than absence of bending. We can try to find a shape of the shell such that the membrane stress σ has the same magnitude at every point and in every direction.

As a first problem of this type we determine the shape of a dome which has to carry its own weight. The problem is a simple one if the dome consists of a plain concrete shell without additional dead load. Then, if γ is the specific weight of the concrete, the load per unit area of the surface is

$$p_\phi = \gamma\, t \sin\phi, \qquad p_r = -\gamma\, t \cos\phi.$$

Introducing this into eq. (9b) and putting

$$N_\phi = N_\theta = -\sigma t$$

(here σ is considered positive when it is a compression, contrary to our usual convention), we get

$$\sigma t \left(\frac{1}{r_1} + \frac{1}{r_2}\right) = \gamma\, t \cos\phi \tag{25}$$

and resolving with respect to r_1:

$$r_1 = \frac{r}{(\gamma/\sigma)\, r \cos\phi - \sin\phi}.$$

By means of the geometric relation (4a) this may be transformed into a simple differential equation for $r(\phi)$ or, better, for $\phi(r)$:

$$\frac{d\phi}{dr} = \frac{\gamma}{\sigma} - \frac{\tan\phi}{r}. \tag{26}$$

It may be solved by numerical integration, beginning at the vertex. There we have $r_1 = r_2$, and hence from (25) and (4a):

$$\frac{d\phi}{dr} = \frac{1}{r_1} = \frac{\gamma}{2\sigma}.$$

We see that there is only one parameter, σ/γ. It has the dimension of a length and determines the size of the shell. When we have found r as a function of ϕ, we determine the meridian in cartesian coordinates by a simple quadrature:

$$z = \int \tan\phi\, dr.$$

The wall thickness follows from eq. (9a), which here assumes the form

$$-\frac{d}{d\phi}(r\,\sigma t) + r_1 \sigma t \cos\phi + \gamma\, t\, r\, r_1 \sin\phi = 0$$

and yields
$$\frac{dt}{d\phi} = \frac{\gamma}{\sigma}\, t \tan\phi \cdot \frac{dr}{d\phi}\,.$$

This equation has a simple solution, when we transform it to rectangular coordinates r, z. We have
$$\frac{d}{d\phi} = r_1 \frac{d}{ds} = r_1 \sin\phi \cdot \frac{d}{dz}, \qquad \frac{dr}{dz} = \cot\phi$$
and therefore
$$\frac{dt}{dz} = \frac{\gamma}{\sigma}\, t$$

which is solved by
$$t = t_0 \exp\frac{\gamma z}{\sigma}\,.$$

The solution is represented in fig. 23. The shell may be extended to greater angles ϕ, but then the exponential growth of t leads to structures which soon cease to be thin-walled and probably are beyond the sphere of technical interest.

For domes of usual sizes, the problem of a shell of constant stress is of no practical importance, because shells of any reason-

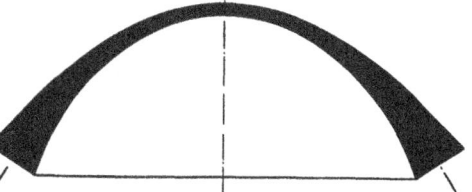

Fig. 23. Dome of constant strength

able shape will have direct stresses far below the admissible limit. But we see, for example, from eq. (13) for a spherical dome, that the stress resultants are proportional to $p\,a$; therefore the stresses caused by the weight of the shell are prop rtional to $\gamma\,a$. This indicates that they increase in proportion to the diameter of the dome, independently of the wall thickness. Therefore, for every shape of the shell there exists a certain size beyond which it can no longer be built in a material of a given σ/γ, and the shell of constant stress is that which allows the biggest dome.

Usually a large dome will have an opening at the top. We may use the previous solution also in this case, if we choose the thickness t so that $N_\phi = -\sigma\,t$, together with a compression ring, will be capable of carrying the loads applied at the upper edge. But this is not the most general solution for a dome having an opening. We find it by numerical integration of eq. (26), beginning at the edge $r = b$ with an arbitrary value $\phi = \phi_0$. This means that we have two parameters and hence a greater variety of shapes. If we choose $\phi_0 = 0$, we find a dome such as that in fig. 24a which abuts against a ring having the compressive force $\sigma\,t\,b$. To avoid a local disturbance, its cross section must have the area $t\,b$ and it will be quite heavy. To carry its weight, we must

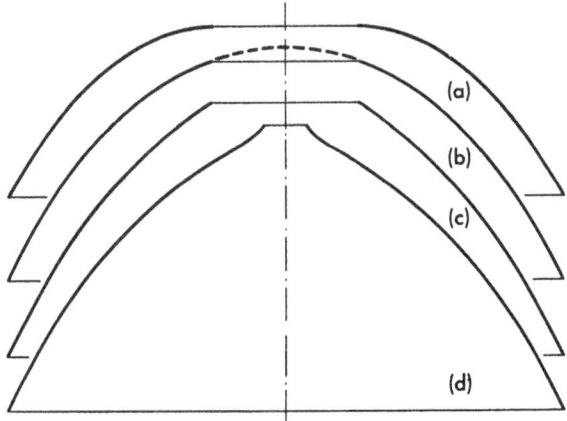

Fig. 24. Domes of constant strength with skylight

choose $\phi_0 > 0$. Among these solutions is the one which we obtained for the closed shell. If we choose ϕ_0 so that

$$\frac{\tan\phi_0}{b} = \frac{\gamma}{\sigma}$$

the meridian begins with a point of inflection, and for still greater values of ϕ_0 we come to shapes as indicated by fig. 24d.

2.4 Loads without Axial Symmetry

2.4.1 General Equations

We shall now drop the assumption that all loads and stress resultants are independent of the coordinate θ. The equations (6a–c) have already been established for the general case. Since one of them, eq. (6c), contains no derivatives, we use it to eliminate N_θ from the other two. Making use of the geometrical relations (1) and (5), we obtain the following set:

$$r_2 \frac{\partial N_\phi}{\partial \phi} \sin\phi + (r_1 + r_2)\, N_\phi \cos\phi + r_1 \frac{\partial N_{\phi\theta}}{\partial \theta} = -r_1 r_2 (p_\phi \sin\phi - p_r \cos\phi),$$

$$r_2 \frac{\partial N_{\phi\theta}}{\partial \phi} \sin\phi + 2 r_1 N_{\phi\theta} \cos\phi - r_2 \frac{\partial N_\phi}{\partial \theta} = -r_1 r_2 \left(p_\theta \sin\phi + \frac{\partial p_r}{\partial \theta} \right). \quad (27)$$

We might go one step further and eliminate $N_{\phi\theta}$. This would lead to a second order differential equation for N_ϕ. We shall come back to this on p. 78 and we shall see then that important conclusions may be drawn from this equation. But for the present purpose it is simpler to use the system (27).

If the load components p_ϕ, p_θ, p_r are arbitrary functions of ϕ, θ, they may always be represented in the form

$$\left.\begin{aligned}
p_\phi &= \sum_0^\infty p_{\phi n} \cos n\,\theta + \sum_1^\infty \bar{p}_{\phi n} \sin n\,\theta, \\
p_\theta &= \sum_1^\infty p_{\theta n} \sin n\,\theta + \sum_0^\infty \bar{p}_{\theta n} \cos n\,\theta, \\
p_r &= \sum_0^\infty p_{r n} \cos n\,\theta + \sum_1^\infty \bar{p}_{r n} \sin n\,\theta,
\end{aligned}\right\} \quad (28)$$

where $p_{\phi n} \ldots \bar{p}_{r n}$ are functions of ϕ only. The first of the two sums in every line represents that part of the load which is symmetric with respect to the plane of the meridian $\theta = 0$ and the second sum represents the antimetric part.

To find the solution of the differential equations (27) which corresponds to such a load, we pick out one of the terms, say

$$p_\phi = p_{\phi n} \cos n\,\theta, \quad p_\theta = p_{\theta n} \sin n\,\theta, \quad p_r = p_{r n} \cos n\,\theta \quad (29)$$

for a fixed but arbitrary integer n. Evidently, the solution may be written in the form

$$N_\phi = N_{\phi n} \cos n\,\theta, \quad N_\theta = N_{\theta n} \cos n\,\theta, \quad N_{\phi\theta} = N_{\phi\theta n} \sin n\,\theta, \quad (30)$$

where $N_{\phi n}, N_{\theta n}, N_{\phi\theta n}$ also are functions of ϕ only. How to find them will be the principal object of this section. Then the general solution for a load which is symmetric with respect to the meridian will be

$$N_\phi = \sum_0^\infty N_{\phi n} \cos n\,\theta, \quad N_\theta = \sum_0^\infty N_{\theta n} \cos n\,\theta, \quad N_{\phi\theta} = \sum_1^\infty N_{\phi\theta n} \sin n\,\theta, \quad (31)$$

and the antimetric part may be found in a similar way.

Now let us introduce (29) and (30) into the differential equations (27). If we do so, we can perform the differentiations with respect to θ and then drop a common factor $\cos n\,\theta$ or $\sin n\,\theta$ from each equation. The result is the following set of equations:

$$\frac{dN_{\phi n}}{d\phi} + \left(1 + \frac{r_1}{r_2}\right) N_{\phi n} \cot\phi + n\,\frac{r_1}{r_2}\,\frac{N_{\phi\theta n}}{\sin\phi} = r_1(-p_{\phi n} + p_{r n} \cot\phi),$$

$$\frac{dN_{\phi\theta n}}{d\phi} + 2\,\frac{r_1}{r_2}\,N_{\phi\theta n} \cot\phi + n\,\frac{N_{\phi n}}{\sin\phi} = r_1\left(-p_{\theta n} + \frac{n}{\sin\phi}\,p_{r n}\right). \quad (32)$$

They are the basic equations of our problem.

2.4.2 Spherical Shell

2.4.2.1 General Solution

Besides the circular cylinder, which we shall treat in the next Chapter with different methods, the sphere is the simplest of all shells of revolution. It therefore yields especially simple results in finite form.

They are a good object for a study of the typical phenomena, which we shall meet again in other shells but then only as the results of lengthy numerical computations.

With $r_1 = r_2 = a$ the left-hand sides of eqs. (32) become very similar to each other:

$$\frac{dN_{\phi n}}{d\phi} + 2\cot\phi \cdot N_{\phi n} + \frac{n}{\sin\phi} N_{\phi\theta n} = a(-p_{\phi n} + \cot\phi \cdot p_{rn}),$$

$$\frac{dN_{\phi\theta n}}{d\phi} + 2\cot\phi \cdot N_{\phi\theta n} + \frac{n}{\sin\phi} N_{\phi n} = a\left(-p_{\theta n} + \frac{n}{\sin\phi} p_{rn}\right).$$

Their sum and their difference are two independent equations for the sum and the difference of the stress resultants,

$$U = N_{\phi n} + N_{\phi\theta n}, \qquad V = N_{\phi n} - N_{\phi\theta n}, \tag{33}$$

namely:

$$\frac{dU}{d\phi} + \left(2\cot\phi + \frac{n}{\sin\phi}\right) U = a\left(-p_{\theta n} - p_{\phi n} + \frac{n + \cos\phi}{\sin\phi} p_{rn}\right),$$

$$\frac{dV}{d\phi} + \left(2\cot\phi - \frac{n}{\sin\phi}\right) V = a\left(p_{\theta n} - p_{\phi n} - \frac{n - \cos\phi}{\sin\phi} p_{rn}\right). \tag{34a, b}$$

Both are linear differential equations of the first order. Now it is well known[1], that the equation

$$\frac{dU}{d\phi} + p(\phi) \cdot U + q(\phi) = 0$$

has the general solution

$$U = \left[C - \int q \exp\left(\int p \, d\phi\right) d\phi\right] \cdot \exp\left(-\int p \, d\phi\right).$$

Applying this formula to (34a, b), we find the following explicit solution:

$$U = \frac{\cot^n\phi/2}{\sin^2\phi} \left[A_n - a\int\left(p_{\phi n} + p_{\theta n} - \frac{n + \cos\phi}{\sin\phi} p_{rn}\right) \sin^2\phi \tan^n\frac{\phi}{2} \, d\phi\right],$$

$$V = \frac{\tan^n\phi/2}{\sin^2\phi} \left[B_n - a\int\left(p_{\phi n} - p_{\theta n} + \frac{n - \cos\phi}{\sin\phi} p_{rn}\right) \sin^2\phi \cot^n\frac{\phi}{2} \, d\phi\right]. \tag{35a, b}$$

These two formulas are the general solution for the sphere, which we shall now discuss.

2.4.2.2 Distributed Load

As an example of the application of the general solution (35) we treat a shell dome subject to a wind load. It has become more or less customary to represent such a load as

$$p_\phi = p_\theta = 0, \qquad p_r = p_{r1}\cos\theta = -p\sin\phi\cos\theta. \tag{36}$$

---

[1] See e. g. PHILLIPS, H. B.: Differential Equations, 3rd ed. New York 1951, p. 44. – MARTIN, W. T., and E. REISSNER: Elementary Differential Equations. Cambridge, Mass. 1956, p. 42 or any other book on differential equations.

This is certainly a somewhat rough approximation, but at least it has the advantage of acknowledging the existence of a large suction on the lee side of the building. Therefore, the pressure p in eq. (36) should be assumed as one half of the normal pressure given by the building codes for a surface at right angles to the wind.

If we introduce $p_{r1} = -p \sin\phi$ into (35) and perform the integrations, we find

$$U = \frac{1 + \cos\phi}{\sin^3\phi} \left[A_1 + p\,a \left(\cos\phi - \frac{1}{3}\cos^3\phi \right) \right],$$

$$V = \frac{1 - \cos\phi}{\sin^3\phi} \left[B_1 - p\,a \left(\cos\phi - \frac{1}{3}\cos^3\phi \right) \right].$$

From this we obtain N_ϕ and $N_{\phi\theta}$ as half the sum and half the difference, multiplied by a factor $\cos\theta$ or $\sin\theta$:

$$N_\phi = \frac{\cos\theta}{\sin^3\phi} \left[\frac{A_1 + B_1}{2} + \frac{A_1 - B_1}{2}\cos\phi + p\,a \left(\cos^2\phi - \frac{1}{3}\cos^4\phi \right) \right],$$

$$N_{\phi\theta} = \frac{\sin\theta}{\sin^3\phi} \left[\frac{A_1 - B_1}{2} + \frac{A_1 + B_1}{2}\cos\phi + p\,a \left(\cos\phi - \frac{1}{3}\cos^3\phi \right) \right].$$

The two constants A_1, B_1 may be determined from the condition that the stress resultants assume finite values at $\phi = 0$. Since the denominator $\sin^3\phi$ has a zero of the third order, the brackets must also have one. The bracket for N_ϕ will be zero, if we put $A_1 = -\frac{2}{3}\,p\,a$. From $N_{\phi\theta}$ comes the same result. The first derivatives of both brackets vanish with any choice of the constants. Vanishing of the second derivatives gives two equations for B_1 which have the identical solution $B_1 = +\frac{2}{3}\,p\,a$. When we introduce these values into the solution and determine the hoop force from (6c), we have the complete solution of our problem:

$$N_\phi = -\frac{p\,a}{3} \frac{(2 + \cos\phi)\,(1 - \cos\phi)\,\cos\phi}{(1 + \cos\phi)\,\sin\phi} \cos\theta,$$

$$N_{\phi\theta} = -\frac{p\,a}{3} \frac{(2 + \cos\phi)\,(1 - \cos\phi)}{(1 + \cos\phi)\,\sin\phi} \sin\theta, \qquad (37)$$

$$N_\theta = -\frac{p\,a}{3} \frac{(3 + 4\cos\phi + 2\cos^2\phi)\,(1 - \cos\phi)}{(1 + \cos\phi)\,\sin\phi} \cos\theta.$$

The distribution of these forces over the meridian is represented in fig. 25.

The dome is supported by the forces N_ϕ and $N_{\phi\theta}$ appearing at the springing line. The normal forces N_ϕ have to equilibrate the moment of the wind loads with respect to the diameter $\theta = \pm\pi/2$ of the springing line. If the shell happens to be a hemisphere, this moment is zero, because all external forces pass through the center of the sphere and hence through the axis of reference. That is the reason that we have $N_\phi = 0$ at $\phi = \pi/2$.

The shearing forces $N_{\phi\theta}$ at the springing line resist the horizontal resultant of the wind forces in so far as it is not resisted by the horizontal components of the meridional forces. They are tangential to the edge of the shell and therefore greatest at those places where the

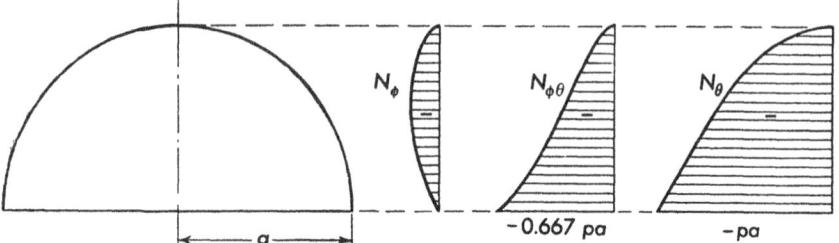

Fig. 25. Spherical dome; stress resultants for wind load

latter is parallel to the direction of the wind, i. e., at $\theta = \pm\pi/2$, although the load there is smallest. In other words, the shell carries the loads from those zones where they are applied (near $\theta = 0$ and $\theta = \pi$) to the sides. This may be recognized very clearly in a picture of the

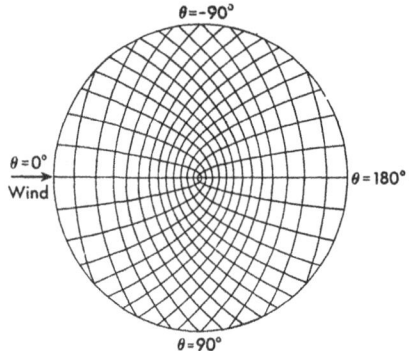

Fig. 26. Spherical dome; stress trajectories for wind load

stress trajectories which in fig. 26 are shown in stereographic projection for a hemisphere. All trajectories which meet the windward meridian $\theta = 0$ have compressive forces; the others have tensile forces. On the wind side, near the springing line, the tension trajectories have almost zero force, since $N_{\phi} = 0$ there. The loads which are applied there are carried away by the vaultlike compressive trajectories toward the sides of the shell. Thus most of the wind pressure is brought to the springing zone lying on both sides between $\theta = \pm\pi/4$ and $\theta = \pm3\pi/4$. The same thing happens to the suction in the lee except that there the tension trajectories do the job as though they were funicular curves. Those loads which are applied in the vicinity of the vertex are first carried by the trajectories with great curvature but soon are transferred to those of the other group, which finally bring them down to the sides of the shell.

The so-called wind load, which we have used here, may be subject to much criticism from an aerodynamic point of view. The formula (36) certainly comes nearer to the truth than most building codes of many countries, which recognize only a pressure on the windward side and ignore the suction. For a hemisphere the pressure distribution should preferably have an axial symmetry to the horizontal diameter parallel to the wind. Such a load distribution, as might be measured in a wind tunnel, may always be represented in the form

$$p_r(\phi, \theta) = \sum_{n=0}^{\infty} p_{rn}(\phi) \cos n\,\theta.$$

To determine the functions $p_{rn}(\phi)$, we only have to subject the values on different parallel circles $\phi = $ const to a harmonic analysis and then collect the numerical values of the n-th Fourier coefficient on different parallel circles as a tabular representation of the function $p_{rn}(\phi)$. Introducing it into (35) leads to $N_{\phi n}$ and $N_{\phi\theta n}$, and if we have made the computation for as many values n as are necessary for convergence. the series (31) give the stress resultants N_ϕ, N_θ, $N_{\phi\theta}$.

2.4.2.3 Edge Load

Let us now consider the homogeneous solution, i. e., that part of (35a, b) which remains when we put $p_\phi \equiv p_\theta \equiv p_r \equiv 0$. It describes the stress resultants in a shell to which loads are applied only at the edges or, perhaps, at the points $\phi = 0$ and $\phi = \pi$. We need it to adapt particular solutions of the inhomogeneous equations to given boundary conditions or to eliminate a singularity from them. We shall study it here in some detail and then apply it to cases where the distributed load has axial symmetry but the boundary conditions have not.

If in (35) we drop all terms containing $p_{\phi n}, p_{\theta n}, p_{rn}$ and determine N_θ from (6c), we have

$$N_{\phi n} = -N_{\theta n} = \frac{1}{2}(U + V) = \frac{1}{2\sin^2\phi}\left(A_n \cot^n \frac{\phi}{2} + B_n \tan^n \frac{\phi}{2}\right),$$

$$N_{\phi\theta n} = \frac{1}{2}(U - V) = \frac{1}{2\sin^2\phi}\left(A_n \cot^n \frac{\phi}{2} - B_n \tan^n \frac{\phi}{2}\right). \tag{38}$$

The solution consists of two independent parts, with the arbitrary factors A_n and B_n.

We see at once that the case $n = 1$ must be treated separately, because there both solutions are infinite at both poles $\phi = 0$ and $\phi = \pi$, whereas for $n \geq 2$ the numerator has a zero of sufficient order to keep the A solution finite (or even zero) at $\phi = \pi$ and the B solution at $\phi = 0$.

For $n = 1$ we have

$$N_\phi = -N_\theta = \frac{\cos\theta}{2\sin^2\phi}\left(A_1 \cot \frac{\phi}{2} + B_1 \tan \frac{\phi}{2}\right),$$

$$N_{\phi\theta} = \frac{\sin\theta}{2\sin^2\phi}\left(A_1 \cot \frac{\phi}{2} - B_1 \tan \frac{\phi}{2}\right). \tag{39}$$

To find out what the singularity at $\phi = 0$ is like, we cut it out by an adjacent parallel circle $\phi = \text{const} \approx 0$ (fig. 27). The forces which act on the small spherical segment that we have cut off are the stress resultants N_ϕ and $N_{\phi\theta}$ and, possibly, an external force acting at the pole.

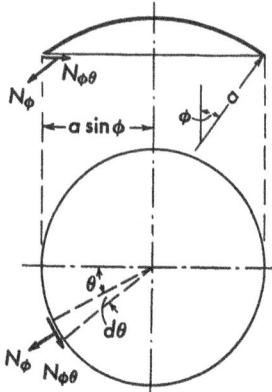

Fig. 27. Spherical cap

The stress resultants have the resultant (positive to the left)

$$\int_0^{2\pi} N_\phi \cos\phi \cdot \cos\theta \cdot a \sin\phi \, d\theta$$

$$- \int_0^{2\pi} N_{\phi\theta} \cdot \sin\theta \cdot a \sin\phi \, d\theta$$

and a resulting moment with respect to the diameter $\theta = \pm\pi/2$ of the circular edge:

$$\int_0^{2\pi} N_\phi \sin\phi \cdot a \sin\phi \cos\theta \cdot a \sin\phi \, d\theta.$$

With $N_\phi = N_{\phi 1} \cos\theta$ and $N_{\phi\theta} = N_{\phi\theta 1} \sin\theta$ the integrals may be evaluated, and if we now go to the limit $\phi = 0$, we find the external actions which must be applied to the point $\phi = 0$ to equilibrate the internal forces. They are a horizontal force (positive to the right)

$$\mathsf{P} = \pi a \lim_{\phi \to 0} (N_{\phi 1} \cos\phi \sin\phi - N_{\phi\theta 1} \sin\phi) \tag{40a}$$

and an external couple (positive as shown in fig. 28)

$$\mathsf{M} = \pi a^2 \lim_{\phi \to 0} (N_{\phi 1} \sin^3\phi). \tag{40b}$$

When we introduce our solution (39) into these formulas, we find an external force

$$\mathsf{P} = \frac{\pi a}{2} (B_1 - A_1)$$

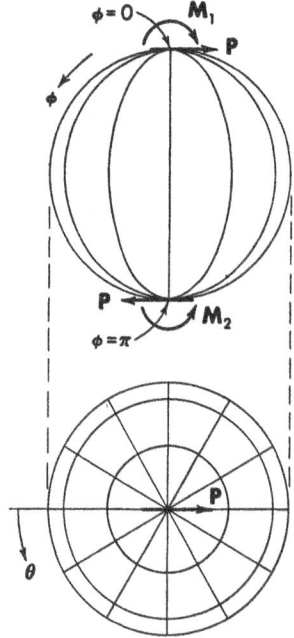

Fig. 28. Concentrated forces and couples acting at the poles of a spherical shell

and an external couple

$$\mathsf{M} = \mathsf{M}_1 = \pi a^2 A_1.$$

For the other pole of the sphere we apply the same formulas (40) and find the same force P, but in the opposite direction, and a couple

$$\mathsf{M} = \mathsf{M}_2 = \pi a^2 B_1.$$

The two forces form a couple too, and we see that the condition of overall equilibrium

$$P \cdot 2a + \mathsf{M}_1 - \mathsf{M}_2 = 0$$

is always fulfilled, whatever the magnitudes of A_1 and B_1. If we chooes $A_1 = 0$, only a force P is applied at $\phi = 0$, and if we choose $A_1 = B_1$, there is only a couple, but no choice is possible where there is nothing at all.

The higher harmonics, $n \geqq 2$, in eqs. (38) have singularities of a different type. There is no external force or couple but a rather complex group of forces having infinite magnitude and canceling each other. We shall not treat these "multi-poles" here in detail, since they do not seem to be of practical interest in the theory of shells.

If the shell is not a complete sphere but ends on a parallel circle $\phi = \phi_0$, we have for every $n \geqq 2$ one solution (that with B_n) which is regular at all points. It corresponds to a combined normal and shear loading of the edge, the ratio of the two parts being fixed by ϕ_0 and n. It may be used to find the stress resultants in such cases as the one illustrated by fig. 29.

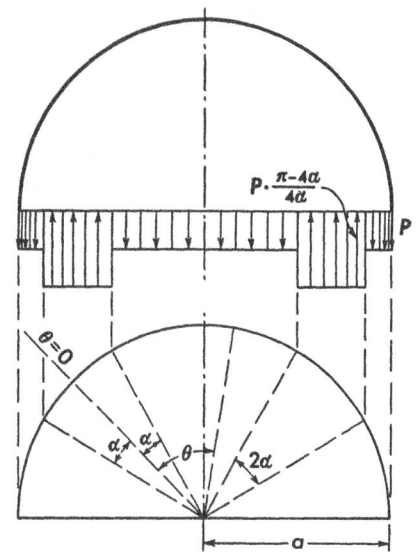

Fig. 29. Edge load at a hemispherical shell

This shell is subject to a discontinuous edge load. In four parts of the circumference it is a compression and on the remainder a tension, and the intensities of both have been so balanced that the external forces are in equilibrium. Such forces will occur if the shell rests on four supports of the angular width 2α and has to carry the edge load P between the supports, which, of course, must be tangential to the sphere.

We develop the edge forces, consisting of the load P and the reaction $-P(\pi - 4\alpha)/4\alpha$, in a Fourier series. Because of the symmetry of the forces, there will appear only the terms with $n = 4, 8, 12, \ldots$;

$$(N_\phi)_{\text{edge}} = -\frac{2P}{\alpha} \sum_{n=4,\,8,\ldots} \frac{\sin n\alpha}{n} \cos n\theta.$$

Now if we drop from (38) the part with A_n and then sum up, we have

$$N_\phi = \frac{1}{2\sin^2\phi} \sum_{n=4,\,8,\ldots} B_n \tan^n\frac{\phi}{2} \cos n\theta,$$

and at the edge of the shell this must become equal to the preceding series. If we assume that, different from fig. 29, this edge is not at $\phi = 90°$, but at some arbitrary angle $\phi = \phi_0$, this will be accomplished if we choose

$$B_n = - \frac{4P}{\alpha} \frac{\sin n \alpha}{n} \frac{\sin^2 \phi_0}{\tan^n \phi_0/2}, \qquad n = 4, 8, 12, \ldots$$

We thus arrive at the following solution:

$$N_\phi = -N_\theta = - \frac{2P}{\alpha} \frac{\sin^2 \phi_0}{\sin^2 \phi} \sum_{n=4, 8, \ldots}^{\infty} \frac{\sin n \alpha}{n} \frac{\tan^n \phi/2}{\tan^n \phi_0/2} \cos n \theta$$

$$N_{\phi\theta} = \frac{2P}{\alpha} \frac{\sin^2 \phi_0}{\sin^2 \phi} \sum_{n=4, 8, \ldots}^{\infty} \frac{\sin n \alpha}{n} \frac{\tan^n \phi/2}{\tan^n \phi_0/2} \sin n \theta.$$

(41)

Because of the quotient of the two tangents the series converges better the farther away we go from the edge. This means that the higher the order n of a harmonic component, the smaller is the zone in which its influence is felt. It also means that the discontinuity of the given boundary values of N_ϕ does not involve a discontinuity in the interior of the shell but that the stress resultants are continuous everywhere except on the boundary. We shall see on p. 78 that this is not a general rule and where its limits are.

The solution (41) may be combined with that for a distributed surface load having axial symmetry, for instance with the solution for the weight of the structure, eq. (13). If we then choose P so that on the edge between the supports the resultant boundary value of N_ϕ is zero, we have the stress resultants of a shell dome resting on four supports of finite width. If there is another number of supports, the only change in the formulas is that the summation must be extended over those harmonics which agree with the symmetry of the structure.

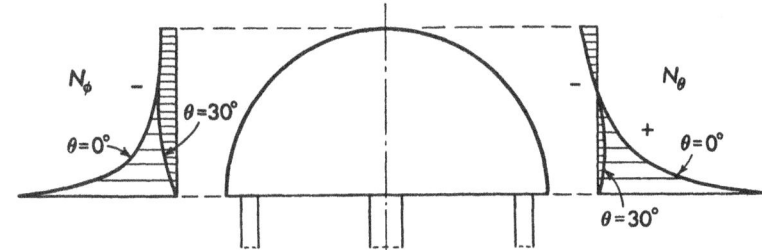

Fig. 30. Hemispherical dome on six supports

Hence, for a shell on six equally spaced supports the harmonics of orders $n = 6, 12, 18, \ldots$ have to be considered. A hemispherical dome of this kind is shown in fig. 30. The width of the supports is $2\alpha = 12°$. The diagrams give N_ϕ and N_θ for the meridians through the center of a support ($\theta = 0°$) and through the center of an opening ($\theta = 30°$). The high value of N_ϕ follows simply from the necessity of carrying the weight of the shell on a limited part of the edge, and N_θ follows then from eq. (6c). The diagrams show that the edge disturbance caused by the supports goes approximately halfway up the meridian before it becomes invisibly small. The major part of it comes from the first harmonic considered, $n = 6$.

The application of eqs. (41) to this problem involves the assumption that the reaction is uniformly distributed over the width of each support. If one wants to have a more exact force distribution, it is necessary to solve a statically indeterminate problem, but since this would essentially affect only the higher harmonics which are not of much importance anyway, this scarcely seems worthwhile.

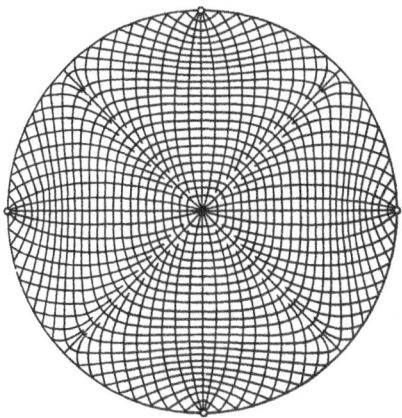

Fig. 31. Hemisphere on four point supports; stress trajectories

The complete solution includes, of course, shearing forces $N_{\phi\theta}$. They are zero on the meridians through the middle of each support and of each span but not elsewhere. In particular they are different from zero along the edge. A ring must be provided there to which this shear can be transmitted. It will be subjected to axial forces and to bending in its own plane, but it does not need to have bending stiffness in the vertical direction. Its weight may be supported by the shell, which then receives additional stresses according to eqs. (41).

In fig. 31 the stress trajectories are shown for a slightly different shell. It has only four supports, and their width is zero. Since such point supports do not occur in a real structure, it is worthwhile to consider them only if this simplifies the computation. This is not the case if eqs. (41) are used, since the Fourier series converge more slowly the smaller α is chosen. But in this case one may use to advantage the complex-variable approach explained on the following pages.

The stress trajectories are, of course, lines on the middle surface of the shell. Fig. 31 is a stereographic projection of these lines. This particular projection was chosen because it preserves the right angles between the curves. It may be seen in the figure how part of the trajectories emanate from the supports, while others leave the free edge at angles of $45°$. At each point of the edge one of these trajectories carries tension and one compression, since there the shell is in a state of pure shear.

The trajectories may convey some idea of the stress pattern, but they may also be misleading. In this particular case they overemphasize the deviation from perfect axial symmetry in the upper part of the shell. Since the forces N_ϕ and N_θ are almost equal there, a rather small shear $N_{\phi\theta}$ makes the directions of principal stress turn through a large angle. Therefore one family of trajectories looks like rounded squares in a region where the stress system is almost exactly that of a continuously supported dome.

If the vertex of the shell is cut away at the parallel circle $\phi = \phi_1$, the A_n terms of (38) are available to fulfill on this edge an additional boundary condition, say $N_\phi = 0$. Every pair of constants A_n, B_n must then be determined from a pair of linear equations, and this is best done numerically.

All these solutions yield the desired distribution of the normal force N_ϕ at the edge or at the edges: but they yield also shearing forces $N_{\phi\theta}$, and their distribution is beyond control, since no further free constants are available. We have to accept them just as they appear and have to provide a stiffening ring of sufficient strength against bending in its own plane. This result is not a deficiency of our method of investigation but corresponds to a real fact. If the ring were missing and thus absence of shearing forces enforced, no equilibrium of the internal forces would be possible without bending moments in the shell. This would mean great stresses and great deformations, the thin-walled shell itself performing the functions of the stiffening ring.

Stiffening members like this ring are necessary at all free boundaries of shells. They are the statical equivalent of the geometric fact that a shell with a free boundary is easily deformable, whereas a closed surface, e. g. a complete sphere or a shell with fixed edges, has a quite remarkable rigidity. We shall discuss this in more detail on p. 91.

The preceding treatment of the shell on isolated supports is possible only if the reactions of these supports are known in advance. This condition is fulfilled if there are only three supports, which may be placed arbitrarily, or if the supports are spaced equally and the shell carries a load which has at least the same degree of symmetry as the arrangement of the supports. In the first case the distribution of the load on the supports is statically determinate, in the second it follows from symmetry. In any other case and also if the width of the individual support is such that the assumption of uniform distribution of its reaction over its width is not justified, the problem becomes statically indeterminate and has to be treated along the lines explained in section 2.5.6.

2.4.2.4 Concentrated Forces and Couples

2.4.2.4.1 Introduction of Complex Variables. There exists another way of solving the stress problem for the spherical shell. It avoids the Fourier series (31) and yields immediate access to a group of singular solutions describing the effect of a concentrated load at an arbitrary point of the shell. For this reason we shall explain it here.

We start from eqs. (27), drop the load terms and specialize for the sphere by putting $r_1 = r_2 = a$:

$$\frac{\partial N_\phi}{\partial \phi} \sin\phi + 2 N_\phi \, \cos\phi + \frac{\partial N_{\phi\theta}}{\partial \theta} = 0,$$

$$\frac{\partial N_{\phi\theta}}{\partial \phi} \sin\phi + 2 N_{\phi\theta} \cos\phi - \frac{\partial N_\phi}{\partial \theta} = 0.$$

After multiplication by $\sin^2\phi$ these equations may be written in the following form:

$$\sin\phi \cdot \frac{\partial}{\partial \phi} (N_\phi \sin^2\phi) + \frac{\partial}{\partial \theta} (N_{\phi\theta} \sin^2\phi) = 0,$$

$$\sin\phi \cdot \frac{\partial}{\partial \phi} (N_{\phi\theta} \sin^2\phi) - \frac{\partial}{\partial \theta} (N_\phi \sin^2\phi) = 0. \tag{42}$$

This suggests to introduce as unknowns the quantities

$$N_1 = N_\phi \sin^2\phi, \qquad N_2 = N_{\phi\theta} \sin^2\phi. \tag{43}$$

In order to remove the factor $\sin\phi$ before the first term of both equations (42), we introduce also a new independent variable

$$\eta = \ln\tan\frac{\phi}{2},$$

$$\frac{\partial}{\partial \eta} = \sin\phi \cdot \frac{\partial}{\partial \phi}. \tag{44}$$

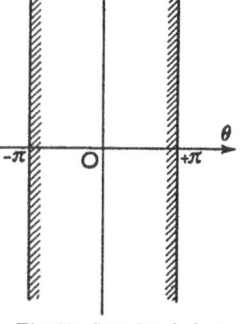

Fig. 32. Complex ζ plane

This transformation may be interpreted as a mapping of the middle surface of the shell in a θ, η plane (fig. 32). The complete sphere is represented by a strip of horizontal width 2π which in the η direction extends both ways to infinity. This mapping is identical with MERCATOR's projection and is conformal.

By the transformation (43), (44) the equations (42) become very simple:

$$\frac{\partial N_1}{\partial \eta} + \frac{\partial N_2}{\partial \theta} = 0,$$

$$\frac{\partial N_2}{\partial \eta} - \frac{\partial N_1}{\partial \theta} = 0. \tag{45}$$

These are the well-known equations of CAUCHY and RIEMANN which exist between the real and imaginary parts of any analytical function

$$N = N_1 + i N_2$$

of the complex variable

$$\zeta = \theta + i\eta.$$

We conclude that any such function describes a possible system of membrane forces in at least a part of the spherical shell. Since our equations have been established under the assumption that the distributed load $p_\phi \equiv p_\theta \equiv p_r \equiv 0$, all these solutions will belong to cases where loads are applied only to the edge of the shell and, perhaps, as concentrated forces and couples at singular points of the function $N(\zeta)$.

When $N(\zeta)$ has the real period 2π, the corresponding membrane forces have the same period and are single-valued on the whole sphere. From LIOUVILLE's theorem[1] and the supposed periodicity, it follows that N has at least one singularity in the strip shaded along its edges in fig. 32, it may be at infinity. At the corresponding point or points of the shell a load must be applied which produces the membrane forces. We shall now consider some solutions of this kind.

2.4.2.4.2 Tangential Point Load. We start with the function

$$N(\zeta) = C \cot \frac{\zeta - \zeta_0}{2}.$$

It has singularities at $\zeta = \zeta_0$ and at $\zeta = \pm i\,\infty$. The corresponding points on the sphere are the poles $\phi = 0$, $\phi = \pi$ and an arbitrary point, which we may place on the meridian $\theta = 0$ by putting

$$\zeta_0 = i\,\eta_0 = i \ln \tan\phi_0/2.$$

The stress resultants follow from N by splitting it into real and imaginary parts:

$$N_1 = C \frac{\sin\theta}{\mathrm{Cosh}\,(\eta - \eta_0) - \cos\theta}, \qquad N_2 = -C \frac{\mathrm{Sinh}\,(\eta - \eta_0)}{\mathrm{Cosh}\,(\eta - \eta_0) - \cos\theta}.$$

From (43), (44) and (6c) we find then

$$N_\phi = -N_\theta = C \frac{\sin\phi_0}{\sin\phi} \frac{\sin\theta}{1 - \cos\phi_0 \cos\phi - \sin\phi_0 \sin\phi \cos\theta},$$

$$N_{\phi\theta} = -C \frac{1}{\sin^2\phi} \frac{\cos\phi_0 - \cos\phi}{1 - \cos\phi_0 \cos\phi - \sin\phi_0 \sin\phi \cos\theta}.$$

At the poles of the sphere the factor $\sin\phi$ in the denominators vanishes, and the second factor does so at the point $\phi = \phi_0$, $\theta = 0$. At these three points the stress resultants assume infinite values, and these singularities correspond to the application of external forces or couples to the shell. To determine their magnitude and direction, we use the

---

[1] CHURCHILL, R. V.: Introduction to Complex Variable and Applications. New York 1948, p. 96.

following method: By a parallel circle $\phi = $ const we cut the shell in two parts (fig. 27) and compute from the forces N_ϕ and $N_{\phi\theta}$ transmitted in this circle the resultant force and the resultant moment with respect to one of the poles. Thus we find the loads acting at the poles, and at the third singular point the load is determined by the overall equilibrium of the sphere.

From the antimetry of all stress resultants with respect to the meridian $\theta = 0$ it follows that the resultant force in the section $\phi = $ const must be perpendicular to the plane of this meridian. It is

$$\mathsf{R} = \int\limits_{-\pi}^{+\pi} (N_\phi \cos\phi \sin\theta + N_{\phi\theta} \cos\theta)\, a \sin\phi \, d\theta.$$

When we introduce here the expressions for N_ϕ and $N_{\phi\theta}$, we arrive after some computation at the following formula

$$\mathsf{R} = 2\pi\, C\, a \cot\phi_0 + \frac{2C\,a}{\sin\phi_0}\left[\arctan \frac{(\cos\phi - \cos\phi_0)\tan\theta/2}{1 - \cos(\phi_0 - \phi)} \right]_{\theta = -\pi}^{\theta = +\pi}.$$

For $\phi < \phi_0$ this yields

$$\mathsf{R} = 2\pi\, C\, a \frac{1 + \cos\phi_0}{\sin\phi_0},$$

independent of ϕ, and this is the force which must be applied in the opposite direction at the pole $\phi = 0$ of the sphere. When we choose $\phi > \phi_0$, the cosine difference under the arc tan changes sign and therefore the resultant becomes

$$\mathsf{R} = -2\pi\, C\, a \frac{1 - \cos\phi_0}{\sin\phi_0}.$$

This is the force which must be applied at the pole $\phi = \pi$ in the direction shown in fig. 33. The force acting at $\theta = 0$, $\phi = \phi_0$ must be equal

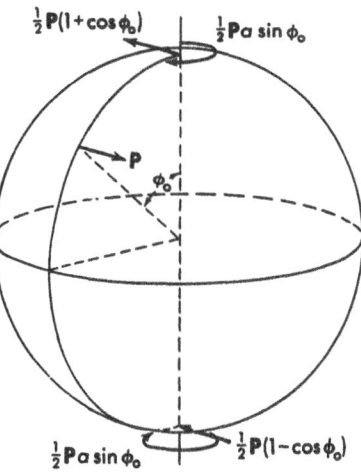

Fig. 33. Loads on a spherical shell corresponding to solution (46)

and opposite in direction to the sum of the two:

$$\mathsf{P} = 2\pi\, C\, a \frac{2}{\sin\phi_0}.$$

This relation allows us to express the constant C in terms of the force P. The stress resultants are then

$$N_\phi = -N_\theta = \frac{\mathsf{P}}{4\pi a}\, \frac{\sin^2\phi_0}{\sin\phi}\, \frac{\sin\theta}{1 - \cos\phi_0\cos\phi - \sin\phi_0\sin\phi\cos\theta},$$

$$N_{\phi\theta} = \frac{\mathsf{P}}{4\pi a}\, \frac{\sin\phi_0}{\sin^2\phi}\, \frac{\cos\phi - \cos\phi_0}{1 - \cos\phi_0\cos\phi - \sin\phi_0\sin\phi\cos\theta},$$

$$(46)$$

and the corresponding loads are those shown in fig. 33.

The equilibrium of the shell still requires external couples, applied at the poles and turning about the vertical axis of the sphere. We find them from the moment of the forces N_ϕ and $N_{\phi\theta}$ in fig. 27. It is

$$M = \int_{-\pi}^{+\pi} N_{\phi\theta}\, a^2 \sin^2\phi\, d\theta = -\frac{Pa}{2\pi}\sin\phi_0 \left[\arctan \frac{(\cos\phi - \cos\phi_0)\tan\theta/2}{1 - \cos(\phi_0 - \phi)}\right]_{-\pi}^{+\pi}.$$

Again this has different values for $\phi < \phi_0$ and $\phi > \phi_0$. In the first case we have $M = +\tfrac{1}{2}P\, a \sin\phi_0$, in the second case $M = -\tfrac{1}{2}P\, a \sin\phi_0$. This leads to the external couples shown in fig. 33. It may easily be checked that there are no external couples about other axes passing through the poles, and then it follows from the equilibrium of the complete sphere that the tangential force P is the only load applied at the point $\theta = 0$, $\phi = \phi_0$.

In a quite similar way the complex function

$$N(\zeta) = \frac{P \sin\phi_0}{4\pi a}\, i \cot \frac{\zeta - \zeta_0}{2}$$

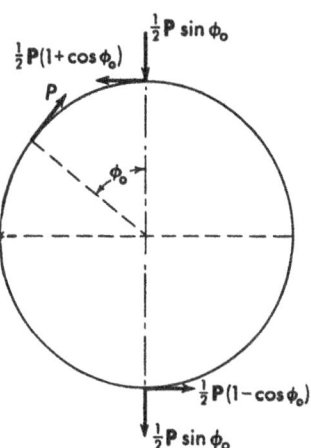

Fig. 34. Loads on a spherical shell corresponding to solution (47)

may be investigated. It will be found that it belongs to a group of loads situated entirely in the plane of the meridian $\theta = 0$ and shown in fig. 34. The corresponding stress resultants are

$$N_\phi = -N_\theta = \frac{P}{4\pi a}\, \frac{\sin\phi_0}{\sin^2\phi}\, \frac{\cos\phi_0 - \cos\phi}{1 - \cos\phi_0 \cos\phi - \sin\phi_0 \sin\phi \cos\theta},$$

$$N_{\phi\theta} = \frac{P}{4\pi a}\, \frac{\sin^2\phi_0}{\sin\phi}\, \frac{\sin\theta}{1 - \cos\phi_0 \cos\phi - \sin\phi_0 \sin\phi \cos\theta}. \tag{47}$$

2.4.2.4.3 Normal Point Load. The solution for a load P normal to the shell (fig. 35) must have a stronger singularity. We might easily establish such a function $N(\zeta)$ and then go through all the formalities just described to find the constant factor C and the reactions at the poles. But we have an easier approach, using the solution (47). Fig. 36 shows two forces P' acting at adjacent points of the meridian $\theta = 0$. If we now write $P'N'(\zeta; \phi_0)$ for the function $N(\zeta)$ corresponding to fig. 34, the function corresponding to fig. 36 will be

$$N(\zeta) = P'N'(\zeta; \phi_0) - P'N'(\zeta; \phi_0 + \Delta\phi_0).$$

If $\Delta\phi_0$ is small, the stress resultants in most parts of the shell will not be much different from those produced by the resultant of the two forces P', a force

$$P = P' \Delta\phi_0,$$

applied halfway between them, and this becomes exact for the whole

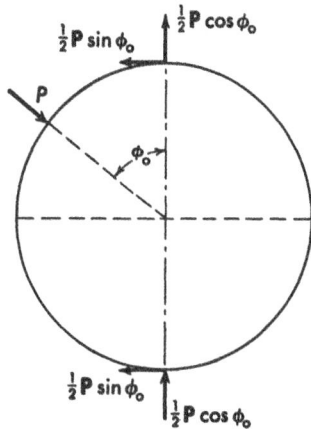

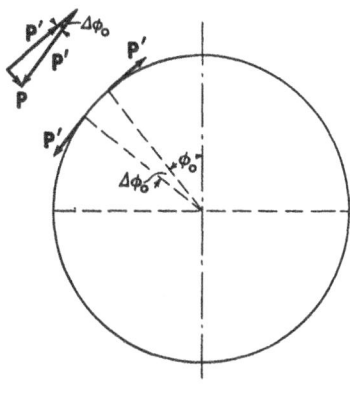

Fig. 35. Loads on a spherical shell corresponding to solution (48)

Fig. 36. Concentrated forces applied at two adjacent points of a meridian

sphere if we go to the limit $\Delta\phi_0 \to 0$ with finite resultant P. In this case we may write

$$N(\zeta) = -P' \frac{dN'(\zeta;\phi_0)}{d\phi_0} \Delta\phi_0 = -P \frac{dN'(\zeta;\phi_0)}{d\phi_0}.$$

Introducing the expression for N', we obtain

$$N(\zeta) = \frac{P}{4\pi a}\left(\frac{1}{2\sin^2\frac{\zeta-\zeta_0}{2}} - i\cos\phi_0\cot\frac{\zeta-\zeta_0}{2}\right)$$

and consequently

$$N_\phi = -N_\theta = -\frac{P}{4\pi a}\left[\frac{\cos\phi_0(\cos\phi_0-\cos\phi)}{\sin^2\phi(1-\cos\phi_0\cos\phi-\sin\phi_0\sin\phi\cos\theta)}\right.$$
$$\left.+\frac{\sin\phi_0}{\sin\phi}\frac{(1-\cos\phi_0\cos\phi)\cos\theta-\sin\phi_0\sin\phi}{(1-\cos\phi_0\cos\phi-\sin\phi_0\sin\phi\cos\theta)^2}\right],$$

$$N_{\phi\theta} = -\frac{P}{4\pi a}\frac{\sin\phi_0\sin\theta}{\sin\phi}\left[\frac{\cos\phi_0}{1-\cos\phi_0\cos\phi-\sin\phi_0\sin\phi\cos\theta}\right.$$
$$\left.+\frac{\cos\phi_0-\cos\phi}{(1-\cos\phi_0\cos\phi-\sin\phi_0\sin\phi\cos\theta)^2}\right].$$

(48)

The reactions at the poles follow from fig. 34 by differentiating with respect to ϕ_0 and changing signs. The whole load system is situated in the plane of the meridian $\theta = 0$ and is shown in fig. 35.

2.4.2.4.4. **Gas Tank on Point Supports.** The formulas given on the preceding pages have many useful applications. One of them is illustrated by fig. 37. This spherical gasholder is supported by six bars, which are situated in planes tangential to the middle surface of the shell. Since the internal pressure of the gas is a self-equilibrating load system, the bars receive forces only from the weight of the shell and from the wind load. Since the support is statically determinate,.

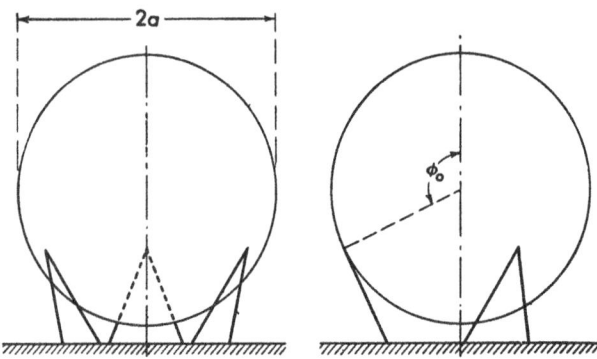

Fig. 37. Spherical gas tank

these forces may, in any case, be found without recourse to shell theory, and then we may apply the preceding formulas to study their influence on the stresses in the shell.

We shall show this here in some detail for the weight of the shell. The forces of two bars meeting at one point may be combined to form a resultant which, for vertical load, lies in the plane of a meridian. Let p be the weight of the shell per unit area of its middle surface, then each of these forces will be

$$P = \frac{4\pi p a^2}{3\sin\phi_0}.$$

We choose their meridians to be $\theta = 0$ and $\theta = \pm 120°$.

We now consider a shell with two fictitious supports at the poles and apply to it the three forces P and the distributed load p. For the load P in the meridian $\theta = 0$ the stress resultants are given by eqs. (47). For the other two loads P we find them from the same formulas by simply replacing θ by $\theta - 120°$ or $\theta + 120°$, respectively. The corresponding reactions at each pole consist of three horizontal forces canceling each other and of three vertical forces which add up to

$$3 \cdot \frac{1}{2} P \sin\phi_0 = 2\pi p a^2.$$

On both poles together they are equal to the weight of the shell. When we now determine the forces due the load p from the integral (10), we must choose the constant C so that these reactions at the poles are compensated. This leads to

$$N_\phi = p\,a\,\frac{\cos\phi}{\sin^2\phi}, \qquad N_\theta = -p\,a\,\frac{(1+\sin^2\phi)\cos\phi}{\sin^2\phi}, \qquad N_{\phi\theta} \equiv 0.$$

The combination of all these solutions looks for N_ϕ like this:

$$N_\phi = p\,a\,\frac{\cos\phi}{\sin^2\phi} + \frac{p\,a}{3\sin^2\phi}\,(\cos\phi_0 - \cos\phi)$$

$$\cdot \left[\frac{1}{1 - \cos\phi_0\cos\phi - \sin\phi_0\sin\phi\cos\theta} \right.$$

$$+ \frac{1}{1 - \cos\phi_0\cos\phi - \sin\phi_0\sin\phi\cos(\theta - 120°)}$$

$$+ \left. \frac{1}{1 - \cos\phi_0\cos\phi - \sin\phi_0\sin\phi\cos(\theta + 120°)} \right].$$

As it is, this formula is not fit for numerical evaluation, because each of its four terms has a strong singularity at each pole, while the sum

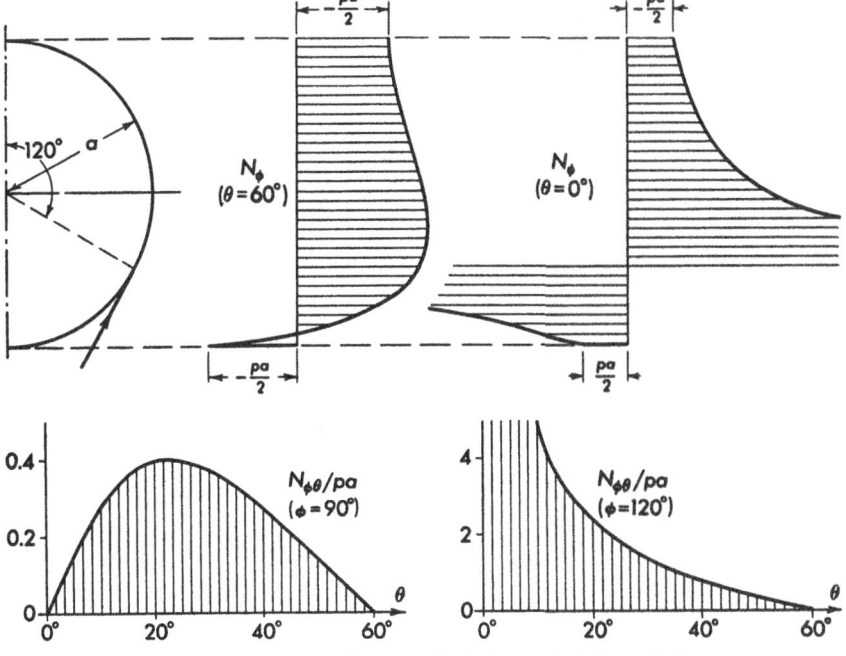

Fig. 38. Diagrams for stress resultants in a spherical gas tank

is regular. We must, therefore, reduce it to a more reasonable appearance. This is a rather tedious procedure and leads to the following result:

$$N_\phi = \frac{p\,a}{\Delta}\,[\cos\phi_0(3 + \cos^2\phi_0)\,(1 + \cos^2\phi) - 2(1 + 3\cos^2\phi_0)\cos\phi$$

$$- \sin^3\phi_0 \sin\phi \cos\phi \cos3\,\theta],$$

$$N_\theta = -\frac{p\,a}{\varDelta}[\cos\phi_0(3+\cos^2\phi_0)(1-2\cos^2\phi-\cos^4\phi)$$
$$-(1+3\cos^2\phi_0)(1-3\cos^2\phi)\cos\phi$$
$$-(1+\sin^2\phi)\sin^3\phi_0\sin\phi\cos\phi\cos3\,\theta],$$

$$N_{\phi\theta}=\frac{p\,a}{\varDelta}\sin^3\phi_0\sin\phi\sin3\,\theta,$$

with

$$\varDelta=(1-\cos\phi_0\cos\phi)[4(\cos\phi_0-\cos\phi)^2+\sin^2\phi_0\sin^2\phi]$$
$$-\sin^3\phi_0\sin^3\phi\cos3\,\theta.$$

From these formulas some diagrams have been computed which are shown in fig. 38. They may give an idea of the distribution of internal forces in this case.

2.4.3 Conical Shell

2.4.3.1 General Solution

For a conical shell (fig. 16) we saw on p. 37, that we have to start from eqs. (7) which, with $r = s\cos\alpha$, $r_1 = \infty$ and $r_2 = s\cot\alpha$ assume the following form:

$$\frac{\partial(N_s\,s)}{\partial s}+\frac{\partial N_{s\theta}}{\partial\theta}\frac{1}{\cos\alpha}-N_\theta+p_s\,s=0,$$
$$\frac{\partial(N_{s\theta}\,s)}{\partial s}+\frac{\partial N_\theta}{\partial\theta}\frac{1}{\cos\alpha}+N_{s\theta}+p_\theta\,s=0, \qquad (49\,\text{a—c})$$
$$N_\theta=p_r\,s\cot\alpha.$$

Again introducing the loads and the stress resultants in the form (29), (30), we find the n-th harmonic of the hoop force $N_{\theta n}$ immediately from (49c):

$$N_{\theta n}=p_{rn}\,s\cot\alpha,$$

independent of all boundary conditions, and we can eliminate it at once from (49a, b), which then read

$$\frac{dN_{sn}}{ds}+\frac{1}{s}N_{sn}+\frac{n}{s\cos\alpha}N_{s\theta n}=-p_{sn}+p_{rn}\cot\alpha.$$
$$\frac{dN_{s\theta n}}{ds}+\frac{2}{s}N_{s\theta n}\qquad\qquad=-p_{\theta n}+p_{rn}\frac{n}{\sin\alpha}. \qquad (50\,\text{a, b})$$

These are two ordinary differential equations for N_{sn} and $N_{s\theta n}$, which may be solved one after the other. Eq. (50b) contains only the shear, and by applying the general formula mentioned on p. 50, we have

$$N_{s\theta n}=-\exp\left(-\int\frac{2ds}{s}\right)\left\{\int\left[\left(p_{\theta n}-p_{rn}\frac{n}{\sin\alpha}\right)\exp\int\frac{2ds}{s}\right]ds-A_n\right\}$$
$$=-\frac{1}{s^2}\left\{\int\left(p_{\theta n}-p_{rn}\frac{n}{\sin\alpha}\right)s^2\,ds-A_n\right\}. \qquad (51\,\text{a})$$

Introducing this in eq. (50a), we find in quite the same way

$$N_{sn} = -\frac{1}{s}\left[\int\left(\frac{n}{\cos\alpha}N_{s\theta n} + s\,p_{sn} - s\,p_{rn}\cot\alpha\right)ds - B_n\right]. \quad (51\,\mathrm{b})$$

As an example of the application of these simple formulas we consider the mushroom-shaped roof of fig. 17 for a kind of wind load which we assume, not very correctly but conventionally, to be

$$p_s = p_\theta = 0, \quad p_r = -p\sin\alpha\,\cos\theta.$$

We have to use our formulas with $n = 1$ and easily find

$$N_\theta = -p\,s\cos\alpha\,\cos\theta,$$

$$N_{s\theta} = -\frac{1}{s^2}\left(p\,\frac{s^3}{3} - A_1\right)\sin\theta.$$

The edge $s = l$ is to be free of external forces. This yields $A_1 = \frac{1}{3}p l^3$ and therefore the shearing force

$$N_{s\theta} = \frac{1}{3}p\,\frac{l^3 - s^3}{s^2}\sin\theta.$$

After the second constant B_1 has been determined by the same argument, the meridional force follows as

$$N_s = \frac{p}{\cos\alpha}\left(\frac{l^3 - s^3}{3s^2} - \frac{l^2 - s^2}{2s}\sin^2\alpha\right)\cos\theta.$$

At $s = 0$ this becomes infinite like s^{-2}. This singularity corresponds to the action of a couple, exerted by the central column in order to equilibrate the moment of the loads. The shear $N_{s\theta}$ at the top has not only to yield a horizontal resultant but also to compensate the resultant of the N_s, and, therefore, it too has a singularity of the second order.

2.4.3.2 Homogeneous Solution

When we drop the terms with p_{sn}, $p_{\theta n}$, p_{rn}, in eqs. (51a, b) we have the solution for a shell which is subjected only to edge-loads

$$N_{s\theta n} = \frac{A_n}{s^2}, \qquad N_{sn} = \frac{n}{\cos\alpha}\frac{A_n}{s^2} + \frac{B_n}{s}, \qquad N_{\theta n} = 0. \quad (52)$$

We see at a glance that there will always be infinite stresses at the vertex $s = 0$, whatever the values of A_n and B_n are. For the first harmonic, this singularity describes the action of a horizontal force P and a couple M with a horizontal axis (fig. 39). Putting $n = 1$ in eq.(52), we find the horizontal resultant of the forces transmitted through an arbitrary parallel circle:

$$\mathsf{P} = \int\limits_{-\pi}^{+\pi} (N_{s1}\cos\alpha\,\cos^2\theta - N_{s\theta 1}\sin^2\theta)\,s\cos\alpha\,d\theta = B_1\,\pi\cos^2\alpha$$

and the moment with respect to an axis through the apex:

$$\mathsf{M} = \int\limits_{-\pi}^{+\pi} N_{s\theta 1}\sin^2\theta \cdot s\cos\alpha \cdot s\sin\alpha \cdot d\theta = A_1\,\pi\cos\alpha\,\sin\alpha.$$

These equations determine A_1 and B_1 when P and M are given. Together with the solution for the vertical force, given on p. 38 and one

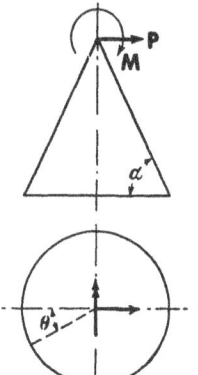

Fig. 39. Loads applied to the apex of a conical shell

for a couple with vertical axis, which may easily be
established, we have the complete solution for an
arbitrary load applied to the top of the cone.

For the higher harmonics, $n \geq 2$, a singularity
at the top does not correspond to an external force
or moment. It is open to discussion whether or not
a solution containing such a singularity has any mechanical significance. If we do not admit it, we must
conclude that a closed conical shell cannot carry an
edge load such as that shown on a spherical shell in
fig. 29.

It is of principal interest in this connection to
study some combinations of parts of spherical and
conical shells. The shell in fig. 40a has a conical top.
From eq. (52) we see that in this part the homogeneous solution must vanish identically if we do not
want it to become infinite. For the spherical part at the base,
the homogeneous solution is represented by eq. (38). On the parallel

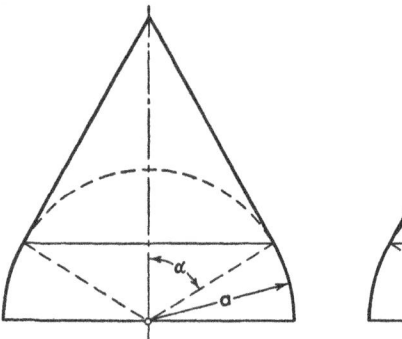
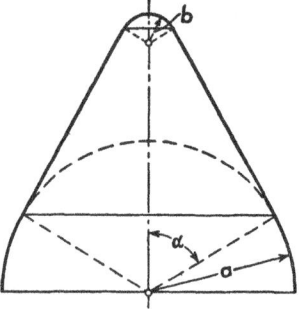

Fig. 40. Shells of revolution, (a) with pointed apex, (b) with rounded apex

circle separating cone and sphere, N_ϕ and $N_{\phi\theta}$ are zero and hence
$A_n = B_n = 0$ in eq. (38), and there is no homogeneous solution at
all. This shell cannot stand any kind of self-equilibrating edge load
without having infinite membrane forces at the apex.

Quite different is the behavior of the shell in fig. 40b. Here the apex is spherical and eqs. (38) yield a regular force system if we only put $A_n = 0$. Writing s instead of ϕ in the subscripts, we have

$$N_{sn} = -N_{s\theta n} = \frac{1}{2} B_n \frac{\tan^n \phi/2}{\sin^2 \phi}.$$

Now we may choose the constants in eq. (52) so that the conical part of the shell has for $s = b \tan\alpha$ the same forces N_s and $N_{s\theta}$. This leads to the formulas

$$N_{sn} = \frac{1}{2} B_n \frac{\tan^n \alpha/2}{\cos^2 \alpha} \left[\frac{n + \cos\alpha}{\sin\alpha} \frac{b}{s} - \frac{n}{\cos\alpha} \frac{b^2}{s^2} \right],$$

$$N_{s\theta n} = -\frac{1}{2} B_n \frac{\tan^n \alpha/2}{\cos^2 \alpha} \frac{b^2}{s^2}, \qquad N_{\theta n} \equiv 0.$$

For the spherical base we again use (38), this time with both constants:

$$N_{sn} = -N_{\theta n} = \frac{b}{4a} B_n \frac{\tan^n \alpha/2}{\sin^2 \phi} \left[\frac{a-b}{a} \left(1 + \frac{n}{\cos\alpha} \right) \tan^n \frac{\alpha}{2} \cot^n \frac{\phi}{2} \right.$$
$$\left. + \left(\frac{a+b}{a} + \frac{a-b}{a} \frac{n}{\cos\alpha} \right) \cot^n \frac{\alpha}{2} \tan^n \frac{\phi}{2} \right],$$

$$N_{s\theta n} = \frac{b}{4a} B_n \frac{\tan^n \alpha/2}{\sin^2 \phi} \left[\frac{a-b}{a} \left(1 + \frac{n}{\cos\alpha} \right) \tan^n \frac{\alpha}{2} \cot^n \frac{\phi}{2} \right.$$
$$\left. - \left(\frac{a+b}{a} + \frac{a-b}{a} \frac{n}{\cos\alpha} \right) \cot^n \frac{\alpha}{2} \tan^n \frac{\phi}{2} \right].$$

An example of these forces is shown in fig. 41. Starting at the edge of the shell, they decrease along the circular part of the meridian, becoming almost insignificant in the case $n = 5$. In the conical part they

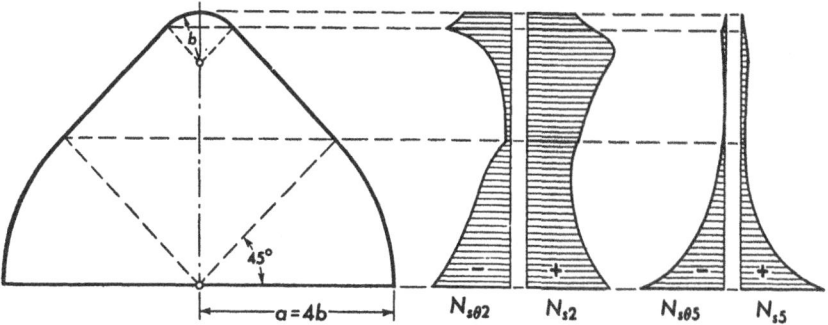

Fig. 41. Stress resultants produced by edge loads

recover. The diagram $n = 2$ shows clearly how they decrease again in the small sphere and end up with finite values at the apex, as may be expected with the second harmonic. In the case $n = 5$, the values in the conical part are too small to show clearly what happens. The shear $N_{s\theta}$ recovers from 0.2% of its boundary value to 3%; the meridional force N_s has a maximum about halfway up and then decreases again. In the small sphere both die out very quickly. If we choose a smaller radius b, the recovery along the straight meridian will be more effective and the state of stress will approach that of the shell with a conical apex. We shall see on p. 73 what conclusions we may draw from this result.

2.4.4 Solution for Shells of Arbitrary Shape

In earlier sections we solved eqs. (32) for the sphere by a trick which is not generally applicable, and for the cone by making use of simplifications arising from the straightness of the meridian. To find a solution of more general applicability, we must look for other methods.

2.4.4.1 Solution by an Auxiliary Variable

We start from the set of differential equations (8) and treat it in the same way as we did with eq. (6) to get the set (32), introducing (29) and (30) and eliminating $N_{\theta n}$. Thus we obtain the following pair of equations:

$$\frac{d}{dz}(rN_{\phi n})\sin\phi + \frac{r_2}{r_1}N_{\phi n}\cos\phi + nN_{\phi\theta n} = -rp_{\phi n} + rp_{rn}\cot\phi,$$

$$\frac{d}{dz}(rN_{\phi\theta n})\sin\phi + N_{\phi\theta n}\cos\phi + n\frac{r_2}{r_1}N_{\phi n} = -rp_{\theta n} + \frac{nr}{\sin\phi}p_{rn}.$$

When we eliminate $N_{\phi\theta n}$, making some use of the geometric relations (1) and (4), we find the following second order differential equation for $N_{\phi n}$:

$$\frac{d^2(rN_{\phi n})}{dz^2}r\sin\phi + 2\frac{d(rN_{\phi n})}{dz}\left(\frac{r}{r_1}+\sin\phi\right)\cot\phi$$

$$+ N_{\phi n}\frac{r}{r_1}\left(2\cot^2\phi - \frac{r}{r_1\sin^3\phi} - \frac{r}{r_1^2}\frac{\cot\phi}{\sin\phi}\frac{dr_1}{d\phi} - \frac{n^2}{\sin^2\phi}\right)$$

$$= \frac{nr}{\sin\phi}\left(p_{\theta n} - \frac{n}{\sin\phi}p_{rn}\right) - (p_{\phi n} - p_{rn}\cot\phi)r\cot\phi$$

$$- \frac{d}{dz}(p_{\phi n}r^2 - p_{rn}r^2\cot\phi).$$

Its left-hand side assumes a very simple form, if we introduce the auxiliary variable

$$U_n = r^2N_{\phi n}\sin\phi.$$

The angle ϕ and the radius r then disappear completely on the left, and with the help of the relations

$$\cot\phi = \frac{dr}{dz}, \qquad \frac{1}{\sin\phi} = \sqrt{1+\left(\frac{dr}{dz}\right)^2}$$

we may make ϕ disappear on the right-hand side also:

$$\frac{d^2 U_n}{dz^2} + (n^2 - 1)\frac{1}{r}\frac{d^2 r}{dz^2} U_n = p_{\theta n}\, n\, r \sqrt{1 + \left(\frac{dr}{dz}\right)^2} - 3 p_{\phi n}\, r\, \frac{dr}{dz} \tag{53}$$

$$- p_{rn}\left[n^2 r + (n^2 - 3)\, r\left(\frac{dr}{dz}\right)^2 - r^2 \frac{d^2 r}{dz^2}\right] - \frac{dp_{\phi n}}{dz}\, r^2 + \frac{dp_{rn}}{dz}\, r^2\, \frac{dr}{dz}.$$

This is now the differential equation of our problem. We shall restrict its discussion here to the action of edge loads on some typical shapes of shells.

In a paraboloid of revolution,

$$r = \sqrt{a z}\,.$$

Introducing this into eq. (53) and putting $p_\phi \equiv p_\theta \equiv p_r \equiv 0$, we have

$$\frac{d^2 U_n}{dz^2} - \frac{n^2 - 1}{4 z^2} U_n = 0\,.$$

The complete solution of this equation is

$$U_n = A z^{(1+n)/2} + B z^{(1-n)/2},$$

as one may easily check by substitution. The second term becomes infinite for $z = 0$ and is therefore not applicable to shells closed at the top.

From U_n we find the n-th harmonic of the meridional force

$$N_{\phi n} = \frac{U_n}{r^2 \sin\phi}\,,$$

of the shear

$$N_{\phi\theta n} = -\frac{1}{n}\left[\frac{d}{dz}\,(r N_{\phi n})\sin\phi + \frac{r}{r_1} N_{\phi n}\cot\phi\right],$$

and of the hoop force

$$N_{\theta n} = -\frac{U_n}{r_1\, r \sin^2\phi}\,.$$

If only the A term of the solution is used, this yields

$$N_{\phi n} = A\,\frac{1}{2a}\,\sqrt{z^{n-2}}\,\sqrt{a + 4z}\,,$$

$$N_{\phi\theta n} = -A\,\frac{\sqrt{z^{n-2}}}{2\sqrt{a}}\,,$$

$$N_{\theta n} = -A\,\frac{\sqrt{z^{n-2}}}{2\sqrt{a + 4z}}\,.$$

If $n = 1$, the stress resultants approach ∞ for $z \to 0$, corresponding to a horizontal concentrated load as shown in fig. 28 for a sphere. For $n = 2$, the stress resultants approach finite limits, and for $n > 2$ they vanish at the top of the shell.

The results given here for a parabolic shell show the same general features as those found on p. 53 for the sphere. In the vicinity of the apex they may be used as an approximation for the stress resultants

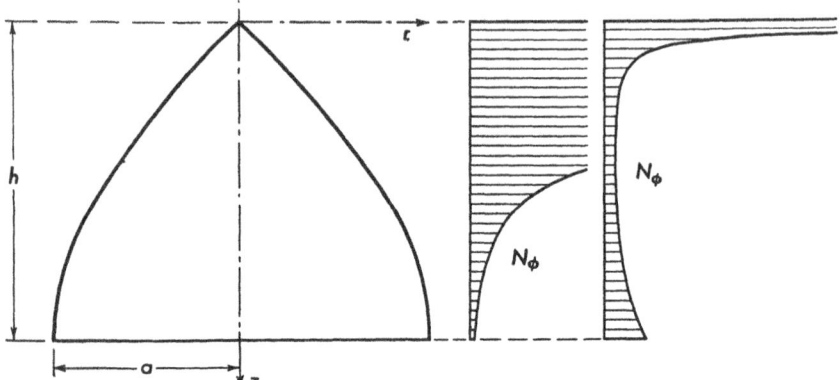

Fig. 42. Pointed shell, (a) meridional section, (b, c) meridional force N_ϕ, A term and B term, for $n = 3$

in any other shell which there has a finite curvature equal to that of the paraboloid.

As a second example we consider a pointed shell. The formulas become particularly simple if the meridian has the equation

$$r = a \sin \frac{\pi z}{2h}$$

(fig. 42). In this case, eq. (53) assumes the form

$$\frac{d^2 U_n}{dz^2} - (n^2 - 1) \frac{\pi^2}{4h^2} U_n = 0$$

and is solved by exponential functions:

$$U_n = A e^{-\lambda z} + B e^{-\lambda(2h - z)}, \qquad \lambda^2 = \frac{n^2 - 1}{4} \frac{\pi^2}{h^2}.$$

Both terms are regular at $z = 0$, but the corresponding stress resultants are not. They are

$$N_{\phi n} = \frac{A e^{-\lambda z} + B e^{-\lambda(2h - z)}}{a^2 \sin^2 \dfrac{\pi z}{2h}} \sqrt{1 + \frac{\pi^2 a^2}{4h^2} \cos^2 \frac{\pi z}{2h}},$$

$$N_{\phi\theta n} = \frac{\pi}{2n a h \sin \dfrac{\pi z}{2h}} \left[\sqrt{n^2 - 1} \, (A e^{-\lambda z} - B e^{-\lambda(2h - z)}) \right.$$

$$\left. + (A e^{-\lambda z} + B e^{-\lambda(2h - z)}) \cot \frac{\pi z}{2h} \right],$$

and they assume, for any choice of A and B, infinite values at $z = 0$ and at $z = 2h$. Nevertheless, there is a considerable difference in the way these two solutions tend toward infinity. If we start at the edge of the shell, say at $z = h$, and follow the meridian toward the apex, the factor $e^{-\lambda z}$ of the A solution produces an accelerated increase of the stress resultants, which at last is reinforced by the vanishing of the factor $\sin^2(\pi z/2h)$ in the denominators (for the shear, the second sine factor is hidden in the cotangent). Quite differently, the factor $e^{+\lambda z}$ of the B solution makes it decrease rapidly, and it may become insignificant before the vanishing of the denominators becomes felt and finally makes it veer to infinity. In fig. 42 the meridional forces for both solutions are represented separately, showing this difference in appearance.

On p. 69 we saw that in a shell with a conical top one of the solutions becomes regular when the shape of the middle surface is but slightly changed. The same will be true in the present case, and it may be presumed that this regular solution approaches the B solution asymptotically as the spherical top is made smaller and smaller. Just as does this rounding of the top, the bending rigidity of any real shell must also have the effect of quelling the weak singularity of the B solution, and we may therefore simply disregard it in all those cases where, in an intermediate zone between the edge and the top, the stress resultants become negligibly small. If this does not happen, the membrane theory is inadequate to solve the stress problem.

2.4.4.2 Solution by Numerical Integration

The use of the function U_n is the appropriate means of solution if it is possible to solve the differential equation (53) by analytical methods. When numerical integration becomes necessary, it will usually be preferable to avoid an auxiliary variable and to start directly from eqs. (32), and when we at once restrict the discussion to edge loads, we know that there are two linearly independent solutions, say

$$N_{\phi n} = f_1(\phi), \qquad N_{\phi\theta n} = g_1(\phi)$$

and

$$N_{\phi n} = f_2(\phi), \qquad N_{\phi\theta n} = g_2(\phi),$$

and the general solution is a linear combination of the two:

$$N_{\phi n} = C_1 f_1(\phi) + C_2 f_2(\phi),$$
$$N_{\phi\theta n} = C_1 g_1(\phi) + C_2 g_2(\phi).$$

If the shell is closed at the top, one of these solutions must be discarded, because it becomes infinite at a rounded apex and strongly infinite at a pointed apex. Let this solution be f_2, g_2, while f_1, g_1 may be the one which either is regular (like the B term in eq. (38)) or has a weak singularity (like the solution plotted in fig. 42c). We have then $C_2 = 0$

for all closed shells and there will be only one boundary condition at the edge, from which C_1 must be determined.

When we want to solve eqs. (32) by numerical integration, we solve them for the derivatives:

$$\frac{dN_{\phi n}}{d\phi} = -\frac{r_1+r_2}{r_2}\cot\phi \cdot N_{\phi n} - n\frac{r_1}{r_2}\frac{N_{\phi\theta n}}{\sin\phi} + r_1 p_{rn}\cot\phi - r_1 p_{\phi n},$$

$$\frac{dN_{\phi\theta n}}{d\phi} = -2\frac{r_1}{r_2}\cot\phi \cdot N_{\phi\theta n} - n\frac{N_{\phi n}}{\sin\phi} + nr_1\frac{p_{rn}}{\sin\phi} - r_1 p_{\theta n}.$$

If we have boundary values $\overline{N}_{\phi n}, \overline{N}_{\phi\theta n}$ for both unknowns, we may simply introduce them on the right-hand side of these equations and compute the first derivatives and, from them, the increments $\Delta N_{\phi n}$ and $\Delta N_{\phi\theta n}$ for a small (negative) increment $\Delta\phi$ of the coordinate. Thus we obtain values of the unknowns at an adjacent point of the meridian and may there repeat the procedure and then again all the way up the meridian. This simple method of numerical integration, the EULER method, may be replaced by any of the more sophisticated methods offered in books on numerical analysis, depending on the degree of accuracy required.

Normally, however, only one of the unknowns will be given at the edge (usually $\overline{N}_{\phi n}$), and if we make an arbitrary choice of the other boundary value, $\overline{N}_{\phi\theta n}$, the result of the integration will be strongly singular at the apex, i. e. it will contain both elementary solutions. This soon becomes apparent as the numerical integration proceeds, since both variables tend to increase rapidly toward $+\infty$ or $-\infty$.

Now let us consider all those solutions which may be obtained for various assumptions of $\overline{N}_{\phi\theta n}$, while $\overline{N}_{\phi n}$ is held at a given value. To each of these assumptions belongs a pair of constants C_1, C_2, and since the differential equations are linear, the constants will be linear functions of the boundary value $\overline{N}_{\phi\theta n}$. Our problem is to find that value of $\overline{N}_{\phi\theta n}$ for which $C_2 = 0$.

Because of the linear relation between C_2 and $\overline{N}_{\phi\theta n}$ this may be done by linear interpolation between two solutions. In that region where the values of $N_{\phi n}$ and $N_{\phi\theta n}$ become large, it may be assumed that f_2, g_2 overshadow the decreasing solution f_1, g_1 and that there

$$N_{\phi n} \approx C_2 f_2(\phi), \qquad N_{\phi\theta n} \approx C_2 g_2(\phi).$$

It is then only necessary to have the results of two numerical integrations, made for different choices of $\overline{N}_{\phi\theta n}$. Picking from them the values of either variable at the same, sufficiently advanced ϕ, one may interpolate or extrapolate that value of $\overline{N}_{\phi\theta n}$, for which this variable would be zero. This is the required boundary value of the shear.

The two trial integrations should preferably be chosen so that the final solution lies between them and so that they are as close to it as feasible. How this is done is best seen from an example.

We want to find the membrane forces which are set up by a uniform dead load $p = 58$ lb/ft^2 in the shell dome represented in fig. 43a. Its meridian is a common parabola with a horizontal axis. The shell rests on eight supports, each of an angular width of $10°$.

First of all, we calculate the stress resultants $N_{\phi 0}$ and $N_{\theta 0}$ for a shell which is uniformly supported along the whole base. This may be done with the help of eqs. (10) and (9b), and we do not need to describe

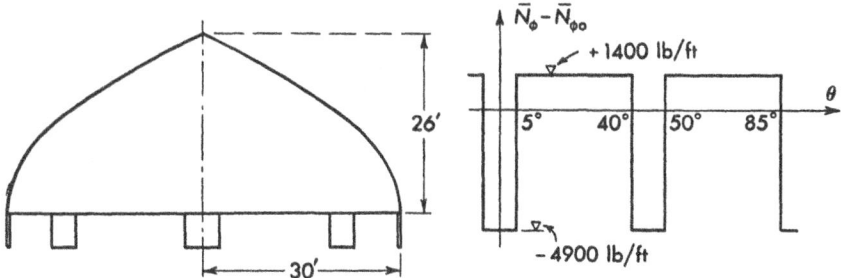

Fig. 43. Pointed shell on eight supports, (a) elevation of the shell, (b) edge load

the details here. The result is an edge value of the meridional force of $\overline{N}_{\phi 0} = -1400$ lb/ft. In order to satisfy the real boundary condition of the shell, we superpose a self-equilibrating edge load which cancels the force $\overline{N}_{\phi 0}$ on the free parts of the edge. Part of this load is shown in fig. 43b. It may be expanded into a Fourier series:

$$\sum_n \overline{N}_{\phi n} \cos n\theta = (-2580 \cos 8\theta - 1972 \cos 16\theta - 1157 \cos 24\theta$$
$$-343 \cos 32\theta + 274 \cos 40\theta + 579 \cos 48\theta$$
$$+564 \cos 56\theta + \ldots) \text{ lb/ft}.$$

For each one of these terms, eqs. (32) must be integrated numerically by the method just described. The term $n = 16$ may serve as an example. A first integration was started from $\overline{N}_{\phi 16} = 1$, $\overline{N}_{\phi \theta 16} = -1.632$. The step $\Delta\phi$ was initially chosen as $-0.5°$, but in the course of the integration it was soon increased to $-1°$ and later to $-2°$. The following values were obtained:

ϕ =	90°	85°	80°	76°	72°	68°	64°
$N_{\phi 16}$ =	1.000	0.4254	0.1770	0.0799	0.0200	−0.0354	−0.1288
$N_{\phi \theta 16}$ =	−1.632	−0.6953	−0.2993	−0.1604	−0.1064	−0.1169	−0.2151

After the change in sign in $N_{\phi 16}$ the values start to grow rapidly. From a close inspection of the details of this computation it was concluded that the solution would tend to $+\infty$ if $|\overline{N}_{\phi\theta 16}|$ were chosen slightly smaller. A second integration was performed with $\overline{N}_{\phi\theta 16} = -1.625$, and the following values were obtained:

ϕ =	90°	85°	80°	76°	72°	68°	64°
$N_{\phi 16}$ =	1.000	0.4298	0.1894	0.1065	0.0782	0.0981	0.1971
$N_{\phi\theta 16}$ =	−1.625	−0.6858	−0.2782	−0.1179	−0.0160	0.0842	0.2584

This solution indeed tends toward the other side, and the true solution must lie between them. It was not thought worthwhile to continue the integration to still smaller values of ϕ, and from the results for $\phi = 64°$ the correct starting value was interpolated. When $N_{\phi 16}$ is used, this yields

$$\overline{N}_{\phi\theta 16} = -1.625 - \frac{(1.632 - 1.625) \times 0.1971}{0.1971 + 0.1288} = -1.629,$$

and from $N_{\phi\theta 16}$ the same result may be obtained. The trial solutions and the final solution lie so close together that it is not worthwhile to do the integration over again, but the final solution may be interpolated between the other two.

In this way the first harmonics $n = 8, 16, 24$ were treated, while the higher ones could be handled with the analytic approximation

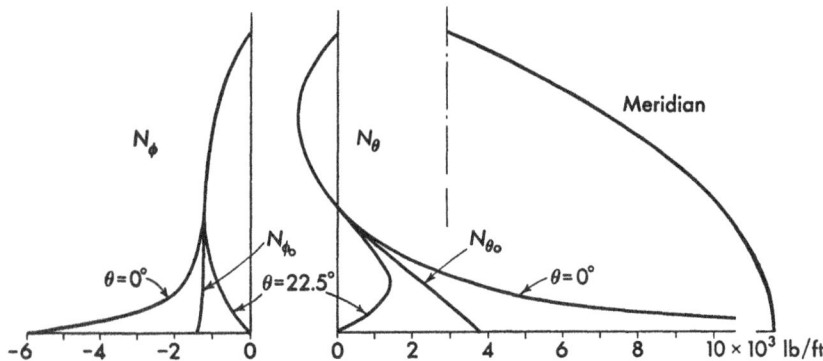

Fig. 44. Stress resultants in the shell shown in Fig. 43

explained on p. 77. The results of these computations yield Fourier series for each of the stress resultants. Some typical results are shown in the diagrams, fig. 44. In these diagrams the curve in the middle gives $N_{\phi 0}$ or $N_{\theta 0}$, and the other two give the final values for the meridian through the center of a support ($\theta = 0°$) and for that one through the center of the free span ($\theta = 22.5°$). It may be seen how rapidly the shell attenuates the high peak values at the supported part of the edge and how the stress system comes closer and closer to axial symmetry as one follows the meridian upward. Theoretically, the contributions of all harmonics $n > 0$ must finally rise to infinity. similarly to fig. 42 c, but when they once have faded out beyond recognition, the bending stiffness even of a thin shell will be sufficient to prevent their revival.

When making numerical integrations of the kind just described, one is always confronted with the difficult question as to how $\overline{N}_{\phi\theta n}$ should be chosen for the first trial. If no other information is at hand, one may consult the solution for the sphere, eq. (38). This will at least give the correct sign and some idea of the order of magnitude, however rough it may be. A better means is to consult the solution for an ellipsoid (see eqs. (IV–35)) which approximates the boundary region of the shell. Still better, one may use any past experience with other shells, including the results given here, or the results just obtained for another harmonic in the same shell. In the present example, for instance, it was easy to make a good guess for $n = 16$ after the computation for $n = 8$ had been carried to a successful completion, and it was still easier for $n = 24$. Another means of getting started is to use the analytical approximation which will now be presented.

The higher the order of the harmonic, the more the solution takes on the character of a local disturbance along the edge of the shell. The engineer's interest is always limited to the zone in which the forces have appreciable magnitude; this zone may be so small that we can safely neglect the variability of the coefficients in the differential equations (32) and replace them by average values, say those at the center of the interesting domain, $\phi \doteq \phi'$. Thus we arrive at equations with constant coefficients, which may solved by exponential functions

$$N_{\phi n} = A\, e^{\alpha\phi}, \qquad N_{\phi\theta n} = B\, e^{\alpha\phi}.$$

Introducing these into eqs. (32), we get two homogeneous linear equations for A and B:

$$A\left[\alpha + \left(\frac{r_1}{r_2} + 1\right)\cot\phi'\right] + B\frac{r_1}{r_2}\frac{n}{\sin\phi'} = 0,$$

$$A\frac{n}{\sin\phi'} + B\left[\alpha + 2\frac{r_1}{r_2}\cot\phi'\right] = 0.$$

These have only the trivial solution $A = B = 0$, unless the determinant of the coefficients vanishes. This yields an equation for α:

$$\alpha^2 + \alpha\left(3\frac{r_1}{r_2} + 1\right)\cot\phi' + 2\frac{r_1}{r_2}\left(\frac{r_1}{r_2} + 1\right)\cot^2\phi' - \frac{n^2 r_1}{r_2\sin^2\phi'} = 0.$$

In the example just treated we have at $\phi' = 80°$:

$$r_1 = 11.78 \text{ ft}, \qquad r_2 = 30.30 \text{ ft},$$

and hence the equation

$$\alpha^2 + 0.3820\,\alpha - 25.60 = 0,$$

of which only the positive root $\alpha \doteq 4.87$ is of interest. Either of the linear equations then gives

$$B = -1.621A,$$

as compared with 1.622 determined as the result of the numerical integration. The approximation, of course, is not always as good and depends on the choice of the representative angle ϕ'. In the present case the extreme choice $\phi' = 90°$ leads to $B/A = -1.632$.

In such shells where it seems more advantageous, the method of numerical integration and the analytic approximation may also be applied to eqs. (7) which use s instead of ϕ as a coordinate.

It will be seen in the next section that the special problem of determining $\overline{N}_{\phi \theta n}$ does not exist for shells of negative curvature. The numerical integration then becomes an extremely simple procedure.

2.4.5 Shell Formed as a One-sheet Hyperboloid

The general properties of all stress systems, which we have found in all previous examples, are bound to one essential supposition: The GAUSSIAN curvature $1/r_1 r_2$ must be positive. Shells of negative curvature, as, for example, the one-sheet hyperboloid of fig. 45, behave quite differently. We shall discuss the reason for this.

We start with eqs. (7). First of all, we use eq. (7c) to eliminate N_θ. With eq. (3a) the other two are

$$r \frac{\partial N_s}{\partial s} + N_s \cos\phi + N_s \frac{r}{r_1}\cot\phi + \frac{\partial N_{s\theta}}{\partial \theta} = -p_\phi r + p_r r \cot\phi,$$

$$\text{(54a, b)}$$

$$r \frac{\partial N_{s\theta}}{\partial s} + 2 N_{s\theta} \cos\phi - \frac{\partial N_s}{\partial \theta} \frac{r_2}{r_1} = -p_\theta r - \frac{\partial p_r}{\partial \theta} r_2.$$

From these two equations we may, if we wish, eliminate $N_{s\theta}$. If we differentiate (54b) with respect to θ, it contains $\partial N_{s\theta}/\partial \theta$ and $\partial^2 N_{s\theta}/\partial s \, \partial \theta$. The first derivative may be eliminated by using eq. (54a). The second derivative may be eliminated with the help of the same equation, if we first differentiate it with respect to s. We thus find a partial differential equation for N_s, which is of the second order. We shall not establish it in detail, because its coefficients are anything but simple and the equation is not of much use, since we cannot find easy boundary conditions for its solution. But it is important for our purpose that in this equation the second derivatives of the unknown appear in the combination

$$\frac{\partial^2 N_s}{\partial s^2} + \frac{1}{r_1 r_2 \sin^2\phi} \frac{\partial^2 N_s}{\partial \theta^2}.$$

In the theory of partial differential equations of the second order[1] it is shown that the sign of the coefficients of the second derivatives determines the essential features of the solution. If both coefficients have the same sign, the equation is called elliptic, and then discontinuities of the given boundary values do not propagate into the interior of the shell, but their influence becomes feebler, the farther away we are from the edge. That is exactly the situation which we found in spherical and other shells, and we see here that we can anticipate it in all shells of positive curvature. But if the shell is one of negative curvature, the coefficient of $\partial^2 N_s/\partial\theta^2$ is negative and hence the differential equation is of the hyperbolic type. Such equations have real characteristics. These are curves along which discontinuities of the given boundary values are propagated into the interior. This is the cause of quite surprising phenomena, which may seriously influence the general layout of a shell structure. On the other hand, the characteristics supply an excellent means of finding the solution of the differential equations for shells with negative curvature. All this we shall study here for the simplest of such shells, one which has

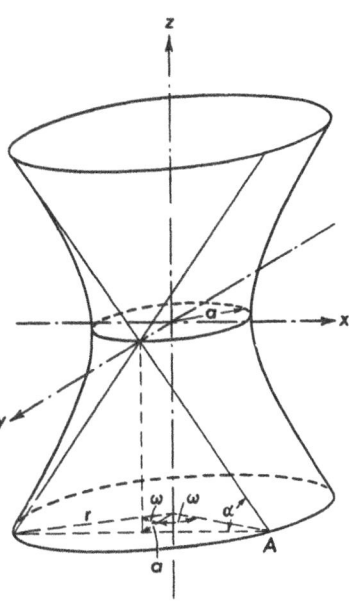

Fig. 45. One-sheet hyperboloid

the shape of a one-sheet hyperboloid. Later we shall meet similar problems and methods when treating another kind of shell with negative curvature in Chapter 4.

Fig. 45 shows a one-sheet hyperboloid. It has the equation

$$\frac{x^2 + y^2}{a^2} - \frac{z^2}{b^2} = 1.$$

Let us intersect this surface with the vertical plane $y = a$. When we introduce this value of y in the preceding equation, we get

$$\frac{x^2}{a^2} = \frac{z^2}{b^2},$$

and this means that the curve of intersection consists of a pair of straight lines having the equations $x = \pm z\, a/b$ and hence the slope

---

[1] HILDEBRAND, F. B.: Advanced Calculus for Engineers. New York 1949, p. 463.

$\tan\alpha = b/a$. Since the hyperboloid is a surface of revolution, the tangent plane to any other point of the waist circle will yield a similar pair of straight lines on the hyperboloid, so that there exist two families of straight lines, each of which covers the surface completely. They are called generators, because the hyperboloid may be generated by any one of them, if we rotate it around the z axis. They are, of course, not meridians, since they do not lie in the same plane as the axis.

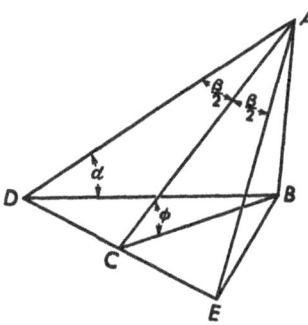

Fig. 46. Tangent plane ADE at point A of a hyperboloid

Before we discuss the equilibrium of a shell element, we need some geometric properties of the generators. Fig. 46 shows the two generators AD and AE which pass through an arbitrary point A of the surface. We want to know the angle β between them. Let the section AC on the tangent to the meridian have the length 1. Then its projection on a vertical has the length $AB = \sin\phi$. From the triangle ABD we find $AD = \sin\phi/\sin\alpha$ and hence from the triangle ADC:

$$\cos\frac{\beta}{2} = \frac{1}{AD} = \frac{\sin\alpha}{\sin\phi}.$$

We shall also need the angle ω between the two meridians which meet the same generator at the waist circle and at the point A, not necessarily on the edge. From fig. 45 it follows that

$$\cos\omega = \frac{a}{r}.$$

For the radii of curvature we have to use the same formulas as for the ellipsoid (see p. 27) except that we must replace b^2 by $-b^2$:

$$r_1 = -\frac{a^2 b^2}{(a^2 \sin^2\phi - b^2 \cos^2\phi)^{3/2}},$$

$$r_2 = \frac{a^2}{(a^2 \sin^2\phi - b^2 \cos^2\phi)^{1/2}}.$$

Since $r = r_2 \sin\phi$, we can now write

$$\cos\omega = \frac{(a^2 \sin^2\phi - b^2 \cos^2\phi)^{1/2}}{a \sin\phi} = \sqrt{1 - \frac{b^2}{a^2}\cot^2\phi}.$$

After these geometric preparations we can begin the investigation of the membrane forces in the shell whose middle surface is a one-sheet hyperboloid. In fig. 47 two adjacent generators are drawn which enclose between them a narrow strip of the shell. This strip is straight but slightly twisted and therefore of variable width, narrowest where it meets the waist circle. Now let us apply to both ends of this strip and in its direction two external forces dP, as shown in fig. 47. We may easily guess that they produce a uniaxial state of tensile stress in the strip, variable in intensity and inversely proportional to the .width, while all the rest of the shell is completely unstressed. Since in such a state of stress every element of the shell will be in perfect equilibrium,

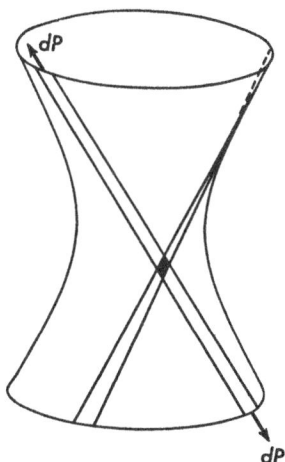

Fig. 47
Shell shaped after a one-sheet hyperboloid

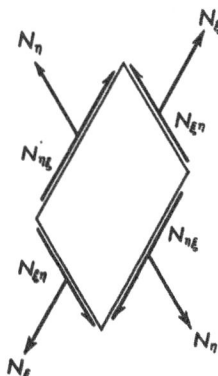

Fig. 48
Shell element shaded in Fig. 47

it must be a solution of the general equations (6), however strange it appears. We shall see now that it is the only possible solution for the given boundary conditions.

To show this, we cut out a particular shell element, which is limited by two pairs of adjacent generators, each pair belonging to one of the two families of such lines (fig. 48). The forces transmitted by its edges are resolved into oblique components, the skew fiber forces N_ξ, N_η and the skew shearing forces $N_{\xi\eta} = N_{\eta\xi}$. The two forces N_ξ on opposite sides of the element lie exactly on the same straight line, one of the generators, and therefore their resultant must do the same. They cannot make any contribution to the equilibrium in the direction normal to the shell, and neither can the forces N_η. To resist a normal load p_r (not shown in fig. 48), only the shearing forces $N_{\xi\eta}$, $N_{\eta\xi}$ are available. These forces on two opposite sides of the element are not strictly parallel, because they have the directions of two different generators which cross each other under an angle of differential magnitude. We shall spare the reader the trouble of finding the exact amount of this angle and of establishing the condition of equilibrium in detail. It is enough to know that it must have the form: $N_{\xi\eta}$ times a geometric coefficient equals p_r. This means that in sections along the generators the shear must be zero, if there is no surface load. Therefore the load dP which is introduced into a strip between two generators at one edge has no chance to leave this strip sideways and must appear on the opposite edge at the end of the strip.

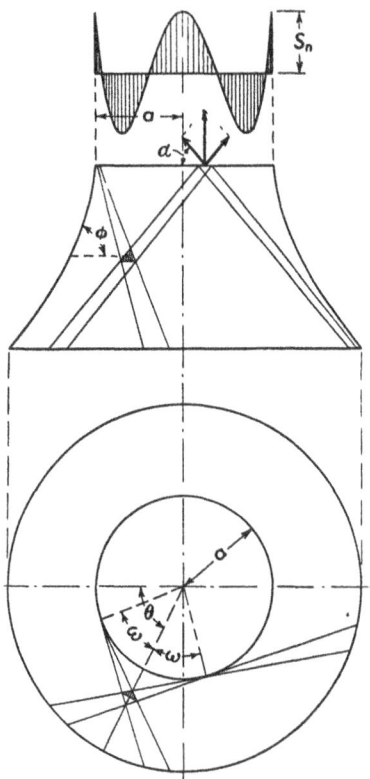

This remarkable fact provides the basis for the construction of the general solution. We cut the shell in the waist circle (fig. 49) and load it there with a harmonic edge load

$$N_\phi = S_n \cos n\,\theta.$$

The vertical force $S_n\,a\,\cos n\theta\,d\theta$, which acts on the line element $a\,d\theta$ of the waist circle, may be resolved into two components

$$\frac{S_n\,a\,\cos n\,\theta\,d\theta}{2\sin\alpha},$$

which have the direction of the two generators passing through the center of this element, as shown in fig. 49 at the point with the coordinate $\theta + \omega$.

Now let us consider an arbitrary point ϕ, θ of the shell. Its two generators meet the waist circle at the meridians $\theta \pm \omega$ and bring from

Fig. 49. Hyperboloid shell loaded at the
waist circle

there the corresponding forces. These act on two sides of the triangular element, fig. 50, which is the half of the one shown in fig. 48. On its right side we have the force

$$dP_1 = \frac{S_n\,a\,\cos n\,(\theta + \omega)}{2\sin\alpha}\,d\theta,$$

and on its left side

$$dP_2 = \frac{S_n\,a\,\cos n\,(\theta - \omega)}{2\sin\alpha}\,d\theta.$$

The force on the horizontal side $r\,d\theta$ of the triangle is equal to the resultant of the other two and has therefore the meridional component

$$N_\phi\,r\,d\theta = \frac{S_n\,a\,d\theta}{2\sin\alpha}\left[\cos n\,(\theta + \omega) + \cos n\,(\theta - \omega)\right]\cos\frac{\beta}{2}$$

and the shear component

$$N_{\phi\theta} r\, d\theta = \frac{S_n a\, d\theta}{2\sin\alpha} \left[\cos n\,(\theta + \omega) - \cos n\,(\theta - \omega)\right] \sin\frac{\beta}{2}.$$

If we make appropriate use of the previously established geometric relations and of (6c), we find from this the following formulas for the stress resultants:

$$N_\phi = \frac{S_n a}{r} \frac{\cos n\,\theta \cos n\,\omega}{\sin\alpha} \frac{\sin\alpha}{\sin\phi} = S_n \frac{\cos\omega \cos n\,\omega}{\sin\phi} \cos n\,\theta,$$

$$N_\theta = -N_\phi \frac{r_2}{r_1} = \frac{S_n a^2}{b^2} \cos^3\omega \sin\phi \cos n\,\omega \cos n\,\theta, \qquad (55)$$

$$N_{\phi\theta} = -\frac{S_n a}{r} \frac{\sin n\,\theta \sin n\,\omega}{\sin\alpha} \sin\frac{\beta}{2} = -\frac{S_n a}{b} \cos^2\omega \sin n\,\omega \sin n\,\theta.$$

By a quite similar reasoning we find for a tangential edge load

$$N_{\phi\theta} = T_n \sin n\,\theta$$

the two strip forces

$$dP_1 = \frac{T_n a \sin n\,(\theta + \omega)}{2\cos\alpha}\, d\theta, \qquad dP_2 = -\frac{T_n a \sin n\,(\theta - \omega)}{2\cos\alpha}\, d\theta,$$

acting on the triangular element, and on its horizontal side the component forces

$$N_\phi r\, d\theta = \frac{T_n a\, d\theta}{2\cos\alpha} \left[\sin n\,(\theta + \omega) - \sin n\,(\theta - \omega)\right] \cos\frac{\beta}{2},$$

$$N_{\phi\theta} r\, d\theta = \frac{T_n a\, d\theta}{2\cos\alpha} \left[\sin n\,(\theta + \omega) + \sin n\,(\theta - \omega)\right] \sin\frac{\beta}{2}.$$

In terms of ϕ, θ and ω we have finally

$$N_\phi = \frac{T_n b}{a} \frac{\cos\omega}{\sin\phi} \sin n\,\omega \cos n\,\theta,$$

$$N_\theta' = \frac{T_n a}{b} \sin\phi \cos^3\omega \sin n\,\omega \cos n\,\theta, \qquad (56)$$

$$N_{\phi\theta} = T_n \cos^2\omega \cos n\,\omega \sin n\,\theta.$$

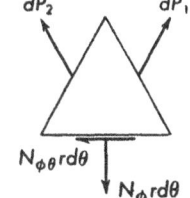

Fig. 50. Triangular shell element shaded in Fig. 49

The two sets of formulas (55), (56) are analogous to the solution (38) for the spherical shell. Both give one term of a FOURIER series which is the general solution for the shell under edge loads. But here the reasoning which led to the formulas (55), (56) opens a second approach to a general solution. We need not submit the given edge load to a harmonic analysis, use our formulas and then sum up again. We resolve the given edge load, not into normal force and shear, but into two components in the directions of the generators, and find from them the forces at any point of the shell from the equilibrium of a triangular element, as we have just done.

But the formulas (55), (56) give still more than that. They show immediately the striking difference in the static behavior of such a shell and one with positive curvature. When a spherical shell has two edges, we can prescribe N_ϕ at both of them and then have to accept the shear $N_{\phi\theta}$ which results. We can also try to determine A_n, B_n so that, for example, at the outer edge both forces N_ϕ and $N_{\phi\theta}$ assume given values. But usually such a solution will lead to unduly high forces in the other boundary zone, because only one of the two homogeneous solutions decreases with increasing distance from the loaded edge and the other one increases. This increase is very pronounced if the order n of the harmonic is high or if the two edges are far from each other. If the shell is closed at the vertex, this increase leads to infinite forces at this point, as we have seen. This indicates that such a set of boundary conditions is not appropriate and that, in any case where it really occurs, bending moments will appear.

Quite to the contrary, eqs. (55), (56) suggest that we choose N_ϕ and $N_{\phi\theta}$ freely at the waist circle and then accept both forces as they result at the other edge. The engineer would certainly prefer to do the same as in the case of the sphere: prescribe N_ϕ at both edges and provide a stiffening ring at each edge to take care of the ensuing shear. Here this procedure is, so to speak, against the nature of the shell. Our formulas show this quite clearly. If $n\omega$ at the lower edge is an integer multiple of π, then a normal force at the waist circle produces only a normal force at the lower edge, and we cannot assume both independently. A shear at one edge produces a pure shear at the other, and we must prescribe one of them to make the problem determinate. Such a result, of course, also appears if we do not choose the waist circle as one of the edges but consider a part of the hyperboloid between any two parallel circles.

If a shell has positive curvature in one part and negative in another, the phenomena described persist. For instance, let the hyperboloid end at the waist circle and be connected there to a hemisphere, as shown in fig. 51. From eqs. (38) for the sphere, it follows

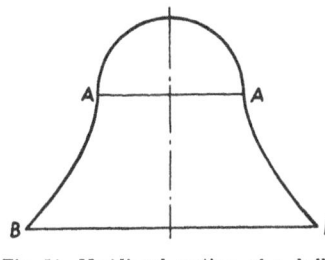

Fig. 51. Meridional section of a bell-shaped shell

that at the circle A–A where $\phi = 90°$, we have $N_{\phi n} = -N_{\phi\theta n}$. The upper edge of the hyperboloid is, therefore, subjected to normal forces $S_n = N_{\phi n}$ and shearing forces $T_n = N_{\phi\theta n} = -S_n$. Eqs. (55) and (56) yield, then, for all points of the hyperboloid

$$N_\phi = S_n \left(\cos n\,\omega - \frac{b}{a} \sin n\,\omega \right) \frac{\cos \omega}{\sin \phi} \cos n\,\vartheta.$$

For a certain hyperboloid b/a is given, and one may find angles ω, for which the factor in parentheses vanishes. If the lower edge of the shell is chosen at a level, where ω assumes one of these special values, then there is always $N_{\phi n} = 0$. For the particular n this shell can resist only a shear load, not a load in the direction of the meridians, and no additional shear and no stiffening ring will help.

The practical meaning of these observations is this: Even if ω at the edge is chosen so that for no integer n does the angle $n\omega$ belong to the series of dangerous values, there will always exist certain harmonics n for which it will very nearly do so. For these harmonics a small load of the dangerous type will produce unduly high stress resultants. No such shell, therefore, can really resist with membrane forces an arbitrary edge load of this kind.

The stresses of other shells of negative curvature may be expected to show the same general features, but the computation of the stress resultants is less simple. Instead of the straight generators of the hyperboloid we have two systems of curvilinear characteristics, and an isolated boundary load P influences not only points of the characteristics, which pass through its point of application, but all the shell between these two lines.

2.5 Deformations

2.5.1 Strains and Displacements

No stress problem is completely solved unless one has also determined the corresponding deformation. In many cases this part of the problem is of no practical interest, but sometimes it is, and in Section 2.4 we met with statically indeterminate problems which require for their solution the analysis of the deformations of the shell.

The deformation of a shell element consists of the elongations of line elements $ds_\phi = r_1 \, d\phi$ (fig. 52) on the meridian and $ds_\theta = r \, d\theta$ on the parallel circle by Δds_ϕ and Δds_θ and of the change of the right angle between these two line elements. We define a meridional strain

$$\epsilon_\phi = \frac{\Delta ds_\phi}{ds_\phi},$$

a hoop strain

$$\epsilon_\theta = \frac{\Delta ds_\theta}{ds_\theta},$$

and a shear strain $\gamma_{\phi\theta}$, which is the decrease of the angle BAC in fig. 52 c.

Between these strains and the stress resultants exists an empirical relation, the elastic law. It depends on the material of the shell. In the mathematical treatment of structural problems, only the linearized form, HOOKE's law, is of importance. Some structural materials, especially steel, follow this law quite perfectly within the limits of the

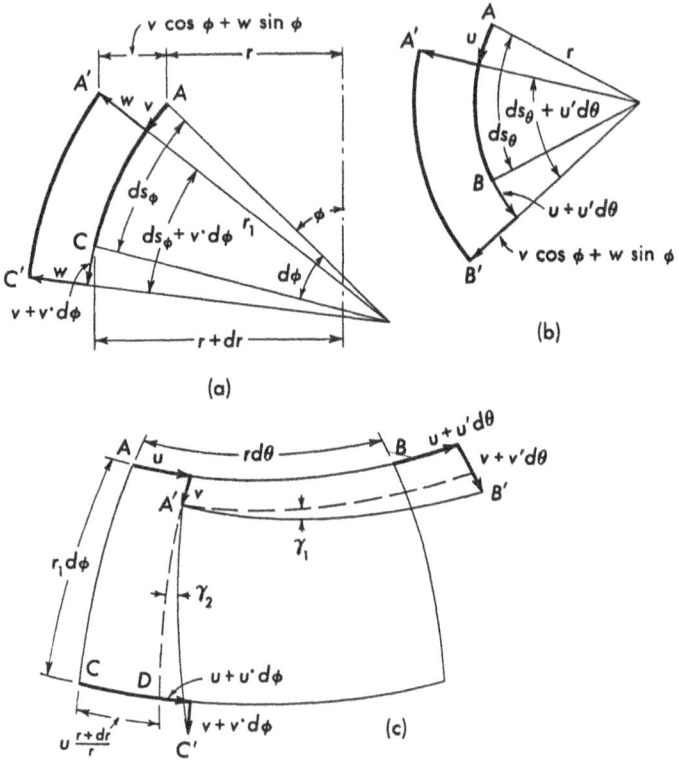

Fig. 52. Line elements before and after deformation,
(a) meridian, (b) parallel circle, (c) both line elements, showing angle between them

usual stresses; for others, such as concrete, it is a linear approximation which, in general, leads to satisfactory results.

Let $\sigma_\phi, \sigma_\theta, \sigma_z$ be the normal stresses acting on sections $\phi, \theta, z = \text{const}$, respectively ($z$ measured as in fig. 1–2). Then HOOKE's law for the two strains is

$$E\,\epsilon_\phi = \sigma_\phi - \nu\,\sigma_\theta - \nu\,\sigma_z,$$

$$E\,\epsilon_\theta = \sigma_\theta - \nu\,\sigma_\phi - \nu\,\sigma_z.$$

The elastic modulus E and POISSON's ratio ν are constants depending on the material. The shear strain $\gamma_{\phi\theta}$ depends only on the shear stress

$$G\,\gamma_{\phi\theta} = \tau_{\phi\theta},$$

where the shear modulus G is connected with the two other elastic constants by the relation

$$G = \frac{E}{2(1 + \nu)} \, .$$

The stress σ_z is of the same order of magnitude as the surface load p_r and, with the exception of the immediate vicinity of concentrated forces, is small compared with σ_ϕ and σ_θ. As in the theory of bars and plates, its influence in HOOKE's law may be neglected. If we introduce the normal and shearing forces instead of the stresses, we have the elastic law of the shell:

$$\epsilon_\phi = \frac{1}{E\,t}\,(N_\phi - \nu\,N_\theta), \qquad \epsilon_\theta = \frac{1}{E\,t}\,(N_\theta - \nu\,N_\phi),$$

$$\gamma_{\phi\theta} = \frac{2\,(1 + \nu)}{E\,t}\,N_{\phi\theta}. \tag{57}$$

Solved for the forces, it takes the following form:

$$N_\phi = D(\epsilon_\phi + \nu\,\epsilon_\theta), \qquad N_\theta = D(\epsilon_\theta + \nu\,\epsilon_\phi),$$

$$N_{\phi\theta} = D\,\frac{1 - \nu}{2}\,\gamma_{\phi\theta}, \tag{58}$$

where

$$D = \frac{E\,t}{1 - \nu^2}$$

is the extensional rigidity of the shell. It is analogous to the product of elastic modulus and cross-sectional area, which gives the extensional rigidity of a bar. Since D contains the wall thickness t, it may be a function of ϕ.

In general, it is not enough to know the strains ϵ_ϕ, ϵ_θ, $\gamma_{\phi\theta}$ for every point of the shell. The quantity which really matters is the displacement of each point of the middle surface. Such displacements are given at the edges where the shell is supported, and when more displacements are prescribed than would be necessary to fix the position of the shell, then the stress problem is statically indeterminate.

The displacement is a vector and may be described by its three components. We choose them in the direction of the tangents to the parallel circle (u) and the meridian (v) and normal to the shell (w). We shall call these components u, v, w the displacements of the point.

The displacements u and v are taken positive in the direction of increasing coordinates θ and ϕ, respectively, and w is considered positive when it points away from the center of curvature of the meridian.

Since u, v, w represent the same geometrical facts as the strains ϵ_ϕ, ϵ_θ, $\gamma_{\phi\theta}$, the displacements and the strains must be connected by a system of equations which is of purely geometric origin. It is our next purpose to establish these equations.

The strains may be expressed in terms of the displacements and their derivatives. It will facilitate the writing of the formulas if we introduce a simple notation for these derivatives. We shall indicate a derivative with respect to θ by a prime and one with respect to ϕ by a dot, for example,

$$\frac{\partial u}{\partial \theta} = u', \qquad \frac{\partial u}{\partial \phi} = u^{\cdot}.$$

We begin with the meridional strain ϵ_ϕ. Fig. 52a shows a meridional element $AC = ds_\phi$. Its ends undergo the tangential displacements v and $v + v^{\cdot} d\phi$, respectively. Their difference $v^{\cdot} d\phi$ is the corresponding elongation of the line element. If there is an additional normal displacement w of the points A and C, it has slightly different directions at the two points and therefore produces an additional elongation of AC. Since the distance r_1 from the center of curvature increases to $r_1 + w$, the length of the arc is increased proportionally to

$$(ds_\phi + v^{\cdot} d\phi) \frac{r_1 + w}{r_1}.$$

An additional radial displacement $w^{\cdot} d\phi$ of the point C' produces an elongation which is small of the second order and therefore without interest. For the same reason we have to drop all products of two displacements, when we compute the elongation

$$\varDelta ds_\phi = (ds_\phi + v^{\cdot} d\phi)\left(1 + \frac{w}{r_1}\right) - ds_\phi = v^{\cdot} d\phi + w \frac{ds_\phi}{r_1}.$$

Dividing by $ds_\phi = r_1 d\phi$, we find the meridional strain

$$\epsilon_\phi = \frac{v^{\cdot} + w}{r_1}. \tag{59a}$$

By similar reasoning we find an expression for the hoop strain ϵ_θ from fig. 52b, where an element ds_θ of a parallel circle is shown. On account of the difference $u' d\theta$ of the tangential displacements of the two ends, the length ds_θ is increased to $ds_\theta + u' d\theta$, and the radial displacement $v \cos\phi + w \sin\phi$ in the plane of the parallel circle (see also fig. 52a) yields a further increase to

$$(ds_\theta + u' d\theta) \frac{r + v \cos\phi + w \sin\phi}{r}.$$

The elongation of the line element is therefore

$$\varDelta ds_\theta = (ds_\theta + u' d\theta)\left(1 + \frac{v}{r} \cos\phi + \frac{w}{r} \sin\phi\right) - ds_\theta$$

$$= u' d\theta + (v \cos\phi + w \sin\phi) \frac{ds_\theta}{r},$$

and hence the hoop strain is

$$\epsilon_\theta = \frac{\Delta\,ds_\theta}{ds_\theta} = \frac{u' + v\cos\phi + w\sin\phi}{r}.$$
(59 b)

To find the corresponding relation for the shear strain we have to consider the whole shell element (fig. 52 c). Its points A, B, C move to A', B', C', and the angle $B'\,A'\,C'$ is smaller than a right angle by $\gamma_1 + \gamma_2 = \gamma_{\phi\theta}$. From the figure we read easily

$$\gamma_1 = \frac{v'\,d\theta}{r\,d\theta + u'\,d\theta} \approx \frac{v'}{r}.$$

To calculate γ_2 we trace the meridian $A'D$. It cuts the arc $CD = u(r + dr)/r$ from the lower parallel circle, where $r + dr$ is the radius of this circle. Subtracting CD from the horizontal displacement $u + u^\cdot\,d\phi$, we find the horizontal projection of DC' and, after division by $r_1\,d\phi + v^\cdot\,d\phi \approx r_1\,d\phi$, the angle

$$\gamma_2 = \frac{u + u^\cdot\,d\phi - u\left(1 + \dfrac{dr}{r}\right)}{r_1\,d\phi} = \frac{u^\cdot}{r_1} - \frac{u}{r_1 r}\frac{dr}{d\phi}.$$

Making use of eq. (4) and then adding γ_1 and γ_2, we find the expression for the shear strain

$$\gamma_{\phi\theta} = \frac{u^\cdot}{r_1} - \frac{u}{r}\cos\phi + \frac{v'}{r}.$$
(59 c)

The relations (59 a–c) enable us to find the strains when the displacements are known functions of the coordinates. Usually we have to deal with the inverse problem, in which the stress resultants of the shell have already been determined and we want to know the displacements. Then HOOKE's law (57) will give us the left-hand sides of (59), and these equations are a set of partial differential equations for the displacements u, v, w. The study of these equations is our next objective.

2.5.2 Inextensional Deformation

2.5.2.1 Differential Equation

We begin with the homogeneous equations

$$\begin{aligned} v^\cdot \quad\quad\ + w \quad\quad\quad &= 0, \\ u' \quad\ + v\cos\phi + w\sin\phi &= 0, \qquad \text{(60 a—c)} \\ u^\cdot \frac{r_2}{r_1}\sin\phi - u\cos\phi + v' \quad\quad &= 0. \end{aligned}$$

They describe a deformation of the shell in which the strains ϵ_ϕ, ϵ_θ, $\gamma_{\phi\theta}$ and hence the stress resultants are all zero. We shall see whether and under what circumstances such a deformation may occur and how we have to fix the edge of the shell to make it impossible.

Since the coefficients in eqs. (60) do not depend on θ, we may write the solution as a Fourier series in θ, and the n-th harmonic will be

$$u = u_n(\phi) \sin n\theta, \quad v = v_n(\phi) \cos n\theta, \quad w = w_n(\phi) \cos n\theta. \quad (61)$$

Introducing this into the partial differential equations (60), we arrive at a system of ordinary differential equations:

$$\dot{v}_n \qquad\qquad + w_n \qquad\qquad = 0,$$
$$n\,u_n \qquad + v_n \cos\phi + w_n \sin\phi = 0, \quad (62\,\text{a—c})$$
$$\dot{u}_n \frac{r_2}{r_1} \sin\phi - u_n \cos\phi - n\,v_n \qquad = 0.$$

We eliminate first the normal component w between the first two of these equations. This yields

$$n\,u_n = \dot{v}_n \sin\phi - v_n \cos\phi \qquad (63)$$

and, after differentiating,

$$n\,\dot{u}_n = \ddot{v}_n \sin\phi + v_n \sin\phi.$$

Introducing this into (62c), we arrive at an equation in which only the meridional component v_n is left:

$$\ddot{v}_n \frac{r_2}{r_1} \sin^2\phi - \dot{v}_n \cos\phi \sin\phi + v_n \left(\frac{r_2}{r_1} \sin^2\phi + \cos^2\phi - n^2 \right) = 0. \quad (64)$$

This is the differential equation of our problem, which we have to solve.

2.5.2.2 Finite Solution for the Spherical Shell

Before we attack the general problem of solving eq. (64) for an arbitrary meridian, we shall apply it to a spherical shell. In this particular case we shall be able to find a simple solution which shows what types of solutions we have to expect in the general case.

For a sphere with $r_1 = r_2 = a$, eq. (64) reads

$$\ddot{v}_n \sin^2\phi - \dot{v}_n \cos\phi \sin\phi + v_n(1 - n^2) = 0,$$

and this may also be written in the following form:

$$\left[\left(\frac{v_n}{\sin\phi} \right)^{\!\cdot} \sin\phi \right]^{\!\cdot} \sin^2\phi - n^2 v_n = 0.$$

This equation has the solution

$$v_n = \sin\phi \left[A \tan^n \frac{\phi}{2} + B \cot^n \frac{\phi}{2} \right]. \qquad (65\,\text{a})$$

Eqs. (63) and (62a) then give

$$u_n = \sin\phi \left[A \tan^n \frac{\phi}{2} - B \cot^n \frac{\phi}{2} \right], \qquad (65\,\text{b, c})$$

$$w_n = -\dot{v}_n = -A \tan^n \frac{\phi}{2} \cdot (n + \cos\phi) + B \cot^n \frac{\phi}{2} \cdot (n - \cos\phi).$$

For $n = 1$ these formulas yield

$$u = [A(1 - \cos \phi) - B(1 + \cos \phi)] \sin \theta,$$

$$v = [A(1 - \cos \phi) + B(1 + \cos \phi)] \cos \theta,$$

$$w = -(A - B) \sin \phi \cos \theta,$$

and they represent two rigid-body rotations of the shell. For the A solution, the axis is the tangent to the meridians $\theta = \pm\pi/2$ at the pole $\phi = 0$, and for the B solution it is the tangent to the same circle at $\phi = \pi$.

For $n \geq 2$ eqs. (65) describe true deformations. Since the strains ϵ_ϕ, ϵ_θ, $\gamma_{\phi\theta}$ are all zero, they are called inextensional deformations. We see from the formulas that there exists no such deformation which is finite on the whole sphere. The A solution becomes infinite at $\phi = \pi$, and the B solution at $\phi = 0$. A complete sphere is therefore not capable of inextensional deformations. For a spherical cap which contains the pole $\phi = 0$, the A solution is regular, and it describes a deformation to which the shell does not offer any resistance as long as we disregard its bending stiffness. We may superpose this solution on any solution of the inhomogeneous equations (69) and so satisfy a boundary condition concerning either u or v, and we must prescribe one of these displacements at the edge in order to make the deformation of the shell determinate. This stituation is the kinematic counterpart to the fact that the equilibrium of the membrane forces requires a ring along the edge of the shell (see p. 58). Instead of prescribing one force and one displacement, say N_ϕ and v, we may also prescribe both displacements but not both forces, as one may easily verify. It is not possible, within the realm of the membrane theory, to prescribe all three components of the displacement. If the edge of the shell is fixed in every direction by the support, a complete fulfillment of all boundary conditions is possible only with the help of bending moments. In some cases it is possible to get reasonable information from the membrane theory by making a judicious choice among the boundary conditions, but this does not always work. Then a study of the bending effects is indispensable in order to determine the membrane forces in the shell.

If the spherical shell is open at the top, both terms in eqs. (65) are regular on the entire shell, just as are both terms in the solution (38) for the membrane forces. We have then twice as many constants for twice as many edges, and all that has just been said for the edge of a spherical cap applies now to both edges of a spherical zone.

Eqs. (65) permit an application which leads to a result of general interest. Each of the two spherical shells shown in fig. 53 has only one pole and therefore admits one inextensional deformation for the n-th harmonic. For the upper sphere we must put $B = 0$ and have

$$u_n = v_n = A \sin\phi_1 \tan^n \frac{\phi_1}{2}, \qquad w_n = -A(n + \cos\phi_1)\tan^n \frac{\phi_1}{2}.$$

For the lower sphere we must drop A and have

$$u_n = -v_n = -B \sin\phi_2 \cot^n \frac{\phi_2}{2}, \qquad w_n = B(n - \cos\phi_2)\cot^n \frac{\phi_2}{2}.$$

When the two spheres are connected at the waist circle $\phi_1 = \alpha$, $\phi_2 = \beta$, their displacements must be the same there. This requirement seems

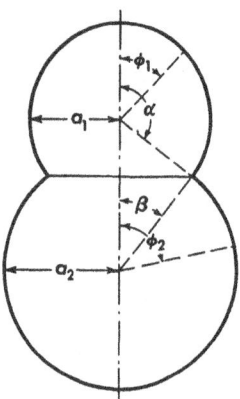

to yield three equations between A and B, but since we have to deal with inextensional deformations, the line element of the waist circle cannot change, and therefore only two equations are independent statements, the third one following from the invariability of the line element. We formulate our conditions of compatibility for the vertical component of the displacement $v_n \sin\phi - w_n \cos\phi$ and for one component in the plane of the waist circle, say u_n. They are

$$A(1 + n \cos\alpha)\tan^n \frac{\alpha}{2} = B(1 - n \cos\beta)\cot^n \frac{\beta}{2},$$

$$A \sin\alpha \tan^n \frac{\alpha}{2} = -B \sin\beta \cot^n \frac{\beta}{2}.$$

Fig. 53. Nonconvex shell, consisting of parts of two spheres

These equations do not admit a solution A, B different from zero unless the determinant of their coefficients vanishes. This condition,

$$(1 + n \cos\alpha)\sin\beta + (1 - n \cos\beta)\sin\alpha = 0,$$

may be brought to the form

$$n = \frac{\sin \dfrac{\alpha + \beta}{2}}{\sin \dfrac{\alpha - \beta}{2}}.$$

Since the numerator is always positive, this can never yield a positive n, if the shell is convex, i. e. if $\alpha < \beta$. But even for shells with $\alpha > \beta$, whose meridian has a reentrant angle, an inextensional deformation is possible only if α and β are such that the quotient becomes an integer. This shows that most non-convex shells of the type under consideration are just as incapable of inextensional deformations as convex shells, for which this property can be proved in general[1]. When shells like

---

[1] BLASCHKE, W.: Vorlesungen über Differentialgeometrie, vol. 1. Berlin 1921, p. 131.

fig. 53 are built, it will be useful to avoid such dimensions where n, as defined by the last formula, is equal or very close to an integer.

2.5.2.3 Solution for Arbitrary Shape of the Meridian

In the general case of an arbitrary meridian it is, of course, not possible to find a simple solution in finite terms for the differential equation (64). We shall establish one in the form of a power series.

The first step toward this is a transformation of the independent variable, which removes all transcendental functions from the coefficients. We put

$$x = 1 - \cos \phi.$$

The equation now assumes the following form

$$\frac{d^2v_n}{dx^2} + \frac{1-x}{x(2-x)} \frac{r_2-r_1}{r_2} \frac{dv_n}{dx} + \left(\frac{1}{x(2-x)} + \frac{(1-x)^2-n^2}{x^2(2-x)^2} \frac{r_1}{r_2} \right) v_n = 0. \quad (66)$$

The singularities of the coefficients are at $x = 0$ and $x = 2$. The properties of the solution in the vicinity of these points depend on these singularities of the coefficients. Here we shall study the singularity at $x = 0$, that is, at $\phi = 0$. The other one may be studied by a series development at $x = 2$, or it may be brought to $x = 0$ by using the substitution $x = 1 + \cos \phi$.

In most shells the two meridians $\theta = 0$ and $\theta = \pi$ will be the two halves of the same simple curve and will be described by a common analytic expression. In this case r_1 and r_2 are necessarily even functions of ϕ. A power series for $r_2 - r_1$ must therefore begin with ϕ^2 and hence with x^1. This x cancels with the x in the denominator, and the coefficient of dv_n/dx has no singularity at all; but the coefficient of v_n has a second order singularity. A differential equation of this type belongs to the FUCHS class. Its solution in the vicinity of the apex $x = 0$ of the shell may be represented in the form

$$v_n = x^\varkappa \sum_{k=0}^{\infty} b_k x^k. \quad (67)$$

The exponent $\varkappa$ determines the type of singularity of v_n and need not be positive or an integer. It is determined along with the coefficients b_k by introducing the series (67) in the differential equation. This procedure is best explained by a concrete example.

We choose a paraboloid as shown in fig. 54. Its radii of curvature are

$$r_1 = \frac{a}{\cos^3\phi}, \qquad r_2 = \frac{a}{\cos\phi},$$

where a is twice the focal distance of the parabola. When we introduce these radii into the differential equation (66), it reads as follows:

$$\frac{d^2 v_n}{dx^2} - \frac{1}{1-x}\,\frac{dv_n}{dx} + \frac{(1+2x-x^2)(1-x)^2 - n^2}{x^2(1-x)^2(2-x)^2}\,v_n = 0\,.$$

We now introduce v_n from eq. (67) and multiply by the denominator of the last term:

$$\begin{aligned} x^2(1-x)^2(2-x)^2 \sum_{k=0}^{\infty}(k+\varkappa)(k+\varkappa-1)\, b_k\, x^{k-2} \\ - x^2(1-x)(2-x)^2 \sum_{k=0}^{\infty}(k+\varkappa)\, b_k\, x^{k-1} \\ + [(1+2x-x^2)(1-x)^2 - n^2] \sum_{k=0}^{\infty} b_k\, x^k = 0\,. \end{aligned}$$

The factor before each of the sums may be written as a polynomial in x. If we multiply every one of its terms separately by the sum, we get as many sums as the factor has terms. In each of them we change the notation of the summation index in such a way that x^k appears everywhere. This yields the equation

$$\begin{aligned} \sum_{k=0}[(2(k+\varkappa)-1)^2 - n^2]\, b_k\, x^k - 4\sum_{k=1}^{\infty}(k+\varkappa-1)\,[3(k+\varkappa)-5]\, b_{k-1}\, x^k \\ + \sum_{k=2}^{\infty}[13(k+\varkappa)^2 - 57(k+\varkappa) + 58]\, k_{k-2}\, x^k \\ - \sum_{k=3}^{\infty}[6(k+\varkappa)^2 - 37(k+\varkappa) + 53]\, b_{k-3}\, x^k \\ + \sum_{k=4}^{\infty}(k+\varkappa-3)(k+\varkappa-5)\, b_{k-4}\, x^k = 0\,. \end{aligned} \tag{68}$$

This equation must be satisfied identically in x, and this requires that for every integer k the sum of all the coefficients in all the sums must be zero. We thus arrive at an infinite set of linear equations for the b_k.

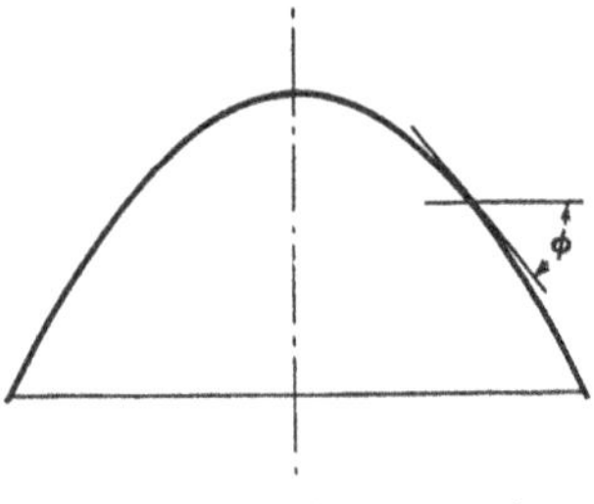

Fig. 54. Paraboloid of revolution

For $k = 0$, only the first sum makes a contribution and leads to the equation

$$[(2\varkappa - 1)^2 - n^2]\, b_0 = 0\,.$$

If we want to get any solution at all, b_0 cannot be zero, and so the other factor must vanish, and that determines $\varkappa$ as

$$\varkappa = \frac{1}{2}(1 \pm n)\,.$$

The two values lead to two solutions of the differential equation, and we see here that one of them [with $\varkappa = \frac{1}{2}(1+n)$] is regular, whereas the other one [with $\varkappa = \frac{1}{2}(1-n)$] has for any $n > 1$ a singularity at $x = 0$, that is, at $\phi = 0$.

For $k \geq 1$, eq. (68) yields recurrence formulas for the coefficients b_k, expressing all the b_k in terms of b_0. The first three of these formulas are somewhat irregular, because only some of the sums in eq. (68) contribute to them:

$$[(1 + 2\varkappa)^2 - n^2] b_1 = -4\varkappa(2 - 3\varkappa) b_0,$$

$$[(3 + 2\varkappa)^2 - n^2] b_2 = 4(1 + \varkappa)(1 + 3\varkappa) b_1$$
$$- [13(2 + \varkappa)^2 - 57(2 + \varkappa) + 58] b_0,$$

$$[(5 + 2\varkappa)^2 - n^2] b_3 = 4(2 + \varkappa)(4 + 3\varkappa) b_2$$
$$- [13(3 + \varkappa)^2 - 57(3 + \varkappa) + 58] b_1$$
$$+ [6(3 + \varkappa)^2 - 37(3 + \varkappa) + 53] b_0.$$

For $k \geq 4$ they all have four terms on the right-hand side:

$$[(2k + 2\varkappa - 1)^2 - n^2] b_k = 4(k + \varkappa - 1)(3k + 3\varkappa - 5) b_{k-1}$$
$$- [13(k + \varkappa)^2 - 57(k + \varkappa) + 58] b_{k-2}$$
$$+ [6(k + \varkappa)^2 - 37(k + \varkappa) + 53] b_{k-3}$$
$$- (k + \varkappa - 3)(k + \varkappa - 5) b_{k-4}.$$

Hence all coefficients depend on the first one, b_0, and this one must necessarily stay undetermined, because each constant multiple of the solution is again a solution of the homogeneous equation (66).

Putting $b_0 = 1$ and introducing either $\varkappa_1$ or $\varkappa_2$ into the recurrence formulas, we obtain two linearly independent solutions v_{n1} and v_{n2}, which may be combined to form the general solution

$$v_n = b_{01} v_{n1} + b_{02} v_{n2},$$

having two free constants b_{01} and b_{02}.

The corresponding displacement w_n follows from (62a):

$$w_n = -\dot{v_n} = -\sin\phi \cdot \frac{d v_n}{d x} = -\sin\phi \cdot x^{\varkappa-1} \sum_{k=0}^{\infty} (k + \varkappa) b_k x^k.$$

For the third component, u_n, no new series need be computed, since eq. (63) yields

$$u_n = -\frac{1}{n} (v_n \cos\phi + w_n \sin\phi).$$

Beginning with $n = 2$, $\varkappa_2$ is negative, and the corresponding solution becomes infinite at $\phi = 0$. It therefore cannot appear in a shell which is closed at its apex. We have then only one constant of integration for every harmonic n, and the inextensional deformation is completely determined if one of the displacements u, v, w is given at the boundary. Of course we can use the constant C which appears in the solution for the stress resultants, to fulfill a second condition for the displacement, if we refrain from prescribing forces at the boundary; but never can all three components u, v, w be controlled within the range of the membrane theory. This is the same situation which we encountered in the case of the spherical shell.

2.5.3 Inhomogeneous Solution

2.5.3.1 General Solution

If the stress resultants N_ϕ, N_θ, $N_{\phi\theta}$ of the shell are known, HOOKE's law (57) gives the strains ϵ_ϕ, ϵ_θ, $\gamma_{\phi\theta}$, and we have to deal with the inhomogeneous equations

$$
\begin{aligned}
v^{\boldsymbol{\cdot}} \quad &+ w \quad\quad = r_1 \epsilon_\phi, \\
u' \quad &+ v \cos\phi + w \sin\phi = r_2 \epsilon_\theta \sin\phi, \\
\frac{r_2}{r_1} u^{\boldsymbol{\cdot}} \sin\phi - u \cos\phi + v' \quad &\quad\quad = r_2 \gamma_{\phi\theta} \sin\phi.
\end{aligned}
\tag{69}
$$

If we can find one particular solution of this set, we only need to combine it with the inextensional deformations and we have the complete solution.

To find such a particular solution, we proceed in the same way as we have for the homogeneous system. We subject the strains to a harmonic analysis and consider the general term of the Fourier series:

$$
\epsilon_\phi = \epsilon_{\phi n}(\phi) \cos n\theta, \quad \epsilon_\theta = \epsilon_{\theta n}(\phi) \cos n\theta, \quad \gamma_{\phi\theta} = \gamma_{\phi\theta n}(\phi) \sin n\theta.
$$

Introducing these and eqs. (61) into the differential equations (69), we obtain a set of ordinary differential equations for the n-th harmonic of the displacement:

$$
\begin{aligned}
v_n \quad &+ w_n \quad\quad = r_1 \epsilon_{\phi n}, \\
n u_n \quad &+ v_n \cos\phi + w_n \sin\phi = r_2 \epsilon_{\theta n} \sin\phi, \\
\frac{r_2}{r_1} u_n^{\boldsymbol{\cdot}} \sin\phi - u_n \cos\phi - n v_n \quad &\quad\quad = r_2 \gamma_{\phi\theta n} \sin\phi.
\end{aligned}
$$

We can again use two of these equations to eliminate u_n and w_n and arrive at a second order equation for $v_n(\phi)$:

$$
\begin{aligned}
&v_n^{\boldsymbol{\cdot\cdot}} \frac{r_2}{r_1} \sin^2\phi - v_n^{\boldsymbol{\cdot}} \cos\phi \sin\phi + v_n \left(\frac{r_2}{r_1} \sin^2\phi + \cos^2\phi - n^2 \right) \\
&= \frac{r_2 - r_1}{r_1} (r_1 \epsilon_{\phi n} - r_2 \epsilon_{\theta n}) \cos\phi \sin\phi + n r_2 \gamma_{\phi\theta n} \sin\phi \\
&\quad + \frac{r_2}{r_1} (r_1 \epsilon_{\phi n} - r_2 \epsilon_{\theta n}) \sin^2\phi + \frac{r_2}{r_1} (r_1 \epsilon_{\phi n}^{\boldsymbol{\cdot}} - r_2^{\boldsymbol{\cdot}} \epsilon_{\theta n}) \sin^2\phi.
\end{aligned}
$$

By the substitution $x = 1 - \cos\phi$ the equation assumes the following form

$$
\frac{d^2 v_n}{d x^2} + \frac{1-x}{x(2-x)} \frac{r_2 - r_1}{r_2} \frac{d v_n}{d x} + \left(\frac{1}{x(2-x)} + \frac{(1-x)^2 - n^2}{x^2(2-x)^2} \frac{r_1}{r_2} \right) v_n = F(x).
$$

It may be solved by developing $F(x)$ in a power series:

$$F(x) = x^{\varkappa} \sum_{k=0}^{\infty} B_k x^k. \tag{70}$$

The exponent $\varkappa$ in this development need not be equal to $\varkappa_1$ or $\varkappa_2$, but because of the symmetry of the shell, it will be half of an odd integer if n is even and an integer if n is odd, just as we found for $\varkappa_{1,2}$. If we now assume v_n in the form (67) but with the same value of $\varkappa$ as appears in $F(x)$, and introduce both into the differential equation, we may find the coefficients b_k for a particular solution by comparing coefficients in the equation. Adding the homogeneous solution with an arbitrary constant factor is equivalent to adding an inextensional deformation. It enables us to fulfill one boundary condition for the displacements.

Such a power series method may be expected to yield fairly good results in the vicinity of the top of the shell. But for greater values of x or ϕ the convergence of the series may or may not be satisfactory, depending on the particular shape of the meridian and on the order n of the harmonic. In such cases the power series may be useful to start a numerical integration, which goes right down the meridian. We have then the advantage of finding at once that solution which is regular at $\phi = 0$. However, if n is great, the regular solution will assume perceptible values only near the edge of the shell, and then it will be preferable to start the numerical integration there and to continue it only as far as is needed. In this case the regular solution must be isolated by the method described on p. 74 for the stress resultants.

2.5.3.2 Axially Symmetric Deformation

If the load and the stresses both have axial symmetry, the deformation need not have it too, but if it does not it will always be possible to split it into a particular solution which has this high degree of symmetry and an inextensional deformation such as we have already treated in detail.

For a deformation with axial symmetry, the general equations (59) assume the simpler form:

$$\epsilon_\phi = \frac{v^{\cdot}}{r_1} + \frac{w}{r_1}, \qquad \epsilon_\theta = \frac{v}{r_2} \cot\phi + \frac{w}{r_2}.$$

The third of them disappears completely, because it becomes trivial. Eliminating w, we arrive at a differential equation for v:

$$v^{\cdot} - v \cot\phi = r_1 \epsilon_\phi - r_2 \epsilon_\theta \equiv q(\phi).$$

It has the solution (see p. 50)

$$v = \left[\int q(\phi) \exp\left(-\int \cot\phi \, d\phi \right) d\phi + C \right] \exp \int \cot\phi \, d\phi,$$

and when we evaluate the simple integrals, we have

$$v = \left[\int \frac{q(\phi)}{\sin\phi} \, d\phi + C \right] \sin\phi. \tag{71}$$

With the help of HOOKE's law the function $q(\phi)$ may be expressed in terms of the stress resultants N_ϕ and N_θ:

$$v = \left[\int \frac{1}{Et} [N_\phi(r_1 + \nu r_2) - N_\theta(r_2 + \nu r_1)] \frac{d\phi}{\sin\phi} + C \right] \sin\phi. \tag{72a}$$

The constant C means simply a rigid-body displacement of the shell in the direction of its axis and must be determined by a boundary condition. When v has been found, we can easily find w from ϵ_θ:

$$w = r_2 \epsilon_\theta - v \cot\phi. \tag{72b}$$

We see that the symmetric deformations of a shell of arbitrary meridian may be computed by a mere quadrature when the stress resultants are known.

As an example, we apply eqs. (72) to the ogival shell of figs. 43 and 44, but with the difference that we now assume that the entire length of the edge is uniformly supported. The membrane forces N_ϕ, N_θ in the shell are then those which are plotted as $N_{\phi 0}$ and $N_{\theta 0}$ in fig. 44.

In eq. (72a) we need the radii of the middle surface. They are

$$r_1 = \frac{h^2}{2b \sin^3\phi}, \qquad r_2 = \frac{4b^2 - h^2 \cot^2\phi}{4b \sin\phi}.$$

When we introduce these expressions into (72a) and assume that the wall thickness t is a constant, we obtain

$$v = \frac{h^2 \sin\phi}{2Ebt} \int_{\phi_0}^{\phi} \frac{N_\phi - \nu N_\theta}{\sin^4\phi} \, d\phi$$
$$- \frac{\sin\phi}{4Ebt} \int_{\phi_0}^{\phi} (N_\theta - \nu N_\phi) \frac{4b^2 - h^2 \cot^2\phi}{\sin^2\phi} \, d\phi + C \sin\phi.$$

We have to choose C such as to make $v = 0$ at the springing line $\phi = \pi/2$. This amounts to replacing the lower limit ϕ_0 of both integrals by $\pi/2$ and dropping the term $C \sin\phi$ at the end. The expressions for N_ϕ and N_θ, which are not given here, are rather clumsy, and the integrals are best evaluated numerically. This has been done for $\nu = 0$, and the result is shown in fig. 55.

At five points of the meridian the resultant displacement is indicated by arrows. It appears that the major part of the shell moves essentially vertically downward. The slight inward component is due to the negative hoop strain ϵ_θ. But near the springing line, where the hoop force is positive, the deflection is outward, and at the springing line

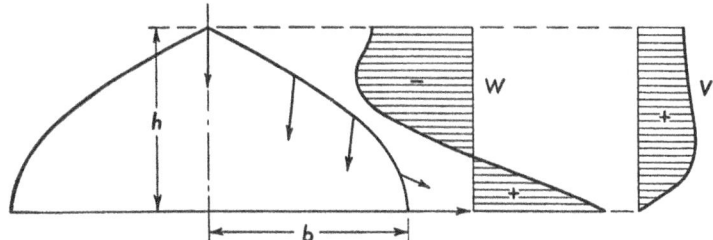

Fig. 55. Displacements for a pointed dome with dead load

it is horizontal. At the apex the components v and w combine, of course, to a vertical deflection w_v. For $p = 58$ lb/ft² it was found that $E\,t\,w_v = 96,800$ lb. This deflection is comparable in magnitude with the horizontal displacement at the springing line.

2.5.4 Toroidal Shell

We have already encountered some difficulties in treating the stress resultants in toroidal shells. Here we shall see that the deformation of these shells also has some peculiarities. On p. 33 we found that a system of membrane forces is not always possible in such a shell, even under conditions which would be sufficient in other shells. But we did at least find membrane forces in some simple and important cases as for instance in that of a toroid of circular cross-section which is subjected to an internal gas pressure. We shall see now that even in this case membrane forces are possible only in restricted parts of the shell, because otherwise impossible consequences for the deformation would result.

To simplify the mathematical representation, we assume here $v = 0$. Then the stress resultants as given on p. 33 produce the strains

$$\epsilon_\phi = \frac{p\,a}{2\,E\,t}\,\frac{2\,R + a\sin\phi}{R + a\sin\phi}, \qquad \epsilon_\theta = \frac{p\,a}{2\,E\,t}.$$

From these we find

$$q(\phi) = a\,\epsilon_\phi - \left(a + \frac{R}{\sin\phi}\right)\epsilon_\theta = -\frac{p\,a}{2\,E\,t}\,\frac{R^2}{(R + a\sin\phi)\sin\phi}.$$

Introducing this into eq. (71) we get the tangential component of the displacement

$$v = \left[-\frac{p\,a\,R^2}{2\,E\,t}\int\frac{d\phi}{(R + a\sin\phi)\sin^2\phi} + C\right]\sin\phi.$$

The integral may be evaluated by elementary means and yields

$$v = \left[\frac{p\, a\, \overset{\scriptscriptstyle\smile}{R}}{2E\,t} \left(\frac{2a^2}{R\sqrt{R^2 - a^2}} \arctan \frac{a + R\tan\phi/2}{\sqrt{R^2 - a^2}} \right. \right.$$
$$\left. \left. - \frac{a}{R} \ln \tan\frac{\phi}{2} - \cot\phi \right) + C \right] \sin\phi.$$

The brackets contain two terms which have singularities: $\ln \tan\phi/2$ and $\cot\phi$. For $\phi \to 0$ the logarithm tends toward $-\infty$, the cotangent toward $+\infty$, and the opposite is true for $\phi \to \pi$. However, the singularity of the cotangent is stronger, and therefore the two do not cancel. In the formula for v both are multiplied by $\sin\phi$, and this factor vanishes strongly enough to keep v finite. But when we now use eq. (72b) to find the normal displacement w, the factor $\sin\phi$ is replaced by $\cos\phi$, and no finite value of the free constant C will help to keep w from becoming infinite at $\phi = 0$ and at $\phi = \pi$. This indicates that even under this load bending must occur in the shell, although the membrane-force system would be sufficient for equilibrium.

This surprising and remarkable result may be explained by the particular geometrical properties of the two circles $\phi = 0$ and $\phi = \pi$. For all points of such a circle the tangential plane of the toroid is the same. To the same degree of approximation to which, usually, adjacent points of the middle surface lie in one plane, here adjacent parallel circles lie in this tangential plane. The state of stress here must therefore be much more akin to a plane state of stress than in other parts of the shell. Indeed, here the difference of the hoop-strains in adjacent circles determines the meridional strain, and through HOOKE's law this gives an additional relation between N_ϕ and N_θ, which may, and in general will, be incompatible with the forces following from the conditions of equilibrium. If we change N_ϕ and N_θ so that they comply with the new relation, the equilibrium of the shell element will be disturbed, and it can be restored only by admitting transverse forces Q_ϕ and consequently bending moments.

2.5.5 Strain Energy

The equations of equilibrium (6) are linear relations between the loads p_ϕ, p_θ, p_r and the stress resultants N_ϕ, N_θ, $N_{\phi\theta}$ which they produce. Therefore, if λ is a constant factor, the load $\lambda\, p_\phi$, $\lambda\, p_\theta$, $\lambda\, p_r$ will produce the forces $\lambda\, N_\phi$, $\lambda\, N_\theta$, $\lambda\, N_{\phi\theta}$. HOOKE's law (57) is a set of linear equations between the forces N_ϕ, N_θ, $N_{\phi\theta}$ and the strains ϵ_ϕ, ϵ_θ, $\gamma_{\phi\theta}$. The load $\lambda\, p_\phi$, $\lambda\, p_\theta$, $\lambda\, p_r$ will therefore produce strains $\lambda\, \epsilon_\phi$, $\lambda\, \epsilon_\theta$, $\lambda\, \gamma_{\phi\theta}$.

Applying the load p_ϕ, p_θ, p_r to the shell means, in our notation, that the factor λ is increased continuously from 0 to 1. During this procedure the forces acting on a shell element (fig. 2) will do some work, because their different points of application are displaced. We shall now calculate this work.

Since the forces on the element are in equilibrium, their work during a rigid-body movement will be zero. The position of the element in space may therefore be fixed arbitrarily, and we have only to ask for the work which is done during the relative displacement of the different sides of the element due to its deformation.

If we assume that the center of the element does not move, the external forces will do no work. When the length of the element $r_1 d\phi$ of the meridian is increased from $r_1 d\phi (1 + \lambda \epsilon_\phi)$ to $r_1 d\phi (1 + \lambda \epsilon_\phi + d\lambda \epsilon_\phi)$, the two forces $\lambda N_\phi \cdot r \, d\theta$ do the work

$$\lambda N_\phi \, r \, d\theta \cdot r_1 \, d\phi \, d\lambda \, \epsilon_\phi,$$

and in the same way we find the work of the forces $\lambda N_\theta \cdot r_1 \, d\phi$ by multiplying them by the relative displacement $r \, d\theta \cdot d\lambda \, \epsilon_\theta$ of their points of application. To find the work done by the shearing forces, let us eliminate an unessential rigid-body movement by assuming that the sides $r \, d\theta$ of the shell element stay horizontal and that the meridional sides $r_1 \, d\phi$ turn by an angle $d\lambda \, \gamma_{\phi\theta}$. Then only one of the forces $\lambda N_{\phi\theta} \cdot r \, d\theta$ does work of amount

$$\lambda N_{\phi\theta} \, r \, d\theta \cdot d\lambda \, \gamma_{\phi\theta} \, r_1 \, d\phi.$$

If we sum up, we see that the whole work is proportional to the area $dA = r_1 \, d\phi \cdot r \, d\theta$ of the shell element and is

$$(N_\phi \, \epsilon_\phi + N_\theta \, \epsilon_\theta + N_{\phi\theta} \, \gamma_{\phi\theta}) \, \lambda \, d\lambda \, dA.$$

This is the work for an increase of λ by a differential $d\lambda$. The total work done when λ increases from 0 to 1 is the integral over the variable λ. Since

$$\int_0^1 \lambda \, d\lambda = \frac{1}{2},$$

this work is

$$dV = \frac{1}{2} (N_\phi \, \epsilon_\phi + N_\theta \, \epsilon_\theta + N_{\phi\theta} \, \gamma_{\phi\theta}) \, dA. \tag{73}$$

When the shell is unloaded, λ decreases from 1 to 0. The forces on the shell element will do the work $-dV$, and the energy which was stored during loading, is set free. It was stored in the deformed shell as potential energy and is called strain energy. The total strain energy V of the shell is the integral of dV, extended over the middle surface:

$$V = \frac{1}{2} \int (N_\phi \, \epsilon_\phi + N_\theta \, \epsilon_\theta + N_{\phi\theta} \, \gamma_{\phi\theta}) \, dA. \tag{74a}$$

When the shell is deformed, the points of application of the external forces move, and these forces do some work. Two kinds of external forces must be considered separately, the distributed loads p_ϕ, p_θ, p_r and the forces applied to the edges of the shell (loads or reactions). The latter are identical with the values $\overline{N}_\phi$ and $\overline{N}_{\phi\theta}$ which the stress resultants assume at the edge.

When the distributed loads are λp_ϕ, λp_θ, λp_r, the displacements of their points of application are λu, λv, λw. If λ again increases from 0 to 1, these forces do the work

$$\frac{1}{2} (p_\theta u + p_\phi v + p_r w) \, dA \, .$$

When the expression is integrated over the middle surface of the shell, it yields the work done by the distributed loads.

The forces at the edge do work on the edge displacements:

$$\frac{1}{2} \int (\overline{N}_\phi \overline{v} + \overline{N}_{\phi\theta} \overline{u}) \, ds \, ,$$

where ds is the line element of the edge and the integral is to be extended over all edges where the integrand is not zero. When we add this integral to the work of the distributed load, we have the total work of the external forces:

$$T = \frac{1}{2} \int (p_\theta u + p_\phi v + p_r w) \, dA + \frac{1}{2} \int (\overline{N}_\phi \overline{v} + \overline{N}_{\phi\theta} \overline{u}) \, ds. \qquad (74\,\mathrm{b})$$

This work must be equal to the strain energy of the shell:

$$V = T \, .$$

Now let us consider two different load systems p_θ, p_ϕ, p_r and p_θ^*, p_ϕ^*, p_r^*. The stress resultants, strains and displacements produced by the combined load $\lambda p_\theta + \mu p_\theta^*, \ldots$ are given by $\lambda N_\phi + \mu N_\phi^*$, $\lambda \epsilon_\phi + \mu \epsilon_\phi^*$, $\lambda u + \mu u^*$ etc.

At first we keep $\mu = 0$ and let λ increase from 0 to 1. During this part of the loading procedure the strain energy

$$V_1 = \frac{1}{2} \int (N_\phi \epsilon_\phi + N_\theta \epsilon_\theta + N_{\phi\theta} \gamma_{\phi\theta}) \, dA$$

is stored. If we now keep $\lambda = 1$ and let μ increase from 0 to 1, the force $N_\phi + \mu N_\phi^*$ does work as the strain increases from $\epsilon_\phi + \mu \epsilon_\phi^*$ to $\epsilon_\phi + (\mu + d\mu) \epsilon_\phi^*$, and this work is

$$(N_\phi + \mu N_\phi^*) \, r \, d\theta \cdot d\mu \, \epsilon_\phi^* \cdot r_1 \, d\phi \, .$$

This and two corresponding terms for N_θ and $N_{\phi\theta}$ must be integrated from $\mu = 0$ to $\mu = 1$ and give, together with the preceding integral, the total strain energy for the full load $p_\theta + p_\theta^*, \dots$:

$$V = \frac{1}{2} \int (N_\phi \epsilon_\phi + N_\theta \epsilon_\phi + N_{\phi\theta} \gamma_{\phi\theta})\, dA + \int (N_\phi \epsilon_\phi^* + N_\theta \epsilon_\theta^* + N_{\phi\theta} \gamma_{\phi\theta}^*)\, dA$$

$$+ \frac{1}{2} \int (N_\phi^* \epsilon_\phi^* + N_\theta^* \epsilon_\theta^* + N_{\phi\theta}^* \gamma_{\phi\theta}^*)\, dA. \tag{75a}$$

The first and the third integral are the strain energies V_1 and V_2, respectively, which are obtained when only one or the other of the two load systems is applied to the shell. The second integral respresents the work V_{12} done by the stress resultants due to the first load system, during the deformation which is produced by the second. The final state of stress and deformation is the same, if the second load system is applied first, and the same total strain energy results. But in this case the second integral in eq. (75a) looks different: the asterisks are attached to the forces instead of to the strains. It follows that both forms of this integral must be equal to each other, which may also be proved by eliminating all stress resultants with the help of HOOKE's law, eq. (57).

During the loading procedure which leads to eq. (75a) for the strain energy, the external forces do the work

$$T = \frac{1}{2} \int (p_\theta u + p_\phi v + p_r w)\, dA + \int (p_\theta u^* + p_\phi v^* + p_r w^*)\, dA$$

$$+ \frac{1}{2} \int (p_\theta^* u^* + p_\phi^* v^* + p_r^* w^*)\, dA$$

$$+ \frac{1}{2} \int (\bar{N}_\phi \bar{v} + \bar{N}_{\phi\theta} \bar{u})\, ds + \int (\bar{N}_\phi \bar{v}^* + \bar{N}_{\phi\theta} \bar{u}^*)\, ds$$

$$+ \frac{1}{2} \int (\bar{N}_\phi^* \bar{v}^* + \bar{N}_{\phi\theta}^* \bar{u}^*)\, ds. \tag{75b}$$

Here, again, the final result must be the same if we transfer all the asterisks in the second and fifth integrals from the displacements to the forces, and therefore

$$\int (p_\theta u^* + p_\phi v^* + p_r w^*)\, dA + \int (\bar{N}_\phi \bar{v}^* + \bar{N}_{\phi\theta} \bar{u}^*)\, ds$$

$$= \int (p_\theta^* u + p_\phi^* v + p_r^* w)\, dA + \int (\bar{N}_\phi^* \bar{v} + \bar{N}_{\phi\theta}^* \bar{u})\, ds. \tag{76}$$

Both sides of this equation must be equal to the second integral in eq. (75a). When we write this equation, it is advisable to use HOOKE's law (57) to express the strains in terms of the stress resultants:

$$\int (p_\theta^* u + p_\phi^* v + p_r^* w)\, dA + \int (\bar{N}_\phi^* \bar{v} + \bar{N}_{\phi\theta}^* \bar{u})\, ds$$

$$= \frac{1}{E t} \int [N_\phi N_\phi^* + N_\theta N_\theta^* - \nu N_\phi N_\theta^* - \nu N_\theta N_\phi^*$$

$$+ 2(1 + \nu) N_{\phi\theta} N_{\phi\theta}^*]\, dA. \tag{77}$$

Eq. (76) contains MAXWELL's theorem of reciprocity. To derive it from this equation, we need consider only two very special load systems. Let the system p_θ, p_ϕ, p_r consist of nothing but a concentrated force P acting normal to the shell at some point 1. At some other point 2 this load may produce a displacement w_{21}. The load system p_θ^*, p_ϕ^*, p_r^* we shall take as a concentrated force P* acting on the shell at point 2, normal to the middle surface. The deflection w^* which it produces at point 1 will be called w_{12}. In this case eq. (76) reduces to the simple statement

$$P\, w_{12} = P^*\, w_{21},$$

and if the two forces are of equal magnitude, we have $w_{12} = w_{21}$, which is exactly MAXWELL's theorem.

Eq. (77) may be used to find the displacement of a point of the shell without solving the complete deformation problem as explained in sections 2.5.2. through 2.5.4.

As an example we calculate the vertical displacement for the top of the shell dome of fig. 43, but under the assumption that the shell is supported continuously along the springing line. The load is the same as before.

The stress resultants N_ϕ, N_θ are the same which on p. 98 were called $N_{\phi 0}$, $N_{\theta 0}$ and which are plotted in fig. 44. In addition we now need the membrane forces N_ϕ^*, N_θ^* for a vertical unit load applied at the top of the shell. These forces may be found easily from eqs. (10) and (9b). They are

$$N_\phi^* = -\frac{2b}{\pi(4b^2 - h^2 \cot^2\phi)\sin\phi}, \qquad N_\theta^* = \frac{b\sin\phi}{\pi h^2}.$$

In eq. (77) the first integral reduces to w_v and the second integral is zero since the edge is supported and does not move. On the right-hand side the shear is zero, and if we assume $\nu = 0$, two more terms drop out, and the integral becomes

$$\frac{4b}{Et}\int\left(\frac{-N_\phi}{(4b^2 - h^2 \cot^2\phi)\sin\phi} + \frac{N_\theta \sin\phi}{2h^2}\right) r\, r_1\, d\phi.$$

When the expressions for the radii are introduced, this yields finally

$$E t\, w_v = \frac{h^2}{4b}\int\left(-\frac{2N_\phi}{\sin^4\phi} + \frac{N_\theta(4b^2 - h^2 \cot^2\phi)}{h^2 \sin^2\phi}\right)d\phi.$$

This integral must be evaluated by SIMPSON's rule. The result is $E t\, w_v = 98{,}700$ lb, which differs by only 2% from the figure obtained on p. 99 from a complete analysis of the deformation of this shell.

If the shell is supported on columns as in fig. 43, the fictitious unit load at the top will give rise to membrane forces N_ϕ^*, N_θ^*, $N_{\phi\theta}^*$ containing harmonic constituents of orders $n = 8, 16, \ldots$, and each of these harmonics will contribute to the integral on the right-hand side of eq. (77) if multiplied by the harmonic of the same order of the actual stress system. On the left-hand side the first integral is still equal to w_v, and in the second integral the total contribution must be zero, since everywhere on the edge either the resultant force or the resultant displacement is zero.

2.5.6 Statically Indeterminate Structures

The shell structures which we studied in the preceding Sections were all statically determinate. Indeed, for determining the three unknown functions N_ϕ, N_θ, $N_{\phi\theta}$, we always had at hand the three equations (6), which are the conditions of equilibrium for an arbitrary shell element. Occasionally we determined a free constant from a condition of regularity of the stress system, but this too can be interpreted as a condition of equilibrium of a particular shell element.

However, cases exist in which equilibrium conditions are not sufficient to determine the stress resultants in a shell. We see this best by an example.

Let a hemispherical shell be subjected to a load

$$p_\phi = p_\theta = 0, \quad p_r = p_{rn}(\phi) \cos n\theta \text{ with } n \geqq 2.$$

The general solution for the stress resultants is given by eqs. (35) in connection with eqs. (33). It contains two constants of integration, one of which, A_n, is determined by a condition of regularity. If either $N_{\phi n}$ or $N_{\phi\theta n}$ is given at the boundary $\phi = \pi/2$, this fact supplies an equation for B_n, as we have seen in the slightly different case of the shell on isolated supports. These problems are statically determinate.

The situation is different if the shell rests with the whole circumference of its boundary circle on an unyielding foundation. The boundary condition which will help us to find B_n is then a condition of zero displacement, and this makes the problem statically indeterminate.

To solve it, we follow the usual method of treating statically indeterminate structures. We can imagine that the edge support consists of two separate structural elements. The first one is a circular stiffening ring, not deformable in its own plane but unresisting to forces normal to this plane. This ring absorbs the shearing forces $N_{\phi\theta} = N_{\phi\theta n} \sin n\theta$ of the shell. The second element consists of an infinite number of vertical bars connecting the edge of the shell with the rigid foundation and transmitting the forces $N_{\phi n} \cos n\theta$.

If we cut through these bars, the forces $N_{\phi n} \cos n\theta$ become a system of external forces, which we can choose as we like, and the shell becomes statically determinate. In this modified structure we compute two systems of stress resultants. The first one is produced by the given load p_r and boundary forces N_ϕ such that $B_n = 0$. From eqs. (35)

we find

$$U^{(0)} = +a\,\frac{\cot^n\phi/2}{\sin^2\phi} \int\limits_0^\phi p_{rn}(\phi)\,(n+\cos\phi)\sin\phi\,\tan^n\frac{\phi}{2}\,d\phi,$$

$$V^{(0)} = -a\,\frac{\tan^n\phi/2}{\sin^2\phi} \int\limits_0^\phi p_{rn}(\phi)\,(n-\cos\phi)\sin\phi\,\cot^n\frac{\phi}{2}\,d\phi,$$

and from eqs. (33):

$$N_{\phi n}^{(0)} = \frac{1}{2}\,(U^{(0)}+V^{(0)}), \qquad N_{\theta n}^{(0)} = -N_{\phi n}^{(0)}+a\,p_{rn},$$

$$N_{\phi\theta n}^{(0)} = \frac{1}{2}\,(U^{(0)}-V^{(0)}).$$

For the second system of stress resultants we remove the load p_r from the shell and apply only an edge load $N_\phi = N_{\phi n}\cos n\theta$ of such magnitude that $B_n = 1$. From eqs. (38) we find

$$N_{\phi n}^{(1)} = -N_{\theta n}^{(1)} = -N_{\phi\theta n}^{(1)} = \frac{\tan^n\phi/2}{2\sin^2\phi}.$$

The real forces, which we want to find, are a linear combination of the two, for example

$$N_{\phi n} = N_{\phi n}^{(0)} + B_n N_{\phi n}^{(1)},$$

and here B_n is the redundant quantity, which must be determined from the condition that the displacements

$$u_n = u_n^{(0)} + B_n u_n^{(1)}, \qquad v_n = v_n^{(0)} + B_n v_n^{(1)}$$

are zero at the edge $\phi = \pi/2$ of the shell.

To find such edge displacements, we have eq. (77). For the forces $N_\phi \ldots$ of this formula we substitute our forces $N_\phi^{(0)} \ldots$, and for the $N_\phi^* \ldots$ our $N_\phi^{(1)} \ldots$. With the simplifying assumption $\nu = 0$ (which, of course, is not essential for the method) we get:

$$\int\limits_0^{2\pi} (\bar{N}_{\phi n}^{(1)}\,\bar{v}_n^{(0)}\cos^2 n\theta + \bar{N}_{\phi\theta n}^{(1)}\,\bar{u}_n^{(0)}\sin^2 n\theta)\,a\,d\theta$$

$$= \frac{1}{Et}\int\limits_0^{\pi/2} (N_{\phi n}^{(0)} N_{\phi n}^{(1)} + N_{\theta n}^{(0)} N_{\theta n}^{(1)} + 2\,N_{\phi\theta n}^{(0)} N_{\phi\theta n}^{(1)})\,a\sin\phi\,d\phi \cdot \int\limits_0^{2\pi}\frac{\cos^2 n\theta}{\sin^2 n\theta}\,a\,d\theta.$$

The integrations over θ may be performed at once, but the ϕ integration on the right which still depends on the function $p_{rn}(\phi)$, will usually have to be done numerically. The boundary forces on the left-hand side of the equation are $\bar{N}_{\phi n}^{(1)} = \frac{1}{2}$, $\bar{N}_{\phi\theta n}^{(1)} = -\frac{1}{2}$, and therefore the formula gives the difference of the two displacements:

$$\bar{v}_n^{(0)} - \bar{u}_n^{(0)} = \frac{2a}{Et}\int\limits_0^{\pi/2} (N_{\phi n}^{(0)} N_{\phi n}^{(1)} + N_{\theta n}^{(0)} N_{\theta n}^{(1)} + 2\,N_{\phi\theta n}^{(0)} N_{\phi\theta n}^{(1)})\sin\phi\,d\phi.$$

If we again apply eq. (77) but this time introduce $N_\phi^{(1)} \dots$ for both $N_\phi \dots$ and $N_\phi^* \dots$, we obtain an analogous result for $\bar{v}_n^{(1)} - \bar{u}_n^{(1)}$:

$$\bar{v}_n^{(1)} - \bar{u}_n^{(1)} = \frac{2a}{Et} \int\limits_0^{\pi/2} \left((N_{\phi n}^{(1)})^2 + (N_{\theta n}^{(1)})^2 + 2(N_{\phi\theta n}^{(1)})^2 \right) \sin\phi \, d\phi$$

$$= \frac{3a}{2Et} \int\limits_0^{\pi/2} \frac{\tan^{2n}\phi/2}{\sin^3\phi} \, d\phi.$$

These formulas do not give information about $\bar{u}_n$ or $\bar{v}_n$, and they cannot even be expected to do so, because the displacements are not completely determined by the stress resultants in the shell. Any one of them may be changed at will at the expense of the other one by adding an inextensional deformation of suitable size. As we see from eqs. (65a, b) with $B = 0$, in an inextensional deformation we always have $\bar{u}_n = \bar{v}_n$ at the edge of the hemisphere. Therefore the difference $\bar{u}_n - \bar{v}_n$ is not affected by inextensional deformations, and so is dependent upon the stress resultants only.

In our problem of a shell resting on an unyielding foundation, both $\bar{v}_n + \bar{u}_n$ and $\bar{v}_n - \bar{u}_n$ must be zero. The first of these conditions determines an inextensional deformation, and the second one serves to determine B_n. We write it in the form

$$\bar{v}_n - \bar{u}_n = \bar{v}_n^{(0)} - \bar{u}_n^{(0)} + B_n(\bar{v}_n^{(1)} - \bar{u}_n^{(1)}) = 0$$

and find from this

$$B_n = -\frac{\bar{v}_n^{(0)} - \bar{u}_n^{(0)}}{\bar{v}_n^{(1)} - \bar{u}_n^{(1)}}$$

as the quotient of two deformations which we have just determined as energy integrals. When we have found B_n, our problem is solved.

Chapter 3

DIRECT STRESSES IN CYLINDRICAL SHELLS

3.1 Statically Determinate Problems

3.1.1 General Theory

3.1.1.1 Differential Equations

A cylinder is generated by moving a straight line along a curve while maintaining it parallel to its original direction. It follows from this definition that through every point of the cylinder one may pass a straight line which lies entirely on this surface. These lines are called the generators. For convenience of language we shall assume here that the generators are horizontal. All planes which are normal to the generators intersect the cylinder in identical curves which are called profiles. The cylinder is named after the shape of the profile, e. g. a circular or a parabolic cylinder.

Generators and profiles suggest themselves as a natural net of coordinate lines. We choose an arbitrary profile as the datum line and

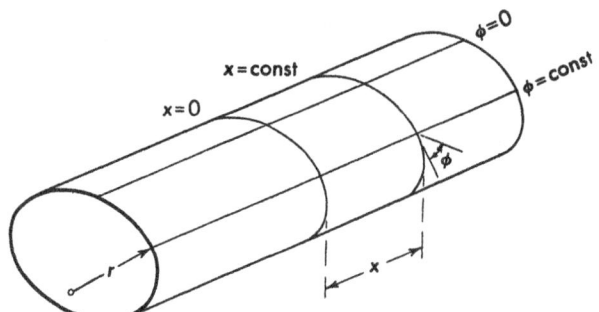

Fig. 1. Coordinates on a cylinder

from this measure the coordinate x along the generators, positive in one direction and negative in the other. The second coordinate must vary from generator to generator. In analogy to the angle ϕ used on

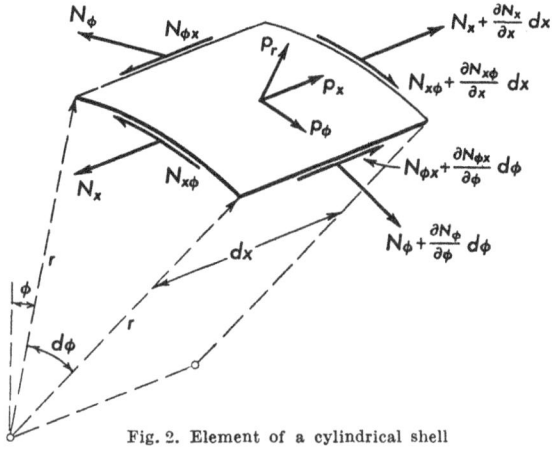

Fig. 2. Element of a cylindrical shell

surfaces of revolution, we introduce here the angle ϕ which a tangent to the profile (or a tangential plane to the cylinder) makes with a horizontal plane (fig. 1).

Now let us consider a shell whose middle surface is a cylinder. We cut from it an element bounded by two adjacent generators ϕ and $\phi + d\phi$ and by two adjacent profiles x and $x + dx$ (fig. 2). The membrane forces which act on the four edges must all lie in tangential planes to the middle surface and may be resolved into normal and shear components as shown. The forces per unit length of section are N_x, N_ϕ (normal forces) and $N_{x\phi} = N_{\phi x}$ (shearing forces). The load per unit area of the shell element has the components p_x, p_ϕ, in the directions of increasing x and ϕ, respectively, and a radial (normal) component p_r, positive outward.

The stress resultants N_x, N_ϕ, $N_{x\phi}$ are of the same kind as those appearing in shells of revolution, and, again, three conditions of equilibrium of the shell element will help us to find them as functions of x and ϕ.

These conditions may easily be read from fig. 2. The equilibrium in the x direction yields the equation

$$\frac{\partial N_x}{\partial x} dx \cdot r \, d\phi + \frac{\partial N_{\phi x}}{\partial \phi} d\phi \cdot dx + p_x \cdot dx \cdot r \, d\phi = 0,$$

and for the forces parallel to a tangent to the profile we have

$$\frac{\partial N_\phi}{\partial \phi} d\phi \cdot dx + \frac{\partial N_{x\phi}}{\partial x} dx \cdot r \, d\phi + p_\phi \cdot dx \cdot r \, d\phi = 0.$$

At right angles to the middle surface we have, besides the external force $p_r \cdot dx \cdot r \, d\phi$, only the resultant of the two forces $N_\phi \, dx$, pointing inward:

$$N_\phi \, dx \cdot d\phi - p_r \cdot dx \cdot r \, d\phi = 0.$$

After division by the two differentials, these three conditions of equilibrium are already the differential equations for the membrane forces of the shell:

$$N_\phi = p_r \, r,$$

$$\frac{\partial N_{x\phi}}{\partial x} = -p_\phi - \frac{1}{r} \frac{\partial N_\phi}{\partial \phi} \tag{1a--c}$$

$$\frac{\partial N_x}{\partial x} = -p_x - \frac{1}{r} \frac{\partial N_{x\phi}}{\partial \phi}.$$

They correspond to eqs. (II–6a–c) of the preceding Chapter. We see at a glance that they are of a much simpler structure; they may be resolved one by one. We shall see later how we have to pay for this mathematical advantage with a mechanical disadvantage.

3.1.1.2 General Solution

From eq. (1a) we obtain N_ϕ. This "hoop force" depends only on the local intensity of the normal load p_r and cannot be influenced by boundary conditions. This is not of great importance for shells whose profiles are closed curves and which have only two profiles as boun-

daries. But for shells like the one in fig. 11, the impossibility of prescribing arbitrary values of N_ϕ at the straight edges leads to the crucial
point in the membrane theory of cylindrical shells. We shall see this
in detail on p. 122.

The other two equations (1b, c) may be solved by simple integrations in the x direction. Each yields a "constant of integration",
which, of course, must be independent of x but may be an arbitrary
function of ϕ:

$$N_{x\phi} = -\int \left(p_\phi + \frac{1}{r} \frac{\partial N_\phi}{\partial \phi}\right) dx + f_1(\phi),$$
$$N_x = -\int \left(p_x + \frac{1}{r} \frac{\partial N_{x\phi}}{\partial \phi}\right) dx + f_2(\phi).$$

(2a, b)

The functions f_1 and f_2 must be determined from two boundary conditions. Each must be of the kind that on a profile $x = $ const (one end
of the shell or a plane of symmetry) one of the forces $N_{x\phi}$ or N_x is given
as an arbitrary function of ϕ.

In some simple and important cases of such boundary conditions
it is possible to introduce them into eqs. (2) and to determine the functions f_1 and f_2 before making any decision regarding the shape of the
profile of the cylinder or the particular kind of loading.

All these cases have in common that $p_x = 0$ and that the other
two load components, p_ϕ and p_r, are independent of x. We may then
perform the simple integrations in eqs. (2) and obtain the following
set:

$$N_{x\phi} = -\left(p_\phi + \frac{1}{r} \frac{dN_\phi}{d\phi}\right) x + f_1(\phi) = -x F(\phi) + f_1(\phi),$$
$$N_x = \frac{x^2}{2r} \frac{dF(\phi)}{d\phi} - \frac{x}{r} \frac{df_1(\phi)}{d\phi} + f_2(\phi).$$

(3a, b)

Here $F(\phi)$ is an abbreviation, defined by eq. (3a), which will be helpful
in writing subsequent formulas.

The different boundary conditions which we now shall consider will
all deal with the support which the shell receives along an edge $x = $ const.
This support is usually supplied by connecting this edge with a ring
or an arch. When we apply loads in the direction of the generators to
this reinforced edge, the ring cannot substantially participate in carrying them, since it could only do so with rather large deflections which
the much stiffer shell will not permit. We conclude that, if no such
load is applied, there must be $N_x = 0$ along the edge, and this must be
used as a boundary condition.

On the other hand, when the ring deflects in its own plane, a thin
shell will follow such deformation, unless it requires a change of the
hoop strain ϵ_ϕ. This latter will not be possible, since N_ϕ is fixed by

eq. (1a) and $N_x = 0$ because of the boundary condition. We conclude that the ring can and will receive shearing forces $N_{x\phi}$ from the shell in any amount which eq. (2a) or (3a) may require and that any discrepancies between the *plane* deformation of the ring and that of the shell can be settled only by the bending theory, since the membrane theory is evidently unable to deal with them.

Depending on the different applications of the theory, the stiffening member may be a ring, an arch, a rib, a truss, or a thin solid wall. In order to cover all these cases with a common expression, we shall

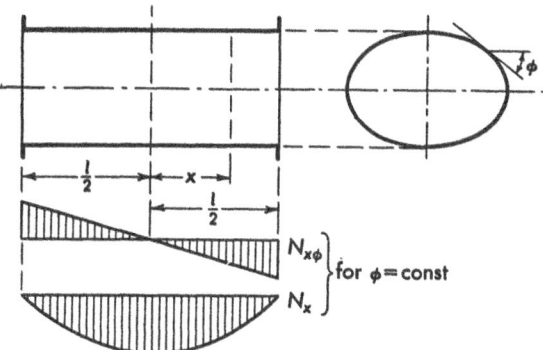

Fig. 3. Cylindrical shell supported by diaphragms at both ends. Spanwise distribution of $N_{x\phi}$ and N_x

henceforth speak of a diaphragm and shall use this word to mean any plane stiffening member which is capable of accepting from the shell any force lying in its plane but which offers no resistance to forces normal to its plane.

The simplest and most important case of boundary conditions is that of a shell of length l which is stiffened and supported by a diaphragm at each end (fig. 3). If we count the coordinate x from the profile halfway between the ends of the cylinder, we have as boundary conditions:

$$N_x \equiv 0 \quad \text{at} \quad x = \pm l/2.$$

When we introduce here N_x from eq. (3b), we find that

$$f_1(\phi) \equiv 0, \qquad f_2(\phi) = -\frac{l^2}{8r}\frac{dF(\phi)}{d\phi}$$

and hence

$$N_{x\phi} = -x F(\phi),$$
$$N_x = -\frac{1}{8r}(l^2 - 4x^2)\frac{dF(\phi)}{d\phi}. \tag{4}$$

From these formulas we see that the lengthwise distribution of the shearing force $N_{x\phi}$ is the same as that of the transverse shearing force of a simple beam of span l carrying a uniformly distributed load. Correspondingly, the forces N_x are distributed in the x direction as the bending moments of such a beam. This indicates that a cylindrical shell with these boundary conditions really acts like such a beam, transmitting all its load to the two diaphragms at the ends of the span l. Of course, the distribution of the forces $N_{x\phi}$ and N_x over the cross section cannot be derived from the beam formulas but is governed by the equations (1) for the equilibrium of the shell element.

Another case of boundary conditions is represented by fig. 4. Here one end, $x = l$, of the shell is completely built in, i. e., the support at

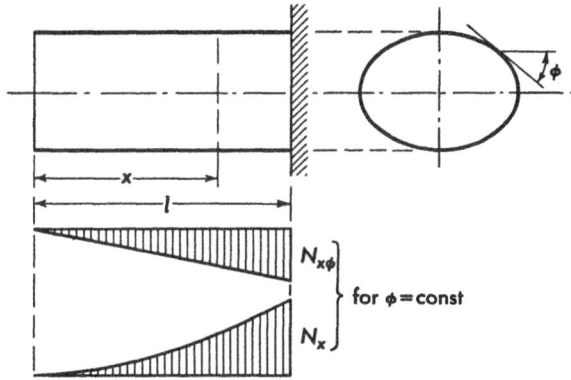

Fig. 4. Cantilever shell. Spanwise distribution of $N_{x\phi}$ and N_x

this side can resist not only shearing forces $N_{x\phi}$ but also normal forces N_x. The other end, $x = 0$, may then be left without any support at all, and we have the boundary conditions

$$x = 0. \qquad N_{x\phi} = 0 \quad \text{and} \quad N_x = 0.$$

A glance at the set (3) shows that in this case f_1 and f_2 must be identically zero, and hence we have

$$N_{x\phi} = -xF,$$
$$N_x = \frac{x^2}{2r}\frac{dF}{d\phi}. \tag{5}$$

This shell is supported like a cantilever beam, and, again, the spanwise distributions of $N_{x\phi}$ and N_x are those of the shear and the bending moment of the beam analogue.

The three-dimensional support of such a cantilever shell will scarcely be accomplished by a solid wall, as shown in fig. 4, but rather by an adjoining span of the same shell (fig. 5). In such a construction we have again two diaphragms of the usual type, which resist only shearing forces but do not accept forces N_x from the shell. The forces N_x coming from the cantilever section must therefore be transmitted across the diaphragm to the adjoining bay of the shell, which therefore has the boundary conditions

$$x = 0: \qquad N_x = \frac{l_1^2}{2r} \frac{dF}{d\phi},$$

$$x = l_2: \qquad N_x = 0.$$

When we determine f_1 and f_2 from these conditions, we arrive at the formulas:

$$N_{x\phi} = \left(\frac{l_1^2 + l_2^2}{2l_2} - x \right) F(\phi),$$

$$N_x = \frac{1}{2r} \left(x^2 - x\frac{l_1^2 + l_2^2}{l_2} + l_1^2 \right) \frac{dF}{d\phi}. \tag{6}$$

Again $N_{x\phi}$ and N_x have the same spanwise distribution as the shearing force and the bending moment of the beam analogue. This coincidence

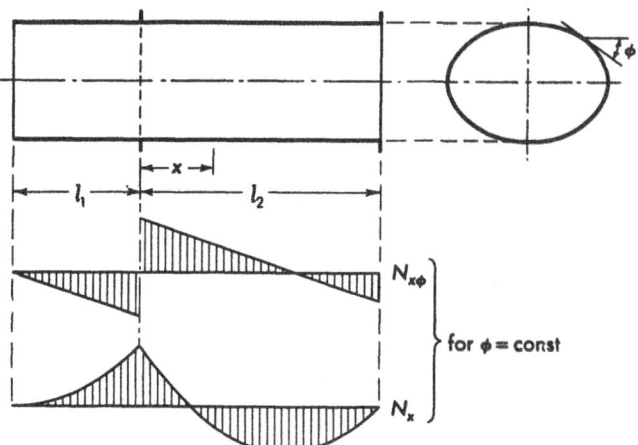

Fig. 5. Cylindrical shell with overhanging end. Spanwise distribution of $N_{x\phi}$ and N_x

will also be found if another cantilever shell is added at the other end of the main span, but the analogy cannot be extended to statically indeterminate cases as, for example, that of a cylindrical shell spanning two bays between three diaphragms. Here the result will be influenced by the deformation of the shell, and this is different from that of a simple beam, as we shall see on p. 138.

In all the preceding cases, N_ϕ is found from equation (1a), which is not affected by the choice of the boundary conditions.

3.1.1.3 Homogeneous Solution

If we put $p_x = p_\phi = p_r = 0$, we obtain the membrane forces in a shell which is subjected to loads only at its boundaries. From eq. (1a) we see that, in this case, $N_\phi \equiv 0$, and eqs. (3) yield

$$N_{x\phi} = f_1(\phi),$$

$$N_x = -\frac{x}{r}\frac{df_1}{d\phi} + f_2(\phi). \tag{7}$$

Let us apply these formulas to a shell supported as shown in fig. 6. Then eqs. (7) describe the stress resultants produced by given edge loads at $x = 0$ and $x = l$. We choose $N_x \equiv 0$ at $x = l$ and an arbitrary distribution $N_x = \overline{N}_x(\phi)$ at $x = 0$. We have then

$$\frac{l}{r}\frac{df_1}{d\phi} = f_2 = \overline{N}_x.$$

The shear follows by an integration:

$$N_{x\phi} = f_1 = \frac{1}{l}\int\limits_0^\phi r f_2 \, d\phi + C.$$

The constant C is still to be determined from a suitable condition, but, in any case, the formula must yield the same value $N_{x\phi}$ for $\phi = 0$ and $\phi = 2\pi$. Therefore, we must have

$$\int\limits_0^{2\pi} r f_2 \, d\phi = 0,$$

and this is a restriction to which the choice of the forces $\overline{N}_x$ is subject. Since $r \, d\phi$ is the line element of the edge, the restriction simply means

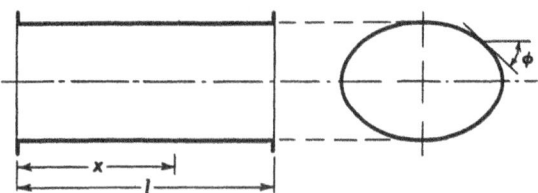

Fig. 6. Cylindrical shell supported by the two rings or diaphragms

that the forces applied to the edge must have the resultant zero. This is a very plausible limitation. If we had admitted similar forces at the other end of the cylinder, it would only be necessary that both loads have equal resultants opposite in direction.

3.1.2 Tubes and Pipes

3.1.2.1 Circular Cylinder

We have not yet discussed what happens if the cylinder has edges ϕ = const, but we are already sufficiently prepared to find the stress resultants in tubelike shells. The simplest case is that of a circular cylinder. Then the radius of curvature $r = a$ is a constant. If such a

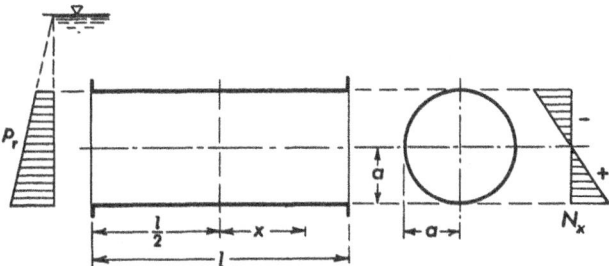

Fig. 7. Circular cylinder filled with water

pipe is filled with water of specific weight γ, the external forces (the water pressure) are

$$p_x = p_\phi = 0, \quad p_r = p_0 - \gamma\, a \cos\phi.$$

Here p_0 is the pressure at the level of the axis of the tube and may be anything $\geqq \gamma\, a$, but not less.

If we choose the boundary conditions of fig. 7, we find at once from eqs. (1a) and (4) the stress resultants:

$$\begin{aligned}
N_\phi &= p_0\, a - \gamma\, a^2 \cos\phi, \\
N_{x\phi} &= -\gamma\, a\, x \sin\phi, \\
N_x &= -\frac{1}{8}\gamma(l^2 - 4x^2)\cos\phi.
\end{aligned}$$ (8a–c)

The average pressure p_0 produces only hoop stresses. The load term with γ represents the weight of the content and produces a kind of over-all bending of the pipe, which acts as a beam carrying this weight between the supports at $x = +l/2$ and $x = -l/2$. We have already seen that therefore the shear $N_{x\phi}$ and the normal force N_x have the same spanwise distribution as the shear and the bending moment of a beam. The distribution of N_x over the profile is shown in the N_x diagram in fig. 7. Incidentally, this is the same linear distribution as that of bending stresses in common beam theory. This result is a peculiarity of the circular cylinder and, even there, is restricted to certain simple loads.

The boundary conditions which we have chosen are not easy ,to realize. They will prevail in a free-spanning section of a pipeline if there are expansion joints at both ends beyond the stiffening rings, which here replace the diaphragms.

If our shell is the wall of a horizontal cylindrical tank, the bulk-heads will also be subjected to the water pressure and will transmit it to the cylinder, where it creates additional forces N_x. If the bulk-heads are plane elastic plates, these forces will be distributed over the circumference according to a law, $A + B \cos\phi$, and from the equilibrium of the bulkhead we find A and B such that we have

$$N_x = \frac{p_0\,a}{2} - \frac{\gamma\,a^2}{4}\cos\phi$$

as the boundary condition at both ends of the cylinder. This will not change N_ϕ and $N_{x\phi}$, but we have to go back to eq. (2) to find f_2. If there we put $f_1 = 0$, corresponding to the symmetry of the shell with respect to the plane $x = 0$, we have

$$N_x = \frac{1}{2}\gamma\,x^2\cos\phi + f_2,$$

and to satisfy the new boundary condition, we have to put

$$f_2 = \frac{1}{2}\,p_0\,a - \frac{1}{8}\,\gamma(2a^2 + l^2)\cos\phi.$$

This yields

$$N_x = \frac{1}{2}\,p_0\,a + \frac{1}{8}\,\gamma(4x^2 - l^2 - 2a^2)\cos\phi.$$

The distribution of N_x in this case varies along the span. It is the same as that of a beam which carries an eccentric axial load corresponding to the pressure on the bulkheads, in addition to the weight of the water in the tank.

When we put $\gamma = 0$ in the last formula, we have the well-known pressure vessel formula with $N_\phi = p_0\,a$ and $N_x = p_0\,a/2$ for a cylindrical vessel with uniform internal pressure p_0. Our complete formula shows, that the cylindrical shell may also resist a variable pressure by a simple system of direct stresses. From the next examples we shall see that this is no peculiarity of the circular cylinder.

3.1.2.2 Elliptic Cylinder

Fig. 8 shows a cylindrical tank of elliptic cross section. It may be subjected to the same hydrostatic pressure as the circular cylinder. Here this pressure is, of course, not proportional to $\cos\phi$, but must be written as

$$p_x = p_\phi = 0, \qquad p_r = p_0 - \gamma\,z,$$

where z is the vertical coordinate in the cross sectional plane. From the equation of the ellipse the following relation may easily be derived:

$$z = \frac{b^2 \cos\phi}{(a^2 \sin^2\phi + b^2 \cos\phi)^{1/2}}$$

which connects z with our coordinate ϕ. For the radius of curvature r the same formula holds which we used already for r_1 in Chapter II, p. 27:

$$r = \frac{a^2 b^2}{(a^2 \sin^2\phi + b^2 \cos^2\phi)^{3/2}}.$$

Upon introducing all this into eq. (1a) we find the hoop-force

$$N_\phi = \frac{p_0 a^2 b^2}{(a^2 \sin^2\phi + b^2 \cos^2\phi)^{3/2}} - \frac{\gamma a^2 b^4 \cos\phi}{(a^2 \sin^2\phi + b^2 \cos^2\phi)^2}$$

and assuming again the somewhat academic boundary conditions of fig. 7, we find from eqs. (4)

$$N_{x\phi} = 3 p_0 (a^2 - b^2) \, x \, \frac{\cos\phi \sin\phi}{a^2 \sin^2\phi + b^2 \cos^2\phi} - \gamma b^2 \, x \, \frac{3(a^2 - b^2) \cos\phi + a^2}{(a^2 \sin^2\phi + b^2 \cos^2\phi)^{3/2}} \sin\phi,$$

$$N_x = -\frac{3}{8} p_0 \frac{a^2 - b^2}{a^2 b^2} (l^2 - 4 x^2) \frac{a^2 \sin^4\phi - b^2 \cos^4\phi}{(a^2 \sin^2\phi + b^2 \cos^2\phi)^{1/2}}$$

$$+ \frac{1}{8} \frac{\gamma}{a^2} (l^2 - 4 x^2) \frac{8 a^4 \sin^2\phi - a^2 b^2 (4 + 5\sin^2\phi) + 3 b^4 \cos^2\phi}{a^2 \sin^2\phi + b^2 \cos^2\phi} \cos\phi.$$

Here again the lengthwise distribution of N_x and $N_{x\phi}$ is the same as that of the bending moment and shearing force of a simple beam

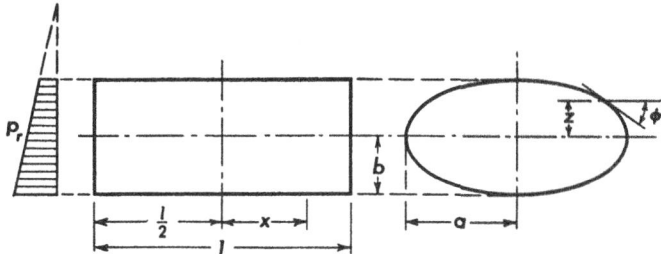

Fig. 8. Elliptic cylinder filled with water

of span l, and N_ϕ does not depend on x. But now the constant pressure p_0 produces also forces $N_{x\phi}$ and N_x. These forces enable the shell to withstand the load without bending stresses, and the bulkheads or stiffening rings at the ends are needed here to receive the shear $N_{x\phi}$, which results from this stress system.

In fig. 9 numerical results are plotted for two different conditions. The diagrams at the right belong to the case $p_0 = \gamma b$, where the pipe is just filled to the top without additional pressure. At the left, the forces produced by a simple gas pressure p_0 are given. They result from our formulas when we put $\gamma = 0$, and they may therefore also be considered as the limiting case of a water content with so high a pressure p_0 that the γ terms become insignificant. The diagrams show that the stress systems are far from simple; however. they exist, and the shell can carry the load.

If we close the ends of the cylinder by plane bulkheads, they will transmit additional forces N_x to the edges of the shell. Their magnitude and distribution follows from the solution of a plate problem, with

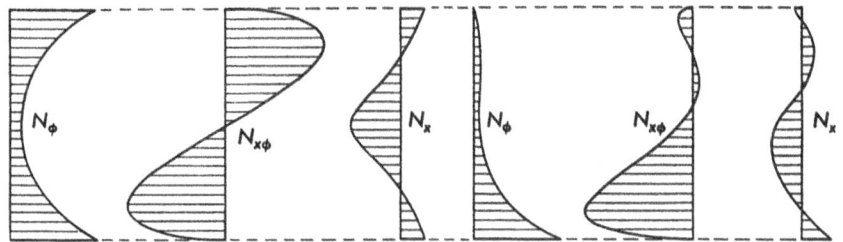

Fig. 9. Stress resultants in an elliptic cylinder. Left half: gas pressure p only. Right half: water pressure, zero pressure at highest point

which we are not concerned here. The other two stress resultants, N_ϕ and $N_{x\phi}$, are not changed. The case of curved bulkheads presents an additional shell problem and may be treated by the method explained in Section 4.4.2.2.

3.1.2.3 Inclined Cylinder

In all examples which we have studied thus far, the external forces were independent of x. We shall add here one case of a more general nature, which may show the possibilities of stress systems in cylindrical shells.

Fig. 10 shows a circular cylinder whose axis is inclined at an angle α from the vertical. The cylinder is partially filled with water. The water pressure is, of course, normal to the wall, so that we again have $p_x = p_\phi \equiv 0$. The normal load is

$$p_r = \gamma(x \cos\alpha - a \sin\alpha \cos\phi) \text{ for } x > a \tan\alpha \cos\phi$$
$$\text{and } p_r = 0 \qquad\qquad \text{for } x < a \tan\alpha \cos\phi.$$

In the part of the shell which lies above the water level, the stress resultants are given by the homogeneous solution (7). If we assume as boundary conditions that the upper edge is completely free, we have for this domain $f_1 \equiv f_2 \equiv 0$, and the shell is here absolutely free of stress.

For the lower part of the shell we have to use the general equations (1a) and (2a, b), where, of course, f_1 and f_2 are not the same functions as before but have to be determined from the condition that the stress resultants are continuous at the water level.

Equation (1a) yields immediately

$$N_\phi = \gamma a(x \cos\alpha - a \sin\alpha \cos\phi),$$

and this fortunately is zero at the level $x = a \tan\alpha \cos\phi$.

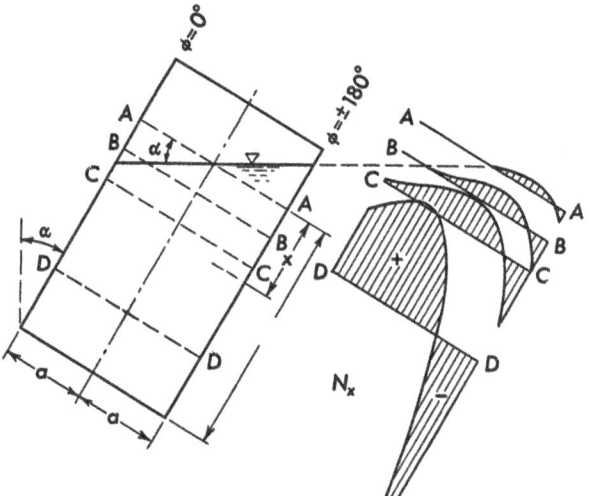

Fig. 10. Inclined cylindrical tank

The partial derivative is

$$\frac{\partial N_\phi}{\partial \phi} = \gamma a^2 \sin\alpha \sin\phi$$

and therefore, according to eq. (2a):

$$N_{x\phi} = -\gamma a x \sin\alpha \sin\phi + f_1(\phi).$$

This vanishes at the water level if we put

$$f_1(\phi) = \gamma a^2 \tan\alpha \sin\alpha \cos\phi \sin\phi,$$

and this yields the shearing force as

$$N_{x\phi} = \gamma a(a \tan\alpha \cos\phi - x) \sin\alpha \sin\phi.$$

Differentiating again with respect to ϕ and applying eq. (2b) yields an expression for the normal force:

$$N_x = \frac{\gamma}{2}[x^2 \cos\phi - 2a x \tan\alpha(\cos^2\phi - \sin^2\phi)] \sin\alpha + f_2(\phi).$$

The function $f_2(\phi)$ has to be determined in such a way that N_x vanishes at the water level, and this leads to the result

$$N_x = \frac{\gamma}{2} [x^2 \cos\phi - 2a\, x \tan\alpha (\cos^2\phi - \sin^2\phi)$$
$$+ a^2 \tan^2\alpha (\cos^2\phi - 2\sin^2\phi) \cos\phi] \sin\alpha.$$

Quite different from all preceding and following examples, the stress resultants here do not appear as products of a function of ϕ and a function of x, and it is therefore not possible to represent them by graphs giving their distribution over the profile and along the span, respectively. In fig. 10 some diagrams for N_x are given, which show how differently this force is distributed over different cross sections.

If $\alpha = 0$, the formulas degenerate into the trivial results for a vertical cylindrical tank. The other limiting case, $\alpha = \pi/2$, does not exist for a cylinder of finite length.

3.1.2.4 Fourier Series Solutions for the Circular Cylinder

Since eqs. (1) can easily be solved by quadratures, it seems unnecessary to employ Fourier series in so simple a problem. However, such solutions are useful on certain occasions, such as the connection of a cylinder with a shell of revolution or the combined use of membrane and bending theories.

In shells with a closed circumference all quantities must be periodic functions of the coordinate ϕ with the period 2π. We may therefore write the stress resultants as Fourier series with this period. This is particularly useful for the homogeneous solution for the circular cylinder. In this case, $N_\phi \equiv 0$, and the other two forces are given by eqs. (7) when we put $r = a$.

We assume $N_{x\phi}$ and N_x of the form

$$N_{x\phi} = \sum_1^\infty N_{x\phi m} \sin m\phi, \qquad N_x = \sum_0^\infty N_{x m} \cos m\phi. \tag{9}$$

Upon comparing this with eqs. (7) we see that the coefficients $N_{x\phi m}$ must be constants, say

$$N_{x\phi m} = -A_m \tag{10a}$$

while $N_{x m}$ must depend linearly on x in the form

$$N_{x m} = A_m \frac{m\, x}{a} + B_m. \tag{10b}$$

For every harmonic, except $m = 0$, there are available two free constants A_m, B_m, and for the zero-order harmonic there is one constant, B_0, which describes a uniform tension or compression.

It follows that we may prescribe N_{x0} at one end of the cylinder and that for each of the other harmonics two boundary conditions are admitted. These may prescribe N_{xm} at both ends, or there may be one condition for N_{xm} and one for $N_{x\phi m}$.

When we write $N_{x\phi}$ as a cosine series and N_x as a sine series, we get another pair of constants for each harmonic $m > 0$. They lead to similar stress systems as before. For $m = 0$ there results a constant shear $N_{x\phi 0}$ which corresponds to BREDT's theory of torsion of a thin-walled tube.

There is another way of using Fourier series for cylindrical shells. In a cylinder of length l the loads and the stresses are defined only within this length and may be extrapolated to form periodic functions of x with any period $\geq l$. In the bending theory of barrel vaults it is necessary to choose $2l$ as the period, and we shall, therefore, write now the loads in the form

$$p_r = \sum_1^\infty p_{rn} \sin\frac{n\pi x}{l}, \qquad p_\phi = \sum_1^\infty p_{\phi n} \sin\frac{n\pi x}{l}, \tag{11}$$

where the Fourier coefficients p_{rn}, $p_{\phi n}$ are independent of x but may and usually do depend on ϕ. The stress resultants must then be assumed as

$$N_\phi = \sum_1^\infty N_{\phi n} \sin\frac{n\pi x}{l}, \qquad N_{x\phi} = \sum_1^\infty N_{x\phi n} \cos\frac{n\pi x}{l},$$

$$N_x = \sum_1^\infty N_{xn} \sin\frac{n\pi x}{l}, \tag{12}$$

where the n-th harmonics $N_{\phi n}$, $N_{x\phi n}$, N_{xn} of the stress resultants are again functions of ϕ alone.

If we measure x from one end of the cylinder, then $N_x \equiv 0$ at both ends, while the shear $N_{x\phi}$ does not vanish. The Fourier series represent, therefore, the solution for a cylinder which is supported at the ends $x = 0$ and $x = l$ (fig. 6).

When we introduce these Fourier series into the differential equations (1), we obtain the following results:

$$N_{\phi n} = p_{rn} r,$$

$$\frac{n\pi}{l} N_{x\phi n} = p_{\phi n} + \frac{1}{r}\frac{dN_{\phi n}}{d\phi}, \tag{13 a–c}$$

$$\frac{n\pi}{l} N_{xn} = -\frac{1}{r}\frac{dN_{x\phi n}}{d\phi}.$$

As an example of the application of these formulas we consider a circular cylinder ($r = a$) of length l which has to carry its own weight. This is the shell shown in fig. 7 but with the coordinate x as shown in fig. 6. If p is the weight per unit of surface, the load components are

$$p_\phi = p \sin\phi, \qquad p_r = -p \cos\phi. \tag{14}$$

They do not depend on x. To bring them into the form (11), we must expand a constant into a Fourier series. The wellknown[1] formula

$$1 \equiv \frac{4}{\pi} \sum_{1,\,3,\,5,\,\ldots}^{\infty} \frac{1}{n} \sin \frac{n \pi x}{l}$$

yields in our case

$$p_{\phi n} = \frac{4p}{\pi n} \sin \phi, \qquad p_{r n} = -\frac{4p}{\pi n} \cos \phi,$$

valid for odd n, while all the even-order coefficients are zero.

When we introduce these load coefficients in eqs. (13), we find (for odd n)

$$N_{\phi n} = -\frac{4pa}{\pi n} \cos \phi, \qquad N_{x \phi n} = \frac{8pl}{\pi^2 n^2} \sin \phi,$$

$$N_{x n} = -\frac{8pl^2}{a \pi^3 n^3} \cos \phi.$$

(15 a–c)

This result is, of course, identical with eqs. (16) below. For the simple purposes of a membrane stress analysis the closed form (16) is preferable, but we shall need the series form (12) on p. 267 when we discuss bending stresses in the shell.

3.1.3 Barrel-Vaults

3.1.3.1 Circular Cylinder

As an introduction to the theory of barrel vaults we consider the case just treated, a tube of circular profile, supported as in fig. 7, and subjected to the load described by eqs. (14). Using the coordinate x as shown in fig. 7, we find from eqs. (1a) and (4):

$$N_{\phi} = -p a \cos \phi, \qquad N_{x \phi} = -2p x \sin \phi,$$

$$N_x = -\frac{p}{4a} (l^2 - 4x^2) \cos \phi.$$

(16)

The most remarkable feature of this force system is that on the generators $\phi = \pm \pi/2$ we have $N_{\phi} \equiv 0$. If we cut away the lower half of the shell, the upper half need not be supported at the straight edges and may carry its weight freely between the diaphragms, just as the tubular shells do. Such barrel vaults have been used as roof structures.

However, the straight edges of a barrel vault are not completely free of external forces. There is a shear $N_{x \phi} = \pm 2p x$, and a structural element must be provided to which it can be transmitted. This so-called edge member is a straight bar, and if properly placed, it is stressed only in tension (fig. 11). Its axial force N is, of course, variable

---

[1] See, for example, MARKS, L. S.: Mechanical Engineers' Handbook. New York 1958, p. 2–69.

along the span. It can easily be found by integrating the shear N_x, beginning at the end $x = -l/2$ where $N = 0$. For the edge $\phi = +\pi/2$ the integration is like this:

$$N = \int_{-l/2}^{x} N_{x\phi}\,dx = -2p \int_{-l/2}^{x} x\,dx = \frac{1}{4}\,p\,(l^2 - 4\,x^2),$$

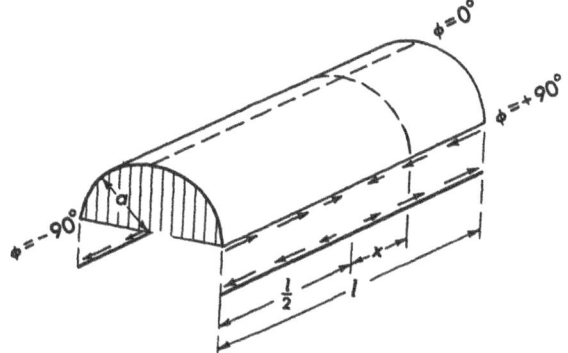

Fig. 11. Barrel-vault shell

and at the other side the same result appears.

The statical necessity of this force may be understood from a look at the N_x diagram in fig. 12. It has only compressive forces, and if we

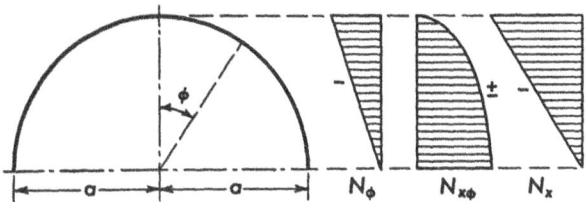

Fig. 12. Stress resultants in a barrel vault with semicircular profile

cut the shell apart in a plane $x = $ const, the horizontal equilibrium of each half requires that tensile forces of the same amount also appear. Now the integral of the forces N_x in the cross section is

$$\int_{-\pi/2}^{+\pi/2} N_x a\,d\phi = -\frac{1}{4}\,p\,(l^2 - 4\,x^2) \int_{-\pi/2}^{+\pi/2} \cos\phi\,d\phi = -\frac{1}{2}\,p\,(l^2 - 4\,x^2).$$

This is exactly the same compressive force as the two tensile forces N in the edge members so that they just maintain the horizontal equilibrium.

The resultant of the compressive forces lies somewhere in the semicircular profile and therefore higher than the tensile force $2N$, and both combine to form a couple. When we consider the barrel vault as a beam of span l, this couple is the bending moment. Since the load of the "beam" per unit length is $\pi a p$, its bending moment is

$$M = \pi a p \frac{l^2 - 4x^2}{8}.$$

To find the moment of the stress resultants N_x and N in the cross section of the shell, the axis of reference may be chosen arbitrarily. We choose the horizontal diameter. Then N makes no contribution, and N_x gives

$$- \int\limits_{-\pi/2}^{+\pi/2} N_x \cdot a \cos\phi \cdot a\, d\phi = \frac{1}{8} \pi p a (l^2 - 4x^2),$$

which is exactly equal to M. In the same way we may check that the vertical resultant of the shearing forces $N_{x\phi}$ in the cross section is equal to the transverse shearing force $-\pi a p x$ of the beam analogue.

This comparison between the barrel vault and its beam analogue gives a good general idea of the stress system in the shell and yields a useful check for computations. It cannot disclose details of the membrane stress distribution, since they depend essentially on the shape of the cylinder. We shall study this now in several examples of technical interest.

3.1.3.2 Elliptic Cylinder

To make a shell suitable for the construction of freespanning barrel vaults, the force N_ϕ must be zero at the straight edges. From eq. (1a) we conclude that this always happens when the normal load component $p_r = 0$ there. Now, the essential load of such a structure is its own weight, and we see from eq. (14) that in this case p_r vanishes for $\phi = \pm\pi/2$. The profile of a free spanning barrel vault must, therefore, terminate with a vertical tangent.

Two simple curves which satisfy this condition are the ellipse and the cycloid, and they have, therefore, been suggested and used as profiles of barrel vaults.

For the ellipse we have already seen how the radius r depends on the coordinate ϕ. Introducing this and the dead load, eq. (14), into eqs. (1a), (4), we find

$$N_\phi = -p a^2 b^2 \frac{\cos\phi}{(a^2 \sin^2\phi + b^2 \cos^2\phi)^{3/2}},$$

$$N_{x\phi} = -p x \frac{2a^2 + (a^2 - b^2)\cos^2\phi}{a^2 \sin^2\phi + b^2 \cos^2\phi} \sin\phi,$$

$$N_x = -\frac{1}{8} p (l^2 - 4x^2) \frac{3 a^2 (b^2 - (a^2 - b^2)\sin^2\phi) - (a^2 \sin^2\phi + b^2 \cos^2\phi)^2}{a^2 b^2 (a^2 \sin^2\phi + b^2 \cos^2\phi)^{1/2}} \cos\phi.$$

The distribution of these forces over the profile is shown in fig. 13. We see that N_ϕ really vanishes at the edge, and that at the edge the shear $N_{x\phi}$ has a finite value $\pm 2 p\, a\, x$, which incidentally is the same as for the circular cylinder of radius a. The force in the edge member therefore is again

$$N = \frac{1}{4}\, p\,(l^2 - 4 x^2).$$

The normal force N_x in the direction of the generators is, of course, a compression for the major part of the profile, but if b/a is small enough, there is a zone with tensile stress near the edges, as shown in fig. 13.

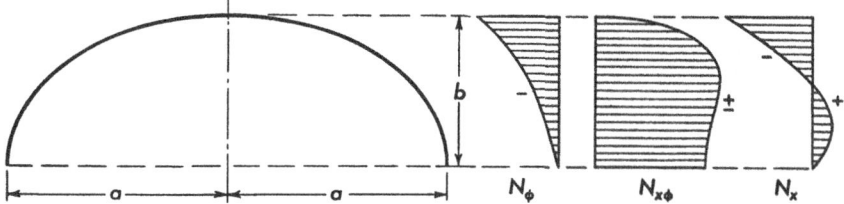

Fig. 13. Stress resultants in a barrel vault with elliptic profile

3.1.3.3 Cycloidal Cylinder and Related Shapes

A cycloid (fig. 14) is described by the equations

$$y = a(2\phi + \sin 2\phi), \qquad z = a(1 + \cos 2\phi),$$

where a is the radius of the generating circle and ϕ the slope angle as defined in fig. 1. The radius of curvature r of the curve is found from the well-known formula

$$r = -\frac{\left[\left(\dfrac{\partial y}{\partial \phi}\right)^2 + \left(\dfrac{\partial z}{\partial \phi}\right)^2\right]^{3/2}}{\dfrac{\partial y}{\partial \phi}\dfrac{\partial^2 z}{\partial \phi^2} - \dfrac{\partial z}{\partial \phi}\dfrac{\partial^2 y}{\partial \phi^2}}$$

in which the minus sign is arbitrary and serves to obtain a positive result:

$$r = 4a \cos\phi.$$

Introducing this and the load (14) into eqs. (1a) and (4), we find the stress resultants for a uniform dead load p:

$$N_\phi = -4p\, a\, \cos^2\phi, \qquad N_{x\phi} = -3p\, x\, \sin\phi,$$

$$N_x = -\frac{3p}{32 a}\,(l^2 - 4 x^2).$$

Here the force N_x is constant over the whole profile. This may sometimes be desirable since it indicates that all material is efficiently used. But in most cases we should rather like to have a tensile force N_x along the edge to avoid the large discrepancy between the lengthwise strain in the shell and in its edge member.

It is relatively easy to improve the properties of the cycloidal cylinder by subjecting it to an affine transformation. When we replace

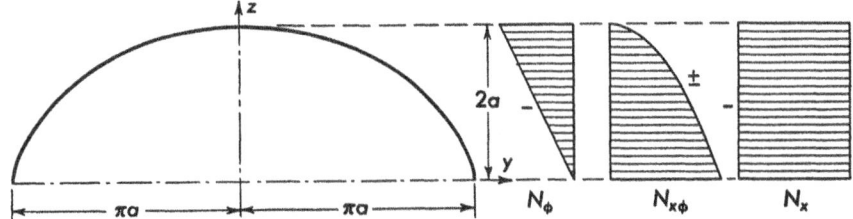

Fig. 14. Stress resultants in a barrel vault with a cycloid as its profile

a by b in the equation for z, we obtain a curve of the same span $2\pi a$ but a height $2b$ (fig. 15). However, the angle ϕ on the right-hand side is then no longer the slope angle and we should use a different notation, writing

$$y = a(2\psi + \sin 2\psi), \qquad z = b(1 + \cos 2\psi).$$

From simple formulas of analytic geometry we find that the real slope angle ϕ is connected with the parameter ψ by the relation

$$a \tan\phi = b \tan\psi$$

and that the radius of curvature may easily be expressed in terms of ϕ:

$$r = \frac{4a^2 b^3 \cos\phi}{(a^2 \sin^2\phi + b^2 \cos^2\phi)^2}.$$

Introducing this into eqs. (1a), (4), we find the following stress resultants for a uniform dead load p:

$$N_\phi = -4p\,a^2 b^3 \frac{\cos^2\phi}{(a^2 \sin^2\phi + b^2 \cos^2\phi)^2},$$

$$N_{x\phi} = -p\,x \frac{3a^2 + (a^2 - b^2) \cos^2\phi}{a^2 \sin^2\phi + b^2 \cos^2\phi} \sin\phi,$$

$$N_x = -\frac{p(l^2 - 4x^2)}{32 a^2 b^3} [3a^2 b^2 - (a^2 - b^2)(4a^2 \sin^2\phi + a^2 \sin^4\phi + b^2 \cos^4\phi)].$$

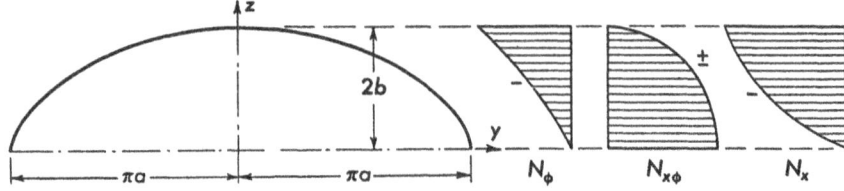

Fig. 15. Stress resultants in a barrel vault whose profile has been obtained by an affine deformation of a cycloid

For $a = b$ all these formulas reduce to those for the common cycloid, but if $b \neq a$, the force N_x is no longer uniformly distributed over the profile of the cylinder. If we choose in particular $b^2 = \frac{5}{8} a^2$ we get $N_x = 0$ at the edges $\phi = \pm 90°$, as shown in fig. 15, and for still flatter shapes there is even a zone with positive N_x which permits us to adapt the strain in the cylinder and in the edge member to each other. In the limiting case (fig. 15) the span of the profile is 3.974 times its height, and cylinders having positive N_x are rather flat.

3.1.3.4 Special Cylinders

It is not always desirable to give a shell as flat a profile as must be done in the preceding case in order to obtain a positive force N_x at the edge. Then another shape is useful which is characterized by the relation

$$r = a_0 + a_1 \cos \phi$$

for its radius of curvature. We find the ordinates y, z in the plane of a profile by simple integration of the projections of the line element $r \, d\phi$:

$$y = \int_0^\phi r \cos\phi \, d\phi = a_0 \sin\phi + \frac{1}{4} a_1 (2\phi + \sin 2\phi),$$

$$z = - \int_{\pi/2}^\phi r \sin\phi \, d\phi = a_0 \cos\phi + \frac{1}{4} a_1 (1 + \cos 2\phi).$$

These formulas show that the curves lie somewhere between a circle ($a_1 = 0$) and a cycloid ($a_0 = 0$), but this does not imply that the stresses lie halfway between these two cases. We see this at once when we use eqs. (1a) and (4) to find the stress resultants for dead load p:

$$N_\phi = -p(a_0 + a_1 \cos\phi) \cos\phi,$$

$$N_{x\phi} = -p \, x \frac{2a_0 + 3a_1 \cos\phi}{a_0 + a_1 \cos\phi} \sin\phi,$$

$$N_x = -\frac{p}{8}(l^2 - 4x^2) \frac{(2a_0 + 3a_1 \cos\phi)(a_0 + a_1 \cos\phi) \cos\phi - a_0 a_1 \sin^2\phi}{(a_0 + a_1 \cos\phi)^3}.$$

In all except the two limiting cases N_x is positive at $\phi = \pi/2$. If the profile is almost circular ($a_0 \gg a_1$), the positive force along the edge is small, and if the profile comes close to a cycloid ($a_0 \ll a_1$), N_x changes quite suddenly from an almost constant negative value to an extremely high positive one. In fig. 16 the case $a_0 = a_1$ has been chosen for plotting. The N_x distribution ends with a positive value of good size, and the transition from negative to positive is rather gradual. Such a shape may be useful to avoid difficulties at the edge member.

A comparison of fig. 16 with the four preceding ones shows impressively how different the N_x diagrams can be for cylindrical shells which do not differ much in shape. The shear diagrams are more like each other, and N_ϕ in all cases comes close to a linear distribution.

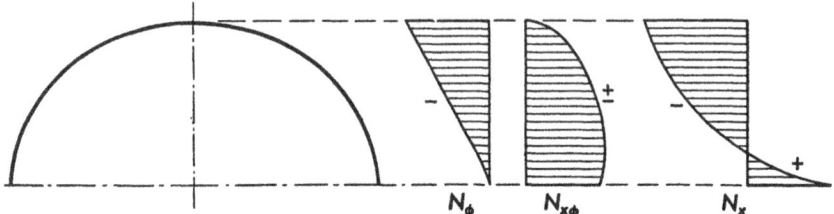

Fig. 16. Stress resultants in a barrel vault of special profile

3.1.3.5 Critical Remarks

Barrel-vault shell roofs are usually built of reinforced concrete. In the edge member of a single-span shell there is always a tensile force of considerable magnitude for which a steel reinforcement must be provided. Its cross section will, of course, be chosen so that the positive strain in it is of appreciable magnitude. On the other hand we see from figs. 12, 14, 15 that the cylindrical shell usually has a negative strain in the x direction, sometimes even at the edges. Of course, this discrepancy in the strains of adjacent fibers cannot exist in reality. To remedy it, an additional stress system will appear, consisting primarily of an additional shear $N_{x\phi}$ but accompanied by bending moments M_ϕ and transverse shearing forces Q_ϕ of considerable magnitude.

There is still another source of bending stresses in the shell. Each edge member has to carry a tensile force equal to the integral of all the compressive stresses across half the profile of the shell. It will have a considerable weight, and this has to be supported somehow. If the edge member is shaped as a slender bar, it will be suspended at the shell, requiring a tensile force N_ϕ. Such a force would be in contradiction to eq. (1a) and therefore is possible only in connection with bending stresses.

If the span l of the shell is greater than the width of the vault ($2a$ for circle and ellipse, $2\pi a$ for the cycloid) the bending stress system will spread over the entire area of the shell, and all the normal and shearing forces will be thoroughly changed by it. Only if l is comparatively small are the bending stresses limited to a border zone.

There exists a way of avoiding both sources of trouble by making the edge member a self-supporting girder and choosing its cross section so that no discrepancy in strain ϵ_x occurs and so that the elastic deflections of both this edge girder and the edge of the shell are the same.

We shall discuss this possibility in the next Section, when we treat the deformations of barrel vaults.

A similar problem arises, when the straight edges of the shell are continuously supported. The edge member will then be built without appreciable bending stiffness, and the support must yield just as far as the deformation of the shell requires but then carry the weight of the edge member. In any other case there will again be additional bending.

Besides the bending stresses which are caused by the presence of the edge member, there will also be bending stresses due to the mere fact that the shell *has* a finite bending stiffness and that there is no reason why the stresses should be uniformly distributed across the wall thickness. These stresses will be more pronounced, the thicker the shell is, and their tendency is to wipe out minor differences of the N_x diagrams. The thinner the shell, the smaller this effect must be, and for sufficiently thin shells the diagrams of the membrane forces must be taken seriously. However, at the present state of the bending theory of cylindrical shells, without much numerical material available, it is hard to say when a shell is *sufficiently* thin. Therefore, much judgment is needed when the results of the membrane theory are to be used in any but the simplest cases.

3.2 Deformations

3.2.1 General Theory

3.2.1.1 Differential Equations

In studying the deformations of a cylindrical shell, we may begin in the same way as we did for shells of revolution. The strains are again two normal strains, ϵ_x and ϵ_ϕ, and a shear strain $\gamma_{x\phi}$. HOOKE's law is the same as given by eqs. (II–57) or (II–58); we just have to replace the subscript θ by x:

$$\epsilon_x = \frac{1}{E\,t}\,(N_x - \nu\,N_\phi), \qquad \epsilon_\phi = \frac{1}{E\,t}\,(N_\phi - \nu\,N_x),$$

$$\gamma_{x\phi} = \frac{2\,(1+\nu)}{E\,t}\,N_{x\phi}. \tag{17}$$

The next step is to find the relations between the strains ϵ_x, ϵ_ϕ, $\gamma_{x\phi}$ and the displacements. These are (fig. 17): the axial displacement u, parallel to the axis of the shell and positive in the direction of increasing x, the circumferential displacement v in the direction of the profile of the middle surface and positive in the direction of increasing ϕ, and the radial displacement w, normal to the middle surface and positive when outward.

The strain ϵ_x represents the stretching of the straight line element dx, caused by the difference between the displacements u and $u + \dfrac{\partial u}{\partial x}\, dx$ of both its ends:

$$\epsilon_x = \frac{\partial u}{\partial x}. \tag{18a}$$

To find the hoop strain ϵ_ϕ, we have to proceed in the same way as we did for the meridional strain of a shell of revolution, and we again find eq. (II–59a) in a slightly different notation:

$$\epsilon_\phi = \frac{1}{r}\left(\frac{\partial v}{\partial \phi} + w\right). \tag{18b}$$

The shear is the sum of the rotations of the two line elements dx and $r\, d\phi$ (fig. 18):

$$\gamma_{x\phi} = \frac{\partial v}{\partial x} + \frac{1}{r}\,\frac{\partial u}{\partial \phi}. \tag{18c}$$

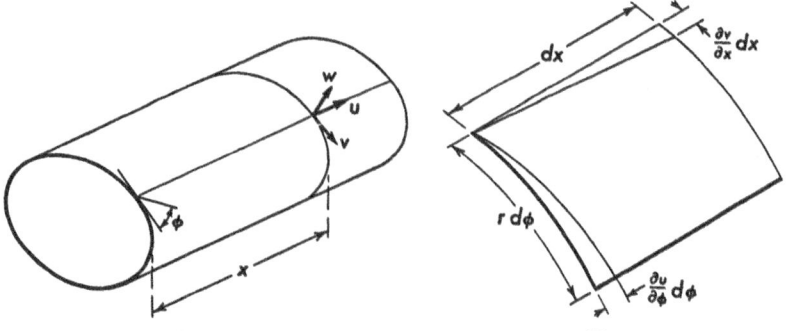

Fig. 17   Fig. 18
Displacements u, v, w for a cylindrical shell   Shear deformation of a shell element

When we eliminate the strains from HOOKE's law (17) and the kinematic relations (18), we obtain the following equations:

$$\frac{\partial u}{\partial x} = \frac{1}{E\,t}\,(N_x - \nu N_\phi),$$

$$\frac{\partial v}{\partial x} + \frac{1}{r}\,\frac{\partial u}{\partial \phi} = \frac{2(1+\nu)}{E\,t}\,N_{x\phi}, \tag{19a–c}$$

$$\frac{w}{r} + \frac{1}{r}\,\frac{\partial v}{\partial \phi} = -\frac{1}{E\,t}\,(N_\phi - \nu N_x).$$

These are the differential equations of the deformation of a cylindrical shell under the influence of direct stresses.

3.2.1.2 Solution in General Terms

These equations may be solved one after the other by simple integrations. For $t = $ const the solution is

$$E\,t\,u = \int (N_x - \nu N_\phi)\,dx + f_3(\phi),$$

$$E\,t\,v = 2(1 + \nu) \int N_{x\phi}\,dx - E\,t\,\frac{1}{r} \int \frac{\partial u}{\partial \phi}\,dx + f_4(\phi), \qquad (20)$$

$$E\,t\,w = r(N_\phi - \nu N_x) - E\,t\,\frac{\partial v}{\partial \phi}.$$

The generalization for variable wall thickness is obvious. In any case, the solution yields as "constants of integration" two more arbitrary functions of ϕ, which may be used to fulfill two boundary conditions at the edges $x = $ const of the shell.

Eqs. (20) may be used to find the deformation in such general cases as the one shown in fig. 10. Usually we shall have to deal with problems where $p_x \equiv 0$ and p_ϕ and p_r are independent of x. Then we may introduce N_ϕ from eq. (1a) and $N_{x\phi}$, N_x from eqs. (3) and perform the integrations:

$$E\,t\,u = \frac{x^3}{6r}\frac{dF}{d\phi} - \nu\,x\,N_\phi - \frac{x^2}{2r}\frac{df_1}{d\phi} + x f_2 + f_3,$$

$$E\,t\,v = -\frac{x^4}{24r}\frac{d}{d\phi}\left(\frac{1}{r}\frac{dF}{d\phi}\right) - (1+\nu)\,x^2 F + \frac{\nu\,x^2}{2r}\frac{dN_\phi}{d\phi} \qquad (21)$$

$$+ \frac{x^3}{6r}\frac{d}{d\phi}\left(\frac{1}{r}\frac{df_1}{d\phi}\right) - \frac{x^2}{2r}\frac{df_2}{d\phi} + x\left[2(1+\nu)\,f_1 - \frac{1}{r}\frac{df_3}{d\phi}\right] + f_4,$$

$$E\,t\,w = \frac{x^4}{24}\frac{d}{d\phi}\left[\frac{1}{r}\frac{d}{d\phi}\left(\frac{1}{r}\frac{dF}{d\phi}\right)\right] + \frac{x^2}{2}\left[(2+\nu)\frac{dF}{d\phi} - \nu\frac{d}{d\phi}\left(\frac{1}{r}\frac{dN_\phi}{d\phi}\right)\right]$$

$$+ r\,N_\phi - \frac{x^3}{6}\frac{d}{d\phi}\left[\frac{1}{r}\frac{d}{d\phi}\left(\frac{1}{r}\frac{df_1}{d\phi}\right)\right] + \frac{x^2}{2}\frac{d}{d\phi}\left(\frac{1}{r}\frac{df_2}{d\phi}\right)$$

$$- x\left[(2+\nu)\frac{df_1}{d\phi} - \frac{d}{d\phi}\left(\frac{1}{r}\frac{df_3}{d\phi}\right)\right] - \nu\,r\,f_2 - \frac{df_4}{d\phi}.$$

Just as we have done with the solution (3) of the stress problem, we may specialize eqs. (21) by introducing certain boundary conditions. We shall do so for the simple case represented by fig. 6. From the condition that $N_x = 0$ at both ends $x = \pm l/2$ we determined the functions f_1 and f_2 which are given on p. 111. We still have f_3 and f_4 which may be used to satisfy two conditions for the displacements. Since we assume that the diaphragms are perfectly flexible in the x direction (hence $N_x = 0$), we have nothing to say about u, but we should, of course, like to have $v = 0$ and $w = 0$ at both ends o the shell because of its connection with the diaphragms. But this is too much for only two free functions, and we have to make a choice. Now there are forces $N_{x\phi}$ at the edge which may enforce a displacement v, but there is no force in the direction of w. It therefore seems most reasonable to determine f_3 and f_4 so as to have $v = 0$ at $x = \pm l/2$ and to leave it to additional bending stresses to fulfill a similar condition for w. In this way we arrive at the following set of formulas:

$$Et\,u = \frac{x}{24r}\left(4x^2 - 3l^2\right)\frac{dF}{d\phi} - v\,x\,N_\phi,$$

$$Et\,v = -\frac{1}{384r}\left(5l^2 - 4x^2\right)\left(l^2 - 4x^2\right)\frac{d}{d\phi}\left(\frac{1}{r}\frac{dF}{d\phi}\right)$$

$$+ \frac{1}{8}\left(l^2 - 4x^2\right)\left[2\left(1 + v\right)F - \frac{v}{r}\frac{dN_\phi}{d\phi}\right] \tag{22}$$

$$Et\,w = \frac{1}{384}\left(5l^2 - 4x^2\right)\left(l^2 - 4x^2\right)\frac{d}{d\phi}\left[\frac{1}{r}\frac{d}{d\phi}\left(\frac{1}{r}\frac{dF}{d\phi}\right)\right]$$

$$- \frac{1}{8}\left(l^2 - 4x^2\right)\left[\left(2 + v\right)\frac{dF}{d\phi} - v\frac{d}{d\phi}\left(\frac{1}{r}\frac{dN_\phi}{d\phi}\right)\right] + r\,N_\phi.$$

Upon introduction of $r(\phi)$, N_ϕ, and $F(\phi)$ for a special case they yield immediately the displacements for the assumed boundary conditions.

3.2.2 Application to Pipes and Barrel Vaults

3.2.2.1 Circular Shell

As a first application of these formulas we shall now calculate the deflection of a pipe of circular cross section $(r = a)$, filled with liquid of specific weight γ and supported by two rings as shown in fig. 7. The stress resultants are given by eqs. (8). Comparison of eqs. (8b) and (3a) yields $F(\phi) = \gamma\,a\,\sin\phi$. Using this and N_ϕ from eq. (8a), we find from (22):

$$Et\,u = -v\,p_0\,a\,x + \gamma\,x\left(\frac{x^2}{6} - \frac{l^2}{8} + v\,a^2\right)\cos\phi,$$

$$Et\,v = \frac{\gamma}{384a}\left(5l^2 - 4x^2\right)\left(l^2 - 4x^2\right)\sin\phi + \frac{\left(2 + v\right)\gamma\,a}{8}\left(l^2 - 4x^2\right)\sin\phi,$$

$$Et\,w = p_0\,a^2 - \frac{\gamma}{384a}\left(5l^2 - 4x^2\right)\left(l^2 - 4x^2\right)\cos\phi \tag{23}$$

$$- \frac{\gamma\,a}{4}\left(l^2 - 4x^2\right)\cos\phi - \gamma\,a^3\cos\phi.$$

The uniform pressure p_0 produces only a uniform increase of the diameter and a shrinkage of the length of the shell, represented by the first term of w and of u, respectively. The γ terms may be interpreted in a similar way as it will be done in the following example. We shall need eqs. (23) on p. 264 when discussing the stresses in a partially filled pipe.

As a second application of eqs. (22) we consider the influence of the weight of the pipe itself on the deflections. The stress resultants for this case are given by eqs. (16). From them and eqs. (22) we find

$$E t u = \frac{p x}{a} \left(\frac{x^2}{3} - \frac{l^2}{4} + \nu a^2 \right) \cos\phi,$$

$$E t v = \frac{1}{8} p(l^2 - 4x^2) \left(\frac{5 l^2 - 4 x^2}{24 a^2} + 4 + 3\nu \right) \sin\phi, \qquad (24)$$

$$E t w = - \frac{1}{8} p(l^2 - 4x^2) \left(\frac{5 l^2 - 4 x^2}{24 a^2} + 4 + \nu \right) \cos\phi - p a^2 \cos\phi.$$

Since in this particular case the distribution of N_x over the cross section is incidentally the same as that assumed by the elementary beam theory, it is interesting to compare the deflection of the shell with that of a beam of span l, carrying the load $q = 2\pi p a$ per unit length and having $I = \pi a^3 t$ as moment of inertia of its cross section. The well-known formula yields

$$w_{\text{beam}} = \frac{q}{384 EI} (5 l^2 - 4 x^2) (l^2 - 4 x^2) = \frac{p}{192 E t a^2} (5 l^2 - 4 x^2) (l^2 - 4 x^2).$$

This might be expected to be equal to v for $\phi = \pi/2$ or to $-w$ for $\phi = 0$, but evidently it is not. For v, the difference lies in the term $4 + 3\nu$ and is due to the fact that in the shell formulas the influence of $N_{x\phi}$ on the deformation is taken into account, which is not done in the common beam formula. For w, an additional source of discrepancy will be found in the deformation of the cross sections, which prohibits having zero displacement on the whole circumference of the circles $x = \pm l/2$. We had to choose the points where we wanted to fulfill this condition exactly, and although our choice of the points $\phi = \pm\pi/2$ is certainly reasonable, it is nevertheless arbitrary. The discrepancy does not depend on the thickness t, but it disappears in the limit $l/a \to \infty$, i. e. if the pipe is so slender that the contribution of the shearing forces to the total deflection and the relative displacements within each cross section become negligible.

Eqs. (24) may equally well be applied to a barrel vault, and we can now answer the question as to how stiff an edge member must be built, in order to fit a semicircular cylinder without any discrepancy in the deformation.

Fig. 19 shows the system we have to investigate. We ask how to choose the cross section of the edge beam in order to satisfy the following conditions along the line where the beam meets the shell:

1) Beam and shell must have the same longitudinal strain ϵ_x.
2) Beam and shell must have the same vertical deflection.

If we follow a practice common in reinforced concrete stress analysis and put $\nu = 0$, we may as well speak of the stress σ_x instead of the strain in the first condition. The stress in the beam comes from two kinds of forces: its own weight and the shear $N_{x\phi}$ transmitted from the shell. Let A be the area, I the moment of inertia of the cross section, and γ the specific weight of the concrete. Then the load per unit length is γA and the corresponding bending moment at midspan is $\frac{1}{8}\gamma A\, l^2$.

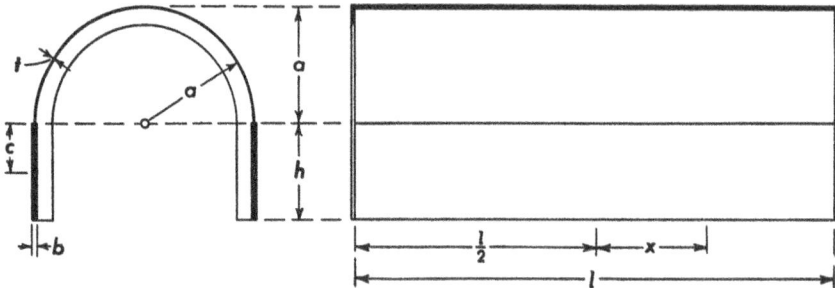

Fig. 19. Barrel-vault shell with high edge beams

The shear $N_{x\phi}$ produces in the beam an eccentric normal force at a distance c above the centroid of the cross section. As we saw on p. 123, it varies according to the same parabolic law as the bending moment due to dead load and has a maximum $N = \frac{1}{4}\,p\,l^2$ at midspan. The combined stress in the top fiber of the edge-beam is therefore

$$\sigma_x = \frac{p\,l^2}{4A} - \frac{c}{I}\left(\frac{1}{8}\,\gamma\,A\,l^2 - \frac{1}{4}\,p\,l^2 c\right).$$

It must be equal to the stress $\sigma_x = N_x/t$ of the shell at $\phi = \pi/2$, and since this is zero, we must have $\sigma_x = 0$. This is our first equation.

The deflection of the beam consists of two parts. One comes from its bending moment and varies as $(5\,l^2 - 4\,x^2)\,(l^2 - 4\,x^2)$ and the other comes from the transverse shear and varies as $(l^2 - 4\,x^2)$ only. The deflection of the shell is v from eq. (24) at $\phi = \pi/2$. It too has two terms with exactly the same functions of x as factors. Our condition 2 will be satisfied for all values of x, if the maximum (at midspan) of each one of the two parts of the deflections is the same for beam and shell.

The deflection of the beam due to bending stresses is at midspan

$$w_b = \frac{5\,l^2}{48\,EI}\left(\frac{1}{8}\,\gamma\,A\,l^2 - \frac{1}{4}\,p\,l^2 c\right).$$

For the shell the corresponding displacement is

$$v = \frac{5p\,l^4}{192\,E\,t\,a^2}.$$

The condition $w_b = v$ is our second equation. Now let us, for a moment, forget the other part of the deflection, which would yield a third equation, just enough to determine the three unknowns c, A, I.

When we want to give the beam a rectangular cross section of height h and width b, the unknowns depend on these two parameters:

$$c = h/2, \quad A = b\,h, \quad I = b\,h^3/12.$$

In this case we may put our two equations in the form

$$\frac{p}{\gamma\,t} = \frac{3h^2}{2a^2} = \frac{3t}{b}.$$

If the barrel vault consists of bare concrete, $p = \gamma\,t$ and hence

$$h = 0.816a, \quad b = 3t.$$

If the shell has a protecting layer, which adds to its weight but does not participate in carrying it, we have $p > \gamma\,t$, say $p = 1.2\,\gamma\,t$. This yields

$$h = 0.894a, \quad b = 2.5t.$$

We see that the height of the beam is not far short of the height of the barrel vault. In such beams the transverse shear makes an appreciable contribution to the total deflection. The shearing force of the edge beam is $\gamma\,A\,x$, and the deflection due to it is at mid-span

$$w_s = \frac{\varkappa}{G\,A} \cdot \frac{1}{8}\,\gamma\,A\,l^2 = \frac{\varkappa\,\gamma\,l^2}{8\,G}$$

with $\varkappa = 1.2$ for a rectangular cross section. Now we have to make this equal to that part of the deflection of the shell which is proportional to $(l^2 - 4x^2)$, and that is $p\,l^2/2\,E\,t$. The ensuing equation may be brought to the form

$$\frac{p}{\gamma\,t} = \frac{\varkappa}{2},$$

and here it becomes evident that a rectangular section will not do, because it is too rigid. The solution may be found by choosing a cross section with flanges or by using a truss as an edge beam. We shall not discuss this question in detail here, because it would lead us far into the problems of reinforced concrete design, which are beyond the scope of this book. The essential result is that it is feasible to combine a barrel vault with edge beams in such a way that there are no discontinuities in deformation and therefore no need of investigating bending stresses in the shell; but that such a structure would become rather high and that there are not many parameters left which the engineer might choose freely to comply with practical needs outside the field of stress analysis. This is the reason that the bending theory of the barrel vault is of such great practical importance.

3.2.2.2 Barrel Vaults of Other than Circular Profile

Eqs. (21) and (22) may be applied to any cylindrical shell. In some cases the repeated differentiations which they require may be rather tedious. We shall therefore give only one sample, the cycloidal cylinder under a uniform dead load p. The stress resultants for this case were given on p. 125.

From $N_{x\phi}$ we obtain $F(\phi)$ by dropping the minus sign and the factor x. When we introduce N_ϕ and F into eqs. (22), we get

$$E\,t\,u = \frac{3\,p\,x}{8\,a}\left(\frac{x^2}{3} - \frac{l^2}{4}\right) + 4\,\nu\,p\,a\,x\cos^2\phi,$$

$$E\,t\,v = \frac{3+2\nu}{4}\,p\,(l^2 - 4\,x^2)\sin\phi,$$

$$E\,t\,w = -\frac{6+\nu}{8}\,p\,(l^2 - 4\,x^2)\cos\phi - 16\,p\,a^2\cos^3\phi.$$

Since N_x is uniformly distributed over the profile, this is essentially a shear deformation modified by the influence of the hoop force N_ϕ. Therefore it is hopeless to try here what we just found to be at the limits of technical possibilities in the case of the circular shell. A cycloidal cylinder may never be combined with a high edge beam without a severe discontinuity of membrane stresses.

3.2.2.3 Fourier Series Solutions for the Circular Cylinder

Just as we did on p. 120 for the stress problem of the circular cylinder, we may also write solutions of the deformation problem as Fourier series.

For a circular cylinder of any length which carries only edge loads, the membrane forces are given by eqs. (9) and (10). We now put

$$u = \sum_0^\infty u_m\cos m\phi, \qquad v = \sum_1^\infty v_m\sin m\phi, \qquad w = \sum_0^\infty w_m\cos m\phi \qquad (25)$$

and find from eqs. (20)

$$E\,t\,u_m = A_m\frac{m\,x^2}{2\,a} + B_m\,x + C_m,$$

$$E\,t\,v_m = A_m\left[\frac{m^2\,x^3}{6\,a^2} - 2(1+\nu)\,x\right] + B_m\frac{m\,x^2}{2\,a} + C_m\frac{m\,x}{a} + D_m,$$

$$E\,t\,w_m = -A_m\left[\frac{m^3\,x^3}{6\,a^2} - (2+\nu)\,m\,x\right] - B_m\left[\frac{m^2\,x^2}{2\,a} + \nu\,a\right] \qquad (26)$$

$$- C_m\frac{m^2\,x}{a} - D_m\,m.$$

The constants A_m, B_m are the same as in eqs. (10) and describe a particular solution of eqs. (19) corresponding to membrane forces which are caused by the edge loads. The constants C_m and D_m represent solutions of the homogeneous equations (19) and describe an inextensional deformation of the shell.

If the edge loads are given, the constants A_m and B_m are already known, and two conditions for the edge displacements are needed to determine C_m and D_m for every m. It is also possible not to give any edge loads at all and to prescribe four conditions for edge displacements, two at each edge. Then all constants are determined by the edge constraints, and the forces which produce the deformation may be found by introducing A_m and B_m in eqs. (10). One may, of course, also consider the intermediate case that one condition refers to an edge load and three to displacements, but it is not possible to have more than two conditions for the forces.

For a cylinder of length l we may write the displacements as a Fourier series in x:

$$u = \sum_1^\infty u_n \cos \frac{n\pi x}{l}, \qquad v = \sum_1^\infty v_n \sin \frac{n\pi x}{l}, \qquad w = \sum_1^\infty w_n \sin \frac{n\pi x}{l}.$$

When we introduce this and the corresponding expressions (12) for the stress resultants into eqs. (20), we find

$$E t u_n = -\frac{l}{n\pi}(N_{xn} - \nu N_{\phi n}),$$

$$E t v_n = \frac{2(1+\nu)l}{n\pi} N_{x\phi n} + \frac{l^2}{n^2\pi^2 a}\left(\frac{dN_{xn}}{d\phi} - \nu\frac{dN_{\phi n}}{d\phi}\right), \qquad (27)$$

$$E t w_n = a(N_{\phi n} - \nu N_{xn}) - \frac{2(1+\nu)l}{n\pi}\frac{dN_{x\phi n}}{d\phi}$$
$$- \frac{l^2}{n^2\pi^2 a}\left(\frac{d^2 N_{xn}}{d\phi^2} - \nu\frac{d^2 N_{\phi n}}{d\phi^2}\right).$$

There is no room for the arbitrary functions $f_3(\phi)$ and $f_4(\phi)$ in these formulas, but the series for v satisfies automatically the condition $v = 0$ at $x = 0$ and $x = l$, while u is not restricted in these cross sections of the cylinder. At first glance one might expect that even the condition $w = 0$ is satisfied at the ends. This, however, is not so. The formula for w contains a term with $N_{\phi n}$, and we see in eq. (15a) that N_ϕ is derived from p_{rn} without a factor n^{-1} appearing anywhere. Therefore, the convergence of the three Fourier series for p_{rn}, $N_{\phi n}$, w_n is of the same quality. Since the load usually does not tend toward zero as $x = 0$ or $x = l$ is approached, the corresponding series is non-uniformly convergent, and so are the series for $N_{\phi n}$ and w_n.

This may readily be seen in an example. We introduce into the general equations (27) the expressions (15) for the stress resultants in a cylinder subjected to its own weight. We find the following expressions for the displacements:

$$E t\, u_n = \frac{4\,p\,l}{n^2\,\pi^2} \left(\frac{2\,l^2}{a\,n^2\,\pi^2} - \nu\,a \right) \cos\phi,$$

$$E t\, v_n = \frac{4\,p\,l^2}{n^3\,\pi^3} \left(4 + 3\nu + \frac{2\,l^2}{a^2\,n^2\,\pi^2} \right) \sin\phi, \qquad (28)$$

$$E t\, w_n = - \frac{4\,p}{n\,\pi} \left(a^2 + \frac{(4+\nu)\,l^2}{n^2\,\pi^2} + \frac{2\,l^4}{a^2\,n^4\,\pi^4} \right) \cos\phi.$$

The terms in u_n have at least n^2 in the denominator and those for v_n not less than n^3, but the first term in w_n has only n, just like the formulas for $p_{\phi n}$ and $p_{r n}$ on p. 122.

3.3 Statically Indeterminate Structures

As we saw on p. 111, the membrane forces of pipes and barrel vaults have the same spanwise distribution as those of simple beams, provided that there are not more than two diaphragms to which the loads are transmitted. If a cylindrical shell is supported by three

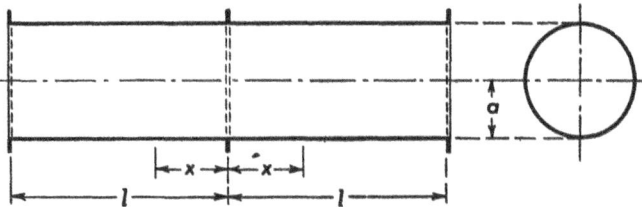

Fig. 20. Cylindrical shell supported by three diaphragms

diaphragms, the stress problem is statically indeterminate. The theory of deformations presented in the preceding Section furnishes the means to solve it if bending stresses are absent in the shell. This is the case for all pipes and those barrel vaults which have a well adapted edge beam.

As an example of this kind we consider a pipe of circular profile, having two equal spans between three diaphragms (fig. 20). We consider the dead load described by eqs. (14). For the hoop force we may use eqs. (16):

$$N_\phi = -p\,a\,\cos\phi,$$

but for $N_{x\phi}$ and N_x we have to go back to the set (3), where the functions f_1 and f_2 have not yet been determined. In our special case we find from them

$$N_{x\phi} = -2p\,x\,\sin\phi + f_1,$$

$$N_x = \frac{p\,x^2}{a}\cos\phi - \frac{x}{a}\frac{df_1}{d\phi} + f_2.$$

Introducing N_ϕ into eq. (21), we find for the displacements the expressions

$$E\,t\,u = \frac{p\,x^3}{3a}\cos\phi + \nu\,p\,a\,x\cos\phi - \frac{df_1}{d\phi}\frac{x^2}{2a} + f_2\,x + f_3,$$

$$E\,t\,v = \frac{p\,x^4}{12a^2}\sin\phi - (4 + 3\nu)\frac{p\,x^2}{2}\sin\phi + \frac{d^2f_1}{d\phi^2}\frac{x^3}{6a^2} - \frac{df_2}{d\phi}\frac{x^2}{2a}$$

$$+ \left[2(1+\nu)\,f_1 - \frac{1}{a}\frac{df_3}{d\phi}\right]x + f_4.$$

Now we have to find four boundary conditions from which to find f_1, f_2, f_3, f_4. Symmetry with respect to the plane $x = 0$ of the middle diaphragm demands that there be $u \equiv 0$. At the other end of each bay, at $x = l$, we must, of course, have $N_x \equiv 0$. The third and fourth conditions follow from the fact that the shell is connected to the diaphragms at $x = 0$ and $x = l$. To a k that both displacements v and w be zero at both diaphragms would evidently be too much. As previously, we must be content with making $v = 0$ at $x = 0$ and at $x = l$.

The two conditions for the end $x = 0$ yield imme liately $f_3 \equiv f_4 \equiv 0$. From the other two we get two linear equations for f_1 and f_2, which may easily be solved. Thus we get finally:

$$N_{x\phi} = -p\sin\phi\left[2x - \frac{l}{4}\frac{5l^2 + 6(4+3\nu)\,a^2}{l^2 + 6(1+\nu)\,a^2}\right],$$

$$N_x = \frac{p}{a}\cos\phi\left[x^2 - \frac{l}{4}\frac{5l^2 + 6(4+3\nu)\,a^2}{l^2 + 6(1+\nu)\,a^2}x + \frac{l^2}{4}\frac{l^2 - 6\nu\,a^2}{l^2 + 6(1+\nu)\,a^2}\right].$$

We see at a glance that the bracketed expressions depend on POISSON's ratio ν and on the ratio l/a. They cannot, therefore, be expected to represent the spanwise distribution of shear and bending moment of the beam analogue. The difference is caused by the shear deformation which is neglected in the common beam formulas but is of importance in a pipe whose diameter is not small compared with its span. In our formulas the shear deformation enters automatically through eq. (19b).

To get a numerical idea of the effect of shear, we may look at the maximum of N_x at the middle support. When we put $x = 0$ and if we assume furthermore $\nu = 0$, we have

$$N_x = \frac{p\,l^2}{4a}\frac{\cos\phi}{1 + 6a^2/l^2}.$$

The beam analogue would yield instead

$$N_x = \frac{p\,l^2}{4a}\cos\phi,$$

and the factor $(1 + 6a^2/l^2)^{-1}$ represents the influence of the shear deformation on the stress distribution in the shell. In fig. 21 this factor has been plotted as a function of l/a, and one may recognize that for $l/a = 5$ the beam formula is still 20% in error but that for $l/a = 10$ the deviation is already rather small. This result may be used as a starting point for a simplified analysis of long cylindrical shells.

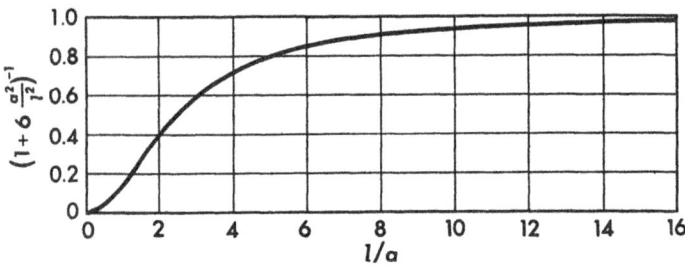

Fig. 21
Influence of the shear deformation on the stress distribution in a statically indeterminate shell

3.4 Polygonal Domes

The cylindrical shells treated thus far have all been limited by two cross-sectional planes. Besides these tubes and barrel vaults, there exists a type of structure in which cylindrical shells are combined to form a polygonal dome. Fig. 22 shows an example. The sectors are cylinders having horizontal generators. Their lines of intersection lie in vertical planes, and we shall see that along these lines the dome must be reinforced by ribs, called hips. Along the polygonal springing line we need a ring which shows features of both the circular foot ring of domes and the edge member of barrel vaults. If the dome is open at the apex, as shown in fig. 22, another polygonal ring must be provided along the upper edge.

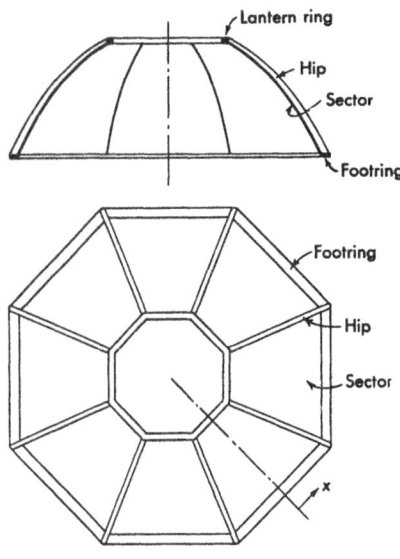

Fig. 22. Regular polygonal dome

Polygonal domes have been built at several places; but beyond their occurrence in practical engineering, their theory is of interest because it illustrates in a striking example the role of ribs provided along edges of a shell.

3.4.1 Regular Dome under Regular Load

We shall call a polygonal dome regular if every horizontal section through its middle surface is a regular polygon. Then all segments are equal and meet at equal angles. We shall call the load regular when it has the same symmetry as the structure (e. g. dead load). The stress pattern will then be the same in all sectors, and we need investigate only one of them.

Since every sector is a cylindrical shell, we use the same notations as before, measuring the coordinate x from the center line of the sector (fig. 22).

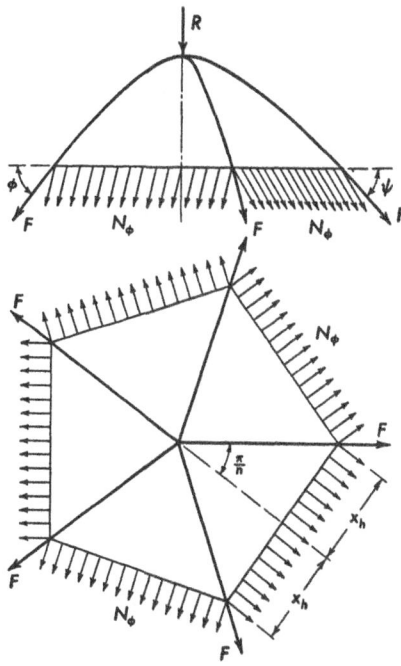

Fig. 23. Top part of a polygonal dome

The stress resultants N_ϕ, $N_{x\phi}$, N_x are given by. eqs. (1a), (3). From the symmetry of structure and load with respect to the vertical plane $x = 0$ we conclude that along this line the shearing force must be zero and hence $f_1(\phi) \equiv 0$, so that we have

$$N_\phi = p_r\, r\,,$$

$$N_{x\phi} = -\left(p_\phi + \frac{1}{r}\, \frac{\partial N_\phi}{\partial \phi}\right) x.$$

(29 a, b)

The function $f_2(\phi)$ is closely connected with the force F in the hips. To find it, we cut the dome along a horizontal plane $\phi = $ const and consider the equilibrium of the cap situated above this plane (fig. 23). The resultant load applied to it is a vertical force $R = R(\phi)$, acting along the axis of the dome. The shearing forces $N_{x\phi}$ transmitted in the section have no resultant and are not shown in fig. 23, and the forces N_ϕ depend only on the local load p_r and, therefore, cannot be expected to be in equilibrium with the resultant load R. The difference must be carried by the hips, and since the forces in all hips are equal, we might find their magnitude if we knew their direction.

The idea suggests itself that it might be possible to establish equilibrium among the internal forces without making use of the bending stiffness of the hips. Then there will be only an axial force F, acting along the axis of the hip. The angle ψ between F and the horizontal plane is a function of ϕ, determined by the relation

$$\tan\psi = \cos\frac{\pi}{n}\tan\phi, \tag{30}$$

where n is the number of sectors or hips of the dome.

The forces F acting at the base of the cap shown in fig. 23 have the vertical resultant $n\,F\sin\psi$, and the force N_ϕ transmitted in n genera-

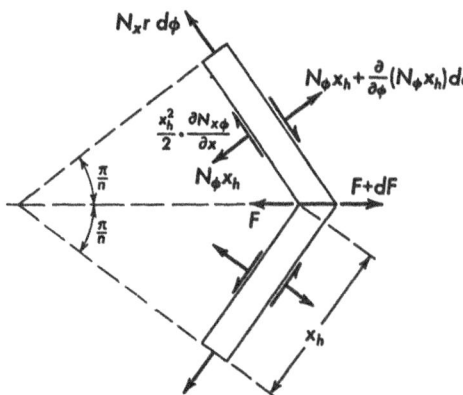

tors each of length $2\,x_h$ contributes $2n\,x_h\,N_\phi\sin\phi$. The equation of equilibrium of vertical forces is therefore

$$n\,(F\sin\psi + 2\,x_h\,N_\phi\sin\phi)$$
$$+\,R = 0$$

and yields the hip force

$$F = -\frac{R + 2n\,x_h\,N_\phi\sin\phi}{n\sin\psi}. \tag{31}$$

Fig. 24. Hip element and adjacent shell strips

We may now find N_x for $x = 0$ by considering the element shown in fig. 24. It is limited by two horizontal planes ϕ and $\phi + d\phi$ and by elements $r\,d\phi$ of the center lines $x = 0$ of two adjacent shell sectors. The forces are indicated in the figure with their true values. When we project them on a horizontal plane and exclude the presence of horizontal loads, we find the following condition of equilibrium:

$$2\,N_x\,r\,d\phi\sin\frac{\pi}{n} = \frac{d}{d\phi}\,(F\cos\psi)\,d\phi + 2\frac{\partial}{\partial\phi}\,(N_\phi\,x_h\cos\phi)\,d\phi\cos\frac{\pi}{n}$$

$$+\,\frac{\partial}{\partial\phi}\left(x_h^2\frac{\partial N_{x\phi}}{\partial x}\right)d\phi\sin\frac{\pi}{n}.$$

Since N_x at $x = 0$ is identical with $f_2(\phi)$, this equation yields

$$f_2(\phi) = \frac{1}{2r\sin\pi/n}\frac{d}{d\phi}\,(F\cos\psi) + \frac{\cot\pi/n}{r}\frac{\partial}{\partial\phi}\,(N_\phi\,x_h\cos\phi)$$

$$+\,\frac{1}{2r}\cdot\frac{\partial}{\partial\phi}\left(x_h^2\frac{\partial N_{x\phi}}{\partial x}\right).$$

Making use of eqs. (30) and (31) and of the geometric relation

$$\frac{dx_h}{d\phi} = r\cos\phi\tan\frac{\pi}{n},$$

one may bring f_2 into the following form which is more convenient for numerical work:

$$f_2(\phi) = -\frac{1}{r}\tan\frac{\pi}{n}\frac{\partial}{\partial\phi}(x_h N_\phi \cos\phi) + x_h\frac{\partial N_{x\phi}}{\partial x}\tan\frac{\pi}{n}\cos\phi$$
$$+ \frac{x_h^2}{2r}\frac{\partial^2 N_{x\phi}}{\partial\phi\,\partial x} - \frac{1}{n\,r\sin(2\pi/n)}\frac{d}{d\phi}(R\cot\phi). \tag{32}$$

When we have f_2, the force N_x at an arbitrary point is given by eq. (3b):

$$N_x = \frac{1}{2r}\frac{d}{d\phi}\left(p_\phi + \frac{1}{r}\frac{dN_\phi}{d\phi}\right)x^2 + f_2(\phi). \tag{33}$$

Interpreting this result, we observe that, along each generator, N_x varies exactly in the same way as if the sector were part of a barrel vault. But the distribution in the ϕ direction is thoroughly different. It is dominated by $f_2(\phi)$. In the barrel vault, this function was determined so as to make N_x vanish at the ends of the span l. In the polygonal dome we determined it so that N_x acts as a kind of hoop force,

keeping the hip forces F centered in the axes of the hips. However, it is not possible to declare the two terms of eq. (33) as due to a "barrel vault action" and a "dome action", since neither of them makes sense without the other. We only may understand the complete stress system as serving two purposes, establishing equilibrium locally in each sector and uniting all sectors in a dome.

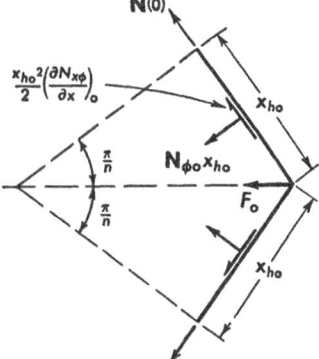

Fig. 25. Part of foot ring of a polygonal dome

At the springing line $\phi = \phi_0$, the forces N_ϕ and $N_{x\phi}$ of the shell and the hip forces F must be transmitted to supports. The horizontal components may be gathered in a foot ring. Because of the shear $N_{x\phi}$, its normal force N will depend on x. For $x = 0$, N may be found from fig. 25 as we found N_x from fig. 24:

$$N(0) = -\frac{F_0\cos\psi_0}{2\sin\pi/n} - N_{\phi 0}\frac{x_{h0}\cos\phi_0}{\tan\pi/n} - \left(\frac{\partial N_{x\phi}}{\partial x}\right)_0\frac{x_{h0}^2}{2}.$$

Here the subscript 0 indicates that the values of all variables have to be taken for $\phi = \phi_0$.

For $x \neq 0$, the term with $N_{x\phi}$ is different, and we have

$$N(x) = -\frac{F_0\cos\psi_0}{2\sin\pi/n} - N_{\phi 0}\frac{x_{h0}\cos\phi_0}{\tan\pi/n} - \left(\frac{\partial N_{x\phi}}{\partial x}\right)_0\frac{x_{h0}^2 - x^2}{2}. \tag{34}$$

The horizontal thrust $N_{\phi 0}\cos\phi_0$ on the foot ring produces bending moments in its plane. In the general case they may be computed from this load by well-known methods of structural theory. If the ring has a constant cross section, the maximum occurs at the corners and is

$$M_{max} = \frac{1}{3} x_{h0}^2 N_{\phi0} \cos\phi_0.$$

If the shell ends with $\phi_0 = 90°$, there will be no thrust and hence no horizontal bending moments in the foot ring.

The vertical load on the ring must be carried to supports which may be arranged arbitrarily. If $\phi_0 = 90°$ and if the local load near the springing line has no horizontal component, we have $N_{\phi0} = 0$, and the dome transmits its total load to the n corner points. In this case it may be supported by n columns which are the prolongations of the hips, and the foot ring has nothing to carry in bending but its own weight.

When the dome has an opening at the top, as shown in fig. 22, a polygonal ring must also be provided at the upper edge. The forces in this ring will be found in the following way: Eq. (29a) yields the thrust N_ϕ of the shell. We resolve it into a vertical and a horizontal component. The former combines with the weight of the ring to form its vertical load which is transmitted by bending and torsion to the corners. There it is in equilibrium with the vertical component of the hip force F, which is determined by this condition. The ensuing horizontal component of F and that of N_ϕ constitute a plane force system producing axial forces and plane bending moments in the ring. They may be handled in the same way as has just been explained for the foot ring.

The stress resultants which we have determined here fulfill the conditions of equilibrium for any possible element which may be cut out of the structure. For the shell sectors and the hips they constitute a kind of generalized membrane force system. Since there is no bending in the hips, there is scarcely a reason for giving them more bending stiffness than that which is connected anyway with a cross section capable of resisting the force F. From the same reasons which justify the application of membrane theory in other cases, we may then expect that also in polygonal domes the internal forces come close to a membrane force system. However, there may be a discrepancy between the deformations of the hip and the shell which leads to bending stresses in both. Little is known at present about them, but it may be assumed that they are of a local character.

It cannot come as a surprise that the polygonal rings at the edges of the shell structure are not free of bending. We saw on p. 58 that the free edge of a shell of revolution needs such a ring when it is to be a stiff structure capable of resisting arbitrary loads.

As an example of the theory presented on the preceding pages, we consider the polygonal dome shown in fig. 26. Its sectors are parts of circular cylinders of radius a. For their number n we shall not fix a definite value. We ask for the stress resultants due to dead load, assuming that both the shell and the hips are of uniform thickness.

It is useful to treat the weight of the sectors and that of the hips separately. If p is the weight per unit area of the middle surface, the components are

$$p_\phi = p \sin\phi,$$

$$p_r = -p \cos\phi.$$

Introducing these into eqs. (29a, b) we find

$$N_\phi = -p\, a \cos\phi,$$

$$N_{x\phi} = -2p\, x \sin\phi,$$

exactly as on p. 122. To find f_2, we need the resultant

$$R = 2n\, p\, a \int_0^\phi x_h\, d\phi.$$

With $x_h = a \sin\phi \tan\pi/n$ we have

$$R = 2n\, p\, a^2 \tan\frac{\pi}{n} \cdot (1 - \cos\phi)$$

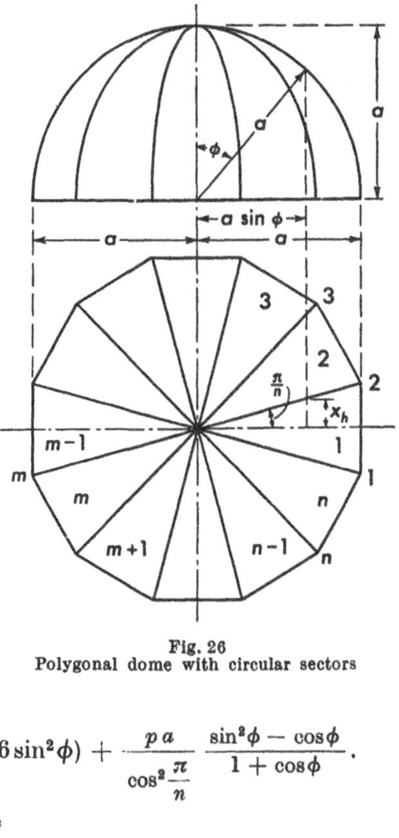

and then from eq. (32):

Fig. 26
Polygonal dome with circular sectors

$$f_2(\phi) = p\, a \tan^2\frac{\pi}{n} \cos\phi \cdot (1 - 6\sin^2\phi) + \frac{p\, a}{\cos^2\dfrac{\pi}{n}} \, \frac{\sin^2\phi - \cos\phi}{1 + \cos\phi}.$$

Eq. (33) now yields the hoop force

$$N_x = p\, a \left[\frac{x^2}{a^2} \cos\phi + \tan^2\frac{\pi}{n} \cos\phi \cdot (1 - 6\sin^2\phi) + \frac{1}{\cos^2\dfrac{\pi}{n}} \, \frac{\sin^2\phi - \cos\phi}{1 + \cos\phi} \right]$$

and eq. (31) the hip force

$$F = -2p\, a^2 \frac{\tan\pi/n}{\sin\psi} (1 - \cos\phi)(\sin^2\phi - \cos\phi).$$

The results are plotted in fig. 27a: the hoop force at $x = 0$, the shear for $x = x_h$, the force N_ϕ, which is independent of x, and the hip force F. The last diagram shows the resultant load R and how it is carried by the resultant $2n\, x_h\, N_\phi \sin\phi$ of the forces in the shell sectors and the resultant $n\, F \sin\psi$ of the hip forces.

The force N_x is very similar to the hoop force N_θ in a hemispherical dome (see eq. (II–13)), with compression in the upper and tension in the lower part of the shell. At the springing line $N_\phi = 0$ and the total load R is carried by the six hips, but toward the apex N_ϕ is so large that the shell carries more load than there is. Therefore the hip force F is positive there.

When the number of sectors, n, is increased, the first two terms of N_x tend toward zero, the first one because $x \leq x_h \to 0$ and the

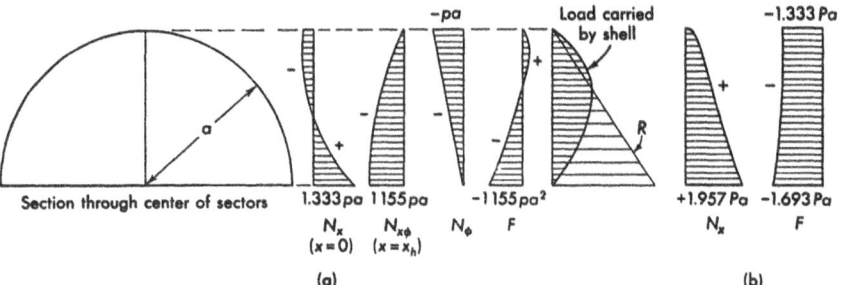

Fig. 27. Stress resultants in a polygonal dome with circular sectors, $n = 6$

second one because of the explicit factor $\tan^2 \pi/n$. In the third term, $\cos \pi/n \to 1$, and we have

$$\lim_{n \to \infty} N_x = p\, a\, \frac{\sin^2\phi - \cos\phi}{1 + \cos\phi}.$$

This is identical with the formula (II–13) for the hoop force N_θ of a spherical shell dome.

The shear $N_{x\phi}$ becomes insignificant with increasing n because of the factor $x \leq x_h$, also in agreement with the stresses in a sphere, but N_ϕ is definitely different. If we want to compare meridional forces, we must average N_ϕ and F. This average is

$$-\frac{R}{2n\, x_h \sin\phi} = -p\, a\, \frac{1 - \cos\phi}{\sin^2\phi},$$

and this really equals N_ϕ as given for the sphere by eqs. (II–13).

For a complete solution of the dead load problem we must still consider the weight of the hips. For this simple example we may assume that they have a constant cross section and therefore a constant weight per unit length, P.

Since the cylindrical sectors now have no load, we have $N_\phi \equiv N_{x\phi} \equiv 0$ and $N_x \equiv f_2(\phi)$. To find f_2 and F, we need the load resultant R. The element of the hip situated between the generators ϕ and $\phi + d\phi$ of the shell sectors has the same vertical projection $a\, d\phi \sin\phi$ as the corresponding meridional element of the cylinders, but the horizontal

projection
$$a\,d\phi\,\frac{\cos\phi}{\cos\pi/n}\,.$$

Its length is therefore

$$ds=\sqrt{\frac{\cos^2\phi}{\cos^2\pi/n}+\sin^2\phi}\;a\,d\phi.$$

Integrating this from 0 to ϕ and multiplying the integral by $n\,P$, we obtain the resultant load

$$R=n\,P\int\limits_0^\phi ds=\frac{n\,P\,a}{\cos\pi/n}\int\limits_0^\phi\sqrt{1-\sin^2\frac{\pi}{n}\sin^2\phi}\;d\phi.$$

The integral in this formula may not be expressed in terms of elementary functions but itself represents a special transcendental function, the elliptic integral of the second kind:

$$E(\alpha,\phi)\equiv\int\limits_0^\phi\sqrt{1-\sin^2\alpha\sin^2\beta}\;d\beta.$$

Its numerical values have been tabulated[1] as functions of the upper limit ϕ and the parameter α. We may therefore write as the final expression

$$R=\frac{n\,P\,a}{\cos\pi/n}\,E\left(\frac{\pi}{n},\phi\right).$$

From eq. (31) we now find

$$F=-\frac{R}{n\sin\psi}=-\frac{P\,a}{\cos\pi/n\,\sin\psi}\,E\left(\frac{\pi}{n},\phi\right)$$

and from (32) and (33):

$$N_x=-\frac{1}{n\,a\sin(2\pi/n)}\frac{d}{d\phi}(R\cot\phi)$$

$$=-\frac{P}{2\cos^2\dfrac{\pi}{n}\sin\dfrac{\pi}{n}}\left(\cot\phi\cdot\sqrt{1-\sin^2\frac{\pi}{n}\sin^2\phi}-\frac{E(\pi/n,\phi)}{\sin^2\phi}\right).$$

These results are plotted in fig. 27 b. The hip force is, of course, a compression throughout and is fairly constant while N_x is positive. When these forces are superposed on those resulting from the weight of the cylindrical sectors, the positive hip force of fig. 27 a will disappear.

The formulas developed on the preceding pages and illustrated here by a simple example are immediately applicable to every dome of arbitrary meridian which is erected over a regular polygon as a basis.

---

[1] For four-place values see: FLüGGE, W.: Four-Place Tables of Transcendental Functions. London 1953; for more digits see the table by A.M.LEGENDRE and its recent reprints.

But the underlying ideas may be used for a more general type of polygonal domes, if we make the necessary changes in some details of the formulas. When we determined $f_1(\phi)$, we had to assume only that

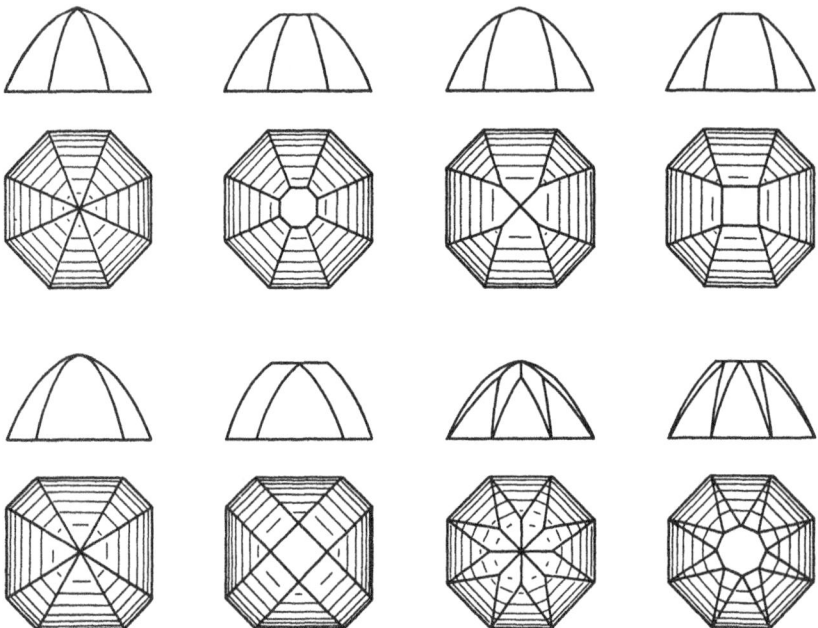

Fig. 28. Polygonal domes to which the theory for regular domes may be applied

there exists in every sector a plane $x = 0$ which is a plane of symmetry for the dome; and when we derived eq. (31) for the hip force, the essential assumption was such a degree of symmetry that we are sure that all hip forces are equal. These two conditions are fulfilled by all the domes shown in fig. 28. It appears that the sectors need not all be equal, but that two different types may alternate. Even a less regular looking structure, the vaulted hip roof shown in fig. 29, fulfills the same conditions and may be treated by the same method. Here the ridge beam does not belong to the hip system but is a degenerated polygonal ring and therefore not free from bending. Of course, one may try to give the structure such dimensions that the weight of this beam is just in equilibrium with the thrust N_ϕ transmitted to it from the two longer shell sectors.

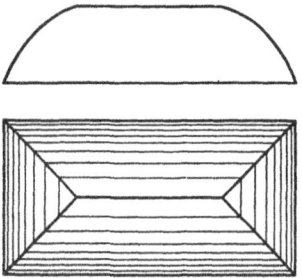

Fig. 29. Vaulted hip roof

3.4.2 Regular Dome under Arbitrary Load

Polygonal domes are particularly suited to large-span roofs. In such structures the wind load is important, and it becomes necessary to consider the general stress problem of a regular dome under arbitrary load.

We must now distinguish between the forces in different hips or in different shell sectors. For this purpose we number the hips and the sectors from 1 to n as shown in fig. 26 and indicate forces referring to the m-th hip or sector by the superscript (m).

In order to keep the formulas from becoming unwieldy, we shall not only assume that the dome is regular but also restrict the discussion to the case that $p_x \equiv 0$ and that p_ϕ and p_r are independent of x within each individual sector. Their variation from sector to sector may be described by specifying the loads in the m-th sector, $p_\phi^{(m)}$, $p_r^{(m)}$, as functions of ϕ for every m from 1 to n. Equivalent to this, but better adapted to our purposes, is the description by FOURIER sums:

$$
\begin{aligned}
p_\phi^{(m)}(\phi) &= \sum_{k=0}^{<n/2} p_{\phi k}(\phi) \cos\frac{2\pi k m}{n} + \sum_{k=1}^{\leqq n/2} \bar{p}_{\phi k}(\phi) \sin\frac{2\pi k m}{n}, \\
p_r^{(m)}(\phi) &= \sum_{k=0}^{<n/2} p_{r k}(\phi) \cos\frac{2\pi k m}{n} + \sum_{k=1}^{\leqq n/2} \bar{p}_{r k}(\phi) \sin\frac{2\pi k m}{n}.
\end{aligned}
\tag{35}
$$

The notations $< n/2$ and $\leqq n/2$ for the upper limits of the sums indicate that, for n odd, all sums must be extended to $k = (n-1)/2$, and, for n even, the cosine sums to $n/2 - 1$ and the sine sums to $n/2$. The two sums together always have n terms.

In order to find the FOURIER coefficient $p_{\phi k}(\phi)$ (the k-th harmonic) of the load from given sector loads $p_\phi^{(m)}(\phi)$, we write in the first eq. (35) $\varkappa$ for k, then multiply both sides by $\cos\dfrac{2\pi k m}{n}$ and make a summation over m:

$$
\begin{aligned}
\sum_{m=1}^{n} p_\phi^{(m)}(\phi) \cos\frac{2\pi k m}{n} = &\sum_{\varkappa=0}^{<n/2} p_{\phi\varkappa}(\phi) \sum_{m=1}^{n} \cos\frac{2\pi\varkappa m}{n} \cos\frac{2\pi k m}{n} \\
&+ \sum_{\varkappa=1}^{\leqq n/2} \bar{p}_{\phi\varkappa}(\phi) \sum_{m=1}^{n} \sin\frac{2\pi\varkappa m}{n} \cos\frac{2\pi k m}{n}.
\end{aligned}
$$

All but one of the sums of cos products and all the sin-cos products are zero. Only for $\varkappa = k$ we have

$$
\sum_{m=1}^{n} \cos^2\frac{2\pi k m}{n} = \frac{n}{2}, \qquad k \neq 0,\ n/2
$$

and hence

$$
\sum_{m=1}^{n} p_\phi^{(m)}(\phi) \cos\frac{2\pi k m}{n} = \frac{n}{2}\, p_{\phi k}(\phi), \qquad k \neq 0,\ n/2.
$$

For $k = 0$ (or for $k = n/2$) we must write n instead of $n/2$ in the last two formulas. However, the harmonic of order 0 is identical with the regular load treated in Section 3.4.1. We shall therefore drop it here and assume that all sums in eq. (35) begin with $k = 1$. The formulas to be developed here are then not all-comprehensive, but together with those for regular loads they cover the whole field.

Formulas for $\bar{p}_{\phi k}$, p_{rk}, $\bar{p}_{rk}$ may be found in a similar way. The coefficient functions $p_{\phi k}$ and p_{rk} represent a load which is symmetric with respect to the plane of symmetry of the sector n, whereas $\bar{p}_{\phi k}$ and $\bar{p}_{rk}$ represent the antimetric part of the load.

We now consider a load

$$p_\phi^{(m)} = p_{\phi k} \cos \frac{2\pi k m}{n}, \qquad p_r^{(m)} = p_{rk} \cos \frac{2\pi k m}{n},$$

where $p_{\phi k}$ and p_{rk} are arbitrary functions of ϕ. The stress resultants in the sectors follow from eqs. (1a) and (3):

$$N_\phi^{(m)} = r \, p_{rk} \cos \frac{2\pi k m}{n} = N_{\phi k} \cos \frac{2\pi k m}{n},$$

$$N_{x\phi}^{(m)} = -\left(p_{\phi k} + \frac{1}{r} \frac{dN_{\phi k}}{d\phi}\right) x \cos \frac{2\pi k m}{n} + f_1^{(m)}(\phi), \tag{36}$$

$$N_x^{(m)} = \frac{1}{r} \frac{d}{d\phi}\left(p_{\phi k} + \frac{1}{r} \frac{dN_{\phi k}}{d\phi}\right) \frac{x^2}{2} \cos \frac{2\pi k m}{n} - \frac{x}{r} \frac{df_1^{(m)}}{d\phi} + f_2^{(m)}(\phi).$$

It may be expected that the functions $f_1^{(m)}$ and $f_2^{(m)}$ will also depend on m in a simple way, and we shall see soon that they must be assumed in the form

$$f_1^{(m)}(\phi) = f_{1k}(\phi) \sin \frac{2\pi k m}{n}, \qquad f_2^{(m)}(\phi) = f_{2k}(\phi) \cos \frac{2\pi k m}{n}. \tag{37}$$

In order to simplify the appearance of the formulas on the next pages, we introduce two abbreviations:

$$G_k = -\left(p_{\phi k} + \frac{1}{r} \frac{dN_{\phi k}}{d\phi}\right), \qquad H_k = -\frac{1}{r} \frac{dG_k}{d\phi}.$$

They represent two functions of ϕ depending on the load and on the parameter k. If we wished, we might compute them numerically at the present stage of the development of the theory, before we have started to determine the free functions f_{1k} and f_{2k}.

The determination of these two functions is the principal problem we have to solve. We combine it with that of finding the forces in the hips. From our experience with regular loads we may hope that also in the general case the hips will be free of bending moments and shearing forces. If we admit this, we can formulate three differential equations with f_{1k}, f_{2k} and the hip force as unknowns. If, for a given shell, these equations have a unique solution, this is proof that a stress system of the assumed kind is possible. For the reasons already explained on p. 144, this stress system will then be a fair representation of what really happens in the shell structure.

Fig. 30 shows an element of the hip m (situated between the sectors $m-1$ and m) and two adjacent shell elements. If the hip is to be free from bending, a first condition is that the sum of all forces perpendicular to the plane of the hip must vanish:

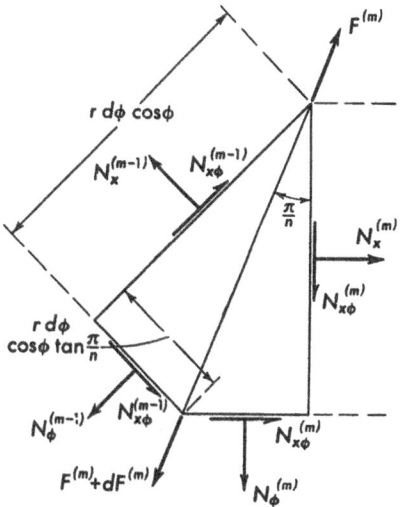

$$(N_\phi^{(m)} - N_\phi^{(m-1)}) \, r \, d\phi \, \cos\phi \tan\frac{\pi}{n}$$
$$\times \cos\phi \sin\frac{\pi}{n}$$
$$+ (N_x^{(m)} - N_x^{(m-1)}) \cdot r \, d\phi \cdot \cos\frac{\pi}{n}$$
$$+ (N_{x\phi}^{(m)} + N_{x\phi}^{(m-1)})$$
$$\times \left(r \, d\phi \cos\phi \tan\frac{\pi}{n} \cos\frac{\pi}{n} \right.$$
$$\left. + r \, d\phi \cdot \cos\phi \sin\frac{\pi}{n} \right) = 0.$$

Fig. 30
Hip element and adjacent shell elements

When we introduce the expressions (36) here, we have to put $x = +x_h$ in the sector $m-1$ and $x = -x_h$ in the sector m.

When we also introduce (37), we may drop a common factor $2 \sin(\pi k(2m-1)/n)$ and arrive at an equation which is independent of m and hence valid for every hip of the dome:

$$\frac{df_{1k}}{d\phi} \frac{x_h}{r} \cot\frac{\pi k}{n} \cot\frac{\pi}{n} + 2f_{1k} \cos\phi \cot\frac{\pi k}{n} - f_{2k} \cot\frac{\pi}{n} =$$
$$= N_{\phi k} \cos^2\phi \tan\frac{\pi}{n} - 2G_k x_h \cos\phi + \frac{1}{2} H_k x_h^2 \cot\frac{\pi}{n}. \tag{38a}$$

This is the first of the three differential equations which we have to find. The other two express the fact that the hip element is in equilibrium in its plane with an axial force $F^{(m)}$ only and without the help of a shearing force. This fact may be cast in equations by simply writing the conditions of equilibrium for vertical and horizontal components in the plane of the hip:

The vertical forces are

$$(N_\phi^{(m)} + N_\phi^{(m-1)}) \cdot r \, d\phi \cos\phi \tan\frac{\pi}{n} \cdot \sin\phi$$
$$+ (N_{x\phi}^{(m)} - N_{x\phi}^{(m-1)}) \cdot r \, d\phi \sin\phi + \frac{d}{d\phi} (F^{(m)} \sin\psi) \, d\phi = 0,$$

and the horizontal forces are:

$$\left(N_\phi^{(m)} + N_\phi^{(m-1)}\right) \cdot r\,d\phi\,\cos\phi\,\tan\frac{\pi}{n} \cdot \cos\phi\,\cos\frac{\pi}{n}$$

$$- \left(N_x^{(m)} + N_x^{(m-1)}\right) r\,d\phi \cdot \sin\frac{\pi}{n}$$

$$+ \left(N_{x\phi}^{(m)} - N_{x\phi}^{(m-1)}\right) \left(r\,d\phi \cdot \cos\phi\cos\frac{\pi}{n} - r\,d\phi\,\cos\phi\,\tan\frac{\pi}{n} \cdot \sin\frac{\pi}{n}\right)$$

$$+ \frac{d}{d\phi}\left(F^{(m)}\cos\psi\right) d\phi = 0.$$

When we introduce the expressions (36) and (37) here, we find it necessary to put

$$F^{(m)} = F_k \cos\frac{\pi\,k(2m-1)}{n}. \tag{39}$$

It is again possible to drop the factor depending on m, and we obtain the following equations:

$$f_{1k}\sin\phi\,\sin\frac{\pi\,k}{n} + \frac{1}{2r}\frac{d}{d\phi}\left(F_k\sin\psi\right)$$

$$= -N_{\phi k}\cos\phi\,\sin\phi\,\cos\frac{\pi\,k}{n}\tan\frac{\pi}{n} + G_k\,x_h\,\sin\phi\,\cos\frac{\pi\,k}{n}, \tag{38b}$$

$$\frac{df_{1k}}{d\phi}\frac{x_h}{r}\sin\frac{\pi\,k}{n}\sin\frac{\pi}{n} + f_{1k}\cos\phi\,\sin\frac{\pi\,k}{n}\frac{\cos 2\pi/n}{\cos\pi/n}$$

$$+ f_{2k}\cos\frac{\pi\,k}{n}\sin\frac{\pi}{n} - \frac{1}{2r}\frac{d}{d\phi}\left(F_k\cos\psi\right)$$

$$= N_{\phi k}\cos^2\phi\,\cos\frac{\pi\,k}{n}\sin\frac{\pi}{n} - G_k\,x_h\,\cos\phi\,\cos\frac{\pi\,k}{n}\frac{\cos 2\pi/n}{\cos\pi/n} \tag{38c}$$

$$- \frac{1}{2}H_k\,x_h^2\cos\frac{\pi\,k}{n}\sin\frac{\pi}{n}.$$

When we study the left-hand side of eqs. (38a–c), we see that f_{2k} may easily be eliminated from the first and the last one. This yields the following equation:

$$\frac{df_{1k}}{d\phi}\frac{x_h}{r}\sin\frac{2\pi}{n} + f_{1k}\cos\phi\left(\cos^2\frac{\pi\,k}{n} - \cos\frac{2\pi}{n}\right)$$

$$- \frac{1}{r}\frac{d}{d\phi}\left(F_k\cos\psi\right)\sin\frac{\pi\,k}{n}\cos\frac{\pi}{n}$$

$$= N_{\phi k}\cos^2\phi\,\sin\frac{2\pi\,k}{n}\tan\frac{\pi}{n} - G_k\,x_h\,\cos\phi\,\sin\frac{2\pi\,k}{n}.$$

From it we may eliminate f_{1k} with the help of eq. (38b) and thus arrive at a differential equation for $F_k \sin \psi$:

$$\frac{d^2}{d\phi^2} (F_k \sin \psi) + \left[\frac{2r}{x_\lambda} \tan \frac{\pi}{n} \cos \phi - \frac{1}{r \sin \phi} \frac{d}{d\phi} (r \sin \phi) \right] \frac{d}{d\phi} (F_k \sin \psi)$$

$$- \frac{2r}{x_\lambda} \frac{\sin^2 \pi k/n}{\sin (2\pi/n) \sin \phi} (F_k \sin \psi)$$

$$= 2r \cos \frac{\pi k}{n} \tan \frac{\pi}{n} \sin \phi \left[\sin \phi - \frac{2r}{x_\lambda} \tan \frac{\pi}{n} \cos^2 \phi \right] N_{\phi k}$$

$$+ 2r^2 \cos \frac{\pi k}{n} \sin \phi \left[(p_{\phi k} + 4G_k) \tan \frac{\pi}{n} \cos \phi - x_\lambda H_k \right].$$

(40)

When we have solved this equation, we may easily find f_{1k} from eq. (38b) and then f_{2k} from eq. (38a) or (38c). Eqs. (36) then yield the stress resultants in the shell sectors.

Of course, we cannot expect to solve the differential equation (40) in general terms. In most cases, it will be necessary to resort to numerical integration. To get an idea of the kind of solutions to be expected, it will be useful to consider a simple example. We choose the dome shown in fig. 26, and we suppose that there is no distributed load on the sectors or the hips, i. e. we ask for the homogeneous solutions. We have then

$$p_{\phi k} \equiv p_{r k} \equiv G_k \equiv H_k \equiv 0,$$

and the right-hand side of eq. (40) vanishes. With the special data of the dome under consideration, the left-hand side assumes the following simple form:

$$\frac{d^2}{d\phi^2} (F_k \sin \psi) + \cot \phi \frac{d}{d\phi} (F_k \sin \psi) - \frac{\lambda^2}{\sin^2 \phi} (F_k \sin \psi) = 0$$

with

$$\lambda = \frac{\sin \pi k/n}{\sin \pi/n}.$$

This equation has the general solution

$$F_k \sin \psi = A \cot^\lambda \frac{\phi}{2} + B \tan^\lambda \frac{\phi}{2}.$$

(41)

Introducing it into (39), we find the individual hip forces

$$F^{(m)} = \frac{1}{\sin \psi} \left(A \cot^\lambda \frac{\phi}{2} + B \tan^\lambda \frac{\phi}{2} \right) \cos \frac{\pi k(2m-1)}{n}.$$

We may now go backwards through our equations and find from (38b)

$$f_{1k}(\phi) = \frac{1}{2a} \frac{1}{\sin \frac{\pi}{n} \sin^2 \phi} \left(A \cot^\lambda \frac{\phi}{2} - B \tan^\lambda \frac{\phi}{2} \right),$$

and from (38a) or (38c):

$$f_{2k}(\phi) = -\frac{1}{a}\frac{\cos\dfrac{\pi k}{n}}{\sin\dfrac{2\pi}{n}\sin^2\phi}\left(A\cot^\lambda\frac{\phi}{2}+B\tan^\lambda\frac{\phi}{2}\right).$$

The stress resultants of the shell sectors are then given by eqs. (36) and (37):

$$N_\phi^{(m)} \equiv 0,$$

$$N_{x\phi}^{(m)} = \frac{1}{2a\sin\pi/n}\frac{1}{\sin^2\phi}\left(A\cot^\lambda\frac{\phi}{2}-B\tan^\lambda\frac{\phi}{2}\right)\sin\frac{2\pi km}{n},$$

$$N_x^{(m)} = -\frac{1}{2a\sin\pi/n}\frac{1}{\sin^2\phi}\left(A\cot^\lambda\frac{\phi}{2}+B\tan^\lambda\frac{\phi}{2}\right) \qquad (42)$$

$$\cdot\left(\frac{\cos\pi k/n}{\cos\pi/n}\cos\frac{2\pi km}{n}+\frac{\lambda-2\cos\phi}{\sin\phi}\cdot\frac{x}{a}\cdot\sin\frac{2\pi km}{n}\right).$$

The formulas (41) and (42) reveal the following facts: The first term of (41) has a singularity at the top $\phi = 0$ of the dome; the second term is regular everywhere unless we extend the shell to the point $\phi = \pi$. For $k = 1$ (first harmonic) we have $\lambda = 1$, independent of n. In this case the B solution corresponds to a loading of the dome by a horizontal force P, the A solution to the application of such a force and an external couple M_1 as shown in fig. II–28 for a spherical shell. These loads may easily be determined by examining the equilibrium of a cap cut from the dome by an arbitrary plane $\phi = $ const. For the B solution one obtains in this way the load

$$P = B\frac{1}{2\cos\pi/n}.$$

For the higher harmonics, $k = 2, 3, \ldots n/2$, the forces $F^{(m)}$ at the top are in equilibrium with each other, and so are the forces $N_{x\phi}^{(m)}$ in a horizontal section through the shell. Then no external force or couple is required at this point.

So far the situation is analogous to that which we found on p. 54 for a spherical shell, but important differences appear when we look at the forces $N_{x\phi}^{(m)}$ and $N_x^{(m)}$ in the cylindrical sectors. Because of the factor $\sin^2\phi$ in the denominator, these forces become infinite at the top even in the B solution, unless $\lambda \geq 2$. For a square dome ($n = 4$), λ is never as great as 2. For a hexagonal dome ($n = 6$) only the highest harmonic $k = 3$ yields λ as large as 2. For domes with more than six edges, there are always some harmonics with finite forces and some with infinite forces. In the case of an octagonal dome ($n = 8$), for example, the harmonics $k = 1$ and $k = 2$ have infinite forces, while $k = 3$ ($\lambda = 2.613$) and $k = 4$ ($\lambda = 2.414$) yield finite values.

For the A solution, of course, everything tends strongly toward infinity when ϕ approaches zero.

We may avoid these singularities either by restricting our attention to the B solution and to those values k for which no singularity occurs or by cutting away the top of the dome. In both cases, the formulas (41) and (42) describe the effect of certain load configurations, applied to the edge (or the edges) $\phi = $ const of the shell. Each one of these configurations consists of n external forces $F^{(m)}$, applied at the ends of the hips and distributed according to a cosine law (39), and of shearing forces $N_{x\phi}^{(m)}$ which will appear automatically, if we provide along the edge a polygonal ring capable of taking care of these forces by bending moments in its plane. By an appropriate superposition of all harmonics from $k = 2$ up to $k = n/2$ (or $k = (n-1)/2$, if n should be odd), any symmetric self-equilibrating group of forces $F^{(m)}$ may be represented. For antimetric groups we must start from the sine terms in eq. (35).

The possibility of these self-equilibrating force systems in the supports of the dome indicates that a polygonal dome which is supported at the foot of every hip is a redundant structure with $n - 3$ redundant forces. The solution which we gave in Section 1 for regular loads is therefore true only, if not only the load is the same in all sectors but also the elastic deformability of the dome inclusive of the foundations has the n-fold symmetry of the structure.

The redundancy becomes apparent when a solution for an arbitrary load, say for an overloading of one shell sector, is sought. It then is not possible to find reasonable boundary conditions for the differential equation (40) without studying the deformations of the dome.

It may be expected that qualitatively similar results will be obtained for regular domes with other profiles when the hips have a horizontal tangent at the apex, and it will be easy to find out how these results are modified in the case of a pointed apex.

3.4.3 Non-regular Domes

When the basic polygon of the dome is not regular, the analysis becomes very involved. The FOURIER sum representation, eqs. (35) and (36), for the loads and the stress resultants is no longer possible, and there is no regular load which might be dealt with more easily than with the general case. Instead of having only one set (38a–c) at a time, we must formulate such equations separately for every hip and then deal with a set of $3n$ simultaneous differential equations.

In simple cases, where symmetry reduces the number of unknowns, it may still be possible to solve the stress problem. For instance, this is the case with the dead load problems of the two structures shown in fig. 31. The first shell, fig. 31 a, has one plane of symmetry and therefore only 3 different hip forces $F^{(m)}$. In the sectors which are intersected by the plane of symmetry, $f_1 \equiv 0$, and there are only 2 functions $f_1^{(m)}$ and 4 functions $f_2^{(m)}$ to be determined. This makes

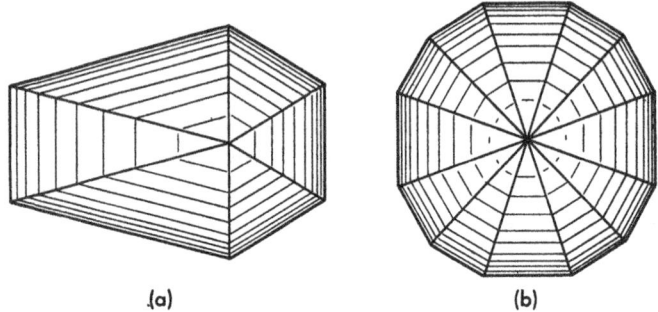

(a)   (b)

Fig. 31. Nonregular polygonal domes

a total of 9 unknown functions. In the other case, fig. 31b, we have 2 different hip forces, 1 function $f_1^{(m)}$ (in the small sectors) and 2 functions $f_2^{(m)}$, i. e., 5 unknowns altogether. Simultaneous systems of this size may still be handled numerically in a reasonable time, once their coefficients have been determined.

3.5 Folded Structures

3.5.1 Uniform Load

There is a type of structure which, although definitely not a shell, is rather similar to the barrel vaults treated in this Chapter. This has been called a folded structure or hipped-plate structure and consists

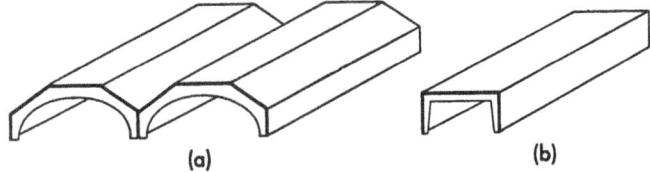

(a)   (b)

Fig. 32. Examples of folded structures

of an assembly of flat plate strips forming a prism. Fig. 32 shows two examples, a roof and a bridge. Like cylindrical shells, these prismatic structures must be stiffened by diaphragms in at least two cross sections.

The similarity in the appearance of folded structures and barrel vaults suggests establishing a kind of membrane theory, neglecting all bending and twisting moments in the plates and admitting only direct stresses which may be represented by the stress resultants N_x, N_y, N_{xy} shown on the plate element $dx \cdot dy_m$ in fig. 33. These membrane forces, as we may call them, cannot be in equilibrium with the loads unless they all also lie in the plane of the plate strip. Within the limits of the membrane theory, therefore, loads which do not fulfill this condition may not be applied. Only at the edges are loads of arbitrary direction permitted, since these may be resolved into two components in the planes of the adjacent strips. This restriction on the

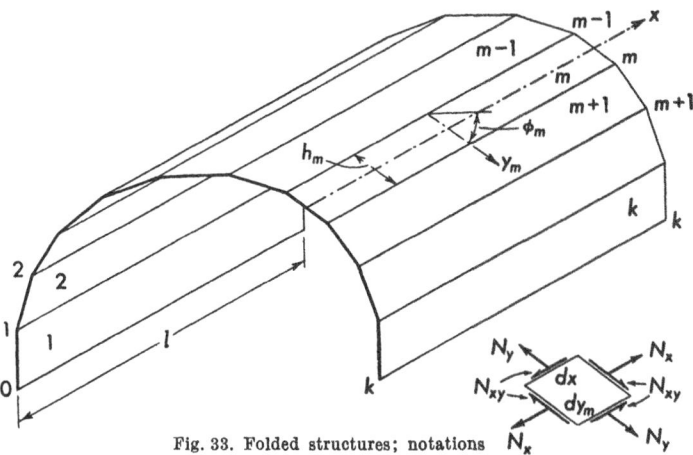

Fig. 33. Folded structures; notations

permissible loads is comparable to the one in trusses where loads are admitted only at the joints. However, in the latter case the requirement is easy to fulfill, while in folded structures the important loads are more or less evenly distributed over the whole surface of each plate strip. Nevertheless, the membrane theory is a useful instrument of analysis, since it describes an essential part of the whole stress system, although it does not tell the entire story.

The structure for which we now shall develop the theory consists of k narrow plate strips (fig. 33) and two diaphragms which we assume to be in the terminal cross sections. The spanwise coordinate x is common to all strips, but there is a special coordinate y_m for each strip. The angle ϕ_m between a horizontal plane and that of the m-th strip corresponds to the angle ϕ of the cylindrical shell, but it cannot be used as a coordinate. In fig. 33 the angles ϕ_m are positive in the right half of the prism.

For simplicity, we shall assume the edge loads vertical and uniformly distributed in the x direction, but the load at every edge may be chosen independently. The force P_m acting on an element of unit length of the m-th edge, may be resolved into two components in the directions y_m and y_{m+1}:

$$S'_m = -P_m \frac{\cos\phi_{m+1}}{\sin\gamma_m}, \qquad S''_m = +P_m \frac{\cos\phi_m}{\sin\gamma_m}, \tag{43}$$

and these loads can be carried by the strips m and $m+1$, respectively (fig. 34).

The plate strip m is bounded by the edges $m-1$ and m and there receives the loads S''_{m-1} and S'_m which, if positive, are both pointing in the same direction. They add up to the resultant load of the strip

$$S_m = S''_{m-1} + S'_m. \tag{44}$$

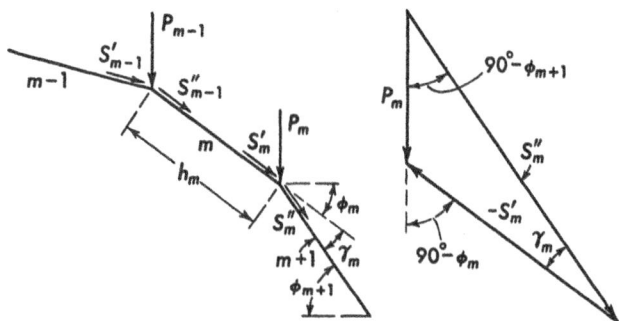

Fig. 34. Cross section through part of a folded structure

If we could separate the strip from its neighbors, it would be a simple beam of span l, depth h_m and width t_m (fig. 35), supported at the diaphragms and subjected to the uniform load S_m. Such a beam has the bending moment

$$M_m^{(0)} = S_m \frac{l^2 - 4x^2}{8} \tag{45a}$$

and the shear force

$$Q_m^{(0)} = -S_m x. \tag{45b}$$

If the plate strip is slender ($h_m \ll l$), and this we shall assume, the bending stress σ_x and the shear stress τ may be found from formulas of elementary beam theory, and so may their products with the thickness t_m, the stress resultants

$$N_x^{(0)} = \frac{12 M_m^{(0)} y_m}{h_m^3}, \qquad N_{xy}^{(0)} = \frac{6 Q_m^{(0)}}{h_m}\left(\frac{1}{4} - \frac{y_m^2}{h_m^2}\right).$$

At the lower edge ($y_m = +h_m/2$) of the strip the normal force $N_x^{(0)}$ produces the strain

$$\epsilon_x^{(0)} = \frac{N_x^{(0)}}{E\,t_m} = +\frac{6 M_m^{(0)}}{E\,t_m h_m^2},$$

while the strain at the upper edge ($y_{m+1} = -h_{m+1}/2$) of the adjacent strip is

$$\epsilon_x^{(0)} = -\frac{6\,M_{m+1}^{(0)}}{E\,t_{m+1}\,h_{m+1}^2}.$$

Since the strips are connected with each other, these strains ought to be equal, but they are by no means so; in general, they are even of

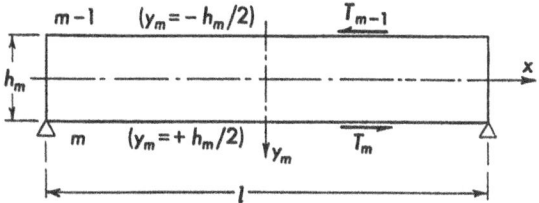

Fig. 35. Beam action on an isolated strip

opposite sign. On the other hand, the two strips will, of course, exert forces upon each other which we have not yet taken into account.

Since such additional forces must lie in the planes of both strips, they can only be shearing forces T_m as shown in fig. 35. Their magnitude and distribution are not known; fig. 35 indicates the direction in which they will be considered positive in agreement with the sign convention for N_{xy} (fig. 33).

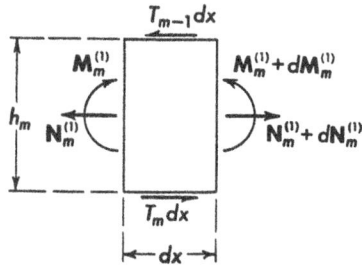

Fig. 36. Strip element

From a strip element (fig. 36) one may see that the edge shears will add a contribution $M_m^{(1)}$ to the beam moment of the strip and that they will also produce an axial force $N_m^{(1)}$. The equilibrium of the strip element yields the relations

$$\frac{d\,N_m^{(1)}}{d\,x} = T_{m-1} - T_m, \qquad \frac{d\,M_m^{(1)}}{d\,x} = -\frac{h_m}{2}\,(T_{m-1} + T_m), \qquad (46)$$

which, of course, cannot be integrated until the shears are known as functions of x.

The force $N_m^{(1)}$ and the moment $M_m^{(1)}$ produce beam bending stresses from which the normal force

$$N_x^{(1)} = \frac{N_m^{(1)}}{h_m} + \frac{12\,M_m^{(1)}\,y_m}{h_m^3}$$

and the edge strain

$$\epsilon_x^{(1)} = \frac{N_m^{(1)}}{E\,t_m\,h_m} \pm \frac{6\,M_m^{(1)}}{E\,t_m\,h_m^2}$$

may be derived. Now, the shearing forces T_m ($m = 1 \ldots k - 1$) must be so chosen that the discrepancy of the deformation at the edges disappears. This means that the sum $\epsilon_x^{(0)} + \epsilon_x^{(1)}$ must be the same, whether it is calculated for the lower edge of the strip m or for the upper edge of the strip $m + 1$:

$$\frac{6\,\mathsf{M}_m^{(0)}}{E\,t_m\,h_m^2} + \frac{\mathsf{N}_m^{(1)}}{E\,t_m\,h_m} + \frac{6\,\mathsf{M}_m^{(1)}}{E\,t_m\,h_m^2} = -\frac{6\,\mathsf{M}_{m+1}^{(0)}}{E\,t_{m+1}\,h_{m+1}^2} + \frac{\mathsf{N}_{m+1}^{(1)}}{E\,t_{m+1}\,h_{m+1}} - \frac{6\,\mathsf{M}_{m+1}^{(1)}}{E\,t_{m+1}\,h_{m+1}^2}\,.$$

$$(47)$$

In order to derive from this relation an equation for the unknown shearing forces, we must first differentiate it with respect to x. Then it is possible, with the help of eq. (46), to express $\mathsf{N}^{(1)}$ and $\mathsf{M}^{(1)}$ on both sides by T_{m-1}, T_m, T_{m+1}, while eq. (45) may be used to express $\mathsf{M}^{(0)}$ in terms of the load:

$$\frac{1}{t_m\,h_m}(2\,T_{m-1} + 4\,T_m) + \frac{1}{t_{m+1}\,h_{m+1}}(4\,T_m + 2\,T_{m+1}) = -\frac{6\,S_m\,x}{t_m\,h_m^2} - \frac{6\,S_{m+1}\,x}{t_{m+1}\,h_{m+1}^2}\,.$$

Such an equation may be written for every edge where two strips meet, $k - 1$ equations altogether for $k - 1$ unknown functions $T_m(x)$. Since all the right-hand sides are proportional to x, the shearing forces must be too, say

$$T_m = T_m'\,x\,,$$

and when we simply differentiate the equations, we obtain a set of ordinary linear equations for the unknown constants T_m':

$$\frac{1}{t_m\,h_m}\,T_{m-1}' + 2\left(\frac{1}{t_m\,h_m} + \frac{1}{t_{m+1}\,h_{m+1}}\right)T_m' + \frac{1}{t_{m+1}\,h_{m+1}}\,T_{m+1}'$$
$$= -\frac{3\,S_m}{t_m\,h_m^2} - \frac{3\,S_{m+1}}{t_{m+1}\,h_{m+1}^2}\,.$$

$$(48)$$

They have the structure of CLAPEYRON's equations for the continuous beam, and all the methods developed for the solution of these equations may be applied here.

When eqs. (48) have been solved, the total stress resultants in a cross section through a plate strip may be obtained from eqs. (45) and (46):

$$\mathsf{M}_m = S_m\frac{l^2 - 4x^2}{8} + (T_{m-1}' + T_m')\,h_m\frac{l^2 - 4x^2}{16}\,,$$

$$\mathsf{Q}_m = -S_m\,x\,,$$

$$\mathsf{N}_m = -(T_{m-1}' - T_m')\frac{l^2 - 4x^2}{8}\,.$$

From these the stress resultants N_x and N_{xy} may be calculated using the formulas for rectangular cross sections:

$$N_x = \left[\frac{12 S_m y_m}{h_m^2} + (T'_{m-1} + T'_m) \frac{6 y_m}{h_m} - (T'_{m-1} - T'_m) \right] \frac{l^2 - 4 x^2}{8 h_m},$$

$$N_{xy} = \left[\frac{3 S_m}{2 h_m} (4 y_m^2 - h_m^2) + \frac{T'_{m-1}}{4} (2 y_m - h_m)(6 y_m + h_m) \right. \tag{49}$$

$$\left. + \frac{T'_m}{4} (2 y_m + h_m)(6 y_m - h_m) \right] \frac{x}{h_m^2}.$$

The normal force N_y in the direction of the y_m axis is rather small and need not be computed.

The foregoing formulas have been derived under the assumption that each edge is uniformly loaded (if at all). One may easily treat the more general case that the load intensity P_m depends on x but that the law of distribution is the same at all loaded edges. It will then be found that M_m, N_m, and N_x depend on x as the bending moment in a beam carrying a similar load and that Q_m, T_m, and N_{xy} have the same spanwise distribution as the shearing force in that beam. The derivative T'_m of the shearing force is no longer a constant, but one may easily define a parameter (e. g. the maximum of T_m) which characterizes the intensity of the shear whose relative distribution is already known, and a system of equations may be set up which takes the place of eqs. (48).

3.5.2 Fourier Series Form of Solution

The solution given on the preceding pages is the simplest one when the loads are uniformly distributed in the x direction. The bending theory explained later (pp. 307—312) makes it desirable to have an alternative form, assuming that the loads P_m are given as FOURIER series in x. For this purpose we measure the coordinate x from one end of the structure (fig. 6) and write

$$P_m = \sum_{n=1}^{\infty} P_{m,n} \sin \frac{n \pi x}{l}, \qquad 0 \leqq x \leqq l. \tag{50}$$

For greater ease in writing we shall consider only a single term of this series and assume that

$$P_m = P_{m,n} \sin \frac{n \pi x}{l}. \tag{51}$$

We have then

$$S_m = S_{m,n} \sin \frac{n \pi x}{l},$$

$$S_{m,n} = P_{m-1,n} \frac{\cos \phi_{m-1}}{\sin \gamma_{m-1}} - P_{m,n} \frac{\cos \phi_{m+1}}{\sin \gamma_m}. \tag{52}$$

The stress resultants for the beam are

$$M_m = M_{m,n} \sin \frac{n \pi x}{l}, \qquad Q_m = Q_{m,n} \cos \frac{n \pi x}{l}, \qquad N_m = N_{m,n} \sin \frac{n \pi x}{l},$$

and for the parts with the superscript (0) we have

$$M_{m,n}^{(0)} = \frac{l^2}{n^2 \pi^2} S_{m,n}, \qquad Q_{m,n}^{(0)} = \frac{l}{n \pi} S_{m,n}. \qquad (53\,a, b)$$

The edge shear also has a cosine distribution:

$$T_m = T_{m,n} \cos \frac{n \pi x}{l},$$

and eqs. (46) assume the form

$$N_{m,n}^{(1)} = \frac{l}{n \pi}(T_{m-1,n} - T_{m,n}), \qquad M_{m,n}^{(1)} = -\frac{h_m l}{2 n \pi}(T_{m-1,n} + T_{m,n}). \quad (54\,a, b)$$

Following the same line of thought, one easily arrives at the following form of the three-shear equation (48):

$$\frac{1}{t_m h_m} T_{m-1,n} + 2\left(\frac{1}{t_m h_m} + \frac{1}{t_{m+1} h_{m+1}}\right) T_{m,n} + \frac{1}{t_{m+1} h_{m+1}} T_{m+1,n}$$
$$= \frac{3l}{n \pi}\left(\frac{S_{m,n}}{t_m h_m^2} + \frac{S_{m+1,n}}{t_{m+1} h_{m+1}^2}\right). \qquad (55)$$

3.5.3 Examples

It is interesting to apply the preceding formulas to some more or less fictitious examples and to compare the results with those obtained for analogous cylindrical shells.

Fig. 37 shows a pipelike structure of octagonal cross section. The best approximation to its own weight which can be made within the framework of the present theory is to assume equal loads $P_1, P_2 \ldots P_8$ at all edges,

$$P_m = 2p\, a \tan 22.5° = 0.828\, p\, a,$$

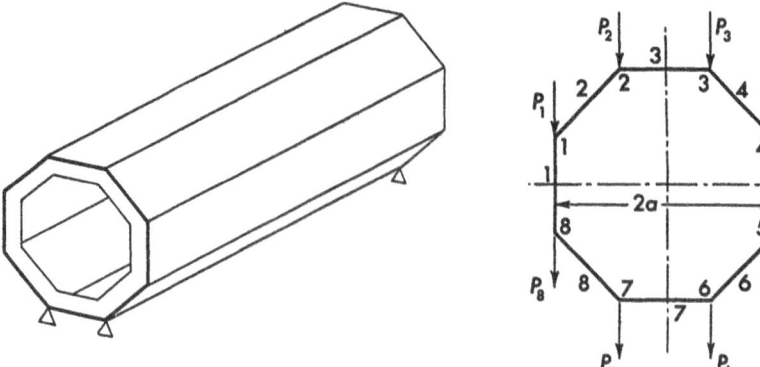

Fig. 37. Tubular folded structure

where p is the weight per unit area of the plates. Eqs. (43) and (45) yield

$$-S_1 = S_5 = 1.656\,pa, \quad -S_2 = S_4 = S_6 = -S_8 = 1.171\,pa, \quad S_3 = S_7 = 0,$$

and since the stresses in each cross section will be distributed symmetrically with respect to a vertical axis and antimetrically with respect to a horizontal axis, it will suffice to consider only one quarter of the structure and to write eqs. (48) for the edges 3 and 4 only. They are

$$T_2' + 4T_3' + T_4' \qquad = -\,4.24\,p,$$
$$T_3' + 4T_4' + T_5' = -10.24\,p,$$

and when we still put $T_2' = -T_3'$, $T_5' = T_4'$ we have two linear equations for T_3' and T_4'. It is not necessary to give more details of the numerical work here. The result is shown in fig. 38. The normal force N_x is constant across the strips 3 (compression) and 7 (tension), and plotted over the vertical diameter it is practically linear and the same as in a cylindrical shell of radius a. The shearing force in a circular cylinder has an elliptic

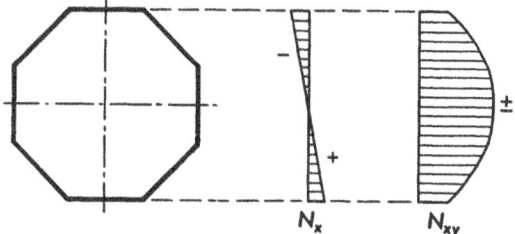

Fig. 38. Stress resultants in the octagonal tube of fig. 37

distribution and almost exactly the same maximum as that of the octagonal prism. One may easily believe that the similarity will be still greater if the number of plate strips is increased.

Results are thoroughly different if one studies structures which look like a prismatic counterpart to barrel vault roofs. A structure of this kind may be obtained from the octagonal tube by cutting away its lower half (fig. 39). The loads $P_1 \ldots P_4$ are the same as before, and

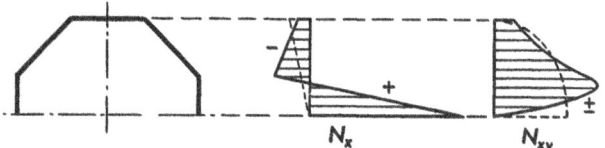

Fig. 39. Prismatic barrel vault

so is the equation for edge 3. The second equation, however, is different. Strip 5 has only one-half the width of the other ones so that the coefficient of T_4' is now $2(1 + 2) = 6$ and the load term is $-16.24\,p$. Edges 0 and 5 are free of any force, hence $T_5' = 0$.

When one again goes through the numerical work, the results plotted in fig. 39 are obtained. They differ widely from those for the cylindrical shell, indicated by dotted lines. In the shell, the shear $N_{x\phi}$ has a maximum at the "free" edge, and we saw that there an edge member *must* be provided to which this shear may be transmitted. The prismatic structure does not need such an edge member, and strips 1 and 5 seem to take its duties, carrying tensile stresses of considerable magnitude.

It is evident that increasing the number of plate strips will not make the stress system approach that of the cylindrical shell. What does occur can be seen from fig. 40. Here the number of edges has been

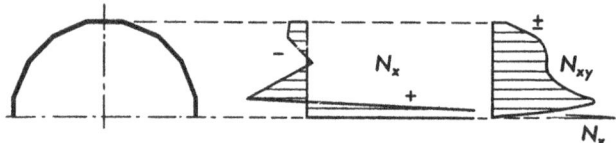

Fig. 40. Prismatic barrel vault with many edges

doubled. The stress diagrams are rather irregular, and it is clear that the membrane theory of folded structures cannot be used as an approximation to the membrane theory of cylindrical shells.

3.5.4 Limitations of the Theory

We saw ·on p. 128 that the membrane theory of the barrel vault roof is subject to severe limitations, resulting from the need of an edge member and the impossibility of incorporating its deformation and its weight in the theory.

In the prismatic roof no edge member is needed, but the limitations of its theory are not less severe. The zigzagging N_x diagram in fig. 40 certainly does not represent a physical reality. The real structure will level the peaks, and it will achieve this with the help of a system of bending moments and transverse shearing forces similar to those in cylindrical shells (see Chapter 5). The bending stresses are caused by incompatibilities in the deformations pertinent to the membrane force system and will be discussed in some detail on p. 307.

Additionally, there is another source of bending stresses. The loads acting on a real folded structure are almost always distributed over the whole surface, not concentrated at the edges. Such loads must necessarily produce plate bending moments M_y which will carry them to the edges. The bending stresses σ_y connected with these moments may be considerable; the *thinner* the plates are, the greater the stresses, and this stress system has no counterpart in cylindrical shells. One may determine the plate moments M_y by cutting a strip of unit width across the prism and treating it as a continuous beam supported at the edges. The forces which it exerts on these fictitious supports are the loads P_m considered in the membrane theory. Since the structure will yield elastically under these loads, the strip is a beam on elastic supports of a peculiar kind, the deflection of each support depending on the reactions on all supports.

Chapter 4

DIRECT STRESSES IN SHELLS OF ARBITRARY SHAPE

4.1 Conditions of Equilibrium

In the two preceding chapters the membrane theory of shells has been developed for two important types: shells of revolution and cylinders. In both cases the theory made use of every advantage which the particular shape of the middle surface offered, thus arriving at the simplest possible solution of many important problems but lacking

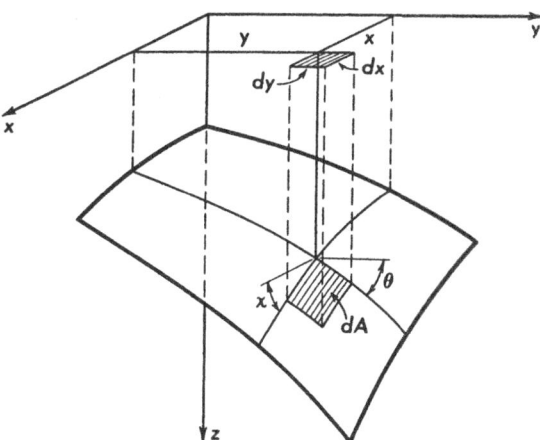

Fig. 1. Shell of arbitrary shape in rectilinear coordinates

generality. It is the purpose of this chapter to develop a general membrane theory for shells of arbitrary shape and then to apply it to some shells which do not fall in one of the special groups considered before.

We describe the middle surface of the shell by a system of rectangular coordinates (fig. 1), assuming that z is given as a function of x and y. Since the latter two coordinates are sufficient to distinguish between the points of the middle surface, they may be used as a pair of GAUSSian (curvilinear) coordinates on the shell. The coordinate lines $x = $ const and $y = $ const on the middle surface are obtained by intersecting this surface with planes normal to the x or y axis. In general, these lines will not meet at right angles, as do their projections on the x, y plane, and it seems reasonable to describe the membrane stresses by a system of skew forces, as explained in Section 1.2.3. A shell element with these forces acting on it is shown in fig. 2. Two of the forces, N_x and N_{yx},

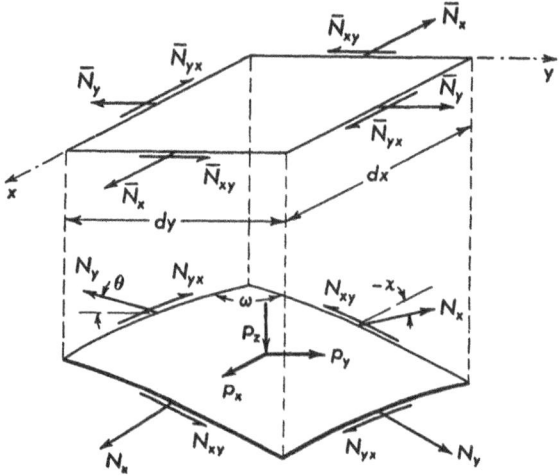

Fig. 2. Shell element and its projection on the x, y plane

are parallel to the x, z plane, while the other two, N_y and N_{xy} have no component parallel to the x axis.

The skew forces N_x, N_y, $N_{xy} = N_{yx}$ are forces per unit length of the line elements through which they are transmitted. The actual forces are obtained by multiplying them by the length of this element, i. e. by $dy/\cos\theta$ or $dx/\cos\chi$, as the case may be. When we multiply by still another $\cos\theta$ or $\cos\chi$, we obtain the horizontal components of these forces, the x components

$$N_x \cdot \frac{dy}{\cos\theta} \cdot \cos\chi = \bar{N}_x\, dy, \qquad N_{yx} \cdot \frac{dx}{\cos\chi} \cdot \cos\chi = \bar{N}_{yx}\, dx$$

and the y components

$$N_y \cdot \frac{dx}{\cos\chi} \cdot \cos\theta = \bar{N}_y\, dx, \qquad N_{xy} \cdot \frac{dy}{\cos\theta} \cdot \cos\theta = \bar{N}_{xy}\, dy.$$

The new quantities

$$\bar{N}_x = N_x \frac{\cos\chi}{\cos\theta}, \qquad \bar{N}_{xy} = \bar{N}_{yx} = N_{xy} = N_{yx}, \qquad \bar{N}_y = N_y \frac{\cos\theta}{\cos\chi} \qquad (1)$$

which we have introduced are the plan projections of the stress resultants, referred to the unit length of the projected line element dx or dy.
We shall use them when we write the conditions of equilibrium of the
shell element. Along with them it is useful to refer the distributed
load to the unit area of the projected shell element $dx \cdot dy$ and to
write $\bar{p}_x\, dx\, dy$, $\bar{p}_y\, dx\, dy$, $\bar{p}_z\, dx\, dy$ for the rectangular components
of the external force acting on the element. The relation between $\bar{p}_x$,
$\bar{p}_y$, $\bar{p}_z$ and the forces p_x, p_y, p_z per unit area of the middle surface is
given by the ratio of the areas of a shell element dA and of its projec-
tion $dx\, dy$. From fig. 2 we read that

$$dA = \frac{dx}{\cos\chi}\, \frac{dy}{\cos\theta}\, \sin\omega = dx\, dy\, \frac{(1 - \sin^2\chi \sin^2\theta)^{1/2}}{\cos\chi \cos\theta},$$

and consequently we have

$$\frac{\bar{p}_x}{p_x} = \frac{\bar{p}_y}{p_y} = \frac{p_z}{p_z} = \frac{dA}{dx\, dy} = \frac{(1 - \sin^2\chi \sin^2\theta)^{1/2}}{\cos\chi \cos\theta}. \tag{2}$$

After these preparations it is easy to write the conditions of equilibrium for the shell element shown in fig. 2. There are three such conditions, one for the forces parallel to each coordinate axis. In the x direc-
tion we have the increments of $\bar{N}_x\, dy$ and $\bar{N}_{yx}\, dx$ and the load $\bar{p}_x\, dx\, dy$:

$$\frac{\partial \bar{N}_x}{\partial x}\, dx \cdot dy + \frac{\partial \bar{N}_{yx}}{\partial y}\, dy \cdot dx + \bar{p}_x \cdot dx\, dy = 0.$$

In the y direction we find a similar equation, and when we drop from
both the factor $dx\, dy$ common to all terms, we have

$$\frac{\partial \bar{N}_x}{\partial x} + \frac{\partial \bar{N}_{xy}}{\partial y} + \bar{p}_x = 0,$$

$$\frac{\partial \bar{N}_{xy}}{\partial x} + \frac{\partial \bar{N}_y}{\partial y} + \bar{p}_y = 0. \tag{3a, b}$$

In the condition for the z components all four stress resultants
appear. The force $N_x \cdot dy/\cos\theta$ has the vertical component

$$N_x \cdot \frac{dy}{\cos\theta} \cdot \sin\chi = \bar{N}_x \tan\chi\, dy = \bar{N}_x \frac{\partial z}{\partial x}\, dy,$$

and the shear N_{xy} on the same side of the shell element gives

$$N_{xy} \cdot \frac{dy}{\cos\theta} \cdot \sin\theta = \bar{N}_{xy} \tan\theta\, dy = N_{xy} \frac{\partial z}{\partial y}\, dy.$$

Similar expressions are obtained for N_y and N_{yx}. The equilibrium
equation involves their differential increments:

$$\frac{\partial}{\partial x}\left(\bar{N}_x \frac{\partial z}{\partial x}\right) + \frac{\partial}{\partial y}\left(\bar{N}_{yx} \frac{\partial z}{\partial x}\right) + \frac{\partial}{\partial x}\left(\bar{N}_{xy} \frac{\partial z}{\partial y}\right) + \frac{\partial}{\partial y}\left(\bar{N}_y \frac{\partial z}{\partial y}\right) + \bar{p}_z = 0.$$

Differentiating the products, we find

$$\overline{N}_x \frac{\partial^2 z}{\partial x^2} + 2\,\overline{N}_{xy} \frac{\partial^2 z}{\partial x\,\partial y} + \overline{N}_y \frac{\partial^2 z}{\partial y^2}$$

$$= -\bar{p}_z - \left(\frac{\partial \overline{N}_x}{\partial x} + \frac{\partial \overline{N}_{xy}}{\partial y}\right)\frac{\partial z}{\partial x} - \left(\frac{\partial \overline{N}_{xy}}{\partial x} + \frac{\partial \overline{N}_y}{\partial y}\right)\frac{\partial z}{\partial y},$$

and making use of eqs. (3a, b):

$$\overline{N}_x \frac{\partial^2 z}{\partial x^2} + 2\,\overline{N}_{xy} \frac{\partial^2 z}{\partial x\,\partial y} + \overline{N}_y \frac{\partial^2 z}{\partial y^2} = -\bar{p}_z + \bar{p}_x \frac{\partial z}{\partial x} + \bar{p}_y \frac{\partial z}{\partial y}. \qquad (3c)$$

The equations (3a–c) are the basic equations of the theory. As in the preceding chapters, they are two differential equations of the first order and one ordinary linear equation. Only the last one has variable coefficients: the second derivatives of z.

Sometimes a direct solution of these equations may be tried. In most cases it will be advantageous to introduce an auxiliary variable, which reduces the system to one second-order equation. As a matter of fact, the set (3a, b) is identical with the conditions of equilibrium of a plane stress system[1], and the method of the stress function which has proved to be a powerful tool for the treatment of that problem, may be applied here as well.

When we put

$$\overline{N}_x = \frac{\partial^2 \Phi}{\partial y^2} - \int \bar{p}_x\,dx, \;\; \overline{N}_y = \frac{\partial^2 \Phi}{\partial x^2} - \int \bar{p}_y\,dy, \;\; \overline{N}_{xy} = -\frac{\partial^2 \Phi}{\partial x\,\partial y}, \qquad (4)$$

we find that two of the conditions of equilibrium, eqs. (3a, b), are identically satisfied. The third one, eq. (3c), yields a differential equation for Φ;

$$\frac{\partial^2 \Phi}{\partial x^2} \frac{\partial^2 z}{\partial y^2} - 2\frac{\partial^2 \Phi}{\partial x\,\partial y} \frac{\partial^2 z}{\partial x\,\partial y} + \frac{\partial^2 \Phi}{\partial y^2} \frac{\partial^2 z}{\partial x^2}$$

$$= -\bar{p}_z + \bar{p}_x \frac{\partial z}{\partial x} + \bar{p}_y \frac{\partial z}{\partial y} + \frac{\partial^2 z}{\partial x^2} \int \bar{p}_x\,dx + \frac{\partial^2 z}{\partial y^2} \int \bar{p}_y\,dy. \qquad (5)$$

In the plane stress problem, where the same stress function is used, an equation of the fourth order is obtained[2]. From comparison with eq. (5) one may understand the essential differences of the two problems. In the plane stress problem the differential equation is

---

[1] TIMOSHENKO, S, and J. N. GOODIER: Theory of Elasticity, 2d edition. New York 1951, p. 22.

[2] Ibid., pp. 26 and 27.

derived from a condition of compatibility between the three strain components ϵ_x, ϵ_y, γ_{xy}; here it stems from the equilibrium of forces in the z direction. In plane stress this third condition of equilibrium is trivial since all forces lie in the x, y plane, and in the case of the shell the compatibility of strains is always assured since there is a third component of the displacement and u, v, w can always be chosen so that a given set of strains will be produced.

In the simplest case the boundary condition for the differential equation (5) will prescribe the value of one of the stress resultants appearing at the boundary. The details depend on whether the differential equation is of the elliptic or of the hyperbolic type, as we shall see later.

4.2 Elliptic Problems

4.2.1 Paraboloid of Revolution, Triangular Shell

A paraboloid of revolution has the equation

$$z = \frac{x^2 + y^2}{h}. \tag{6}$$

When we calculate from it the derivatives of z and introduce them into eq. (5), we see that the second term on the left-hand side vanishes and that the coefficients of the other two become constant. Restricting our attention to vertical loads ($p_x \equiv p_y \equiv 0$), we have then

$$\frac{\partial^2 \Phi}{\partial x^2} + \frac{\partial^2 \Phi}{\partial y^2} = -\frac{1}{2} h \, \bar{p}_z. \tag{7}$$

This is the plane-harmonic equation which is well known in mathematical literature.

For the first approach, we use the simplest type of vertical loading, assuming $p_z = p = $ const. If we interpret this load as the weight of the shell, the wall thickness t would have to be greatest at the top and to decrease as the slope increases. This does not make much sense, but if the slope is not great, t will not be far from constant, and the results will give a reasonable first orientation. We shall see later what to do in more realistic cases.

Under the assumption $\bar{p}_z = p$ we have to deal with the equation

$$\frac{\partial^2 \Phi}{\partial x^2} + \frac{\partial^2 \Phi}{\partial y^2} = -\frac{1}{2} p \, h. \tag{8}$$

There exists one simple solution of this equation, which is probably without practical application but shows some features of general interest. It is:

$$\varPhi = -\frac{1}{8}\,p\,h\left[x^2 + y^2 + \frac{1}{a}\,(3x\,y^2 - x^3)\right].$$

Here a is an arbitrary length.

One may easily check that $\varPhi$ assumes a constant value along all three sides of the triangle shown in the x, y plane of fig. 3, i. e.

$$\text{for}\quad x = -\frac{a}{3}\quad\text{and for}\quad y\,\sqrt{3} = \pm\left(\frac{2a}{3} - x\right).$$

For that side which is parallel to the y axis this implies that $\partial^2\varPhi/\partial y^2 \equiv 0$ and hence from eq. (4) that $\overline{N}_x = 0$. This means that this edge of the shell must be supported by a thin vertical wall (a diaphragm) which will accept only a shear N_{xy} but not a thrust. For the other two edges the situation is similar. All partial derivatives of $\varPhi$ in the direction of the edge are zero and hence there is no thrust normal to the edge. Therefore the formula describes the membrane forces in the triangular shell shown in fig. 3 when its edges are supported by vertical arches or walls which can resist only tangential forces transmitted to them.

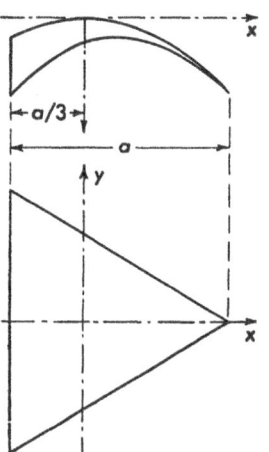

Fig. 3. Triangular shell

The complete formulas for the stress resultants may easily be found, applying eqs. (1) and (4):

$$N_x = -\frac{p\,h}{4}\left(1 + 3\frac{x}{a}\right)\sqrt{\frac{h^2 + 4x^2}{h^2 + 4y^2}},$$

$$N_y = -\frac{p\,h}{4}\left(1 - 3\frac{x}{a}\right)\sqrt{\frac{h^2 + 4y^2}{h^2 + 4x^2}},$$

$$N_{xy} = \frac{3}{4}\,p\,h\,\frac{y}{a}.$$

Some numerical results for a shell with $a = h$ are shown in fig. 4. At the top $x = y = 0$, we have $N_x = N_y$ and $N_{xy} = 0$. Here the normal force is the same in every direction and would also be the same if the shell were limited by a circular springing line. At $x = -a/3$, $y = 0$, the shear $N_{xy} = 0$ from symmetry and $N_x = 0$ from the boundary condition. Here N_y alone carries the load as in an arch. At the corners, a rhomboidal element with edges parallel to those of the shell is in a state of pure shear. The shearing forces are really able to carry the local load, since they have different slopes at opposite sides of the element and therefore different vertical components. We shall see later that this is not always so and that then a disagreeable phenomenon may appear.

It may be mentioned that the use of cylindrical coordinates r, ψ, z permits finding solutions for similar shells limited in plan view by any regular polygon. The solution then is not obtained in finite form but as a FOURIER series in ψ.

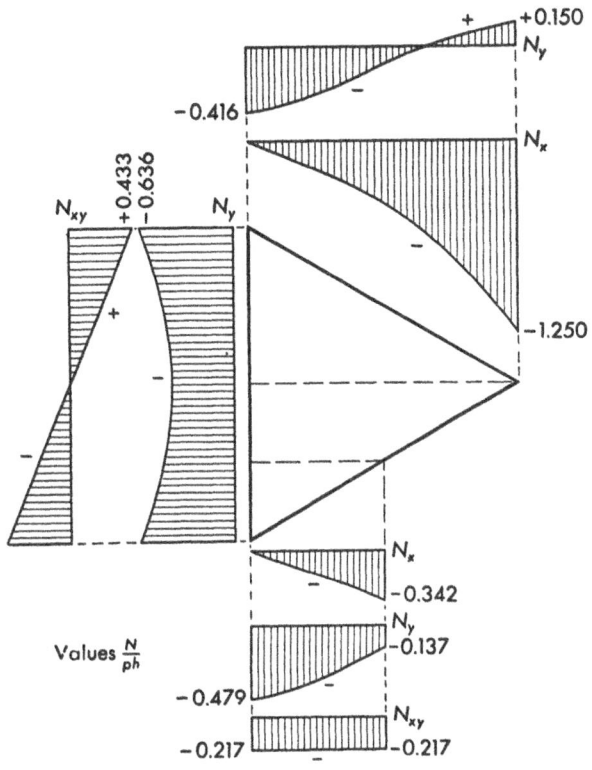

Fig. 4. Stress resultants in a triangular shell

4.2.2 Elliptic Paraboloid

For roof construction, those shells are of the greatest interest which permit covering rectangular areas. If the rectangle is not too different from a square, we may cut a convenient piece out of the paraboloid (6), but for more elongated rectangles it is better to use an elliptic paraboloid

$$z = \frac{x^2}{h_1} + \frac{y^2}{h_2}. \tag{9}$$

Its intersections with planes $x = \text{const}$ or $y = \text{const}$ are again parabolas but of two different sizes.

We now consider a shell which is cut out of such a surface by two planes $x = \pm a/2$ and two planes $y = \pm b/2$ (fig. 5). If we choose the same load as before, $\bar{p}_x \equiv \bar{p}_y \equiv 0$, $\bar{p}_z = p$, the differential equation is

$$\frac{1}{h_2} \frac{\partial^2 \Phi}{\partial x^2} + \frac{1}{h_1} \frac{\partial^2 \Phi}{\partial y^2} = -\frac{p}{2}. \tag{10}$$

Let the edges of the shell be stiffened by four arches. Then the boundary conditions are

$$\text{for} \quad x = \pm \frac{a}{2}: \quad N_x = 0, \quad \text{hence} \quad \frac{\partial^2 \Phi}{\partial y^2} = 0,$$

$$\text{for} \quad y = \pm \frac{b}{2}: \quad N_y = 0, \quad \text{hence} \quad \frac{\partial^2 \Phi}{\partial x^2} = 0.$$

From this we conclude that on each side of the rectangular boundary, Φ must be a linear function of one of the coordinates, x or y. Since the load is distributed symmetrically with respect to both coordinate axes, we may expect that Φ is a constant along the whole boundary, and since we are interested only in second derivatives of Φ, the constant boundary value may arbitrarily be assumed to be zero.

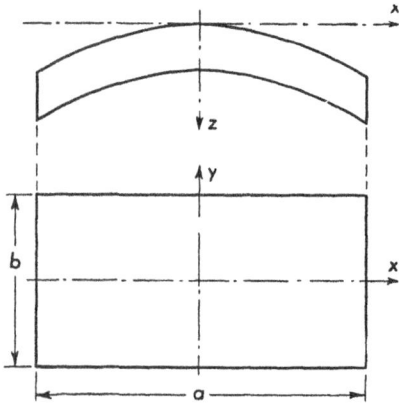

Fig. 5. Elliptic paraboloid

In cases where there is less symmetry one may still take advantage of the fact that the stress resultants are not changed when a linear function of x and y is added to Φ. One may therefore always assume Φ to be zero along two adjacent edges of the rectangle, and then symmetry with respect to one axis is enough to make $\Phi = 0$ along the whole boundary.

In the particular case under consideration, we start from the particular solution

$$\Phi \equiv \Phi_0 = \frac{p h_1}{4} \left(\frac{b^2}{4} - y^2 \right)$$

which not only satisfies the differential equation but assumes the value $\Phi = 0$ on the two longer edges $y = \pm b/2$. To fulfill the same condition on the other two edges we add another stress function

$$\Phi \equiv \Phi_1 = \sum_{n=1, 3, \ldots}^{\infty} C_n \operatorname{Cosh} \frac{n \pi x}{c} \operatorname{ccs} \frac{n \pi y}{b}. \tag{11}$$

This series satisfies term by term the homogeneous differential equation if we put $c = b(h_1/h_2)^{1/2}$. Furthermore, each of its terms vanishes for $y = \pm b/2$, and if we choose the coefficients C_n so that $\Phi_0 + \Phi_1 = 0$

for $x = \pm a/2$, this sum of the two stress functions will represent the solution of our problem.

Now, the boundary values for $x = \pm a/2$ are

$$\Phi \equiv \Phi_0 + \Phi_1 = \frac{p\,h_1}{4}\left(\frac{b^2}{4} - y^2\right) + \sum_{1,\,3,\,5,\ldots}^{\infty} C_n \mathrm{Cosh}\,\frac{n\,\pi\,a}{2\,c}\cos\frac{n\,\pi\,y}{b}.$$

The first term of this expression may be expanded into a FOURIER series

$$\frac{p\,h_1}{4}\left(\frac{b^2}{4} - y^2\right) = \frac{p\,h_1}{4}\frac{8\,b^2}{\pi^3}\left(\cos\frac{\pi\,y}{b} - \frac{1}{3^3}\cos\frac{3\,\pi\,y}{b} + \frac{1}{5^3}\cos\frac{5\,\pi\,y}{b} - + \cdots\right),$$

and we see that the condition $\Phi = 0$ will be fulfilled if we choose

$$C_n = -\frac{(-1)^{\frac{n-1}{2}}\cdot 2\,p\,h_1\,b^2}{\pi^3\,n^3\,\mathrm{Cosh}\,\dfrac{n\,\pi\,a}{2\,c}}.$$

With this expression for the general coefficient the complete solution assumes the following form

$$\Phi = \frac{p\,h_1}{4}\left[\frac{b^2}{4} - y^2 - \frac{8\,b^2}{\pi^3}\sum_{1,\,3,\,5,\ldots}^{\infty}(-1)^{\frac{n-1}{2}}\cdot\frac{1}{n^3}\frac{\mathrm{Cosh}\,n\,\pi\,x/c}{\mathrm{Cosh}\,n\,\pi\,a/2\,c}\cos\frac{n\,\pi\,y}{b}\right].$$

$$(12)$$

Differentiation according to eq. (4) and application of eq. (1) yields the formulas for the stress resultants:

$$N_x = -\frac{p\,h_2}{2}\sqrt{\frac{h_1^2 + 4\,x^2}{h_2^2 + 4\,y^2}}\left[1 + \frac{4}{\pi}\sum_{1,\,3,\,5,\ldots}^{\infty}(-1)^{\frac{n+1}{2}}\frac{1}{n}\frac{\mathrm{Cosh}\,n\,\pi\,x/c}{\mathrm{Cosh}\,n\,\pi\,a/2\,c}\cos\frac{n\,\pi\,y}{b}\right],$$

$$N_{xy} = \frac{2\,p}{\pi}\sqrt{h_1\,h_2}\sum_{1,\,3,\,5\ldots}^{\infty}(-1)^{\frac{n+1}{2}}\cdot\frac{1}{n}\frac{\mathrm{Sinh}\,n\,\pi\,x/c}{\mathrm{Cosh}\,n\,\pi\,a/2\,c}\sin\frac{n\,\pi\,y}{b}, \qquad (13)$$

$$N_y = 2\,\frac{p\,h_1}{\pi}\sqrt{\frac{h_2^2 + 4\,y^2}{h_1^2 + 4\,x^2}}\sum_{1,\,3,\,5,\ldots}^{\infty}(-1)^{\frac{n+1}{2}}\cdot\frac{1}{n}\frac{\mathrm{Cosh}\,n\,\pi\,x/c}{\mathrm{Cosh}\,n\,\pi\,a/2\,c}\cos\frac{n\,\pi\,y}{b}.$$

The convergence of these series shows an interesting peculiarity. Since the terms have alternating signs, the factor $1/n$ is sufficient to assure at least a feeble convergence. But there is still the quotient of the two hyperbolic functions. For a fixed value of x, $0 < x < a/2$, and large values of n it is approximately equal to $\exp[n\,\pi(2\,x - a)/2\,c]$ and decreases exponentially. This gives an excellent convergence, if we do not go all too close to the corner $x = a/2$, $y = b/2$. The forces N_x and N_y there are zero, but in the series for the shear N_{xy} the factor $\sin n\,\pi\,y/b$ is alternately $+1$ and -1 and thus cancels the alternating sign coming from the factor $(-1)^{(n+1)/2}$, and we have there exactly the sum of the odd terms of the harmonic series, which is divergent. This is not a failure of the applied method but indicates a real singularity of the stress system, which we can easily explain by mechanical considerations.

The equation (9) of the elliptic paraboloid has the form

$$z = f(x) + g(y).$$

When a surface is described by an equation of this kind, all its cross sections $x = $ const are congruent to each other, and so are the curves $y = $ const. The surface may therefore be generated by subjecting one of these curves to a transverse translation. Such surfaces are called surfaces of translation, and the curves from which they may be generated are their generators.

When a shell is formed as a surface of translation, a shell element bounded by two pairs of generators is an exact parallelogram. Therefore the shearing forces N_{xy} at opposite edges are exactly parallel, and they cannot contribute to the vertical equilibrium as do the longitudinal forces N_x and N_y. For this reason the second term of eq. (5) is missing in eq. (10). Now, at the corners, the boundary condition requires that N_x and N_y both be zero, and nothing is left to carry the vertical load. This is the reason that the shear tends toward infinity when one approaches this point.

The physical interpretation of this singularity is, of course, this: In the vicinity of the rectangular corner, where the membrane forces cannot carry the load, transverse shearing forces of substantial magnitude will appear which, in turn, will produce bending and twisting moments in the shell.

4.2.3 Solution by Relaxation Method

It is not always possible to solve the stress problem by such simple formulas as we used in the preceding sections. As an example, we may think of the shell shown in fig. 5, but with constant wall thickness. Then the load p_z, produced by its weight, is a constant, but $\bar{p}_z$ increases toward the edges and still more toward the corners, and eq. (2), which describes this increase, is in no way tempting for analytical work.

In this and other cases a numerical method is needed, and it is the particular merit of the stress function that it opens the way for the application of the relaxation method, at least for shells of positive curvature.

It is not the purpose of this book to teach this method, and it will be explained here only so far as is necessary in order to understand its application to shell problems. For the mathematical background of the method, as well as for special tricks and possible refinements, the reader is referred to the special literature on the subject.

In principle, the relaxation method is a method to solve linear algebraic equations. It may be applied to differential equations when it is possible to approximate them by a set of linear algebraic equations. In our case this is done by considering only the values $\Phi_{m,n}$ of Φ in the nodal points $x = m\,d$, $y = n\,d$ of a rectangular network (fig. 6) and by writing

$$\left(\frac{\partial^2\Phi}{\partial x^2}\right)_{m,n} \approx \frac{1}{d^2}\left(\Phi_{m-1,n} - 2\Phi_{m,n} + \Phi_{m+1,n}\right),$$

$$\left(\frac{\partial^2\Phi}{\partial y^2}\right)_{m,n} \approx \frac{1}{d^2}\left(\Phi_{m,n-1} - 2\Phi_{m,n} + \Phi_{m,n+1}\right). \tag{14}$$

When these approximations are introduced in the differential equation (7), it becomes an algebraic equation:

$$\Phi_{m-1,n} + \Phi_{m+1,n} + \Phi_{m,n-1} + \Phi_{m,n+1} - 4\Phi_{m,n} = -\frac{1}{2}\,h d^2\,\overline{p}_{z\,m\,n}. \tag{15}$$

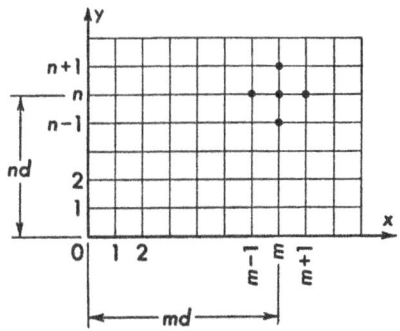

We have such an equation for every grid point in the interior of the domain covered by the shell. In these equations there will appear the values $\phi_{m,n}$ at just these points and at points on the boundary. Since the latter are known to be zero, we have exactly as many unknowns as we have equations.

If we had to solve these equations by the usual elimination procedure, we should have to

Fig. 6. Network in the x, y plane used for applying the relaxation method

limit the number of unknowns to a very few, and this would restrict the use of the finite-difference method to rather coarse nets. This limitation is overcome by the relaxation method. Its essential idea is this: We write eq. (15) in the form

$$\Phi_{m-1,n} + \Phi_{m+1,n} + \Phi_{m,n-1} + \Phi_{m,n+1} - 4\Phi_{m,n} + \frac{1}{2}\,h d^2\,\overline{p}_{z\,m\,n} = R_{m,n}, \tag{16}$$

introducing the residuals $R_{m,n}$ which will be zero if the $\Phi_{m,n}$ represent the correct solution of eqs. (15). When we insert values $\Phi_{m,n}$ which are only approximately correct, the magnitude of the residuals indicates how good or bad the approximation is, and their distribution in the m, n plane shows where improvements are most needed. Such improvements can be made in successive steps, changing one single value $\Phi_{m,n}$ at a time.

When we increase the stress function at a certain point m, n by unity, eq. (16) shows that $R_{m,n}$ will decrease by 4, and when we write eq. (16) for one of the four points surrounding the point m, n, we see that there the residual will increase by 1. In this way five residuals are affected when one Φ value is changed, and we may represent these changes by the "relaxation pattern"

$$
\begin{array}{ccc}
 & 1 & \\
1 & -4 & 1 \\
 & 1 &
\end{array}
$$

By a great number of well-chosen changes of this kind the approximate figures $\Phi_{m,n}$ are made to approach the correct values, i. e. those which satisfy the linear equations (15). To perform this work successfully, certain bookkeeping is necessary. We shall see in an example how this may best be done.

Fig. 7 shows a quadratic shell with $h/a = 1.40$. When the wall thickness is constant, the weight per unit of shell surface is also a constant, say $p_z = p$, and the load per unit of projected area follows from eq. (2):

$$
\bar{p}_z = p \, \frac{(1 - \sin^2\chi \sin^2\theta)^{1/2}}{\cos\chi \cos\theta} . \tag{17}
$$

Since both the shell and its load have four planes of symmetry, it suffices to consider the triangular group of grid points marked in fig. 7. For these points we may easily find approximate values of $\Phi_{m,n}$

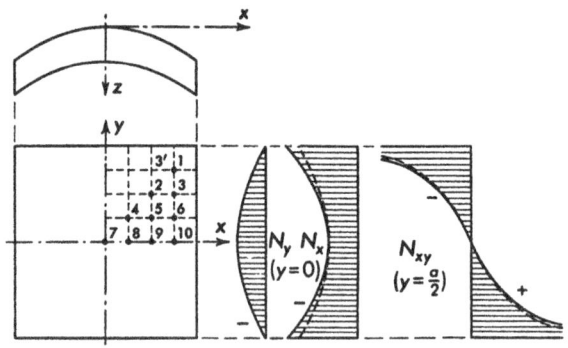

Fig. 7. Paraboloid of revolution; example for the use of the relaxation method

from eq. (12) which assumes that $\bar{p}_z = p$. Using these values and the correct load $\bar{p}_{zmn}$ from eq. (17) we compute the residuals $R_{m,n}$ from eq. (16).

Now we set up the relaxation table, p. 177, in which the systematic correction of the stress function will be done. Each of the rectangular panels in this table represents one of the grid points, and the panels are arranged exactly in the same order as the grid points in the x, y plane. The numbers in the lower right corner of each panel are therefore not necessary for identification and are usually omitted.

Each panel is divided into two compartments, a smaller one for the stress function and a wider one for the residual. At the top of the left compartment we put down the starting value of Φ and at the top

Relaxation Table

		910 45 (1) : : : : **1023**	179 −1 (1) 71 (2) : : : : −1 **1**
2270 36 (3) : : : **2456**	74 146 (2) 2 (3) : : : 0 **2**	1420 36 (2) : : : **1056**	99 144 (1) 0 (2) 36 (3) 72 (4) : 1 **3**
3300 : : : : : **3503**	25 : : : : : 1 **4**	2720 : : : : : **2919**	70 106 (3) 142 (4) : : : 2 **5**

Panel 6 (continuing the staircase):

1670 36 (4) : : **1817**	107 143 (2) −1 (4) : : −1 **6**

3690 : : : : : **3880**	−19 : : : : : 1 **7**	3490 : : : : : **3685**	−7 : : : : : 1 **8**	2870 : : : : : **3062**	29 : : : : : −1 **9**	1750 : : : : : **1895**	95 167 (4) : : : 1 **10**

of the right compartment the corresponding residual R. In practical computation work it is advisable to multiply both by a suitable factor, in general a power of 10, so that no decimal fractions occur, and in computations which aim at results of some general validity, it may even be useful to multiply the unknowns and the residuals by a factor which makes them dimensionless. This has been done in the present case, values $10^5 \Phi/p\,h\,a^2$ and $10^5 R/p\,h\,a^2$ being entered in the relaxation table.

The first four steps of the relaxation are reproduced in the table and marked by the small numbers in parentheses. The computation begins at the point where the residual is largest, namely at point 1 where it is 179. When we add 45 to the stress function, the residual will change by $-4 \times 45 = -180$ and will assume the value $179 - 180 = -1$. This new value has been noted under the old one, and the increment of the stress function has been noted under the starting value of 910. Now the residuals in the adjacent points must also be changed, each by $+45$. Two of the adjacent points lie on the boundary where we do not keep track of the residuals. Another neighbor is point 3, and there the new residual $99 + 45 = 144$ has been noted at the appropriate place. The fourth neighbor is the point 3' which lies beyond the axis of symmetry. The situation there will always be identical with that at point 3 and need not be computed separately.

Looking at the new distribution of residuals, one recognizes that 144 at point 3 is the largest one and should now be improved. When we add 36 to the stress function there, the residual becomes exactly zero. This has been noted in panel no. 3. When we now make the necessary changes in the residuals of the adjacent points, we must keep in mind that, together with 3 its correspondent 3' on the other side of the axis of symmetry also gets 36 added to its stress function. Points 1 and 2 are influenced by both changes, and therefore their residuals will be increased by $2 \times 36 = 72$ while at point 6 the increase is only 36. We may express this situation by saying that the point 3 has an irregular relaxation pattern. In a similar way we find that three other points also have irregular patterns. The irregular patterns are

$$
\begin{array}{cc}
& 2 \\
\text{point 3:} \quad 2 \;\; -4 & \text{point 4:} \qquad\qquad -4 \;\; 1 \\
& 1 \qquad\qquad\qquad\qquad\qquad 2 \\
& 2 \qquad\qquad\qquad\qquad\qquad 2 \\
\text{point 5:} \quad 2 \;\; -4 & \text{point 8:} \quad 4 \;\; -4 \;\; 1 \\
& 2
\end{array}
$$

We may now easily follow the third and fourth operations, which also are reproduced in the relaxation table.

The computation must be continued in this way until the residuals have become sufficiently small. This has been done, and work was terminated when the residuals reached the values given at the ends of the dotted lines. It is evident that further improvement is not possible unless changes of less than unity are applied to the Φ values in the table.

Since for the stress function only the starting values and the increments have been noted, it is still necessary to find the final values of Φ by adding everything in each Φ compartment. This sum is found under the horizontal line and represents the final value of $10^5\,\Phi/p\,h\,a^2$. As a check one may compute the final residuals from the final stress function. This check is very important, and if it is not satisfactory, the relaxation must be continued after replacing wrong residuals by their correct values.

The result of the relaxation table may be used either to compute the stress resultants of the shell or to interpolate values of Φ for a finer net and to start with them a new and bigger relaxation table. Whether the one or the other is done depends largely on the accuracy required for the results.

At the four corners the load per unit projected area is $\bar{p}_z = 1.420\,p$, and since at these points $N_x = N_y = 0$, there again $N_{xy} = \infty$. A finite-difference computation cannot reproduce this singularity, and a rather fine grid is needed to obtain good results in the vicinity of the corners. It is therefore advisable to use a slightly different procedure for the stress analysis, splitting the total of $\bar{p}_z$ in a constant part $1.420\,p$ and the difference between this and the actual values, which varies with x and y. For the constant part one may use eqs. (13), and the variable part has the agreeable property that it is zero at the corners. It may therefore be handled by relaxation without running into a singularity. In this way the results have been obtained from which the diagrams in fig. 7 have been drawn. The stress resultants may be compared with those which were found from eqs. (13) for a shell with constant $\bar{p}_z = p$, given by broken lines in the diagrams.

There is an interesting group of shell problems which may readily be solved by the relaxation method, the shells with a central cut-out. In fig. 8 four such shells are shown in plan projection. Around the hole the shell must, of course, be reinforced by a ring. In the first two examples this ring consists of a frame of four arches in vertical planes. It seems plausible to assume that each arch can resist loads lying in its plane but that the shell will not exert a thrust normal to the edge. By the same line of thought which we applied already to the outer edge, one may then recognize that along the inner edge the stress function must also be a constant, say $\Phi = C$, but, of course, this constant may not simply be set equal to zero. It must be determined from the condition that the sum of all vertical forces transmitted from the shell to the frame must be equal to the weight of the frame. To

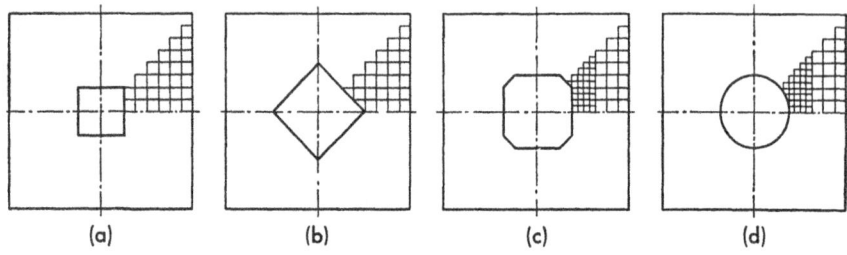

Fig. 8. Square shell roof with skylight

satisfy this condition, two relaxation tables must be set up, one for the load p_z and a reasonable guess of C, yielding a solution Φ_0, and another table for $p_z \equiv 0$ and $C = 1$, yielding the stress function Φ_1. The final solution is then the linear superposition

$$\Phi = \Phi_0 + C' \Phi_1,$$

for which the forces at the inner edge have the required vertical result-ant.

When the numerical work is carried out, it is found that there is a singularity at each reentrant corner. It represents essentially a concentrated thrust in the prolongation of each arch; however, a detailed analytic description of this singularity has not yet been given.

Since the paraboloid, eq. (6), is a surface of revolution, the hole in fig. 8d is not only circular in plan projection but is really bounded by a circle in a horizontal plane. In this case, it does not seem reasonable to require that the shell not lean against the stiffening ring; but it seems reasonable to ask that all forces transmitted to the ring lie in its plane, except for a uniform thrust needed to carry the weight of the ring. In cases like fig. 8c it is not easy to find a plausible boundary condition, and when the hole is oval and bounded by a space curve it becomes entirely impossible to stipulate anything about the boundary condition without first studying the bending theory of the shell.

The use of the relaxation method may easily be extended to the elliptic paraboloid, eq. (9). The only difference from the paraboloid of revolution is that the coefficients of the second derivative in eq. (10) are not the same and that, therefore, the relaxation pattern looks different. In all other shells of positive curvature the coefficients depend on x and y and there is a different relaxation pattern for every grid point. The presence of $\partial^2 z/\partial x\,\partial y$, of course, also influences the relaxation pattern. Except for these circumstances, which make the work somewhat more tedious, everything runs along the same lines and there is no need for explaining more details.

4.3 Hyperbolic Problems

4.3.1 Hyperbolic Paraboloid, Edges Parallel to Generators

The surface which spans in the simplest way a twisted quadrangle (fig. 9) has the equation

$$z = \frac{x\,y}{c} = \frac{x\,y}{a\,b}\,h \qquad (18)$$

and is called a hyperbolic paraboloid. Its intersections with vertical planes $x =$ const or $y =$ const are straight lines, the generators. The quantity $1/c$, the reciprocal of a length, is the twist $\partial^2 z/\partial x\,\partial y$ of the surface, i. e. the difference of slope of two generators, which are a unit length apart.

The hyperbolic paraboloid yields a type of shells which has often been used for roof structures. Its stress problem is best treated with the help of the differential equation (5). Introducing there the relation (18), we find

$$-\frac{2}{c}\,\frac{\partial^2 \varPhi}{\partial x\,\partial y} = -\overline{p}_z + \overline{p}_x\,\frac{y}{c} + \overline{p}_y\,\frac{x}{c}\,.$$

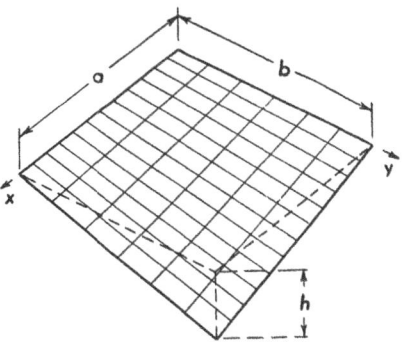

Fig. 9. Hyperbolic paraboloid

$$(19)$$

When we have only vertical loads $\overline{p}_z = p$, constant per unit of horizontal projection of shell surface, the equation is simply

$$\frac{\partial^2 \varPhi}{\partial x\,\partial y} = \frac{1}{2}\,c\,p\,,$$

which, with eq. (4), yields

$$N_{xy} = -\frac{1}{2}\,c\,p \qquad (20\,\text{a})$$

and has the general solution

$$\varPhi = \frac{1}{2}\,c\,p\,x\,y + f_1(x) + f_2(y)$$

with two arbitrary functions f_1 and f_2. Introducing this into eqs. (4) and (1), we find the fiber forces

$$N_x = \sqrt{\frac{c^2 + y^2}{c^2 + x^2}}\,\frac{d^2 f_2}{dy^2}\,, \qquad N_y = \sqrt{\frac{c^2 + x^2}{c^2 + y^2}}\,\frac{d^2 f_1}{dx^2}\,. \qquad (20\,\text{b, c})$$

This is a very simple stress-system. The shear is constant throughout the shell, and the projections $\bar{N}_x$, $\bar{N}_y$ of the fiber forces N_x, N_y are each constant along those generators which have the direction of that force and may vary only from one such generator to the other. From this situation it follows for the force N_x that we can arbitrarily prescribe its values on one of the edges $x = $ const of the shell but that we have no means of influencing the ensuing values on the opposite edge; the same is true for N_y with respect to the edges $y = $ const.

Fig. 10 shows a roof constructed by a combination of four shells of the type just studied. This structure has two vertical planes of symmetry. At the gables there must be edge members to take care of the shear N_{xy} according to eq. (20a). Their axial force F_E must be zero at the top and therefore is

$$F_E = N_{xy}\frac{x}{\cos\alpha} = -\frac{ab}{2h}\frac{px}{\cos\alpha}$$

in the domain AB in fig. 10 and similar in the other parts. At the corner $x = a$ the horizontal component of this force is balanced by

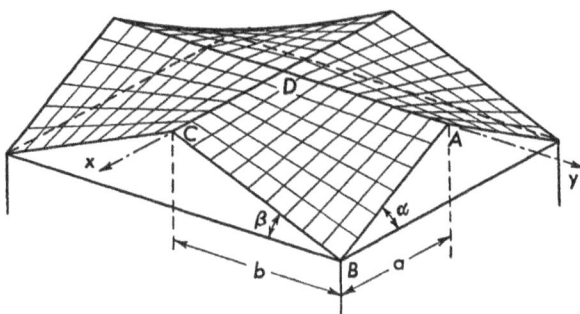

Fig. 10. Shell roof consisting of four hyperbolic paraboloids

means of a tie against a similar component appearing at the next corner. The vertical components of the F_E in two adjoining gables combine to a resultant

$$R = \frac{ab}{2h}\frac{pa}{\cos\alpha}\sin\alpha + \frac{ab}{2h}\frac{pb}{\cos\beta}\sin\beta = pab,$$

which must be carried by a support. It is, of course, equal to one quarter of the total load carried by the structure.

Since the edge member will not be stiff enough to resist a horizontal thrust of the shell, there should be $N_y = 0$ on the edge $y = b$ and $N_x = 0$ on $x = a$. From eqs. (20b, c) it follows that $N_x = N_y = 0$ for the whole shell.

Along the ridges there must be another system of ribs. These receive shearing forces N_{xy} from both sides, which produce a rib force F_R. In the ridge CD it is

$$F_R = 2(a - x) N_{xy} = -c(a - x) p,$$

beginning with $F_R = 0$ at the gable. Beyond the point D, F_R decreases symmetrically.

The stress system described here seems to be extremely simple. But as soon as we look for further details, many difficulties appear. One of them comes from the weight of the ridges. Since these bars are placed along lines where two shells meet at an angle, one might think that the weight should be carried by those shells. This is not possible, because the forces N_x or N_y needed for this purpose cannot exist without giving new trouble at the opposite edges. Therefore, it is necessary that the ridge beams take care of themselves and sustain their own weight as beams supported at the gables.

The situation becomes worse when we try to apply a load to only a part of the roof. Consider, for example, the case that one of the four panels, say $ABCD$, has a uniform load of snow, p,

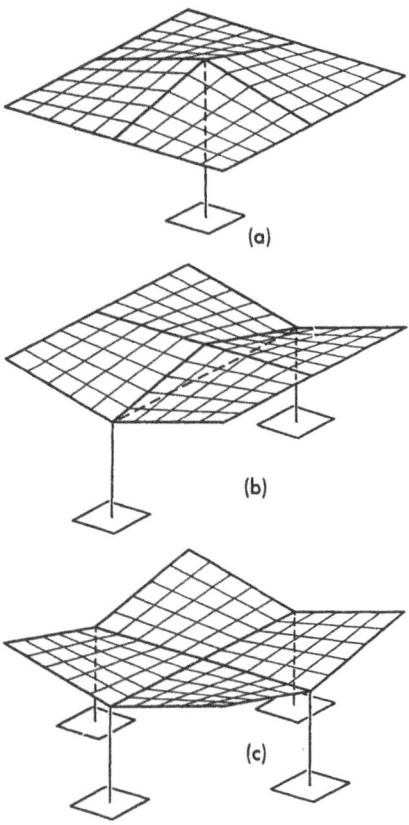

Fig. 11. Examples of shell roofs built up from hyperbolic paraboloids

and that the other three are bare. Then the stress resultants in the first panel will be exactly the same as before,

$$N_{xy} = -\frac{1}{2} a p, \qquad N_x = N_y = 0,$$

and the other three panels must be completely free of stress. The ridge CD would then have exactly one half of the force F_R which we found before, but beyond the point D there is now no shear from the shell to make this force decrease to zero at the far end of the ridge.

It is therefore impossible to find any kind of equilibrium in the structure without resorting to considerable lateral bending of the ribs.

The figures 11 show some other roof structures built up of the same element, fig. 9. They all present the same simple stress problem, if the load is perfectly symmetric, and they all have the same shortcomings under less favorable loading conditions.

4.3.2 Hyperbolic Paraboloid, Edges Bisecting the Directions of the Generators

If we rotate the coordinate system x, y by 45°, the equation of the hyperbolic paraboloid assumes the form

$$z = \frac{x^2 - y^2}{2c}.$$

If we stretch this surface in the x direction, we arrive at a more general type having the equation

$$z = \frac{x^2}{h_1} - \frac{y^2}{h_2}. \tag{21}$$

A rectangular part of such a shell is shown in fig. 12. Its stress resultants will be studied in this Section.

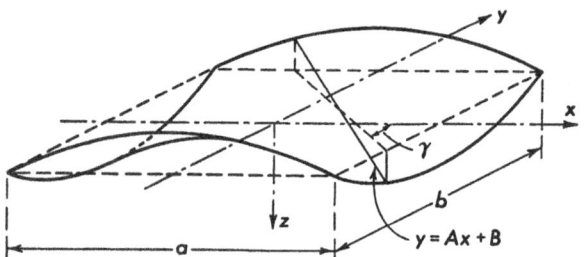

Fig. 12. Hyperbolic paraboloid

Introducing eq. (21) into the general equation (5), we find the differential equation of our particular problem:

$$\frac{2}{h_2}\frac{\partial^2 \Phi}{\partial x^2} - \frac{2}{h_1}\frac{\partial^2 \Phi}{\partial y^2} = \bar{p}_z - \frac{2x}{h_1}\bar{p}_x + \frac{2y}{h_2}\bar{p}_y - \frac{2}{h_1}\int \bar{p}_x\,dx + \frac{2}{h_2}\int \bar{p}_y\,dy. \tag{22}$$

It looks very much like eq. (10) for the elliptic paraboloid, but the second term at the left has a minus sign here, indicating that this equation belongs to the hyperbolic type. This has important consequences for the methods to be used in its solution and for the properties of the stress systems which will be found.

We may easily find a particular solution of eq. (22). For the simplest type of vertical loading, $\bar{p}_x = \bar{p}_y = 0$, $\bar{p}_z = p = \text{const}$, we may choose among

$$\Phi = \frac{1}{4} p h_2 x^2 \quad \text{or} \quad \Phi = -\frac{1}{4} p h_1 y^2 \quad \text{or} \quad \Phi = \frac{1}{8} p (h_2 x^2 - h_1 y^2). \quad (23)$$

These solutions permit a simple mechanical interpretation. We find it by introducing them into eqs. (4) for the stress resultants. For the second solution, only the forces N_x are $\neq 0$ and represent a state of stress in which the shell carries its load like a series of arches parallel to the x, z plane, bringing the load to abutments at the edges $x = \pm a/2$. In the case of the first solution there are similar arches parallel to the y, z plane, but they have tensile forces N_y; and in the third solution both arch systems act jointly in carrying the load. Similar solutions exist for the elliptic paraboloid, but there both arch systems are in compression. There is, however, a difference of much greater importance between these two types of shells. In the elliptic problem we added a homogeneous solution which eliminated the thrust on all four edges and replaced it by tangential forces which could be transmitted to simple diaphragms. Here, in the hyperbolic case, a similar stress system is not generally possible. To make this clear, we must first discuss some more geometric properties of the hyperbolic paraboloid (fig. 12).

We consider a vertical plane

$$y = A + B x.$$

Eliminating y from this equation and from (21), we find a relation valid for all points of the intersection of the plane and the paraboloid:

$$z = \frac{x^2}{h_1} - \frac{A^2 + 2 A B x + B^2 x^2}{h_2}.$$

If we choose $B^2 = h_2/h_1$, the terms with x^2 will cancel, and then the two surfaces will intersect in a straight line. There are two families of such straight lines on the paraboloid, corresponding to $B = \pm \sqrt{h_2/h_1}$. They are the generators of the surface. Their projections on the x, y plane meet the x axis at an angle $\pm \gamma$, where

$$\tan \gamma = \sqrt{\frac{h_2}{h_1}}.$$

When $h_1 = h_2$, the projections meet each other at right angles and the generators are identical with those which were used as coordinate lines in Section 4.3.1. Those shells were bounded by four generators, but here two of them pass through every point on the edges, with the exception of the four corners. From each corner only one generator emanates. It may happen that it traverses the shell diagonally and ends at the opposite corner, but, in general, the generators from the corners will meet one pair of opposite sides of the shell. These sides

will be called the principal sides and the other two, the secondary
sides. With the notations of fig. 12, the sides a are the principal sides,
if the parameter λ, defined by

$$\lambda^2 = a^2 h_2/b^2 h_1,$$

is greater than unity, and it is this parameter which determines essentially the features of the stress system set up in the shell.

Now let us consider the shell in fig. 13 and suppose that, for a
given load system $\bar{p}_x$, $\bar{p}_y$, $\bar{p}_z$, we have found a particular solution of the
differential equation (22). It will yield certain forces on the edges,
which must be applied there as external forces. Some of them may be
undesirable for practical applications, and we are, therefore, interested
in finding solutions of the homogeneous equation which, when added,

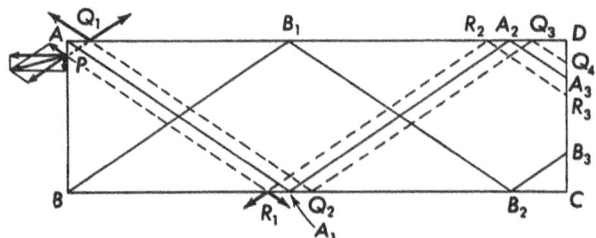

Fig. 13. Plan projection of a hyperbolic paraboloid, showing the generators

cancel those edge forces and adapt the stresses to suitable boundary
conditions, allowing a simple support of the shell.

We start at the secondary edge AB. In a line element ds situated
at an arbitrary point P of this side we have forces $N_x\,ds$ and $N_{xy}\,ds$
which we may combine to form a resultant, lying in the tangential
plane of the shell. To rid the shell of this external force, we apply
a force of the same size and in the opposite direction as an additional
load. We may resolve it into components in the directions of the two
generators PQ_1 and PR_1 passing through P. In the same way, as we
saw for the hyperboloid of revolution on p. 82, these forces will produce stresses only in two straight strips along the generators and will
need for equilibrium external forces of the same size at the opposite
ends Q_1 and R_1 of the strips.

If we do the same at all points of the edge AB, we can rid it completely of all external forces, so that it does not need any support at
all. In exchange we get additional forces on the sections AB_1 and BA_1
of the principal edges AD and BC. They combine with the forces
already present from the inhomogeneous solution.

We may now try to enforce some boundary condition on the lines AB_1
and BA_1 by applying there new external forces. But there we are no

longer completely free in our choice. Consider for instance the point Q_1. The new force to be applied there should have the direction of the generator $Q_1 Q_2$. Otherwise one component would run back to P and there disturb the order we have just created. Hence only the magnitude, not the direction of the new force is free, and we may choose it so that either its y component cancels the already existing thrust N_y or that its x component cancels the shear N_{xy}. In either case, a certain force is introduced into a strip along $Q_1 Q_2$ and reappears at its far end Q_2 as an external force. In the same way, we proceed at all points on AB_1 and BA_1 and get additional forces on $B_1 A_2$ and $A_1 B_2$. This may be continued until we arrive on the secondary edge CD. Then we are through because any additional force applied there would run back to other edges of the shell, where it would not be welcome. The edge CD must, therefore, have a complete support which can resist a thrust N_x as well as a shear N_{xy}.

By the procedure described here we have found a set of boundary conditions which can always be realized on this kind of a shell, irrespective of its length: one secondary edge AB completely free of external forces, the other one completely supported, and the principal edges either resting on plane diaphragms resisting only shear, or, if we should prefer, supported by abutments which can resist only a thrust N_y, but not a shearing force N_{xy}.

The method which we applied to throw boundary forces from one edge to another may be applied again to modify these boundary conditions. It is, for instance, possible to make the edge CD free of forces, if we admit an edge thrust N_y in addition to the shear on certain parts of the principal edges AD and BC. Fig. 14 shows some possibil-

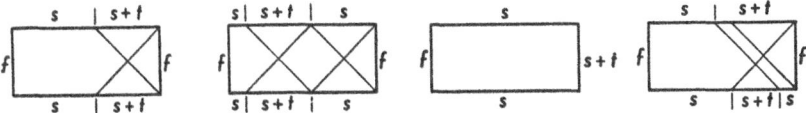

Fig. 14. Plan view of a hyperbolic shell, showing different examples of boundary conditions that may be imposed; f = free edge, s = shear edge (no thrust), $s + t$ = shear and thrust admitted

ities of this kind. Most of them look rather queer. Whether or not we find among them one which is readily applicable to a practical problem depends on the chain of generators $A A_1 A_2 \ldots$, starting at one of the corners. We see this at once when we raise the question whether it would be possible to exchange the thrust on CD against a shear on AB, arriving thus at a set of boundary conditions corresponding to a support of the shell by four diaphragms along its four edges.

Let us study this question on the two shells shown in fig. 15a, b. In fig. 15a the chain of generators starting at A ends at another corner C. In this case the two chains starting at any point E on the left-hand side meet at one point F on the right-hand side. To cancel the thrust at point E, we must there apply an additional thrust, as shown in the figure. If we transfer it through the shell, admitting only additional shear on the principal edges, we see that the forces resulting at F combine exactly to a thrust of the same magnitude. It is hence impossible to exchange the thrust at the left for a shear at the right. Quite different is the behavior of the shell in fig. 15b. Here the two chains of generators emanating from the point E_1 end at two different

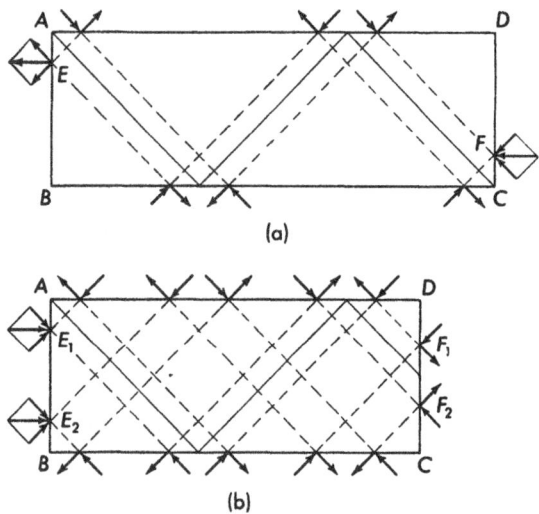

Fig. 15. Transfer of edge loads in hyperbolic shells, depending on the relation of the edges to the generators

points F_1 and F_2, having equal distances from the corners C and D, respectively. But there is another point E_2, which sends its chains of generators to the same two points F_1, F_2. If the load on this shell is symmetric with respect to its horizontal center line, equal thrusts will have to be applied at E_1 and E_2, and when we transfer them across the shell, the forces resulting at F_1 and F_2 combine exactly to two shearing forces of equal magnitude and opposite direction. It follows that this shell may, for a load of a certain symmetry, be supported by four diaphragms. A roof constructed in this way may easily carry its own weight. But it cannot carry unsymmetric loads without transverse bending of the arch ribs on its edges. If we are willing to admit the ensuing stresses, then the simple and intuitive method of solving the stress problem which this kind of shell permits is certainly in their favor.

We shall now illustrate this method by some examples. We begin with the shell shown in plan projection in fig. 16, assuming that there is a vertical load such that $\bar{p}_z = p = \text{const}$. Then the inhomogeneous solutions (23) are applicable, and we choose the second one. It yields the stress resultants

$$\bar{N}_x = \frac{1}{2}\,p\,h_1, \qquad \bar{N}_y = \bar{N}_{xy} = 0. \tag{24}$$

Since we do not want the thrust on the edges AB and CD, we apply here tensile forces of the same magnitude and resolve them in components parallel to the generators. Projected on the x, y plane, and per unit length of line element dy, these components are

$$\frac{p\,h_1}{4\cos\gamma}\,.$$

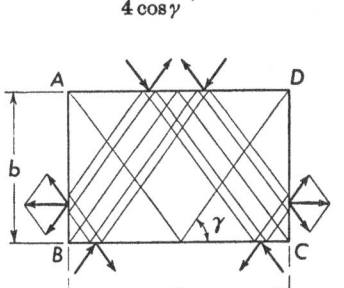

Fig. 16. Transfer of edge loads in a special hyperbolic shell

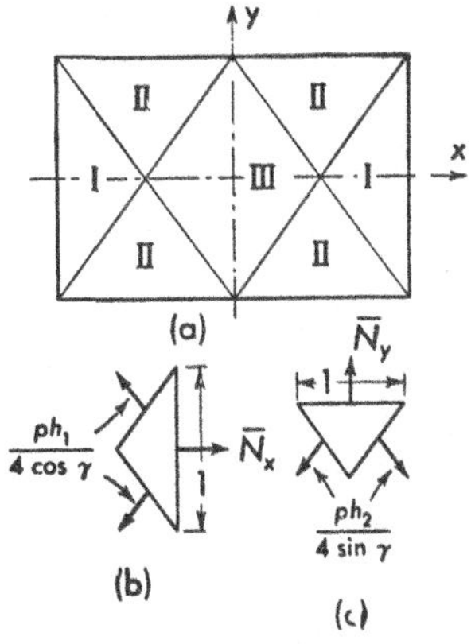

Fig. 17. Calculation of the stress resultants in the shell of Fig. 16

They reappear on the longer sides a, as indicated in the figure. There they are distributed over line elements $ds = dy \cot\gamma$ and hence have the intensity

$$\frac{p\,h_1}{4\cos\gamma} \cdot \frac{1}{\cot\gamma} = \frac{p\,h_2}{4\sin\gamma}\,.$$

To keep these edges free of thrust, we add a compressive force of equal intensity in the direction of the second generator. It happens that these forces equilibrate each other on each generator, and we have found a solution which leaves the sides b free of all external forces and yields only shear on the sides a, certainly an ideal result for practical application.

Until now, we know the stress resultants only for sections parallel to the generators. It will, of course, be useful to know the corresponding forces $\bar{N}_x$, $\bar{N}_y$, $\bar{N}_{xy}$. To find them, we divide the shell into seven zones as shown in fig. 17a. In the zones marked I, we have tensile forces along both generators, as indicated on the triangular shell elements, fig, 17b, c. The equilibrium of these elements yields equations from which we may find the projected forces

$$\text{I:} \; \bar{N}_x = +\frac{1}{2}\,p\,h_1, \quad \bar{N}_{xy} = 0, \quad \bar{N}_y = +\frac{1}{2}\,p\,h_2.$$

In zone III all forces have the opposite sign:

$$\text{III:} \; \bar{N}_x = -\frac{1}{2}\,p\,h_1, \quad \bar{N}_{xy} = 0, \quad \bar{N}_y = -\frac{1}{2}\,p\,h_2.$$

In the zones marked II, one of the generators has tension, the other one compression, and the equilibrium of triangular elements yields

$$\text{II:} \; \bar{N}_x = \bar{N}_y = 0, \quad \bar{N}_{xy} = \pm\frac{1}{2}\,p\,h_1\tan\gamma = \pm\frac{1}{2}\,p\,\sqrt{h_1\,h_2}\,,$$

the positive sign referring to the zones II in the upper left and the lower right of the shell. To all these forces the particular solution (24) must still be added.

The resulting stress system which is represented by the diagrams in fig. 18 has severe discontinuities. When we cross one of the four generators shown in fig. 17a, the fiber force parallel to this line changes its magnitude abruptly, and the corresponding strain does the same. In shells of positive curvature we found such discontinuities only along the edges, but when treating the hyperboloid of revolution (p. 81) we encountered the same phenomenon which we see here: A discontinuity on the edge (there a discontinuous load, here a corner) produces a discontinuity in the stress resultants which is propagated along certain lines right across the shell. This is a general feature of

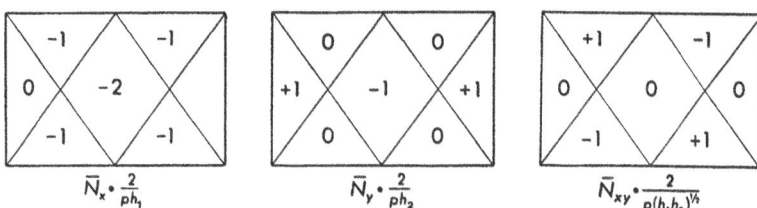

Fig. 18. Distribution of the stress resultants in the shell of Fig. 16

all stress problems governed by differential equations of the hyperbolic type, and we see from eq. (5) that the membrane stress problem is exactly of this type if the shell has negative curvature. This indicates that for all such shells the results of the membrane theory are to be applied with much caution.

As we may already expect from the discussion of fig. 15, the stress resultants will be quite different if we shorten the shell in fig. 16 by one quarter of its length. Starting from the same particular solution as before and applying at the left edge the same additional forces, we find that they will not cancel the thrust at the right edge but lead there to additional shearing forces, and vice versa, elimination of the thrust at the right yields additional shear at the left. When we work out the details we get the stress resultants shown in fig. 19. Along the edges there is no thrust, only shear, just as there was in the case of the elliptic paraboloid. But what a difference in the details! On one half

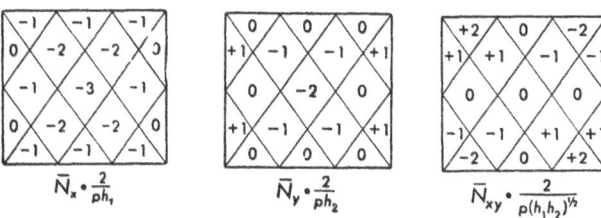

Fig. 19. Stress-distribution patterns similar to those of Fig. 18 but in a shorter shell

of a short side the shear is positive and at the center it suddenly changes to a negative value of the same magnitude, and there are eight lines of discontinuity of stress crossing the interior of the shell.

For any other value of the parameter λ, as defined on p. 186, the stress system in the hyperbolic paraboloid shows more or less different features, which the reader may easily find out by himself.

4.4 Membrane Forces in Affine Shells

4.4.1 General Theory

The left-hand side of the differential equation (5) is not only linear in Φ but also in z. Therefore, if we multiply the stress function Φ by a constant factor and the ordinates z by its reciprocal, the equation will still be satisfied. This hints at a close relationship between the stress resultants in certain families of shells. We shall see how this relationship may be used to solve stress problems.

We start from an arbitrary shell S^*. When we multiply the rectangular coordinates x^*, y^*, z^* of every point of its middle surface by constant factors λ_1, λ_2, λ_3, the new coordinates

$$x = \lambda_1 x^*, \quad y = \lambda_2 y^*, \quad z = \lambda_3 z^* \tag{25}$$

describe the middle surface of another shell S. The two shells are said to be affine to each other, and the set (25) is called an affine transformation.

Now let Φ^* be the stress function which satisfies eq. (5) with the coordinates x^*, y^*, z^* and a certain system of loads $\bar{p}_x^*$, $\bar{p}_y^*$, $\bar{p}_z^*$. We consider the corresponding equation for the shell S, choosing the stress function

$$\Phi = \lambda_1 \lambda_2 \, \Phi^*. \tag{26}$$

The first term of the equation will be

$$\frac{\partial^2 \Phi}{\partial x^2} \frac{\partial^2 z}{\partial y^2} = \frac{\lambda_3}{\lambda_1 \lambda_2} \frac{\partial^2 \Phi^*}{\partial x^{*2}} \frac{\partial^2 z^*}{\partial y^{*2}}.$$

The same constant $\lambda_3/\lambda_1 \lambda_2$ is also the ratio of the other terms on the left-hand sides of both equations and therefore should also be the ratio of the right-hand sides. This condition will be fulfilled if the loads satisfy the relations

$$\lambda_2 \, \bar{p}_x = \bar{p}_x^*, \qquad \lambda_1 \, \bar{p}_y = \bar{p}_y^*, \qquad \frac{\lambda_1 \lambda_2}{\lambda_3} \, \bar{p}_z = \bar{p}_z^*. \tag{27}$$

The quantities $\bar{p}_x^*$, $\bar{p}_y^*$, $\bar{p}_z^*$ were defined as loads per unit of the projected area $dx^* \cdot dy^*$ of a shell element. The total force acting on an element of S^* has therefore the components

$$\bar{p}_x^* \, dx^* \, dy^*, \qquad \bar{p}_y^* \, dx^* \, dy^*, \qquad \bar{p}_z^* \, dx^* \, dy^*,$$

and it follows from eqs. (25) and (27) that the load components on the corresponding element of S have the magnitude

$$\bar{p}_x \, dx \, dy = \lambda_1 \, \bar{p}_x^* \, dx^* \, dy^*, \qquad \bar{p}_y \, dx \, dy = \lambda_2 \, \bar{p}_y^* \, dx^* \, dy^*,$$
$$\bar{p}_z \, dx \, dy = \lambda_3 \, \bar{p}_z^* \, dx^* \, dy^*,$$

that is, they are obtained by applying the same factors which also apply to the coordinates to which these forces are parallel. The loads per unit of the areas dA^* and dA of the real shell elements are then connected by the relations

$$p_x = \lambda_1 \, p_x^* \, \frac{dA^*}{dA}, \qquad p_y = \lambda_2 \, p_y^* \, \frac{dA^*}{dA}, \qquad p_z = \lambda_3 \, p_z^* \, \frac{dA^*}{dA}. \tag{28}$$

We now turn to the stress resultants. They are connected with the stress function by eqs. (4). When we introduce there the relations (25), (26), and (27) we find the corresponding relations for the projected forces:

$$\bar{N}_x = \frac{\lambda_1}{\lambda_2} \, \bar{N}_x^*, \qquad \bar{N}_{xy} = \bar{N}_{xy}^*, \qquad \bar{N}_y = \frac{\lambda_2}{\lambda_1} \, \bar{N}_y^*. \tag{29}$$

According to our former definition, the force $N_x^* \, ds_y^*$ acting on the line element ds_y^* of the shell S^* has the x component $\bar{N}_x^* \, dy^*$. From eqs. (25) and (29) we find its relation to the corresponding force $\bar{N}_x \, dy$ in the shell S:

$$\bar{N}_x \, dy = \lambda_1 \, \bar{N}_x^* \, dy^*.$$

For the vertical component the relation is

$$\bar{N}_x \, dy \cdot \frac{\partial z}{\partial x} = \lambda_1 \, \bar{N}_x^* \, dy^* \cdot \frac{\lambda_3}{\lambda_1} \frac{\partial z^*}{\partial x^*} = \lambda_3 \, \bar{N}_x^* \, dy^* \, \frac{\partial z^*}{\partial x^*}.$$

When we examine all forces in this way, we find that the corresponding components of every internal or external force, applied to corresponding elements of both shells, are always in the same ratio as the coordinates to which they are parallel. And when we put the components of a force together, we see that every pair of corresponding forces, acting on or in the shells S^* and S, has the same ratio as has a pair of (real or hypothetic) line elements of the two shells which would be parallel to these forces. We have, therefore, the following relations for the forces N_x, N_{xy}, N_y of the two shells:

$$N_x = N_x^* \frac{ds_x}{ds_x^*} \frac{ds_y^*}{ds_y}, \qquad N_{xy} = N_{xy}^*, \qquad N_y = N_y^* \frac{ds_y}{ds_y^*} \frac{ds_x^*}{ds_x}. \tag{30}$$

It is easily seen that the last statement and the formulas (30) are not restricted to the forces transmitted in the particular sections along lines $x = $ const or $y = $ const. When we want to apply eqs. (30) to the stress resultants for another reference system, we only need to introduce the line element parallel to the force under consideration and that one in which this force is transmitted.

Together with all other forces in the two shells, the forces transmitted to edge members also are subject to the law of affine transformation. If an edge member is statically determinate, the same is true for the axial and shearing forces therein. To find the rule for the bending moment, let us consider a horizontal ring, e. g. the foot ring of a dome. Its bending moments are the sum of moments of x forces at y lever arms and of y forces at x lever arms. In both cases the factor $\lambda_1 \lambda_2$ applies:

$$M = \lambda_1 \lambda_2 M^*. \tag{31}$$

However, it should be borne in mind that most edge members are statically indeterminate by themselves, and since redundant quantities are not derived from conditions of equilibrium, they are not subject to affine transformation but must be computed for each case according to its own merits.

4.4.2 Applications

4.4.2.1 Vertical Stretching of a Shell of Revolution.

We consider two shells of revolution (fig. 20) whose rectangular coordinates are connected by the simple affine transformation

$$x = x^*, \qquad y = y^*, \qquad z = \lambda z^*, \tag{32}$$

which transforms one shell into the other simply by stretching it in the direction of its axis.

When we use angular coordinates ϕ^*, θ^* and ϕ, θ on both shells, they are connected by the relations

$$\theta = \theta^*, \qquad \tan\phi = \lambda \tan\phi^*. \tag{33}$$

Corresponding parallel circles on both shells have the same radius and therefore equal line elements

$$ds_\theta = ds_\theta^*.$$

The meridional element ds_ϕ^* of the shell S^* has the components $ds_\phi^* \cos \phi^*$ (horizontal) and $ds_\phi^* \sin \phi^*$ (vertical). The latter one is increased by the transformation in the ratio $1 : \lambda$, and the corresponding element on S is therefore

$$ds_\phi = ds_\phi^* \sqrt{\cos^2 \phi^* + \lambda^2 \sin^2 \phi^*}.$$

With this information about the line elements, we may use eqs. (30) to establish relations between the stress resultants in the shells. They are:

$$N_\phi = N_\phi^* (\cos^2 \phi^* + \lambda^2 \sin^2 \phi^*)^{1/2}, \quad N_{\phi\theta} = N_{\phi\theta}^*,$$

$$N_\theta = N_\theta^* (\cos^2 \phi^* + \lambda^2 \sin^2 \phi^*)^{-1/2} \tag{34}$$

The area of the shell elements limited by meridians and parallel circles is, $dA = ds_\phi \cdot ds_\theta$ and $dA^* = ds_\phi^* \cdot ds_\theta^*$, respectively. When their

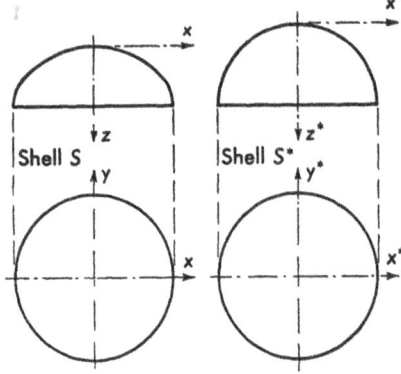

ratio is introduced in eqs. (28), the following relations between the loads on the shells are obtained:

$$p_x = p_x^* (\cos^2 \phi^* + \lambda^2 \sin^2 \phi^*)^{-1/2},$$

$$p_y = p_y^* (\cos^2 \phi^* + \lambda^2 \sin^2 \phi^*)^{-1/2},$$

$$p_z = p_z^* \lambda (\cos^2 \phi^* + \lambda^2 \sin^2 \phi^*)^{-1/2}.$$

We have now all necessary formulas for this particular type of affine transformation and may discuss them and apply them to special problems.

Fig. 20
Vertical stretching of a shell of revolution

When the shell S is a tank or a pressure vessel, there is little merit in applying an affine transformation. The horizontal and vertical components of the pressure p would be multiplied by different factors, and the resulting load on the shell S^* would not be perpendicular to the wall. The advantage gained from substituting, say a sphere S^* for an ellipsoid S is lost through the more complicated load distribution.

When the shell S is a dome, the situation is slightly more in favor of the affine transformation, because the important loads (dead load and snow) are all vertical. But still there is not much advantage in using this detour to the solution of the stress problem, since the straightforward method developed in 2.1 is easily applicable to shells of any meridian.

The real importance of the affine transformation (32) lies in its application to the solutions given in 2.4.2. There we had some simple and important formulas for spherical shells, and we may now adapt them with little more than a stroke of the pen to ellipsoids of revolution. We shall show this here for the formulas (II–38) which describe the effect of an edge load applied to a spherical shell at two edges $\phi = $ const.

We suppose the shell S^* to be part of a sphere, bounded by two parallel circles $\phi^* = \alpha$ and $\phi^* = \beta$. The formulas (II–38) then read as follows:

$$N_{\phi n}^* = -N_{\theta n}^* = \frac{1}{\sin^2\phi^*}\left(A_n \cot^n \frac{\phi^*}{2} + B_n \tan^n \frac{\phi^*}{2}\right),$$

$$N_{\phi\theta n}^* = \frac{1}{\sin^2\phi^*}\left(A_n \cot^n \frac{\phi^*}{2} - B_n \tan^n \frac{\phi^*}{2}\right).$$

The only thing to do is to introduce these expressions on the right-hand sides of eqs. (34), and we already have the desired result:

$$N_{\phi n} = \frac{(\cos^2\phi^* + \lambda^2\sin^2\phi^*)^{1/2}}{\sin^2\phi^*}\left(A_n \cot^n \frac{\phi^*}{2} + B_n \tan^n \frac{\phi^*}{2}\right),$$

$$N_{\theta n} = -\frac{1}{\sin^2\phi^*(\cos^2\phi^* + \lambda^2\sin^2\phi^*)^{1/2}}\left(A_n \cot^n \frac{\phi^*}{2} + B_n \tan^n \frac{\phi^*}{2}\right), \quad (35)$$

$$N_{\phi\theta n} = \frac{1}{\sin^2\phi^*}\left(A_n \cot^n \frac{\phi^*}{2} - B_n \tan^n \frac{\phi^*}{2}\right).$$

Here the stress resultants of the ellipsoid are given in terms of the angle ϕ^* measured on the sphere, which is, of course, as good a coordinate as any. If one prefers to use the angle ϕ measured on the ellipsoid itself, he may use eqs. (33) to introduce it, but this is scarcely worthwhile.

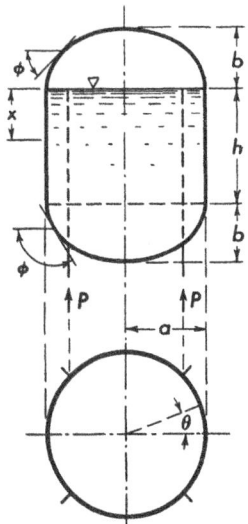

As an example of the application of eqs. (35), we consider the watertank shown in fig. 21. It consists of a cylindrical part closed by two half ellipsoids as roof and bottom, and it is supported by four columns which are riveted to the cylinder over the full length h. When the tank is filled as indicated, the reaction in each column, due to the water weight, will be

$$p = \frac{\pi \gamma a^2}{12}(3h + 2b).$$

We solve the stress problem in two steps. In the first step we try to find simple stress systems in each of the three parts of the tank, disregarding the discrepancies in the shearing forces at the joints. In the second step we use the formulas (35) to make the shearing forces match.

Fig. 21. Water tank, supported by four columns attached to four generators of the cylindrical wall

Since no load is applied directly to the roof shell, we may assume, for the start, that the stress resultants in this half ellipsoid are zero. For the bottom, we found a solution on p. 36. This solution would describe the stress resultants perfectly, if the cylinder could supply the force

$$N_\phi = \frac{1}{6} \gamma a (3h + 2b)$$

needed at the edge, without requiring more.

In the cylinder we have certainly the hoop force

$$N_\theta = \gamma a x$$

due to the water pressure on the shell. In addition, there must be forces N_x which respond to the force N_ϕ from the bottom, and there are four forces P from the supports. We assume that the latter are introduced smoothly along the generators by a uniform line load $P = \mathsf{P}/h$. Each quarter of the shell, cut out between two supports,

$$N_{x\theta} = -\frac{\gamma a^2 \theta}{6h}(3h+2b)$$

$$N_{x\theta} = -\frac{x \gamma a^2}{24h}(3h+2b)$$

$$N_x = \frac{1}{6}\gamma a(3h+2b)$$

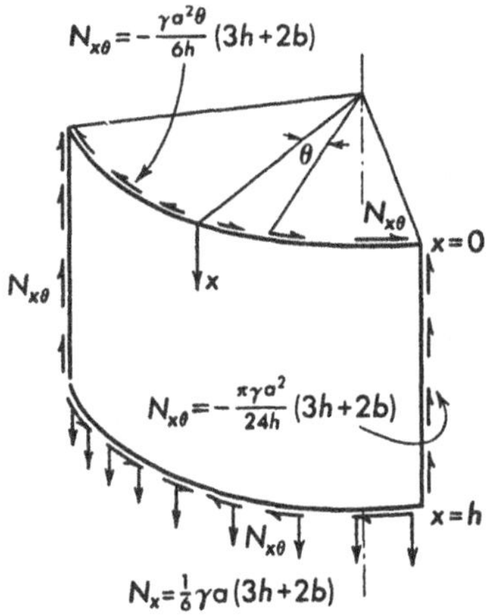

Fig. 22. Forces acting on one quarter of the cylindrical part of Fig. 21

will therefore be subjected to the vertical loads shown in fig. 22. When we compare these forces with the formulas (III–3) for the membrane forces in a cylinder, from which, of course, the load terms must be dropped, we find

$$f_2(\theta) \equiv 0, \qquad \frac{df_1}{d\theta} = -\frac{\gamma a^2}{6h}(3h + 2b)$$

and hence the stress resultants in the cylinder

$$N_{x\theta} = -\frac{\gamma a^2}{6h}(3h + 2b)\theta,$$

$$N_x = \frac{\gamma a}{6h}(3h + 2b)x.$$

At $x = h$, the axial force N_x is in equilibrium with the force N_ϕ coming from the tank bottom. But there are shearing forces $N_{x\theta}$ on both edges of the cylinder which have, so far, no counterpart in the ellipsoids. Since we do not want to apply them as external forces to the structure, we apply additional forces to the edges of the cylinder and of the two half ellipsoids, and we choose them so that the sum of all external forces is everywhere equal to zero.

When we want to apply edge loads to the half ellipsoids, we must use eqs. (35). Since they give the general term of a FOURIER series in θ, we must first write the unbalanced shear of the cylinder as a FOURIER series. The shear distribution along the developed edge of the cylinder is shown in fig. 23. This discontinuous function has the following FOURIER representation:

$$N_{x\theta} = \frac{\gamma a^2}{3h} (3h + 2b) \sum_n \frac{(-1)^{n/4}}{n} \sin n\theta \equiv \sum_n T_n \sin n\theta.$$

The summations in this formula, like all those on the following pages, are to be extended over the values $n = 4, 8, 12, 16, \ldots$ The abbreviation T_n has been introduced to keep the following formulas compact, but we shall get rid of it before we write the final result.

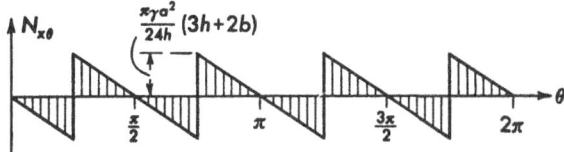

Fig. 23. Edge shear on the cylinder, resulting from Fig. 22

For the roof ellipsoid we must put $A_n = 0$ in order to avoid a meaningless singularity at $\phi = 0$, but B_n may still be chosen freely. At the edge $\phi = 90°$ we have then the shear $N_{\phi\theta} = -B_n \sin n\theta$ and the meridional force $N_\phi = B_n \lambda \cos n\theta$. At the upper edge of the cylinder the n-th harmonic of the shear is $N_{x\theta} = T_n \sin n\theta$. We could easily remove the discrepancy in the shear by putting $B_n = -T_n$, but this would not dispose of the discrepancy in the meridional forces. We must therefore still postpone the final decision on the value of B_n and remove all discrepancies by applying an additional shear

$$N_{x\theta} = -(B_n + T_n) \sin n\theta$$

and an additional axial force

$$N_x = B_n \lambda \cos n\theta$$

to the upper edge of the cylinder.

At the connection of the cylinder with the bottom we have to proceed in a similar way. We apply the solution (35) to the bottom shell, this time putting $B_n = 0$ and leaving A_n undetermined until later. The edge load of the half ellipsoid at $\phi = 90°$ will then be a shear $N_{\phi\theta} = A_n \sin n\theta$ and a meridional force $N_\phi = A_n \lambda \cos n\theta$. At the edge $x = h$ of the cylinder we must then apply the shear

$$N_{x\theta} = (A_n - T_n) \sin n\theta$$

and the axial force

$$N_x = A_n \lambda \cos n\theta.$$

As we have already seen in Chapter 3, it is not possible to apply at both edges of the cylinder arbitrary normal and shearing forces, but we have only the choice of two among them. The relations which exist between the four edge loads just specified will be found when we go with them into eqs. (III–10). When we do so, we not only must replace there m by n, but we must distinguish between the constants in those equations, which we now shall write as A'_n and B'_n, and our present constants A_n and B_n.

We see from eq. (III–10a) that the shear must be the same at both edges $x = 0$ and $x = h$, namely equal to what is now called $-A'_n$. This yields the equation

$$A_n = -B_n$$

and additionally

$$A'_n = B_n + T_n.$$

From eq. (III–10b), applied to the edge $x = 0$, we now find

$$B'_n = B_n \lambda,$$

and when we apply the equation to the other edge $x = h$, we find

$$A'_n \frac{n h}{a} + B'_n = A_n \lambda$$

whence

$$A'_n \frac{n h}{a} = -2 B_n \lambda.$$

When this is compared with the expression just obtained for A'_n, there results

$$B_n = -\frac{n h}{n h + 2 \lambda a} \, T_n = -\frac{\gamma a^2}{3} (3 h + 2 b) \frac{(-1)^{n/4}}{2 b + n h}.$$

We have now the complete solution of our problem. Collecting everything which belongs to the same part of the shell, we find for the roof shell ($0° \leq \phi^* \leq 90°$):

$$N_\phi = -\frac{\gamma a^2}{3} (3 h + 2 b) \frac{(\cos^2\phi^* + \lambda^2 \sin^2\phi^*)^{1/2}}{\sin^2\phi^*} \sum_n \frac{(-1)^{n/4}}{2 b + n h} \tan^n \frac{\phi^*}{2} \cos n\theta,$$

$$N_\theta = \frac{\gamma a^2}{3} (3 h + 2 b) \frac{(\cos^2\phi^* + \lambda^2 \sin^2\phi^*)^{-1/2}}{\sin^2\phi^*} \sum_n \frac{(-1)^{n/4}}{2 b + n h} \tan^n \frac{\phi^*}{2} \cos n\theta,$$

$$N_{\phi\theta} = \frac{\gamma a^2}{3} (3 h + 2 b) \frac{1}{\sin^2\phi^*} \sum_n \frac{(-1)^{n/4}}{2 b + n h} \tan^n \frac{\phi^*}{2} \sin n\theta;$$

for the cylinder ($0 \leq x \leq h$):

$$N_x = \frac{\gamma a x}{6 h} (3 h + 2 b) - \frac{\gamma a b}{3} (3 h + 2 b) \left(1 - \frac{2 x}{h}\right) \sum_n \frac{(-1)^{n/4}}{2 b + n h} \cos n\theta,$$

$$N_{x\theta} = \frac{\gamma a^2}{3} (3 h + 2 b) \sum_n \frac{(-1)^{n/4}}{2 b + n h} \sin n\theta,$$

$$N_\theta = \gamma a x;$$

for the bottom shell ($90° \leqq \phi^* \leqq 180°$):

$$N_\phi = N_{\phi 0} + \frac{\gamma\, a^2}{3}(3h+2b)\frac{(\cos^2\phi^* + \lambda^2 \sin^2\phi^*)^{1/2}}{\sin^2\phi^*}\sum_n \frac{(-1)^{n/4}}{2b+nh}\cot^n\frac{\phi^*}{2}\cos n\theta,$$

$$N_\theta = N_{\theta 0} - \frac{\gamma\, a^2}{3}(3h+2b)\frac{(\cos^2\phi^* + \lambda^2 \sin^2\phi^*)^{-1/2}}{\sin^2\phi^*}\sum_n \frac{(-1)^{n/4}}{2b+nh}\cot^n\frac{\phi^*}{2}\cos n\theta,$$

$$N_{\phi\theta} = \frac{\gamma\, a^2}{3}(3h+2b)\frac{1}{\sin^2\phi^*}\sum_n \frac{(-1)^{n/4}}{2b+nh}\cot^n\frac{\phi^*}{2}\sin n\theta.$$

In fig. 24 some numerical results are shown for $b/a = 0.60$, $h/a = 1.50$. The diagrams give the meridional force (N_x for the cylinder, N_ϕ for the roof and bottom) and the hoop force N_θ in three meridional sections. As in many other cases, the hoop force is discontinuous at the

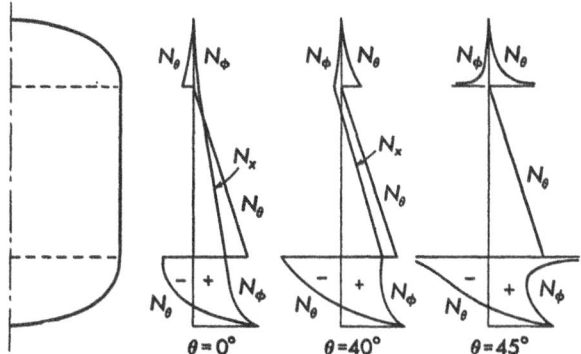

Fig. 24. Stress resultants in the tank shown in Fig. 21

transition from the cylinder to the ellipsoids, indicating that at these joints bending stresses must be expected. The meridional force is, of course, continuous. The roof shell is not much stressed, and in the bottom, which carries the weight of the water, the stress distribution is not too far from being axially symmetric. Only in those meridians which contain the supporting columns are there very localized singularities at the edges of the ellipsoids. Because of these singularities, the force N_x of the cylinder is not defined for $\theta = 45°$, but the other two diagrams show almost identical values of N_x. On the whole, the plots illustrate how easily the shell distributes the concentrated reactions of the columns.

4.4.2.2 Horizontal Stretching of a Shell of Revolution

We consider now (fig. 25) a shell of revolution S^* and a shell S which is derived from S^* by the affine transformation

$$x = \lambda x^*, \quad y = y^*, \quad z = z^*. \tag{36}$$

Through this transformation, every parallel circle of the shell S^* becomes a horizontal ellipse on S. Every meridian of S^* is transformed into a curve which again lies in a vertical plane and may also be called a meridian, but these meridians of S do not all have the same shape. Therefore, the slope is not the same at points which lie on the same level on different meridians, and it is not suitable as a coordinate.

On the shell S^* we may define coordinates ϕ^*, θ^* in the usual way. On the shell S, we use the elliptic parallels and the meridians as coordinate lines, and we attribute to each parallel the value ϕ^* of the corresponding parallel of S^* and to each meridian the value θ^* of the corresponding meridian of S^* (see fig. 25).

We may easily express the length of a line element ds_θ on a parallel or ds_ϕ on a meridian of S by the length of the corresponding element

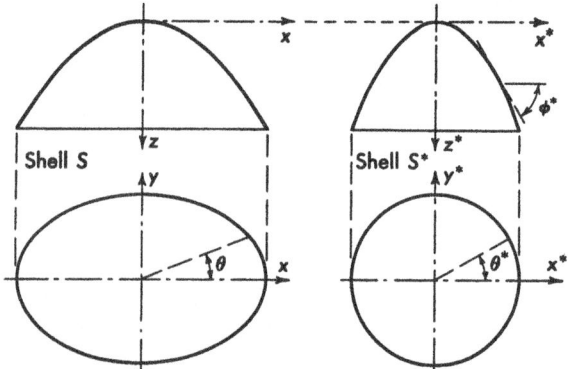

Fig. 25. Horizontal stretching of a shell of revolution

ds_θ^* or ds_ϕ^* on the shell of revolution simply by composing it from its three orthogonal projections on the axes x, y, z:

$$ds_\theta = ds_\theta^* \sqrt{\lambda^2 \sin^2\theta^* + \cos^2\theta^*},$$
$$ds_\phi = ds_\phi^* \sqrt{\lambda^2 \cos^2\phi^* \cos^2\theta^* + \cos^2\phi^* \sin^2\theta^* + \sin^2\phi^*}.$$

On a shell of revolution, the parallels and the meridians meet at right angles. On the shell S, the angle between ds_θ and ds_ϕ will no longer be a right angle. The stress resultants which result from the affine transformation of the forces in S^* will therefore be skew forces as we defined them on p. 15 and as we have already used them in earlier parts of this chapter.

With the help of eqs. (30) the skew forces may be expressed in terms of the stress resultants in the shell S^*:

$$N_\phi = N_\phi^* \sqrt{\frac{\lambda^2 \cos^2\phi^* \cos^2\theta^* + \cos^2\phi^* \sin^2\theta^* + \sin^2\phi^*}{\lambda^2 \sin^2\theta^* + \cos^2\theta^*}},$$
$$N_\theta = N_\theta^* \sqrt{\frac{\lambda^2 \sin^2\theta^* + \cos^2\theta^*}{\lambda^2 \cos^2\phi^* \cos^2\theta^* + \cos^2\phi^* \sin^2\theta^* + \sin^2\phi^*}}, \qquad (37)$$
$$N_{\phi\theta} = N_{\phi\theta}^*.$$

When we want to compute the maximum stresses in the shell, we must apply to these forces the formulas developed in Section 1.2.3. To do so, we need the angle ω between the line elements. The simplest way to find it is to find first the area dA of the shell element which has ds_θ and ds_ϕ as sides. We find it from its projections on the coordinate planes:

$$dA = ds_\theta^* \, ds_\phi^* \, \sqrt{\sin^2\phi^* \cos^2\theta^* + \lambda^2 \sin^2\phi^* \sin^2\theta^* + \lambda^2 \cos^2\phi^*}.$$

Since this area must also be

$$dA = ds_\theta \, ds_\phi \, \sin\omega,$$

we have

$$\sin\omega = \sqrt{\frac{\sin^2\phi^* \cos^2\theta^* + \lambda^2 \sin^2\phi^* \sin^2\theta^* + \lambda^2 \cos^2\phi^*}{(\lambda^2 \cos^2\phi^* \cos^2\theta^* + \cos^2\phi^* \sin^2\theta^* + \sin^2\phi^*)(\cos^2\theta^* + \lambda^2 \sin^2\theta^*)}}.$$

A glance at a sketch will show whether the angle ω is greater or smaller than a right angle.

With the expression just given for dA, we may find from eqs. (28) the transformation of the load components:

$$p_x = p_x^* \, \lambda (\sin^2\phi^* \cos^2\theta^* + \lambda^2 \sin^2\phi^* \sin^2\theta^* + \lambda^2 \cos^2\phi^*)^{-1/2},$$

$$p_y = p_y^* (\sin^2\phi^* \cos^2\theta^* + \lambda^2 \sin^2\phi^* \sin^2\theta^* + \lambda^2 \cos^2\phi^*)^{-1/2}, \quad (38)$$

$$p_z = p_z^* (\sin^2\phi^* \cos^2\theta^* + \lambda^2 \sin^2\phi^* \sin^2\theta^* + \lambda^2 \cos^2\phi^*)^{-1/2}.$$

The preceding formulas permit many useful applications. Since we .can solve any reasonable stress problem for a shell of revolution, we can do the same for every shell that may be derived from it by the affine transformation (36). If the shell S is a pressure vessel or a tank with a vertical z axis, the load p will be perpendicular to the wall and independent of θ^*. The corresponding pressure p^* in the shell of revolution will not have these properties. We may find its components from eqs. (38), solve the stress problem for the shell S^* and compute the forces in the tank from eqs. (37). We shall see on p. 203 the best way of doing this.

When the shell S is a dome, the essential load is vertical ($p_x = p_y = 0$). It consists of the weight of the shell and all the material attached to it for weatherproofing, sound absorption and other purposes. If the thickness of the shell is constant, the load per unit area, p_z, will also be a constant, say p. The corresponding load p_z^* on the shell S^* is then a rather involved function of ϕ^* and θ^*:

$$p_z^* = p_z \sqrt{\lambda^2 + (1 - \lambda^2) \sin^2\phi^* \cos^2\theta^*}.$$

We must resolve it into its components p_ϕ^* and p_r^* (fig. II–2), determine their harmonics and then apply the methods explained in 2.2 and 2.4. As we saw there, the harmonic of order zero requires a foot ring which can resist a tensile force N, due to the thrust of the shell, but no foot ring at all if the meridian happens to end with a vertical tangent. The higher harmonics, however, require a foot ring which can resist a bending load in its plane (p. 57).

When we return to the elliptic dome S, corresponding bending moments appear in its elliptic foot ring and require a heavy design of this ring. The details of their distribution along the circumference may not, of course, be found by affine transformation from the circular dome S^*, because they depend on redundant quantities and hence on deformations.

The elliptic dome may have the same advantage as the dome of revolution, if we distribute the load in a convenient way. We find this distribution when we start from an axially symmetric distribution on the dome S^*. If the load on this shell does not depend on the coordinate θ^*, the forces transmitted to the foot ring consist only in a radial thrust which produces a simple hoop force N^* in the ring. Since this hoop force is determined by an equation of equilibrium it obeys the rules of affine transformation and leads to a hoop force N in the elliptic foot ring which varies in such a way along the springing line that it is at every point in equilibrium with the thrust of the dome without the need of bending moments.

The function p_z^* may in this case be chosen as an arbitrary function of ϕ^*. The third equation (38) then yields the corresponding p_z as a function of both coordinates ϕ^* and θ^*. From this variable load we have to subtract the weight of all accessory material, and the remainder indicates how the wall thickness of the shell must be chosen at each point. If $\lambda > 1$, as in fig. 25, the shell S is thickest at $\theta = 0$ and $\theta = \pi$, thinnest at $\theta = \pm\pi/2$.

4.4.2.3 The General Ellipsoid

The general ellipsoid with three different half axes a, b, c may be derived by the transformation

$$x = \lambda_1 x^*, \qquad y = \lambda_2 y^*, \qquad z = z^*$$

from a sphere of radius c, when we put

$$\lambda_1 = a/c, \qquad \lambda_2 = b/c.$$

On the sphere, we define angular coordinates ϕ^*, θ^* as usual (fig. 26), and we use the same values as coordinates of the corresponding point on the ellipsoid. The relations between the line elements of the two shells are

$$
\begin{aligned}
ds_\theta &= ds_\theta^* \sqrt{\lambda_1^2 \sin^2\theta^* + \lambda_2^2 \cos^2\theta^*}\,, \\
ds_\phi &= ds_\phi^* \sqrt{\lambda_1^2 \cos^2\phi^* \cos^2\theta^* + \lambda_2^2 \cos^2\phi^* \sin^2\theta^* + \sin^2\phi^*}\,.
\end{aligned}
\tag{39}
$$

The element $dA^* = ds^*_\phi \cdot ds^*_\theta$ of the sphere has the following projections on the coordinate planes:

on the y, z plane: $dA^* \sin\phi^* \cos\theta^*$,

on the z, x plane: $dA^* \sin\phi^* \sin\theta^*$,

on the x, y plane: $dA^* \cos\phi^*$.

The projections of the corresponding element dA of the ellipsoid are obtained by multiplying these three quantities by λ_2, λ_1, $\lambda_1\lambda_2$, respectively.

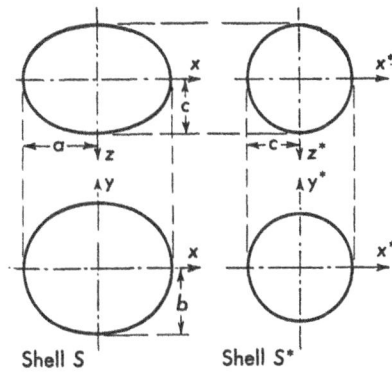

When the ellipsoid is subjected to a constant internal pressure p, a force $p\,dA$ acts on the shell element dA. Its components parallel to the coordinate axes are the products of p with the orthogonal projections of dA, namely

$$p_x\,dA = p\,\lambda_2\,dA^* \sin\phi^* \cos\theta^*,$$
$$p_y\,dA = p\,\lambda_1\,dA^* \sin\phi^* \sin\theta^*,$$
$$p_z\,dA = -p\,\lambda_1\,\lambda_2\,dA^* \cos\phi^*.$$

Shell S Shell S^*

Fig. 26
General ellipsoid S and generating sphere S^*

According to the rule given on p. 192, we find the corresponding forces on the sphere when we divide by λ_1, λ_2, 1, respectively:

$$p_x^*\,dA^* = p\,\frac{\lambda_2}{\lambda_1}\,dA^* \sin\phi^* \cos\theta^* = p\,\frac{b}{a}\,dA^* \sin\phi^* \cos\theta^*,$$
$$p_y^*\,dA^* = p\,\frac{\lambda_1}{\lambda_2}\,dA^* \sin\phi^* \sin\theta^* = p\,\frac{a}{b}\,dA^* \sin\phi^* \sin\theta^*,$$
$$p_z^*\,dA^* = -p\,\lambda_1\,\lambda_2\,dA^* \cos\phi^* = -p\,\frac{ab}{c^2}\,dA^* \cos\phi^*.$$

We have to solve the stress problem of the sphere for this load.

We want to use the formulas of Chapter 2, and so we must express the load by the components p_θ^*, p_ϕ^*, p_r^* as defined by fig. II–2. We find from simple geometric considerations the following formulas:

$$p_\theta^* = -p_x^* \sin\theta^* + p_y^* \cos\theta^*,$$
$$p_\phi^* = (p_x^* \cos\theta^* + p_y^* \sin\theta^*) \cos\phi^* + p_z^* \sin\phi^*,$$
$$p_r^* = (p_x^* \cos\theta^* + p_y^* \sin\theta^*) \sin\phi^* - p_z^* \cos\phi^*.$$

When we introduce here the expressions for p_x^*, p_y^*, p_z^* from the preceding set, we may write the loads as the sum of two harmonic components of orders 0 and 2:

$$p_\theta^* = \frac{p}{2}\left(\frac{a}{b} - \frac{b}{a}\right)\sin\phi^* \sin 2\theta^* \equiv p_{\theta 2}^* \sin 2\theta^*,$$

$$p_\phi^* = \frac{p}{2}\left(\frac{a}{b} + \frac{b}{a} - \frac{2ab}{c^2}\right)\cos\phi^* \sin\phi^* - \frac{p}{2}\left(\frac{a}{b} - \frac{b}{a}\right)\cos\phi^* \sin\phi^* \cos 2\theta^*$$

$$\equiv p_{\phi 0}^* + p_{\phi 2}^* \cos 2\theta^*,$$

$$p_r^* = p\,\frac{ab}{c^2} + \frac{p}{2}\left(\frac{a}{b} + \frac{b}{a} - \frac{2ab}{c^2}\right)\sin^2\phi^* - \frac{p}{2}\left(\frac{a}{b} - \frac{b}{a}\right)\sin^2\phi^* \cos 2\theta^*$$

$$\equiv p_{r0}^* + p_{r2}^* \cos 2\theta^*.$$

The harmonics of order zero, $p_{\phi 0}^*$, p_{r0}^*, may be handled with the integral (II–10) and eq. (II–6 c). They yield the following stress resultants:

$$N_{\phi 0}^* = \frac{p}{2}\,\frac{ab}{c}, \qquad N_{\theta 0}^* = \frac{p}{2}\,\frac{ab}{c} + \frac{p}{2}\left(\frac{ac}{b} + \frac{bc}{a} - 2\frac{ab}{c}\right)\sin^2\phi^*,$$

$$N_{\phi\theta 0}^* = 0.$$

For the second harmonic, $p_{\theta 2}^*$, $p_{\phi 2}^*$, p_{r2}^*, we must use eqs. (II–35) with $n = 2$. They yield the following forces:

$$N_{\phi 2}^* = -\frac{pc}{2}\left(\frac{a}{b} - \frac{b}{a}\right), \qquad N_{\theta 2}^* = \frac{pc}{2}\left(\frac{a}{b} - \frac{b}{a}\right)\cos^2\phi^*,$$

$$N_{\phi\theta 2}^* = \frac{pc}{2}\left(\frac{a}{b} - \frac{b}{a}\right)\cos\phi^*.$$

We are now ready to return to the ellipsoid, using the general formulas (30) with the special expressions (39) for the line elements. The result are formulas for the skew forces N_ϕ, N_θ, $N_{\phi\theta}$ in the ellipsoid:

$$N_\phi = \frac{pc}{2}\left[\frac{ab}{c^2} - \left(\frac{a}{b} - \frac{b}{a}\right)\cos 2\theta^*\right]$$
$$\times \sqrt{\frac{c^2 + (b^2 - c^2)\cos^2\phi^* + (a^2 - b^2)\cos^2\phi^* \cos^2\theta^*}{a^2 + (b^2 - a^2)\cos^2\theta^*}},$$

$$N_\theta = \frac{pc}{2}\left[\frac{ab}{c^2} + \left(\frac{a}{b} + \frac{b}{a} - 2\frac{ab}{c^2}\right)\sin^2\phi^* + \left(\frac{a}{b} - \frac{b}{a}\right)\cos^2\phi^* \cos 2\theta^*\right]$$
$$\times \sqrt{\frac{a^2 + (b^2 - a^2)\cos^2\theta^*}{c^2 + (b^2 - c^2)\cos^2\phi^* + (a^2 - b^2)\cos^2\phi^* \cos^2\theta^*}},$$

$$N_{\phi\theta} = \frac{pc}{2}\left(\frac{a}{b} - \frac{b}{a}\right)\cos\phi^* \sin 2\theta^*.$$

These formulas solve the stress problem for an ellipsoidal shell with constant internal pressure p. Since there is no denominator which might vanish at some point, the stress resultants cannot become infinite, and a membrane stress system is really possible for any choice of the radii a, b, c. This result demonstrates clearly that pressure vessels need not necessarily have a circular cross section. This statement, of course, does not imply that an ellipsoid is better than a sphere, but it indicates that the ellipsoidal tank is feasible at a comparable expense if other circumstances should be in its favor.

In order to give an idea of the stress distribution, some diagrams are shown in fig. 27. They give the forces N_ϕ and N_θ along the ellipses

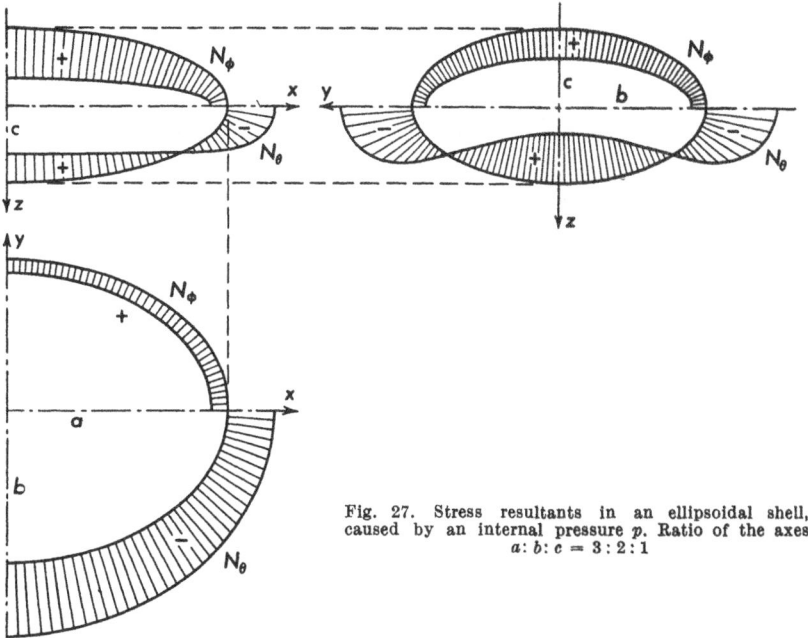

Fig. 27. Stress resultants in an ellipsoidal shell, caused by an internal pressure p. Ratio of the axes $a: b: c = 3 : 2 : 1$

which lie in the planes of symmetry of the shell. On these lines N_ϕ and N_θ are genuine normal forces, and the shear is $N_{\phi\theta} = 0$.

4.4.2.4 Polygonal Domes

A vertical stretching of a polygonal dome according to the transformation (32) may occasionally yield some advantages for the numerical computation. However, most computations on regular domes consist of numerical integrations, and they are best done on the actual structure and not on some affine substitute.

A horizontal stretching according to eqs. (36) transforms a regular dome into a non-regular structure (fig. 28). If the sectors of this oblong dome S all have the same thickness, the load per unit area will be different in the sectors of the corresponding regular dome S^*. With the theory for arbitrary loads developed in 3.4.2 we are prepared to find the forces in S^* and hence in S. The outcome is similar to that which we found for the horizontal stretching of a shell of revolution: The foot ring is subject to heavy bending in its plane, and this makes the structure rather expensive. For large domes it therefore is wise to arrange the dead load so that it corresponds to a regular load on the affine

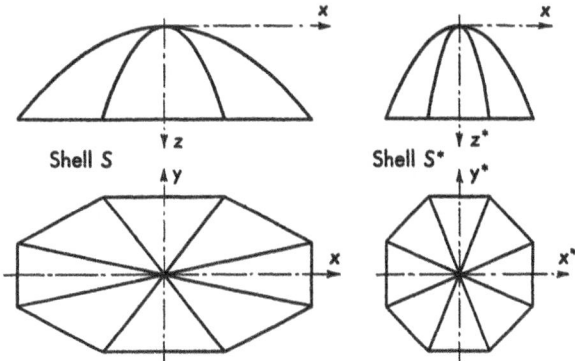

Fig. 28. Horizontal stretching of a polygonal dome

regular dome. The wall thickness must then be different in different sectors and even has a different dependence upon the coordinate ϕ^* in each sector.

4.4.2.5 Cylindrical Shells

If we were to stretch a cylinder in the direction of its generators or in any direction perpendicular to the generators, nothing of interest

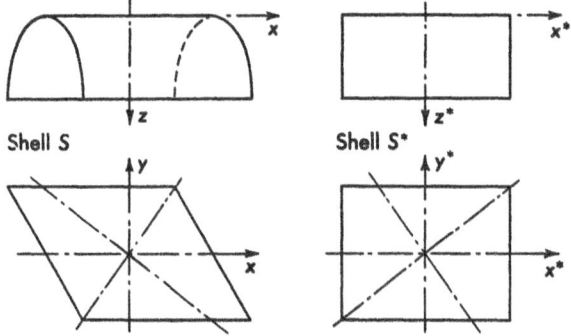

Fig. 29. Affine distorsion of a cylindrical shell

would happen. But when we apply the transformation

$$x = x^* + \lambda y^*, \quad y = y^*, \quad z = z^*, \tag{40}$$

the transformed shell S is a skew cylinder (fig. 29). The equations (40) also represent an affine transformation and not even one of a more general type than eqs. (25). Indeed, we may find an orthogonal system of three axes for which the geometric relation between the two shells would assume the form (25). Two of these axes are indicated by dash-dot lines in fig. 29; the third one is the z axis.

Now suppose that the shell S^* carries that vertical load for which its profile is the funicular curve. If the profile is a common parabola $(r = a/\cos^3 \phi^*)$, this load is $p_z^* = p \cos \phi^*$, if the profile is a catenary $(r = a/\cos^2 \phi^*)$, it is $p_z^* = p = \text{const.}$ In such cases the shell does not need diaphragms or stiffeners on the curved edges, and the stress resultants are

$$N_x^* \equiv N_{x\phi}^* \equiv 0, \qquad N_\phi^* \equiv - p\, a/\cos \phi^*$$

in both cases.

The area dA^* of the shell element is not changed by the transformation (40). The load per unit area is therefore the same on the

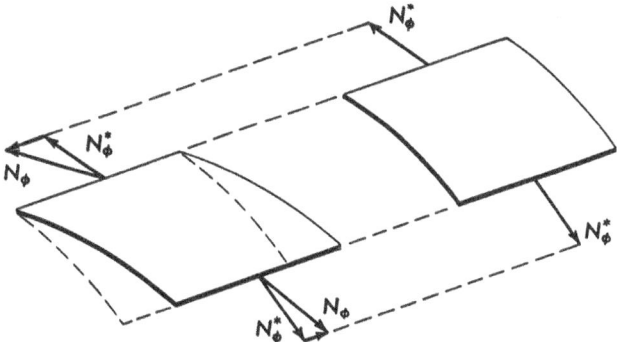

Fig. 30. Corresponding elements of the cylinders S and S^*

skew vault S. On the curved edges and in all sections parallel to them there is no stress, and in sections along the generators there is only the skew force N_ϕ as shown in fig. 30. It may be resolved into an ordinary normal force and an ordinary shear, and from the general rule for the transformation of stress resultants it follows that the normal force is equal to the force N_ϕ^* in S^* and that the shear is $-\lambda\, p\, a$.

This result indicates that a skew vault has the same thrust in the direction of the shortest span as has a straight vault and that there is an additional thrust parallel to the springing line and of such magnitude that the resultant thrust lies in a plane parallel to the faces of the vault.

It may be kept in mind that this simple reasoning is not applicable to arches with bending moments, since these are not subject to the laws of affine transformation. However, since arches are so shaped that they carry most of their load by direct stresses, the affine transformation as applied here must show an essential feature of the force system in a skew arch.

Chapter 5

BENDING OF CIRCULAR CYLINDRICAL SHELLS

In the preceding Chapters dealing with the membrane theory of shells, we often met questions which this theory could not answer. This indicates that in certain cases the bending stiffness of the shell, although small, cannot be disregarded and that it is necessary to develop a bending theory. In such a theory all the stress resultants defined by eqs. (I–1a–j) (pp. 5–6) will appear, and, as one may easily imagine, the mathematical analysis of such stress systems is far from simple. Therefore, solutions have been obtained for only a few of the simplest types of shells. They will be presented in this Chapter and the next.

5.1 Differential Equations

5.1.1 Equilibrium

We use here the same coordinates x and ϕ as we did for the membrane theory of cylinders (see fig. III–1), x being the distance of the point under consideration from a datum plane normal to the generators (here usually coinciding with one edge of the shell) and ϕ measuring the angular distance of the point from a datum generator (here not necessarily the topmost one). The derivatives with respect to the dimensionless coordinates x/a and ϕ will here be indicated by primes and dots:

$$a\,\frac{\partial(\,)}{\partial x} = (\,)', \qquad \frac{\partial(\,)}{\partial\phi} = (\,)^{\cdot}.$$

The shell element as determined by the choice of coordinates is shown in fig. 1a, b. The first of these figures contains all the external and internal forces acting on this element and the second one contains the moments, represented in the usual way by arrows.

These forces and moments have to satisfy six conditions of equilibrium, three of them concerning the force components and the other three, the moments.

The condition for the forces in the x direction is exactly the same as in the membrane theory, eq. (III–1c):

$$N_x' + N_{\phi x}^{\cdot} + p_x\,a = 0. \tag{1a}$$

Eq. (III–1b) deals with the forces in the ϕ direction. Here a term must be added which represents the contribution of the transverse shear Q_ϕ. The two forces $Q_\phi\,dx$ make an angle $d\phi$ with each other and have the tangential resultant $Q_\phi\,dx \cdot d\phi$ which points in the direction of decreasing ϕ. The condition of equilibrium is, therefore,

$$N_\phi^{\cdot} + N_{x\phi}' - Q_\phi + p_\phi\,a = 0. \tag{1b}$$

The third equation refers to the radial components of forces. In the membrane theory it is extremely simple; here it contains contributions of both transverse shears, viz. their increments $(\partial Q_\phi/\partial\phi)\,d\phi \cdot dx$ and $(\partial Q_x/\partial x)\,dx \cdot a\,d\phi$. We have, therefore, the equation

$$Q_\phi^\cdot + Q_x' + N_\phi - p_r\,a = 0, \tag{1c}$$

which no longer requires that the hoop force N_ϕ be proportional to the local load p_r. Thus the restriction that caused most of the trouble we had in the membrane theory of cylinders is eliminated.

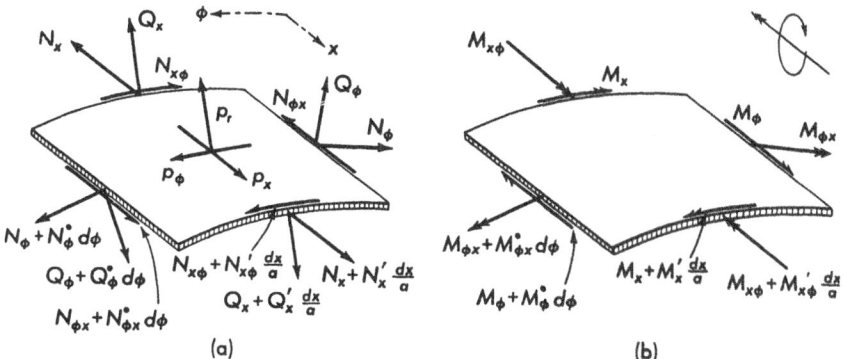

Fig. 1. Shell element

The equations for the equilibrium of moments are easily explained. For an axis of reference, coinciding with the vector p_x in fig. 1a, we have the increments of the bending moment M_ϕ and of the twisting moment $M_{x\phi}$ and the couple formed by the two forces $Q_\phi\,dx$:

$$M_\phi^\cdot + M_{x\phi}' - a\,Q_\phi = 0. \tag{1d}$$

Quite similar is the condition for the moments having the vector p_ϕ as an axis:

$$M_x' + M_{\phi x}^\cdot - a\,Q_x = 0. \tag{1e}$$

The sixth condition of equilibrium contains the moments about a radius of the cylinder. We find there the two couples formed by the forces $N_{x\phi}\,a\,d\phi$ and $N_{\phi x}\,dx$, respectively, and the resultant of the two twisting moments $M_{\phi x}\,dx$ including the angle $d\phi$:

$$a\,N_{x\phi} - a\,N_{\phi x} + M_{\phi x} = 0. \tag{1f}$$

We may easily eliminate the transverse shears Q_x and Q_ϕ from eqs. (1) by means of eqs. (1d, e). In this way we obtain the system:

$$N_x' + N_{\phi x}^\cdot + p_x\,a = 0,$$
$$a\,N_\phi^\cdot + a\,N_{x\phi}' - M_\phi^\cdot - M_{x\phi}' + p_\phi\,a^2 = 0,$$
$$M_\phi^{\cdot\cdot} + M_{x\phi}'^\cdot + M_{\phi x}'^\cdot + M_x'' + a\,N_\phi - p_r\,a^2 = 0, \tag{2a–d}$$
$$a\,N_{x\phi} - a\,N_{\phi x} + M_{\phi x} = 0.$$

Since this set of 4 equations still contains 8 unknown stress resultants, the problem is not statically determinate, and it is necessary to study the deformation of the shell.

5.1.2 Deformation

5.1.2.1 Exact Relations

The deformation of the cylinder may be described by the three components of the displacement of an arbitrary point A of the shell (fig. 2), having the coordinates x, ϕ and the distance z from the middle surface (positive so that $a + z$ is the distance of the point from the axis of the cylinder). For these components we use the following notation:

$u_A =$ displacement along the generator, positive in direction of increasing x,

$v_A =$ displacement along a circle of radius $a + z$, positive in direction of increasing ϕ,

$w_A =$ radial displacement, positive outward.

Determining u_A, v_A, w_A as functions of the three coordinates x, ϕ, z requires the solution of a three-dimensional stress problem. It becomes a shell problem when we establish simple kinematic relations between the displacements u_A, v_A, w_A of an arbitrary point and the corresponding values u, v, w for that point of the middle surface which has the same coordinates x, ϕ. Such relations must be derived from the fact that the shell is thin.

This may be done in different ways. We may start with the fundamental equations of three-dimensional elasticity and investigate which terms become unimportant when t is made small compared with the dimensions of the middle surface; or we may try to use the basic assumptions made in the theories of straight and curved bars and of flat plates. Which way we may prefer will depend essentially on our point of view. If we consider shell theory as a part of the general theory of elasticity, we shall certainly prefer the first way. If we consider shell theory as part of the theory of structures, we may be more inclined toward the second approach, which is nearer the engineer's way of thinking. Of course, results obtained by one approach must agree with those obtained by the other, if both approaches are sound.

We choose the second way and assume:

1) that all points lying on one normal to the middle surface before deformation, do the same after deformation,

2) that for all kinematic relations the distance z of a point from the middle surface may be considered as unaffected by the deformation

of the shell, but that for all considerations of the stress system, the stress σ_z in the z direction may be considered negligible compared with the stresses σ_x and σ_ϕ.

Both assumptions would be exact if the shell were made of a (non-existent) anisotropic material for which the modulus of elasticity in the z direction and the shear moduli for the strains γ_{xz} and $\gamma_{\phi z}$ are infinite, while two of the POISSON constants are zero. In this case all conclusions drawn from the assumptions would be exact. Applied to a real shell, the first assumption means that we neglect the deformations due to the transverse shears Q_x and Q_ϕ. The second assumption means that whatever happens in the z direction, stress or strain, is without significance. This is obviously good if the shell is thin.

To the two basic assumptions we must add a third one, which is needed to keep our equations linear. It is

3) that all displacements are small, i. e. that they are negligible compared with the radii of curvature of the middle surface and that their first derivatives, the slopes, are negligible compared with unity.

From these three assumptions we establish the kinematic relations of the cylindrical shell. Fig. 2a shows a section along a generator. The

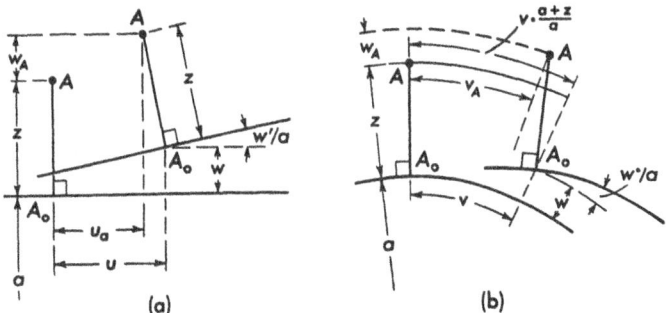

Fig. 2. Displacements of the points A_0 and A

heavy horizontal line is the middle surface before deformation. After-wards, it has the slope $\partial w/\partial x = w'/a$. At the point A_0 the normal $A_0 A$ is erected having the length $z < t/2$. From assumption 1 it follows that this line rotates during deformation by the same angle w'/a. The displacement u_A of point A is therefore equal to the displacement u of point A_0 minus the distance A is shifted back by this rotation of $A_0 A$

$$u_A = u - \frac{z}{a} w'. \tag{3a}$$

To find a similar formula for v_A, we use fig. 2b which shows a trans-verse section through the shell. The point A_0 is displaced by v along the middle surface. Since the normal $A_0 A$ stays normal to this surface, the point A is displaced by $v(a + z)/a$. The rotation of the normal, which is now w'/a, produces an additional displacement $-z\, w'/a$. Together these yield the displacement

$$v_A = \frac{a+z}{a}\, v - \frac{z}{a}\, w^{\cdot}. \tag{3b}$$

Because of our second assumption, the length $A_0 A$ does not change. The difference of the normal displacements w and w_A is then due only to the rotations w'/a and w'/a and proportional to $1 - \cos$ of these angles. Because of the third assumption, this is negligible, and we have

$$w_A = w. \tag{3c}$$

The next step is to find the strains ϵ_x, ϵ_ϕ, $\gamma_{x\phi}$ at the point A. They describe the deformation of an element on the cylindrical surface passing through A. We may therefore apply the formulas (III–18), if we replace the radius r there by $a + z$, and the displacements u, v, w by u_A, v_A, w_A respectively:

$$\epsilon_x = \frac{\partial u_A}{\partial x} = \frac{u'_A}{a},$$

$$\epsilon_\phi = \frac{1}{a+z}\left(\frac{\partial v_A}{\partial \phi} + w_A\right) = \frac{v^{\cdot}_A + w_A}{a+z}, \tag{4a–c}$$

$$\gamma_{x\phi} = \frac{\partial v_A}{\partial x} + \frac{1}{a+z}\,\frac{\partial u_A}{\partial \phi} = \frac{v'_A}{a} + \frac{u^{\cdot}_A}{a+z}.$$

Introducing eqs. (3 a—c) here, we obtain the strains at A as functions of the displacements of A_0:

$$\epsilon_x = \frac{u'}{a} - z\,\frac{w''}{a^2},$$

$$\epsilon_\phi = \frac{v^{\cdot}}{a} - \frac{z}{a}\,\frac{w^{\cdot\cdot}}{a+z} + \frac{w}{a+z}, \tag{5a–c}$$

$$\gamma_{x\phi} = \frac{u^{\cdot}}{a+z} + \frac{a+z}{a^2}\, v' - \frac{w'^{\cdot}}{a}\left(\frac{z}{a} + \frac{z}{a+z}\right).$$

The third step is to find the stresses σ_x, σ_ϕ, $\tau_{x\phi}$ by introducing these expressions into HOOKE's law. Since the second part of the second assumption requires that we neglect σ_z, we have the following formulas, which are equivalent to eqs. (III–17) of the membrane theory:

$$\sigma_x = \frac{E}{1 - \nu^2}\,(\epsilon_x + \nu\,\epsilon_\phi),$$

$$\sigma_\phi = \frac{E}{1 - \nu^2}\,(\epsilon_\phi + \nu\,\epsilon_x), \tag{6a–c}$$

$$\tau_{x\phi} = \frac{E}{2(1 + \nu)}\,\gamma_{x\phi}.$$

When we introduce here the strains from eqs. (5), we have the stresses at A as functions of the displacements of the point A_0 on the middle surface and their derivatives.

The last step is to introduce these expressions in the definitions (I–1) of the stress resultants N and M. In the case of a circular cylinder, we must replace in these equations the subscript y by ϕ and must put the radii $r_x = \infty$, $r_y = a$. We thus obtain the following form of these definitions:

$$N_x = \int_{-t/2}^{+t/2} \sigma_x \left(1 + \frac{z}{a}\right) dz, \qquad N_\phi = \int_{-t/2}^{+t/2} \sigma_\phi \, dz,$$

$$N_{x\phi} = \int_{-t/2}^{+t/2} \tau_{x\phi} \left(1 + \frac{z}{a}\right) dz, \qquad N_{\phi x} = \int_{-t/2}^{+t/2} \tau_{\phi x} \, dz,$$

$$\text{(7a–h)}$$

$$M_x = - \int_{-t/2}^{+t/2} \sigma_x \left(1 + \frac{z}{a}\right) z \, dz, \qquad M_\phi = - \int_{-t/2}^{+t/2} \sigma_\phi \, z \, dz,$$

$$M_{x\phi} = - \int_{-t/2}^{+t/2} \tau_{x\phi} \left(1 + \frac{z}{a}\right) z \, dz, \qquad M_{\phi x} = - \int_{-t/2}^{+t/2} \tau_{\phi x} \, z \, dz.$$

When the stresses from eqs. (6) expressed by the strains from eqs. (5) are introduced in eqs. (7), the integrations with respect to z can be performed. For N_x we find in this way

$$N_x = \frac{E}{1 - \nu^2} \int_{-t/2}^{+t/2} (\epsilon_x + \nu \epsilon_\phi) \frac{a + z}{a} \, dz$$

$$= \frac{E}{a(1 - \nu^2)} \left[(u' + \nu \dot{v} + \nu w) t - w'' \frac{t^3}{12 a^2} \right].$$

On the right-hand side every term contains as a factor either the extensional rigidity

$$D = \frac{E t}{1 - \nu^2}, \tag{8a}$$

which we encountered earlier on p. 87, or the flexural rigidity (bending stiffness)

$$K = \frac{E t^3}{12(1 - \nu^2)}. \tag{8b}$$

Using these notations, the expression for N_x assumes the form (9b) given below. Some of the stress resultants may be treated in exactly the same way; some need an additional explanation. For the other normal force we get e. g.

$$N_\phi = \frac{E}{1 - \nu^2} \int_{-t/2}^{+t/2} (\epsilon_\phi + \nu \epsilon_x) \, dz$$

$$= \frac{E}{a(1 - \nu^2)} \left[(\dot{v} + \nu u') t - \ddot{w} \left(t - a \ln \frac{2a + t}{2a - t} \right) + a w \ln \frac{2a + t}{2a - t} \right].$$

To use here the rigidities D and K, we expand the logarithms in powers of t/a and drop the fifth and higher powers. Thus we get

$$N_\phi = \frac{E}{a(1-\nu^2)} \left[(v^{\cdot} + w + \nu u')\, t + (w^{\cdot\cdot} + w)\, \frac{t^3}{12 a^2} \right],$$

and this may easily be brought into the form (9a). Neglecting the higher powers of t/a evidently means only that the rigidity K appearing at different places in these formulas is not strictly the same but has slightly different values, the differences being of the order t^2/a^2.

Treating all the forces N and the moments M along these lines, we obtain the following set of relations which represent the elastic law for the cylindrical shell:

$$N_\phi = \frac{D}{a}\,(v^{\cdot} + w + \nu u') + \frac{K}{a^3}\,(w + w^{\cdot\cdot}),$$

$$N_x = \frac{D}{a}\,(u' + \nu v^{\cdot} + \nu w) - \frac{K}{a^3}\,w'',$$

$$N_{\phi x} = \frac{D}{a}\,\frac{1-\nu}{2}\,(u^{\cdot} + v') + \frac{K}{a^3}\,\frac{1-\nu}{2}\,(u^{\cdot} + w'^{\cdot}),$$

$$N_{x\phi} = \frac{D}{a}\,\frac{1-\nu}{2}\,(u^{\cdot} + v') + \frac{K}{a^3}\,\frac{1-\nu}{2}\,(v' - w'^{\cdot}),$$

$$M_\phi = \frac{K}{a^2}\,(w + w^{\cdot\cdot} + \nu w''),$$

$$M_x = \frac{K}{a^2}\,(w'' + \nu w^{\cdot\cdot} - u' - \nu v^{\cdot}),$$

$$M_{\phi x} = \frac{K}{a^2}\,(1-\nu)\left(w'^{\cdot} + \frac{1}{2}\,u^{\cdot} - \frac{1}{2}\,v'\right),$$

$$M_{x\phi} = \frac{K}{a^2}\,(1-\nu)\,(w'^{\cdot} - v').$$

(9 a–h)

At first sight these equations appear rather complicated, but it is quite possible to give them a detailed mechanical interpretation. To do this, we first cast them into another form by introducing a set of quantities describing the deformation of an element of the middle surface. They are:

the extensions

$$\bar{\epsilon}_\phi = \frac{v^{\cdot} + w}{a}, \qquad \bar{\epsilon}_x = \frac{u'}{a}, \tag{10a, b}$$

the shear strain

$$\bar{\gamma}_{x\phi} = \frac{u^{\cdot} + v'}{a}, \tag{10c}$$

the changes of curvature

$$\varkappa_\phi = \frac{w^{\cdot\cdot} + w}{a^2}, \qquad \varkappa_x = \frac{w''}{a^2}, \tag{10d, e}$$

the twist

$$\varkappa_{x\phi} = \frac{w'^{\cdot}}{a^2} + \frac{u^{\cdot} - v'}{2a^2}. \tag{10f}$$

The meaning of $\bar{\epsilon}_\phi$, $\bar{\epsilon}_x$ and $\bar{\gamma}_{x\phi}$ is evident, the formulas being identical with eqs. (III–18) used in the membrane theory. In $\varkappa_\phi$ the term w/a^2 needs an explanation. If all points of a shell element undergo a radial displacement w, the radius of curvature is increased from a to $a + w$ and there will be an increase in curvature

$$\frac{1}{a+w} - \frac{1}{a} = \frac{1}{a}\left(1 - \frac{w}{a}\right) - \frac{1}{a} = -\frac{w}{a^2},$$

although there is no rotation of sections or tangents. As formula (9e) shows, the shell really responds to this change of curvature with a bending moment. It is due to the fact that the same increase of length of all hoop fibers, $2\pi w$, produces different strains, the fibers on the inside of the shell being shorter than those on the outside. This leads to a slightly non-uniform stress distribution and hence to a bending moment M_ϕ.

The second term of the twist $\varkappa_{x\phi}$ also needs some comment. It represents the effect of the rotation of the shell element about a normal to the middle surface. It may best be understood by cutting a small rectangle from a piece of paper and placing it on the outside of a cylindrical waterglass. When we rotate the element about its normal and require that it maintain everywhere its contact with the glass (i. e. $w = 0$), two opposite corners will come closer to a tangential plane, while the other corners are moving away from it. Such a deformation of a rectangle is a twist, and this is what the second term in $\varkappa_{x\phi}$ represents.

When we introduce the deformations defined by eqs. (10) into the elastic law (9), it assumes the following form:

$$N_\phi = D(\bar{\epsilon}_\phi + \nu\bar{\epsilon}_x) + \frac{K}{a}\varkappa_\phi,$$

$$N_x = D(\bar{\epsilon}_x + \nu\bar{\epsilon}_\phi) - \frac{K}{a}\varkappa_x,$$

$$N_{\phi x} = \frac{D(1-\nu)}{2}\bar{\gamma}_{x\phi} + \frac{K(1-\nu)}{2a}\left(\varkappa_{x\phi} + \frac{\bar{\gamma}_{x\phi}}{2a}\right),$$

$$N_{x\phi} = \frac{D(1-\nu)}{2}\bar{\gamma}_{x\phi} - \frac{K(1-\nu)}{2a}\left(\varkappa_{x\phi} - \frac{\bar{\gamma}_{x\phi}}{2a}\right),$$

$$M_\phi = K(\varkappa_\phi + \nu\varkappa_x),$$

$$M_x = K\left(\varkappa_x + \nu\varkappa_\phi - \frac{\bar{\epsilon}_x + \nu\bar{\epsilon}_\phi}{a}\right),$$

$$M_{\phi x} = K(1 - \nu)\varkappa_{x\phi},$$

$$M_{x\phi} = K(1 - \nu)\left(\varkappa_{x\phi} - \frac{\bar{\gamma}_{x\phi}}{2a}\right).$$

(11 a–h)

We are now prepared to give a mechanical interpretation of the elastic law. If we put $K = 0$, the moments vanish altogether and for the forces we obtain the simple formulas of the membrane theory: each normal force proportional to the sum of the corresponding strain and ν times the other one and the shearing forces equal to each other and proportional to the shear strain in the middle surface.

When we now consider the complete formulas, we understand easily the terms with $\varkappa$ in the moments. But there is a term with $\bar\epsilon_x$ in M_x. It is due to the fact that the faces $x = $ const of the shell element are trapezoids (see fig. I–2). Therefore the resultant of a uniform distribution of stresses σ_x across these faces does not lie on the middle surface but has a slight eccentricity yielding a contribution to the bending moment M_x. The term $\nu\,\bar\epsilon_\phi$ arises from the fact that $v^{\textbf{·}}$ leads to a uniform distribution of strains ϵ_ϕ and through lateral contraction to uniformly distributed stresses σ_x to which the same reasoning again applies, while the deformation w produces just such a nonuniformity of stresses σ_x that the effect of the trapezoidal shape of the section is compensated. For the terms with K in the normal and shearing forces similar explanations may be found, if we keep in mind that the values of the stresses at $z = 0$ are not necessarily the average values of the stresses across the thickness of the shell.

5.1.2.2 Approximate Relations

The formulas (9) or, what amounts to the same, the combination of eqs. (10) and (11), are a sure foundation for the bending theory of circular cylinders, since they are derived from a clear set of assumptions without losing anything on the way. We shall therefore need them in all such cases where doubts may occur as to whether or not a term may be neglected.

However, they contain a number of terms which in many cases are without importance for the numerical result. It is therefore useful to consider also a simplified version of these equations. There are two sources from which many small terms are derived. One is the trapezoidal shape of the faces $x = $ const of the shell element as expressed by the factor $(1 + z/a)$ in half of the equations (7). If the shell is thin enough we may neglect z/a compared with 1 and simply drop the factor. On the other hand, we have at many places in the kinematic relations (5) the denominator $(a + z)$ which represents the fact that the hoop fibers at different levels z have different lengths. Here also we should consequently neglect z and so simplify the relations. When this is done, the following, much simpler elastic law is obtained:

$$N_\phi = \frac{D}{a}\,(v^{\cdot} + w + \nu\,u'),$$

$$N_x = \frac{D}{a}\,(u' + \nu\,v^{\cdot} + \nu\,w),$$

$$N_{\phi x} = N_{x\phi} = \frac{D(1-\nu)}{2a}\,(u^{\cdot} + v'),$$

$$M_\phi = \frac{K}{a^2}\,(w^{\cdot\cdot} + \nu\,w''),$$

$$M_x = \frac{K}{a^2}\,(w'' + \nu\,w^{\cdot\cdot}),$$

$$M_{\phi x} = M_{x\phi} = \frac{K(1-\nu)}{a^2}\,w'^{\cdot}.$$

(12a–f)

How good or bad this approximation is, we may judge later, when we have a chance to compare results. But there is one point of fundamental interest which may be discussed at once. In the simplifided formulas the difference between the shearing forces $N_{\phi x}$ and $N_{x\phi}$ has disappeared. The sixth condition of equilibrium, (1f), is therefore no longer satisfied if $M_{\phi x} \neq 0$, which is generally the case. This violation of one of the fundamental principles of mechanics which is inseparable from the simplified equations (11), is a serious drawback for all theory founded thereon. In most cases small and otherwise insignificant changes of $N_{\phi x}$ and $N_{x\phi}$ will be sufficient to adjust the equilibrium, but during the mathematical handling of the equations it may happen that the large terms cancel and just the small ones become decisive.

5.1.2.3 Secondary Stresses in Membrane Theory

When discussing membrane forces in Chapter III, we simply made the assumption that all bending and twisting moments were zero. We found that often at the borders such conditions prevail that this assumption cannot be true, but we could not check how good it might be in cases where boundary conditions are favorable.

We now have the means to do this and we shall show here by a simple example the extent to which the membrane forces may represent the real state of stress in a cylindrical shell.

On p. 122 we found the formulas (III–16) for the membrane forces N_ϕ, N_x, $N_{x\phi}$ in a cylindrical pipe supported at both ends by diaphragms and subjected to its own weight. On p. 133 we found the formulas (III–24) which give the corresponding displacements. They describe the deformation of a shell of extensional rigidity $D = E\,t/(1 - \nu^2)$ and of bending rigidity $K = 0$. Now the real shell of thickness t has a finite rigidity K. To ask how much the existence of this bending rigidity will change the deformation would be equivalent to asking for a complete solution of the bending problem, i. e. of the differential equations (13) on p. 219. We set ourselves the simpler problem of asking for the stress resultants which would produce the deformations (III–24), assuming that the loads can be adjusted to satisfy the conditions of equilibrium.

The answer is found by introducing these displacements in the elastic law (9). We simplify this procedure without losing something essential by assuming $\nu = 0$. Then we find that the moments $M_\phi = M_{x\phi} = 0$ and

$$M_x = \frac{p\,t^2}{24\,a^2}\,(8a^2 + l^2 - 4x^2)\cos\phi,$$

$$M_{\phi x} = -\frac{p\,t^2}{6a}\,x\sin\phi.$$

These moments may be introduced into the conditions of equilibrium (1 d, e) which yield the transverse forces

$$Q_\phi = 0, \qquad Q_x = -\frac{p\,t^2}{2a^2}\,x\cos\phi.$$

These shearing forces have been neglected in the conditions of equilibrium used in the membrane theory. That we did well in doing so, we see best in the following way: The existence of a bending moment M_x besides a normal force N_x indicates that the resultant of the stresses σ_x is eccentric with respect to the middle surface. This eccentricity and that of the shearing force $N_{\phi x}$ may be found by dividing the moments by the respective membrane forces:

$$\frac{M_x}{N_x} = -\frac{t^2}{a}\,\frac{8a^2 + l^2 - 4x^2}{6(l^2 - 4x^2)}, \qquad \frac{M_{\phi x}}{N_{\phi x}} = \frac{t^2}{12a}.$$

They are of the order $t \cdot t/a$, i. e. small compared with the thickness of the shell, as this thickness is small compared with the radius a of the cylinder. This proves the reliability of the results of the membrane theory if the boundary conditions are such that no bending of greater order of magnitude is enforced there.

5.1.3 Differential Equations for the Displacements

The conditions of equilibrium in the form (2) and the elastic law (9) together are 12 equations for 11 unknowns: 8 stress resultants and 3 displacements. At first sight it seems that we have at last got one equation too many and that the problem is overdetermined. But the surplus of one equation is not real. In fact, eq. (2d), the "sixth condition of equilibrium", as we called it, because it is the sixth in the original set (1), is an immediate consequence of the relation $\tau_{x\phi} = \tau_{\phi x}$. It therefore becomes an identity when we use eqs. (9) to express the stress resultants by the displacements. When we do the same with eqs. (2a–c), they become three differential equations for u, v, w, the displacements of the middle surface:

$$u'' + \frac{1-\nu}{2} u^{\cdot\cdot} + \frac{1+\nu}{2} v'^{\cdot} + \nu w'$$

$$+ k\left[\frac{1-\nu}{2} u^{\cdot\cdot} - w''' + \frac{1-\nu}{2} w'^{\cdot\cdot}\right] + \frac{p_x a^2}{D} = 0,$$

$$\frac{1+\nu}{2} u'^{\cdot} + v^{\cdot\cdot} + \frac{1-\nu}{2} v'' + w^{\cdot}$$

$$+ k\left[\frac{3}{2}(1-\nu)v'' - \frac{3-\nu}{2} w''^{\cdot}\right] + \frac{p_\phi a^2}{D} = 0, \quad (13\text{a–c})$$

$$\nu u' + v^{\cdot} + w + k\left[\frac{1-\nu}{2} u'^{\cdot\cdot} - u''' - \frac{3-\nu}{2} v''^{\cdot}\right.$$

$$\left. + w^{IV} + 2w''^{\cdot\cdot} + w^{\cdot\cdot\cdot\cdot} + 2w^{\cdot\cdot} + w\right] - \frac{p_r a^2}{D} = 0.$$

Here the dimensionless quantity k stands as an abbreviation for

$$k = \frac{K}{D a^2} = \frac{t^2}{12 a^2}. \tag{14}$$

It is a rather small number.

Eqs. (13) are the differential equations of the bending theory of a circular cylindrical shell. Their solution will be treated in Sections 5.2 through 5.5.

In many cases eqs. (13) may be replaced by a simpler set, derived from the approximate elastic law (12). Since the last of the conditions of equilibrium, (2d), does not contribute to eqs. (13), the fact that it is irreconcilable with eqs. (12) has no immediate consequences. From eqs. (2a–c) and (12) we find

$$u'' + \frac{1-\nu}{2} u^{\cdot\cdot} + \frac{1+\nu}{2} v'^{\cdot} + \nu w' + \frac{p_x a^2}{D} = 0,$$

$$\frac{1+\nu}{2} u'^{\cdot} + v^{\cdot\cdot} + \frac{1-\nu}{2} v'' + w^{\cdot} - k(w''^{\cdot} + w^{\cdot\cdot\cdot}) + \frac{p_\phi a^2}{D} = 0,$$

$$\nu u' + v^{\cdot} + w + k(w^{IV} + 2w''^{\cdot\cdot} + w^{\cdot\cdot\cdot\cdot}) - \frac{p_r a^2}{D} = 0.$$

When we compare these equations with the exact set (13) we see that the terms without the factor k (the membrane terms) are exactly the same but that there are great changes in the k terms (the bending terms). In the brackets of eq. (13c) all the terms with u and v have disappeared, and of the w terms only those with the highest derivatives have survived. The k terms have completely disappeared from eq. (13a), and they have so thoroughly changed in eq. (13b) that they cannot be of much importance. We may therefore feel inclined to neglect them altogether. We find a confirmation of this idea when we check their origin. They stem from the moments in eq. (2b), and these stand for Q_ϕ in eq. (1b). It seems plausible that this term can be neglected, since in a thin shell the transverse shearing forces must be rather small. This argument however, may not be applied in eq. (1c), since Q_x and Q_ϕ may and really do change rather rapidly from one point to another and, therefore, have large derivatives.

As a result of all these considerations we now write the simplified differential equations of the cylindrical shell in the following final form:

$$u'' + \frac{1-\nu}{2}\, u^{\cdot\cdot} + \frac{1+\nu}{2}\, v^{\cdot\prime} + \nu\, w' + \frac{p_x\, a^2}{D} = 0,$$

$$\frac{1+\nu}{2}\, u^{\prime\cdot} + v^{\cdot\cdot} + \frac{1-\nu}{2}\, v'' + w^{\cdot} + \frac{p_\phi\, a^2}{D} = 0, \qquad (15\,\text{a–c})$$

$$\nu\, u' + v^{\cdot} + w + k(w^{\mathrm{IV}} + 2w''^{\cdot\cdot} + w^{\cdot\cdot}) - \frac{p_r\, a^2}{D} = 0.$$

These equations are so simple that it is easy to eliminate u and v from them and so derive a single differential equation for the radial deflection w. We shall show this for the homogeneous equations with $p_x \equiv p_\phi \equiv p_r \equiv 0$.

We recognize in the w terms of eq. (15c) the square of the LAPLACian operator

$$\nabla^2 = a^2 \frac{\partial^2}{\partial x^2} + \frac{\partial^2}{\partial \phi^2} = (\ )'' + (\ )^{\cdot\cdot},$$

and we shall now use the ∇^2 notation.

We begin by eliminating u between eqs. (15a, b). This is done by applying the operator $(\ )^{\prime\cdot}$ to eq. (15a) and then using eq. (15b) to express $u'''^{\cdot}$ and $u'^{\cdot\cdot\cdot}$ in terms of v, w and their derivatives. The result is the equation

$$\nabla^4 v = -(2 + \nu)\, w''^{\cdot} - w^{\cdot\cdot\cdot}. \qquad (16\,\text{a})$$

In a similar way we may eliminate v between the same equations and obtain

$$\nabla^4 u = -\nu\, w''' + w'^{\cdot\cdot}. \qquad (16\,\text{b})$$

When we now apply the operator ∇^4 to eq. (15c), we may use eqs. (16a, b) to express $\nabla^4 u'$ and $\nabla^4 v^{\cdot}$ in terms of w. This yields the differ-

ential equation
$$(1 - \nu^2)\, w^{IV} + k \, \nabla^8 \, w = 0. \tag{17}$$

It is of the eighth order. When it has been solved, eqs. (16a, b) may be solved for u and v. However, since these equations have been obtained by differentiating the original equations (15), not every solution of eqs. (16a, b) is acceptable but only those which also satisfy eqs. (15a,b).

The equations (15) are the result of such sweeping simplifications that they should not be relied upon too much; but since the bending of shells is a rather complicated subject, it will often be useful to have them.

5.2 Solution of the Inhomogeneous Problem

We saw on p. 218 that for thin shells the stress resultants and displacements as computed from the membrane theory are a very good approximation of the solution to be expected from the bending theory. The main objective of the bending theory therefore is not an improvement of these membrane solutions but a study of the stresses produced by certain edge loads which do not fit into the general pattern of the membrane theory. This will be done in much detail in Sections 5.3 through 5.5.

There exist, however, occasions where it is desirable to have a particular solution of eq. (13) for a given surface load p_x, p_ϕ, p_r. We may easily find such a solution if the loads are distributed according to the following formulas:

$$\left.\begin{aligned}
p_x &= p_{xmn} \cos m\,\phi \cos \frac{\lambda\,x}{a}, \\[4pt]
p_\phi &= p_{\phi mn} \sin m\,\phi \sin \frac{\lambda\,x}{a}, \qquad \lambda = n\,\frac{\pi\,a}{l}, \\[4pt]
p_r &= p_{rmn} \cos m\,\phi \, \sin \frac{\lambda\,x}{a},
\end{aligned}\right\} \tag{18}$$

where p_{xmn}, $p_{\phi mn}$, p_{rmn} are three constants which may be given independently. When we introduce eqs. (18) in the differential equations (13), we see that there exists a particular solution in the form

$$\left.\begin{aligned}
u &= u_{mn} \cos m\,\phi \cos \frac{\lambda\,x}{a}, \\[4pt]
v &= v_{mn} \sin m\,\phi \, \sin \frac{\lambda\,x}{a}, \\[4pt]
w &= w_{mn} \cos m\,\phi \sin \frac{\lambda\,x}{a},
\end{aligned}\right\} \tag{19}$$

with three unknown constants u_{mn}, v_{mn}, w_{mn}. They, of course, have to be determined from the differential equations. Introducing (19) into (13), we may drop the trigonometric factors and arrive at the following set of three linear equations for u_{mn}, v_{mn}, w_{mn}:

$$\left[\lambda^2 + \frac{1-\nu}{2} m^2(1+k)\right] u_{mn} + \left[-\frac{1-\nu}{2} \lambda m\right] v_{mn}$$

$$+ \left[-\nu\lambda - k\left(\lambda^3 - \frac{1-\nu}{2}\lambda m^2\right)\right] w_{mn} = \frac{a^2}{D} p_{xmn},$$

$$\left[-\frac{1+\nu}{2}\lambda m\right] u_{mn} + \left[m^2 + \frac{1-\nu}{2}\lambda^2(1+3k)\right] v_{mn}$$

$$+ \left[m + \frac{3-\nu}{2} k\lambda^2 m\right] w_{mn} = \frac{a^2}{D} p_{\phi mn}, \qquad (20)$$

$$\left[-\nu\lambda - k\left(\lambda^3 - \frac{1-\nu}{2}\lambda m^2\right)\right] u_{mn} + \left[m + \frac{3-\nu}{2} k\lambda^2 m\right] v_{mn}$$

$$+ \left[1 + k(\lambda^4 + 2\lambda^2 m^2 + m^4 - 2m^2 + 1)\right] w_{mn} = \frac{a^2}{D} p_{rmn}.$$

From them the numerical values of u_{mn}, v_{mn}, w_{mn} may be found in any concrete case. To obtain the stress resultants we only have to introduce eqs. (19) into the elastic law (9) and the two equations (1d, e). This yields the following set of formulas:

$$N_\phi = \frac{D}{a} [m v_{mn} + (1 + k - k m^2) w_{mn} - \nu\lambda u_{mn}] \cos m\phi \sin\frac{\lambda x}{a},$$

$$N_x = \frac{D}{a} [-\lambda u_{mn} + \nu m v_{mn} + (\nu + k\lambda^2) w_{mn}] \cos m\phi \sin\frac{\lambda x}{a},$$

$$N_{\phi x} = \frac{D(1-\nu)}{2a} [-(1 + k) m u_{mn} + \lambda v_{mn} - k\lambda m w_{mn}] \sin m\phi \cos\frac{\lambda x}{a},$$

$$N_{x\phi} = \frac{D(1-\nu)}{2a} [-m u_{mn} + (1 + k)\lambda v_{mn} + k\lambda m w_{mn}] \sin m\phi \cos\frac{\lambda x}{a},$$

$$M_\phi = -\frac{K}{a^2} (m^2 - 1 + \nu\lambda^2) w_{mn} \cos m\phi \sin\frac{\lambda x}{a},$$

$$M_x = -\frac{K}{a^2} [(\lambda^2 + \nu m^2) w_{mn} - \lambda u_{mn} + \nu m v_{mn}] \cos m\phi \sin\frac{\lambda x}{a}, \qquad (21)$$

$$M_{\phi x} = -\frac{K(1-\nu)}{a^2} \left[\lambda m w_{mn} + \frac{1}{2} m u_{mn} + \frac{1}{2}\lambda v_{mn}\right] \sin m\phi \cos\frac{\lambda x}{a},$$

$$M_{x\phi} = -\frac{K(1-\nu)}{a^2} [\lambda m w_{mn} + \lambda v_{mn}] \sin m\phi \cos\frac{\lambda x}{a},$$

$$Q_\phi = +\frac{K}{a^2} [m(m^2 + \lambda^2 - 1) w_{mn} + (1 - \nu)\lambda^2 v_{mn}] \sin m\phi \sin\frac{\lambda x}{a},$$

$$Q_x = -\frac{K}{a^2} \left[\lambda(\lambda^2 + m^2) w_{mn} + \left(\frac{1-\nu}{2} m^2 - \lambda^2\right) u_{mn} + \frac{1+\nu}{2}\lambda m v_{mn}\right] \cdot$$

$$\cos m\phi \cos\frac{\lambda x}{a}.$$

This, of course, is not the complete solution of the bending problem since it does not have free constants to adapt it to arbitrary boundary conditions, but it fulfills a certain set of such conditions which frequently occurs. All stress resultants and displacements which vary as $\sin \lambda x/a$ vanish at $x = 0$ and $x = l$. Among them are v, w, N_x, and M_x, and they are exactly those which must be zero if the edge of the shell is supported by a plane diaphragm, i. e. a plane structure which is indeformable in its own plane but offers no resistance against displacements perpendicular to this plane. In membrane theory, we used the conditions $v = 0$ and $N_x = 0$. We did not need to care for M_x, of course, but we were bothered by the fact that the membrane solution does not have constants enough to make $w = 0$. Here we have a solution which, for the particular load (18), is better than the membrane solution and not at all complicated. To understand its significance and the limits of its usefulness, we shall discuss some numerical results.

We apply our formulas to the cylinder shown in fig. 3. At both ends of the span $l = \pi a$ it is supported by diaphragms. We assume $\nu = 0$ and ask for the deformations and stress resultants set up by the load

$$p_x = 0, \qquad p_\phi = p_n \sin\phi \sin\frac{n\pi x}{l}, \qquad p_r = -p_n \cos\phi \sin\frac{n\pi x}{l}.$$

Let us first consider a rather thick-walled shell with $t/a = 0.10$, and assume $n = 1$. From eqs. (20) we find with $m = 1$:

$$u_{1,1} = 1.987\, p_1\, a^2/D, \quad v_{1,1} = 5.968\, p_1\, a^2/D, \quad w_{1,1} = -6.957\, p_1\, a^2/D.$$

This comes rather close to the figures which the membrane formulas (III–28) yield in this case. They are

$$u_{1,1} = 2p_1\, a^2/D, \quad v_{1,1} = 6p_1\, a^2/D, \quad w_{1,1} = -7p_1\, a^2/D.$$

A similarly good approximation is found for the normal and shearing forces. They are:

exact	membrane theory
$N_{\phi 1,1} = -0.989\, p_1\, a$	$N_{\phi 1,1} = -p_1\, a$
$N_{x 1,1} = -1.993\, p_1\, a$	$N_{x 1,1} = -2p_1\, a$
$N_{\phi x 1,1} = 1.993\, p_1\, a$	$N_{\phi x 1,1} = 2p_1\, a$
$N_{x \phi 1,1} = 1.990\, p_1\, a$	

This surprisingly good agreement in a rather thick shell may give much confidence in the results of the membrane theory when applied to shells with smoothly distributed loads and appropriate boundary conditions. Moments and transverse shearing forces are, of course, very small. Here are the figures:

$$M_{\phi 1,1} = 0, \qquad\qquad M_{x 1,1} = 7.44 \times 10^{-3}\, p_1\, a^2,$$
$$M_{\phi x 1,1} = 2.480 \times 10^{-3} p_1\, a^2, \qquad M_{x \phi 1,1} = 0.823 \times 10^{-3}\, p_1\, a^2,$$
$$Q_{\phi 1,1} = -0.823 \times 10^{-3}\, p_1\, a, \qquad Q_{x 1,1} = 9.93 \times 10^{-3}\, p_1\, a$$

The influence of the "small" terms in eqs. (9) is considerable here but is nevertheless unimportant because of the unimportance of the moments. When we compute the eccentricities M/N as we did on p. 218,

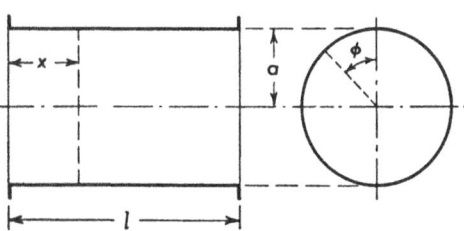

Fig. 3. Circular cylinder

we find them to be a few percent of t.

Now let us consider a shell of the same overall dimensions, but much thinner. We choose $k = 10^{-4}$, corresponding to $t/a = 3.46 \times 10^{-2}$. For $n = 1$ the normal and shearing forces will equal those given by the membrane theory with far better accuracy than can be determined by slide rule, and the moments will be even smaller than in the preceding example.

But when we put $n = 10$, we obtain the following displacements:

<table>
<tr><td>bending theory:</td><td>membrane theory:</td></tr>
<tr><td>$u_{1,10} = +0.980 \times 10^{-3} p_{10} a^2/D$</td><td>$u_{1,10} = +2 \times 10^{-3} p_{10} a^2/D$</td></tr>
<tr><td>$v_{1,10} = +29.85 \times 10^{-3} p_{10} a^2/D$</td><td>$v_{1,10} = +40.2 \times 10^{-3} p_{10} a^2/D$</td></tr>
<tr><td>$w_{1,10} = -510 \times 10^{-3} p_{10} a^2/D$</td><td>$w_{1,10} = -1040 \times 10^{-3} p_{10} a^2/D$</td></tr>
</table>

These figures show a pronounced reduction of the deformation due to the bending stiffness of the shell. The normal and shearing forces are also reduced:

<table>
<tr><td>bending theory:</td><td>membrane theory:</td></tr>
<tr><td>$N_{\phi 1,10} = -0.480 \, p_{10} a$</td><td>$N_{\phi 1,10} = -p_{10} a$</td></tr>
<tr><td>$N_{x 1,10} = -0.00470 \, p_{10} a$</td><td>$N_{x 1,10} = -0.02 \, p_{10} a$</td></tr>
<tr><td>$N_{\phi x 1,10} = +0.1490 \, p_{10} a$</td><td>$N_{\phi x 1,10} = +0.20 \, p_{10} a$</td></tr>
<tr><td>$N_{x \phi 1,10} = +0.1485 \, p_{10} a$</td><td></td></tr>
</table>

These figures may easily be understood, if we keep in mind that for $n = 10$ the shell is divided by 9 circles into 10 strips which carry alternately positive and negative loads. The width of such a strip here is approximately $10t$. Under these circumstances it seems reasonable that the normal load is no longer carried around the shell by hoop forces N_ϕ, but that part of it, here slightly more than one half, is carried by transverse forces Q_x to the adjacent zone with opposit loading. Correspondingly, the bending moments are of greater impore tance here. In fact, they have approximately the same values as in the previous case, although the shell is much thinner and the direct stresses are smaller. The moments are:

$$M_{\phi 1,10} = 0, \qquad\qquad M_{x 1,10} = 5.10 \times 10^{-3} p_{10} a^2,$$
$$M_{\phi x 1,10} = 0.495 \times 10^{-3} p_{10} a^2, \qquad M_{x \phi 1,10} = 0.480 \times 10^{-3} p_{10} a^2$$

Three of the eccentricities are unimportant, but for the longitudinal force we find

$$M_x/N_x = -31.4\, t.$$

One may easily imagine that a heavy bending in the hoop direction will occur if we choose a load pattern having a large m and $n = 1$ and that both bending moments and the twisting moment will be important, if both m and n are large.

This is exactly what limits the usefulness of the solution presented in this section. Loads such as are described by eqs. (18) will not often occur in practical shell problems, but we may represent any load by a double series of such terms:

$$p_x = \sum_{m=0}^{\infty} \sum_{n=0}^{\infty} p_{x\,mn} \cos m\phi \cos\frac{\lambda x}{a},$$

$$p_\phi = \sum_{m=1}^{\infty} \sum_{n=1}^{\infty} p_{\phi\,mn} \sin m\phi \sin\frac{\lambda x}{a},$$

$$p_r = \sum_{m=0}^{\infty} \sum_{n=1}^{\infty} p_{rmn} \cos m\phi \sin\frac{\lambda x}{a}.$$

For every term of this series, eqs. (19) are a solution, and by super-position we find

$$u = \sum_{m=0}^{\infty} \sum_{n=0}^{\infty} u_{mn} \cos m\phi \cos\frac{\lambda x}{a},$$

$$v = \sum_{m=1}^{\infty} \sum_{n=1}^{\infty} v_{mn} \sin m\phi \sin\frac{\lambda x}{a},$$

$$w = \sum_{m=0}^{\infty} \sum_{n=1}^{\infty} w_{mn} \cos m\phi \sin\frac{\lambda x}{a}.$$

and analogous expressions for the stress resultants. Since even discontinuously distributed loads may be represented by such double FOURIER series, it appears that we have here a fairly general solution of the bending problem, at least for a certain useful set of boundary conditions. From a purely mathematical point of view this is certainly true, but for technical applications it is not sufficient that a series converges eventually. It must converge so well that its sum may be obtained with slide rule accuracy from a rather limited number of terms. The solution treated in this section fulfills this condition only for thick-walled shells. If t/a is small, only the series for the N-forces converge quickly, but in those for the M and the Q the coefficients first increase considerably because of the phenomenon just explained in the numerical example, and quite a few terms must be computed until they decrease enough to become negligible. In these cases it is more convenient to avoid the FOURIER series by a skillful combination of membrane solutions with the homogeneous solution presented in the following section.

5.3 Loads Applied to the Edges $x = $ const
5.3.1 General Solution

A circular cylinder may extend from $x = -\infty$ to $x = +\infty$, but a shell must necessarily have an end somewhere. In the simplest case, it will be limited by two planes $x = $ const. We then have to consider the possibility that loads are applied to these edges. As we have seen on p. 120, there exist a few systems of edge loads to which the shell may respond by membrane forces alone, but the general solution of the edge load problem must be found from eqs. (13) of the bending theory.

When we follow a circle $x = $ const around the cylinder, we return at last to the starting point, but ϕ has increased by $2\pi = 360°$. Since at the same point we must always find the same stresses, strains, and displacements, they all must be periodic functions of ϕ with the period 2π and, therefore, may be written as FOURIER series. Since the differential equations (13) have constant coefficients, each term of these series is itself a solution, provided that we choose a judicious combination of sines and cosines. From the symmetry of the shell with respect to the diametral plane $\phi = 0$ it may be expected that the following choice of sines and cosines fits together:

$$
\left.\begin{aligned}
&u = \sum u_m \cos m\phi, \quad v = \sum v_m \sin m\phi, \quad w = \sum w_m \cos m\phi, \\
&N_\phi = \sum N_{\phi m} \cos m\phi, \quad N_x = \sum N_{xm} \cos m\phi, \\
&N_{\phi x} = \sum N_{\phi xm} \sin m\phi, \quad N_{x\phi} = \sum N_{x\phi m} \sin m\phi, \\
&M_\phi = \sum M_{\phi m} \cos m\phi, \quad M_x = \sum M_{xm} \cos m\phi, \\
&M_{\phi x} = \sum M_{\phi xm} \sin m\phi, \quad M_{x\phi} = \sum M_{x\phi m} \sin m\phi, \\
&Q_\phi = \sum Q_{\phi m} \sin m\phi, \quad Q_x = \sum Q_{xm} \cos m\phi.
\end{aligned}\right\} \quad (22)
$$

The coefficients u_m, v_m, w_m, ..., Q_{xm} of these series are, of course, not constants but functions of x.

We now take the general term of the first three series, putting

$$u = u_m(x) \cos m\phi, \quad v = v_m(x) \sin m\phi, \quad w = w_m(x) \cos m\phi,$$

and introduce this into eqs. (13). Since we want to treat the edge load problem, we set $p_x = p_\phi = p_r = 0$. All other terms in each equation will have a common factor $\sin m\phi$ or $\cos m\phi$, which we may drop. In this way we arrive at a set of three simultaneous differential equations with only one independent variable x:

$$
\left.
\begin{aligned}
u_m'' &- \frac{1-\nu}{2}\, m^2 u_m + \frac{1+\nu}{2}\, m v_m' + \nu w_m' \\
&\qquad - k\left(\frac{1-\nu}{2}\, m^2 u_m + w_m''' + \frac{1-\nu}{2}\, m^2 w_m'\right) = 0, \\
-\frac{1+\nu}{2}\, m u_m' &- m^2 v_m + \frac{1-\nu}{2}\, v_m'' - m w_m \\
&\qquad + k\left(\frac{3}{2}(1-\nu) v_m'' + \frac{3-\nu}{2}\, m w_m''\right) = 0, \\
\nu u_m' + m v_m + w_m &+ k\left(-\frac{1-\nu}{2}\, m^2 u_m' - u_m''' - \frac{3-\nu}{2}\, m v_m''\right. \\
&\left. \qquad + w_m^{\mathrm{IV}} - 2m^2 w_m'' + m^4 w_m - 2m^2 w_m + w_m\right) = 0.
\end{aligned}
\right\}
\tag{23}
$$

These equations have constant coefficients and may be solved by exponential functions:

$$
u_m = A\, e^{\lambda x/a}, \qquad v_m = B\, e^{\lambda x/a}, \qquad w_m = C\, e^{\lambda x/a}. \tag{24}
$$

After introducing eqs. (24) into (23), we may drop the exponential factor and then have three ordinary linear equations for the constants A, B, C:

$$
\begin{aligned}
\left[\lambda^2 - \frac{1-\nu}{2}\, m^2(1+k)\right] A &+ \left[\frac{1+\nu}{2}\, \lambda m\right] B \\
&+ \left[\nu\lambda - k\left(\lambda^3 + \frac{1-\nu}{2}\, \lambda m^2\right)\right] C = 0,
\end{aligned}
$$

$$
\begin{aligned}
\left[\frac{1+\nu}{2}\, \lambda m\right] A &+ \left[-\frac{1-\nu}{2}\, \lambda^2 + m^2 - \frac{3}{2}(1-\nu) k \lambda^2\right] B \\
&+ \left[m - \frac{3-\nu}{2}\, k\lambda^2 m\right] C = 0,
\end{aligned} \tag{25a—c}
$$

$$
\begin{aligned}
\left[\nu\lambda - k\left(\lambda^3 + \frac{1-\nu}{2}\, \lambda m^2\right)\right] A &+ \left[m - \frac{3-\nu}{2}\, k\lambda^2 m\right] B \\
+ [1 + k(\lambda^4 - 2\lambda^2 m^2 &+ m^4 - 2m^2 + 1)] C = 0.
\end{aligned}
$$

Since these equations are homogeneous, they can have a solution A, B, C different from zero only if the determinant formed from their nine coefficients vanishes. This condition yields an equation for λ. We obtain it by expanding the determinant and arranging the terms by descending powers of λ. The coefficients may be simplified by neglecting everywhere the small number k in comparison to 1, and then we have

$$
\begin{aligned}
\lambda^8 - 2(2m^2 - \nu)\lambda^6 &+ \left[\frac{1-\nu^2}{k} + 6m^2(m^2 - 1)\right]\lambda^4 \\
&- 2m^2[2m^4 - (4-\nu)m^2 + (2-\nu)]\lambda^2 + m^4(m^2 - 1)^2 = 0.
\end{aligned} \tag{26}
$$

This is a fourth-degree equation for λ^2. It may be shown that its four roots are all complex and hence two pairs of conjugate complex numbers. The 8 roots λ may therefore be written in the following form with real $\varkappa$ and μ:

$$
\begin{aligned}
\lambda_1 &= -\varkappa_1 + i\,\mu_1, & \lambda_5 &= +\varkappa_1 + i\,\mu_1, \\
\lambda_2 &= -\varkappa_1 - i\,\mu_1, & \lambda_6 &= +\varkappa_1 - i\,\mu_1, \\
\lambda_3 &= -\varkappa_2 + i\,\mu_2, & \lambda_7 &= +\varkappa_2 + i\,\mu_2, \\
\lambda_4 &= -\varkappa_2 - i\,\mu_2, & \lambda_8 &= +\varkappa_2 - i\,\mu_2.
\end{aligned}
$$

There are different methods to find these values numerically. The best among them seems to be that one given by W. S. Brown[1] which yields the complex values λ^2.

Each of the 8 values λ_j yields one solution of eqs. (23), and the complete solution is the sum of all them with 8 independent sets of constants A_j, B_j, C_j:

$$
\left.
\begin{aligned}
u_m &= e^{-\varkappa_1 x/a}(A_1 e^{i\mu_1 x/a} + A_2 e^{-i\mu_1 x/a}) \\
&\quad + e^{-\varkappa_2 x/a}(A_3 e^{i\mu_2 x/a} + A_4 e^{-i\mu_2 x/a}) \\
&\quad + e^{+\varkappa_1 x/a}(A_5 e^{i\mu_1 x/a} + A_6 e^{-i\mu_1 x/a}) \\
&\quad + e^{+\varkappa_2 x/a}(A_7 e^{i\mu_2 x/a} + A_8 e^{-i\mu_2 x/a}), \\
v_m &= e^{-\varkappa_1 x/a}(B_1 e^{i\mu_1 x/a} + B_2 e^{-i\mu_1 x/a}) \\
&\quad + e^{-\varkappa_2 x/a}(B_3 e^{i\mu_2 x/a} + B_4 e^{-i\mu_2 x/a}) \\
&\quad + e^{+\varkappa_1 x/a}(B_5 e^{i\mu_1 x/a} + B_6 e^{-i\mu_1 x/a}) \\
&\quad + e^{+\varkappa_2 x/a}(B_7 e^{i\mu_2 x/a} + B_8 e^{-i\mu_2 x/a}), \\
w_m &= e^{-\varkappa_1 x/a}(C_1 e^{i\mu_1 x/a} + C_2 e^{-i\mu_1 x/a}) \\
&\quad + e^{-\varkappa_2 x/a}(C_3 e^{i\mu_2 x/a} + C_4 e^{-i\mu_2 x/a}) \\
&\quad + e^{+\varkappa_1 x/a}(C_5 e^{i\mu_1 x/a} + C_6 e^{-i\mu_1 x/a}) \\
&\quad + e^{+\varkappa_2 x/a}(C_7 e^{i\mu_2 x/a} + C_8 e^{-i\mu_2 x/a}).
\end{aligned}
\right\} \tag{27}
$$

For every j, the three constants A_j, B_j, C_j are related among each other by the linear equations (25). Since the determinant of these equations is zero, we may use any two of them to determine A_j and B_j as multiples of C_j, introducing the corresponding value of λ_j into the coefficients:

$$
A_j = \alpha_j\,C_j, \qquad B_j = \beta_j\,C_j.
$$

The α_j, β_j are complex numbers, but we need to solve only two pairs of equations to find them, since they are so interconnected that we have all of them when we have the real and imaginary parts of α_1, β_1, α_3, β_3. Indeed, by inspecting the coefficients of eqs. (25) one may

---

[1] Brown, W. S.: Solution of biquadratic equations. Aircraft Engg. Vol. 16 (1944) p. 14.

easily verify that the following relations must hold:

$$\alpha_1 = -\alpha_6 = \bar{\alpha}_1 + i\,\bar{\alpha}_2, \qquad \alpha_3 = -\alpha_8 = \bar{\alpha}_3 + i\,\bar{\alpha}_4,$$
$$\alpha_2 = -\alpha_5 = \bar{\alpha}_1 - i\,\bar{\alpha}_2, \qquad \alpha_4 = -\alpha_7 = \bar{\alpha}_3 - i\,\bar{\alpha}_4,$$
$$\beta_1 = \beta_6 = \bar{\beta}_1 + i\,\bar{\beta}_2, \qquad \beta_3 = \beta_8 = \bar{\beta}_3 + i\,\bar{\beta}_4,$$
$$\beta_2 = \beta_5 = \bar{\beta}_1 - i\,\bar{\beta}_2, \qquad \beta_4 = \beta_7 = \bar{\beta}_3 - i\,\bar{\beta}_4.$$

Since the α_j, β_j depend only on the dimensions of the shell, the C_j are the only free constants of our problem which must be determined from 8 boundary conditions, 4 at each of the two edges $x = $ const. Such a problem involving the determination of 8 constants, although simple in its mathematical structure, is rather tedious in numerical execution. For practical applications of the theory it is therefore important that the number of free constants may be reduced in special cases.

5.3.2 Semi-infinite Cylinder

One half of the eight elementary solutions contained in the formulas (27) have a factor $e^{-\varkappa x/a}$ and therefore decrease more or less rapidly with increasing x; the other four do exactly the contrary and increase beyond all limits when x increases.

On a cylinder which begins at $x = 0$ and which reaches very far in the direction of positive x, we cannot expect that loads applied to the edge $x = 0$ will produce stresses and displacements which, beginning with moderate values at $x = 0$, increase exponentially with x. We may, therefore, entirely disregard the particular solutions $j = 5 \ldots 8$, when fulfilling the boundary conditions at $x = 0$.

The other four may be written in real form by combining two by two the exponential functions of the same imaginary argument:

$$u_m = e^{-\varkappa_1 x/a}\left[(A_1 + A_2)\cos\frac{\mu_1 x}{a} + i(A_1 - A_2)\sin\frac{\mu_1 x}{a}\right]$$
$$+ e^{-\varkappa_2 x/a}\left[(A_3 + A_4)\cos\frac{\mu_2 x}{a} + i(A_3 - A_4)\sin\frac{\mu_2 x}{a}\right],$$
$$v_m = e^{-\varkappa_1 x/a}\left[(B_1 + B_2)\cos\frac{\mu_1 x}{a} + i(B_1 - B_2)\sin\frac{\mu_1 x}{a}\right]$$
$$+ e^{-\varkappa_2 x/a}\left[(B_3 + B_4)\cos\frac{\mu_2 x}{a} + i(B_3 - B_4)\sin\frac{\mu_2 x}{a}\right],$$
$$w_m = e^{-\varkappa_1 x/a}\left[(C_1 + C_2)\cos\frac{\mu_1 x}{a} + i(C_1 - C_2)\sin\frac{\mu_1 x}{a}\right]$$
$$+ e^{-\varkappa_2 x/a}\left[(C_3 + C_4)\cos\frac{\mu_2 x}{a} + i(C_3 - C_4)\sin\frac{\mu_2 x}{a}\right].$$

In these formulas $(A_1 + A_2)$, $i(A_1 - A_2)$ etc. are real quantities, and when we put

$$C_1 + C_2 = \bar{C}_1, \qquad C_3 + C_4 = \bar{C}_3,$$
$$i(C_1 - C_2) = \bar{C}_2, \qquad i(C_3 - C_4) = \bar{C}_4, \tag{28}$$

we have

$$A_1 + A_2 = \bar{\alpha}_1 \bar{C}_1 + \bar{\alpha}_2 \bar{C}_2, \qquad B_1 + B_2 = \bar{\beta}_1 \bar{C}_1 + \bar{\beta}_2 \bar{C}_2,$$
$$i(A_1 - A_2) = \bar{\alpha}_1 \bar{C}_2 - \bar{\alpha}_2 \bar{C}_1, \qquad i(B_1 - B_2) = \bar{\beta}_1 \bar{C}_2 - \bar{\beta}_2 \bar{C}_1, \tag{29}$$

and four similar relations for the subscripts 3 and 4. Introducing the expressions for u_m, v_m, w_m into the elastic law (9) and passing from there to eqs. (1d, e), we may find similar expressions for all the stress resultants listed in eqs. (22). They may all be written in the general

Table 1. Semi-infinite Cylinder

f	c	a_1
w	1	1
w'	1	$-\varkappa_1$
u	1	$\bar{\alpha}_1$
v	1	$\bar{\beta}_1$
N_ϕ	D/a	$1 + k - k\,m^2 + m\,\bar{\beta}_1 - \nu\,(\varkappa_1\bar{\alpha}_1 + \mu_1\bar{\alpha}_2)$
N_x	D/a	$\nu - (\varkappa_1\bar{\alpha}_1 + \mu_1\bar{\alpha}_2) + \nu\,m\,\bar{\beta}_1 - k(\varkappa_1^2 - \mu_1^2)$
$N_{\phi x}$	$\dfrac{D(1-\nu)}{2a}$	$-(1+k)\,m\,\bar{\alpha}_1 - (\varkappa_1\bar{\beta}_1 + \mu_1\bar{\beta}_2) + k\,m\,\varkappa_1$
$N_{x\phi}$	$\dfrac{D(1-\nu)}{2a}$	$-m\,\bar{\alpha}_1 - (1+k)(\varkappa_1\bar{\beta}_1 + \mu_1\bar{\beta}_2) - k\,m\,\varkappa_1$
T_x	$\dfrac{D(1-\nu)}{2a}$	$-m\,\bar{\alpha}_1 - (1+3k)(\varkappa_1\bar{\beta}_1 + \mu_1\bar{\beta}_2) - 3k\,m\,\varkappa_1$
M_ϕ	K/a^2	$1 - m^2 + \nu\,(\varkappa_1^2 - \mu_1^2)$
M_x	K/a^2	$(\varkappa_1^2 - \mu_1^2) + (\varkappa_1\bar{\alpha}_1 + \mu_1\bar{\alpha}_2) - \nu\,m(m + \bar{\beta}_1)$
$M_{\phi x}$	$\dfrac{K(1-\nu)}{2a^2}$	$m(2\varkappa_1 - \bar{\alpha}_1) + (\varkappa_1\bar{\beta}_1 + \mu_1\bar{\beta}_2)$
$M_{x\phi}$	$\dfrac{K(1-\nu)}{a^2}$	$m\,\varkappa_1 + (\varkappa_1\bar{\beta}_1 + \mu_1\bar{\beta}_2)$
Q_ϕ	K/a^3	$m(m^2 - 1) - m(\varkappa_1^2 - \mu_1^2) - (1 - \nu)[(\varkappa_1^2 - \mu_1^2)\bar{\beta}_1 + 2\varkappa_1\mu_1\bar{\beta}_2]$
Q_x	$\dfrac{K}{2a^3}$	$2\varkappa_1(m^2 - \varkappa_1^2 + 3\mu_1^2) - [(1-\nu)\,m^2 + 2(\varkappa_1^2 - \mu_1^2)]\,\alpha_1$ $-\,4\varkappa_1\mu_1\bar{\alpha}_2 + (1 + \nu)\,m(\varkappa_1\bar{\beta}_1 + \mu_1\bar{\beta}_2)$
S_x	$\dfrac{K}{2a^3}$	$2\varkappa_1[(2 - \nu)\,m^2 - \varkappa_1^2 + 3\mu_1^2] - [(1 - \nu)\,m^2 + 2(\varkappa_1^2 - \mu_1^2)]\,\bar{\alpha}_1$ $-\,4\varkappa_1\mu_1\bar{\alpha}_2 + (3 - \nu)\,m(\varkappa_1\bar{\beta}_1 + \mu_1\bar{\beta}_2)$

form

$$f = c \left[e^{-\varkappa_1 x/a} \left((a_1 \overline{C}_1 + a_2 \overline{C}_2) \cos\frac{\mu_1 x}{a} + (a_1 \overline{C}_2 - a_2 \overline{C}_1) \sin\frac{\mu_1 x}{a} \right) \right.$$
$$\left. + e^{-\varkappa_2 x/a} \left((a_3 \overline{C}_3 + a_4 \overline{C}_4) \cos\frac{\mu_2 x}{a} + (a_3 \overline{C}_4 - a_4 \overline{C}_3) \sin\frac{\mu_2 x}{a} \right) \right] {\cos \atop \sin} \, m\phi. \tag{30}$$

The coefficients a_1, a_2 are given in Table 1 for the various displacements and stress resultants. The other two, a_3, a_4, are found by changing in the formulas of the Table the subscript of $\varkappa$ and μ from 1 to 2 and the subscripts of $\bar{\alpha}$, $\bar{\beta}$ from 1 and 2 to 3 and 4.

Table 1. (Continued)

a_1	ϕ factor
0	cos
μ_1	cos
$\bar{\alpha}_2$	cos
$\bar{\beta}_2$	sin
$m\bar{\beta}_2 - \nu(\varkappa_1\bar{\alpha}_2 - \mu_1\bar{\alpha}_1)$	cos
$-(\varkappa_1\bar{\alpha}_2 - \mu_1\bar{\alpha}_1) + \nu m\bar{\beta}_2 + 2k\varkappa_1\mu_1$	cos
$-(1+k)m\bar{\alpha}_2 - (\varkappa_1\bar{\beta}_2 - \mu_1\bar{\beta}_1) - km\mu_1$	sin
$-m\bar{\alpha}_2 - (1+k)(\varkappa_1\bar{\beta}_2 - \mu_1\bar{\beta}_1) + km\mu_1$	sin
$-m\bar{\alpha}_2 - (1+3k)(\varkappa_1\bar{\beta}_2 - \mu_1\bar{\beta}_1) + 3km\mu_1$	sin
$-2\nu\varkappa_1\mu_1$	cos
$-2\varkappa_1\mu_1 + (\varkappa_1\bar{\alpha}_2 - \mu_1\bar{\alpha}_1) - \nu m\bar{\beta}_2$	cos
$-m(2\mu_1 + \bar{\alpha}_2) + (\varkappa_1\bar{\beta}_2 - \mu_1\bar{\beta}_1)$	sin
$-m\mu_1 + (\varkappa_1\bar{\beta}_2 - \mu_1\bar{\beta}_1)$	sin
$2m\varkappa_1\mu_1 - (1-\nu)[(\varkappa_1^2 - \mu_1^2)\bar{\beta}_2 - 2\varkappa_1\mu_1\bar{\beta}_1]$	sin
$-2\mu_1(m^2 - 3\varkappa_1^2 + \mu_1^2) - [(1-\nu)m^2 + 2(\varkappa_1^2 - \mu_1^2)]\bar{\alpha}_2$ $+ 4\varkappa_1\mu_1\bar{\alpha}_1 + (1+\nu)m(\varkappa_1\bar{\beta}_2 - \mu_1\bar{\beta}_1)$	cos
$-2\mu_1[(2-\nu)m^2 - 3\varkappa_1^2 + \mu_1^2] - [(1-\nu)m^2 + 2(\varkappa_1^2 - \mu_1^2)]\bar{\alpha}_2$ $+ 4\varkappa_1\mu_1\bar{\alpha}_1 + (3-\nu)m(\varkappa_1\bar{\beta}_2 - \mu_1\bar{\beta}_1)$	cos

When Table 1 is used for numerical work, it will be found that many terms are negligibly small. They have all been kept in the formulas, because it depends to some extent on the special nature of the problem whether a term is important or not. Of course, in each individual case everything should be dropped which does not make a contribution of reasonable magnitude.

We are now prepared to solve specific problems. At the start of any such computation we have to decide for which harmonics m we want to work out the solution. In the FOURIER series (22), the order m of the terms runs from 0 or 1 to ∞, but, practically speaking, we need only a certain choice, as we shall see in an example on p. 234. For a chosen m, we begin by solving eq. (26), which will yield $\varkappa_1$, $\varkappa_2$, μ_1, μ_2 as real and imaginary parts of the solutions λ. Then we find from the first two of eqs. (25) the complex numbers α_1, α_3, β_1, β_3 as the values of A and B for $C = 1$. The next step is to select from the coefficient table those displacements or forces which appear in the four boundary conditions and to write their values at $x = 0$ as functions of $\overline{C}_1$, $\overline{C}_2$, $\overline{C}_3$, $\overline{C}_4$. This will yield 4 linear equations for these 4 unknowns. When they have been solved, the coefficient table will give numerical expressions for all the displacements and stress resultants we want to know. When we have done all this for several m, a FOURIER synthesis of the results according to eqs. (22) will conclude the work.

One point in this procedure still needs some explanation, namely the formulation of the boundary conditions. Let us consider a simply supported edge, as may be realized by those diaphragms which we used in the membrane theory to support the edges of cylindrical shells. The connection of the shell to such a diaphragm indeformable in its own plane means that the displacements v and w must be zero, and the normal force N_x and the bending moment M_x may be given arbitrarily. The full set of boundary conditions is therefore:

$$M_x = \text{given}, \quad N_x = \text{given}, \quad v = 0, \quad w = 0. \tag{31}$$

The values which the forces and moments $N_{x\phi}$, $M_{x\phi}$, Q_x assume at the edge, will be found from the solution of the bending problem of the shell as we just described it. It is interesting to investigate their influence on the diaphragm.

Fig. 4a is a side view of the edge $x = 0$, looking in the direction of increasing x. It shows two adjacent elements of length $ds = a\,d\phi$ each. On the left one a twisting moment $M_{x\phi}\,ds$ is acting, on the right a moment $(M_{x\phi} + M'_{x\phi}\,d\phi)\,ds$, and the same moments act in opposite direction on the supporting diaphragm. In fig. 4b each moment has been replaced by an equivalent group of three forces. The two forces F_n on the left element are almost parallel to each other and must have a moment equal to $M_{x\phi}\,ds$, hence

$$F_n \cdot ds = M_{x\phi}\,ds.$$

But since they are slightly divergent, they have a horizontal resultant $F_n\,d\phi$, pointing to the left, which is compensated by the third
force $F_t = F_n\,d\phi$, so that the three forces F_n, F_n, F_t are statically
equivalent to the distributed shearing stresses which yield the twisting
moment $M_{x\phi}$. Therefore their effect on
the shell and on the support cannot be
much different from that of $M_{x\phi}$, appreciable differences appearing only in a
zone whose width is of the same order
as the thickness t of the shell. If we
disregard them, we can no longer dis-
criminate between F_t and a genuine
shearing force and may combine both
into a resultant force per unit length,
the effective shear

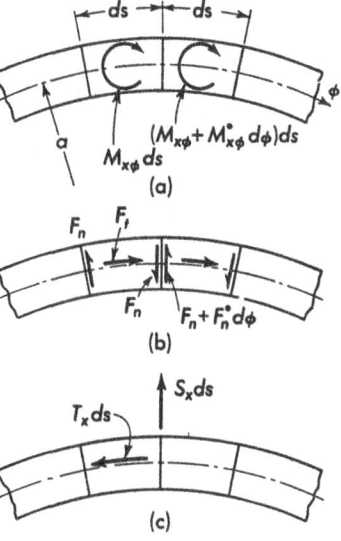

$$T_x = N_{x\phi} - \frac{F_t}{ds} = N_{x\phi} - \frac{M_{x\phi}}{a}. \quad (32\,\text{a})$$

A similar reasoning may be applied
to the forces F_n. At the right end of
the left element in fig. 4b we have a
force F_n pointing inward, and adjacent
to it at the left end of the right element
the force $F_n + F_n'\,d\phi$ pointing outward.

Their difference, $F_n'\,d\phi = M_{x\phi}'\,d\phi$, may
be combined with the transverse shearing force $Q_x\,ds$. This leads to
the effective transverse force

$$S_x = Q_x + \frac{M_{x\phi}'}{a}. \quad (32\,\text{b})$$

It is the same force which appears under similar circumstances in the
theory of plane plates and is known there under the name of KIRCH-
HOFF's force.

As we have seen here, the static effect of the three stress result-
ants $N_{x\phi}$, $M_{x\phi}$, Q_x may be expressed by two quantities, S_x and T_x.
This is essential when we have to formulate the boundary conditions
for a completely unsupported edge. We might expect at first that all
the edge forces and moments may be arbitrarily given. But they are
five, N_x, $N_{x\phi}$, M_x, $M_{x\phi}$, Q_x, whereas we have only four constants of
integration, C_1, C_2, C_3, C_4. Here the forces S_x and T_x will help us. Since
they are equivalent to the combined action of three of the original

stress resultants, they may replace them in the boundary conditions, and then we have only four essential external forces and moments, which may be arbitrarily prescribed: N_x, T_x, M_x, S_x, and they will lead to 4 boundary conditions, determining the 4 constants C_j.

It becomes evident that the introduction of the effective edge forces S_x and T_x is an essential feature of the bending theory as it is represented in this book. It is beyond the scope of this theory to study local stress problems in a boundary zone of the order t. They may be of interest when a small hole is drilled in the shell. For this problem, we have to abandon our basic assumptions and to replace them by something better. In all other cases boundary conditions on structures and machine parts are not defined exactly enough to make such local border problems a possible object of investigation. When the edge of the shell is riveted, welded, or glued to an edge member, or if there is a smooth transition between them, as it frequently happens in concrete, cast iron, and plastics, we often cannot even tell exactly where the shell ends and something else begins.

5.3.3 Cooling Tower

The theory developed here may be applied to structures like the cooling tower shown in fig. 5. If the shell were put directly on a continuous foundation, there would be no problem at all. The weight of the shell would cause compressive forces N_x, increasing from zero at the top to, say, $N_x = -P$ at the base and distributed uniformly over any horizontal cross section of the tower. Now, in order to allow the cool air to enter, the shell stands on a number of columns, and in the space between them no force N_x is allowed to act on the edge. Therefore, an edge load must be super imposed on the simple stress system just described, and this edge load must be self-equilibrating and must cancel the edge load $N_x = -P$ along the free parts of the edge. A part of this load is shown in fig. 6. It may be expanded in a FOURIER series (22) in which only the terms $m = 8, 16, 24, \ldots$ appear. This series respresents the values which N_x must assume for $x = 0$. Besides N_x, three other quantities must be known along the edge to make the problem determinate. Since there is certainly not much that might restrict

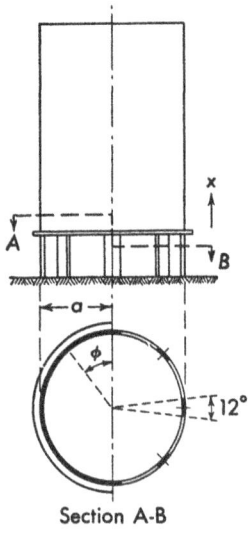

Section A-B

Fig. 5
Cylindrical cooling tower

the rotation of a vertical tangent to the cylinder, at least not between the supports, it is reasonable to choose $M_x = 0$ as a second
boundary condition. The other two depend on the size of the stiffening
ring which will be provided at the edge. For a numerical example we
consider the extreme case that the ring is very stiff in its plane, and
prescribe $v = w = 0$. When the shell is high enough, the solution (30)

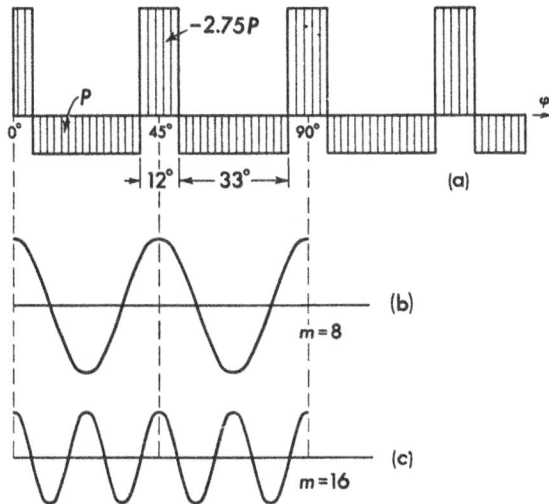

Fig. 6. Edge load applied to the cooling-tower shell, (a) total edge load, (b, c) first and
second harmonics of this load

for the semi-infinite cylinder may be applied, and then it is not necessary
to have another set of four boundary conditions for the upper end.

The first thing to be done on the way toward a numerical solution
is to solve eq. (26) and to determine $\varkappa_1$, $\varkappa_2$, μ_1, μ_2 from its solutions.
This must be done for every m. Under the assumption $a/t = 150$ the
following figures have been obtained:

m	8	16	24	32	40
$\varkappa_1$	18.22	24.55	32.24	40.17	48.17
$\varkappa_2$	2.14	8.39	16.05	23.95	31.89
μ_1	14.42	12.00	10.76	10.12	9.74
μ_2	1.69	4.12	5.39	6.08	6.51

The next step is to find the ratios α_j and β_j for $j = 1$ and $j = 3$,
using two of the three equations (25). This must also be done separately
for every m.

Thus far, the computation does not depend on the particular set
of boundary conditions, but now it is time to introduce them. According

to eq. (30), every quantity needed at the edge $x = 0$ has there the amplitude $c(a_1 \overline{C}_1 + a_2 \overline{C}_2 + a_3 \overline{C}_3 + a_4 \overline{C}_4)$, the quantities c, $a_1, \ldots, a_4$ in this expression to be taken from the appropriate line of Table 1. This may now be done for N_{xm}, M_{xm}, v_m, w_m, and for every m a set of four equations with real coefficients may be set up whose unknowns are the real quantities $\overline{C}_1$, $\overline{C}_2$, $\overline{C}_3$, $\overline{C}_4$ introduced by eqs. (28). When these equations have been solved, Table 1 together with eq. (30) will provide any required information.

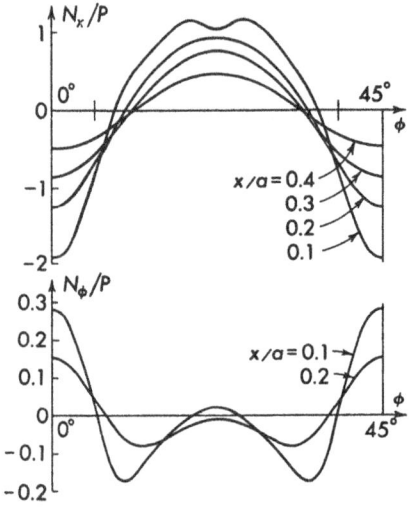

Fig. 7. Distribution of N_x and N_ϕ at different levels of the cooling tower in fig. 5

While this work was being done, it was found necessary to use 6-digit machine computation up to the determination of the α_j's and β_j's; afterwards careful sliderule work was sufficient.

Some of the results are shown in figs. 7 and 8. At the edge $x = 0$ the longitudinal force is given (fig. 6). When we proceed to higher cross sections (fig. 7), the peaks at $\phi = 0°$, $45°, \ldots$ become less and less pronounced, and at $x/a = 0.4$ the distribution is practically sinusoidal. This development is due to the fact that $\varkappa_2$, the smaller one of the two damping exponents, increases substantially from $m = 8$ to $m = 16$ and 24 and that, therefore, at some distance from the edge only the lowest harmonic survives. In fig. 8 the vertical distribution of N_x is shown for the generators $\phi = 0°$ (centerline of a column) and $\phi = 22.5°$ (midspan section). It can be seen that the stresses decay with increasing x and diminish most rapidly where they are highest at the edge.

A cylindrical shell put on columns may be compared with a plane wall supported and loaded in the same way. The horizontal forces in such a wall, which would correspond to the hoop forces N_ϕ in the shell, may readily be understood as the bending stresses in a continuous beam of unusual height. Since our boundary conditions $v = w = 0$ imply the presence of a sturdy edge member, it may be expected that this edge member has compressive stresses at the supports and tensile stresses at midspan and that the lower part of the wall has stresses of opposite sign which higher up decrease to zero, with or without change of sign. In the plane wall this is really what happens, but in the shell the hoop forces are differently distributed. In fig. 8 both are shown for $\phi = 0°$ and $22.5°$, and one may recognize that the stresses in the shell are much lower and – surprisingly – do not even have the expected sign at midspan. We may understand this result when we consider the typical difference between a plane and a cylindrical wall.

In the shell the presence of a hoop force requires either a radial load p_r or transverse forces, preferably Q_ϕ [see eq. (1c)]. Since in our example we assumed $p_r \equiv 0$, we face the alternatives of either having hoop forces N_ϕ and shearing forces Q_ϕ and then necessarily also bending

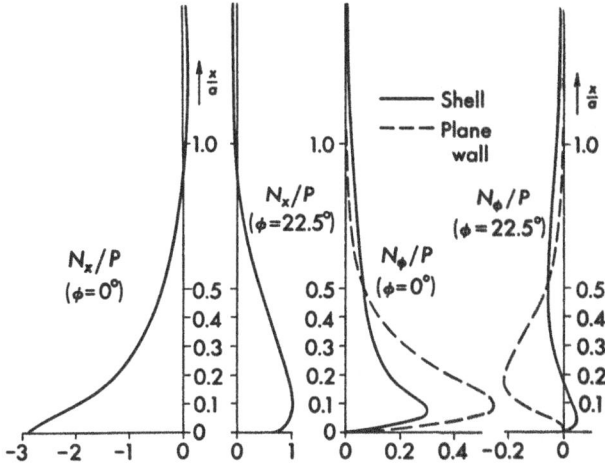

Fig. 8. Distribution of N_x and N_ϕ along two generators of the cooling tower in fig. 5

moments M_ϕ [see eq. (1d)] or having neither shear nor hoop force and hence a stress pattern which comes very close to that of the membrane theory. For the lower harmonics, a certain amount of Q_ϕ leads to a much larger bending moment than for the higher harmonics, and in the same way bending moments produce large deflections w for the lower harmonics and small wrinkles for higher m. Since large deflections are not compatible with our boundary condition $w = 0$, we find that for $m = 8$ our solution has membrane character and that $m = 16$ makes a much larger contribution to the hoop force N_ϕ. This is clearly visible in fig. 7, which shows the distribution of N_ϕ along two typical cross sections.

5.3.4 Simplified Theory

When m is not large, the coefficient of λ^4 in eq. (26) is much greater than all the other coefficients because of the term $(1 - \nu^2)/k$, the parameter k being a small number. The solution must then be either very great or very small. In the first case, the terms with λ^2 and λ^0 are negligible, in the second case those with λ^6 and λ^8. This simplifies the equation. For the roots of greater magnitude, say those expressed in $\varkappa_1$, μ_1, we use the equation

$$\lambda^4 - 2(2m^2 - \nu)\,\lambda^2 + \left(\frac{1-\nu^2}{k} + 6m^2(m^2-1)\right) = 0 \qquad (33\,\text{a})$$

and for the smaller roots, those expressed in terms of $\varkappa_2$, μ_2:

$$\left(\frac{1-\nu^2}{k} + 6m^2(m^2-1)\right)\lambda^4$$

$$- 2m^2\left(2m^4 - (4-\nu)\,m^2 + (2-\nu)\right)\lambda^2 + m^4(m^2-1)^2 = 0. \qquad (33\,\text{b})$$

This splitting of the λ equation leads to a complete splitting of the problem. From eqs. (33) it is seen that the large roots λ are of the order $k^{-1/4}$, the small ones of the order $k^{1/4}$. Assuming $\nu = 0$ and then solving eqs. (25 a, b) for $\alpha_j = A_j/C_j$ and $\beta_j = B_j/C_j$ shows that α_j is of orders $k^{1/4}$ and $k^{-1/4}$ for the large and the small roots, respectively, and β_j of orders $k^{1/2}$ and 1. We therefore have $\bar{\alpha}_3, \ldots, \bar{\beta}_4 \gg \bar{\alpha}_1, \ldots, \bar{\beta}_2$, but $\varkappa_1, \mu_1 \gg \varkappa_2, \mu_2$. Let us now have a look in Table 1 for the values of the functions at $x = 0$. We find

$$w_m(0) = \bar{C}_1 + \bar{C}_3, \quad w'_m(0) = (-\varkappa_1\bar{C}_1 + \mu_1\bar{C}_2) + (-\varkappa_2\bar{C}_3 + \mu_2\bar{C}_4),$$

$$u_m(0) = (\bar{\alpha}_1\bar{C}_1 + \bar{\alpha}_2\bar{C}_2) + (\bar{\alpha}_3\bar{C}_3 + \bar{\alpha}_4\bar{C}_4),$$

$$v_m(0) = (\bar{\beta}_1\bar{C}_1 + \bar{\beta}_2\bar{C}_2) + (\bar{\beta}_3\bar{C}_3 + \bar{\beta}_4\bar{C}_4).$$

If the constants $\bar{C}_1 \ldots \bar{C}_4$ are all of the same order of magnitude, the first term of w'_m will be much larger than the second, while the opposite is true for u_m and v_m. What this difference means, becomes clear when we have a look at the stress resultants. We find that the terms with $\varkappa_1$, μ_1 are markedly preponderant in M_x, Q_x, S_x, but that the terms with $\varkappa_2$, μ_2 dominate in N_x and, if C_3, C_4 are large enough, also in T_x. The second half of the solution (30) is therefore expecially fit to satisfy a pair of boundary conditions concerning N_x or u and T_x or v, and those are exactly the conditions we already were able to impose on the membrane solution. This part of the bending solution is not more than an improved form of the membrane solution, and since the gain is but slight, the extra effort needed to obtain it is hardly justified, if we use membrane forces as an approximation for the inhomogeneous solution.

The first half of the solution (30), to the contrary, is suitable for satisfying a pair of boundary conditions concerning M_x or w' and S_x

or w. If we replace w here by the hoop strain $w + v^{\cdot}$, these are exactly the conditions which the membrane theory cannot fulfill, and this part of the solution is therefore the essential complement to the membrane theory.

The situation changes for greater values of m. There the splitting of eq. (26) into the pair (33 a, b) is no longer possible and the values $\varkappa_1$, $\varkappa_2$ and μ_1, μ_2 come closer together. In those cases both parts of the solution work together in fulfilling all four boundary conditions, and it is not possible to anticipate part of these conditions when writing the membrane solution for the given loads.

5.4 Loads Applied to the Edges ϕ = const

5.4.1 Exact Solution

5.4.1.1 General Theory

Only in the case of tubular shells is the solution developed in Section 5.3 a complete solution of the bending problem of the circular cylinder. If the perimeter of the shell covers less than 360°, as it does in the case of barrel vaults, this solution is not applicable, because it does not permit prescribing arbitrary boundary conditions along edges ϕ = const. It has so far not been possible, and probably never will be with simple mathematical means, to find a solution which can satisfy any desired set of boundary conditions along all four edges of a rectangular panel cut from a cylindrical shell. However, we may exchange the roles of the coordinates x and ϕ and find a solution which does the same services at two edges ϕ = const that eqs. (22) did at two edges x = const.

We assume that the shell has a finite length l, or at least that all forces are periodic in x with the period $2l$. Then we may write

$$u = \sum_{n=0}^{\infty} u_n \cos\frac{n\pi x}{l}, \quad v = \sum_{n=1}^{\infty} v_n \sin\frac{n\pi x}{l}, \quad w = \sum_{n=1}^{\infty} w_n \sin\frac{n\pi x}{l}, \quad (34)$$

where u_n, v_n, w_n are functions of ϕ only. When we introduce this in the differential equations (13) and again drop the load terms, we find a set of three ordinary differential equations for the n-th harmonics u_n, v_n, w_n of the displacements. With the abbreviation

$$\lambda = \frac{n\pi a}{l},$$

they are:

$$-\lambda^2 u_n + \frac{1-\nu}{2} u_n'' + \frac{1+\nu}{2} \lambda v_n' + \nu \lambda w_n$$
$$+ k\left[\frac{1-\nu}{2} u_n'' + \lambda^3 w_n + \frac{1-\nu}{2} \lambda w_n''\right] = 0,$$

$$-\frac{1+\nu}{2} \lambda u_n' + v_n'' - \frac{1-\nu}{2} \lambda^2 v_n + w_n'$$
$$+ k\left[-\frac{3}{2}(1-\nu)\lambda^2 v_n + \frac{3-\nu}{2}\lambda^2 w_n'\right] = 0,$$

$$-\nu\lambda u_n + v_n' + w_n + k\left[-\frac{1-\nu}{2}\lambda u_n'' - \lambda^3 u_n\right.$$
$$\left.+\frac{3-\nu}{2}\lambda^2 v_n' + (\lambda^4 + 1) w_n - 2(\lambda^2 - 1) w_n'' + w_n''''\right] = 0.$$

$$\tag{35}$$

They may be solved by putting

$$u_n = A\, e^{m\phi}, \quad v_n = B\, e^{m\phi}, \quad w_n = C\, e^{m\phi}, \tag{36}$$

where m is not an integer but a still unknown quantity which will play the same role here as λ did in Section 5.3. Introducing eqs. (36) into eqs. (35), we arrive at three linear equations for the constants A, B, C:

$$\left[\lambda^2 - \frac{1-\nu}{2} m^2(1+k)\right]A + \left[-\frac{1+\nu}{2}\lambda m\right]B$$
$$+ \left[-\nu\lambda - k\left(\lambda^3 + \frac{1-\nu}{2}\lambda m^2\right)\right]C = 0,$$

$$\left[-\frac{1+\nu}{2}\lambda m\right]A + \left[m^2 - \frac{1-\nu}{2}\lambda^2 - \frac{3}{2}(1-\nu)k\lambda^2\right]B$$
$$+ \left[m + \frac{3-\nu}{2}k\lambda^2 m\right]C = 0,$$

$$\left[-\nu\lambda - k\left(\lambda^3 + \frac{1-\nu}{2}\lambda m^2\right)\right]A + \left[m + \frac{3-\nu}{2}k\lambda^2 m\right]B$$
$$+ [1 + k(\lambda^4 - 2\lambda^2 m^2 + m^4 + 2m^2 + 1)]C = 0.$$

$$\tag{37}$$

The only formal difference between this set and eqs. (25) lies in the sign of some of the terms, but we must keep in mind that here λ is known and m is not. Therefore, the condition that the determinant of the nine coefficients vanishes now yields an equation for m;

$$m^8 - 2(2\lambda^2 - 1) m^6 + [6\lambda^4 - 2(4-\nu)\lambda^2 + 1] m^4$$
$$- 2\lambda^2[2\lambda^4 - 3\lambda^2 + (2-\nu)] m^2 + \left[\frac{1-\nu^2}{k}\lambda^4 + \lambda^6(\lambda^2 - 2\nu)\right] = 0. \tag{38}$$

The eight solutions of this equation are all complex and may be written as follows:

$$
\begin{aligned}
m_1 &= -\varkappa_1 + i\,\mu_1, & m_5 &= +\varkappa_1 + i\,\mu_1, \\
m_2 &= -\varkappa_1 - i\,\mu_1, & m_6 &= +\varkappa_1 - i\,\mu_1, \\
m_3 &= -\varkappa_2 + i\,\mu_2, & m_7 &= +\varkappa_2 + i\,\mu_2, \\
m_4 &= -\varkappa_2 - i\,\mu_2, & m_8 &= +\varkappa_2 - i\,\mu_2.
\end{aligned}
\tag{39}
$$

The $\varkappa$, μ may be found by the same numerical methods as the $\varkappa$, μ in Section 5.3. Each of the 8 values m_j yields one particular solution of the differential equations (35):

$$
u_n = A_j\, e^{m_j\phi}, \qquad v_n = B_j\, e^{m_j\phi}, \qquad w_n = C_j\, e^{m_j\phi},
$$

where the three constants A_j, B_j, C_j of each set are related by the linear equations (37), in which the corresponding value m_j has to be used when computing the coefficients. Since the determinant is zero, we may use any pair out of the three equations to determine A_j, B_j or a given C_j. The result may be written as

$$
A_j = \alpha_j\, C_j, \qquad B_j = \beta_j\, C_j,
$$

where α_j, β_j are complex constants derived from the coefficients of eqs. (37). From closer inspection of these equations one may recognize that for α_j and β_j the following relations must hold:

$$
\begin{aligned}
\alpha_1 &= \alpha_6 = \bar{\alpha}_1 + i\,\bar{\alpha}_2, & \alpha_3 &= \alpha_8 = \bar{\alpha}_3 + i\,\bar{\alpha}_4, \\
\alpha_2 &= \alpha_5 = \bar{\alpha}_1 - i\,\bar{\alpha}_2, & \alpha_4 &= \alpha_7 = \bar{\alpha}_3 - i\,\bar{\alpha}_4, \\
\beta_1 &= -\beta_6 = \bar{\beta}_1 + i\,\bar{\beta}_2, & \beta_3 &= -\beta_8 = \bar{\beta}_3 + i\,\bar{\beta}_4, \\
\beta_2 &= -\beta_5 = \bar{\beta}_1 - i\,\bar{\beta}_2, & \beta_4 &= -\beta_7 = \bar{\beta}_3 - i\,\bar{\beta}_4,
\end{aligned}
$$

in which the quantities marked with a bar represent real numbers.

The complete solution for any one of the displacements is the sum of the eight particular solutions, e. g.:

$$
\begin{aligned}
w_n = {}& e^{-\varkappa_1\phi}(C_1 e^{i\mu_1\phi} + C_2 e^{-i\mu_1\phi}) + e^{-\varkappa_2\phi}(C_3 e^{i\mu_2\phi} + C_4 e^{-i\mu_2\phi}) \\
& + e^{+\varkappa_1\phi}(C_5 e^{i\mu_1\phi} + C_6 e^{-i\mu_1\phi}) + e^{+\varkappa_2\phi}(C_7 e^{i\mu_2\phi} + C_8 e^{-i\mu_2\phi}),
\end{aligned}
$$

which may as well be written in the following form:

$$
\begin{aligned}
w_n = {}& e^{-\varkappa_1\phi}[(C_1 + C_2)\cos\mu_1\phi + i(C_1 - C_2)\sin\mu_1\phi] \\
& + e^{-\varkappa_2\phi}[(C_3 + C_4)\cos\mu_2\phi + i(C_3 - C_4)\sin\mu_2\phi] \\
& + e^{+\varkappa_1\phi}[(C_5 + C_6)\cos\mu_1\phi + i(C_5 - C_6)\sin\mu_1\phi] \\
& + e^{+\varkappa_2\phi}[(C_7 + C_8)\cos\mu_2\phi + i(C_7 - C_8)\sin\mu_2\phi].
\end{aligned}
$$

Since all formulas, though simple in structure, become very clumsy in appearance, we shall not continue the general treatment but shall specialize on two important particular cases.

From general experience in the bending theory of shells, one may expect that for thin cylinders the forces and moments applied to one boundary ϕ = const will produce a localized disturbance but will not influence very much the situation at the opposite edge. It may easily be checked how far this is true in the present case, just by solving

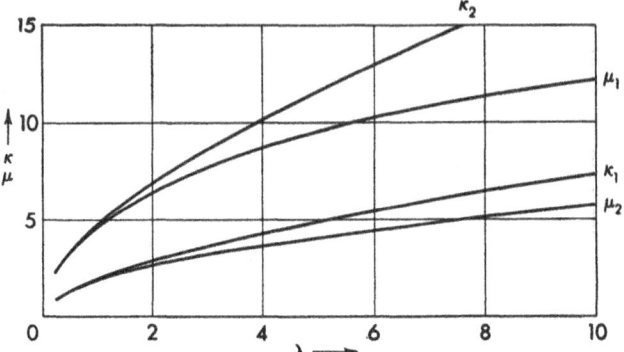

Fig. 9. Real and imaginary parts of $m = \varkappa + i\mu$ from eq. (38) for $k/(1 - \nu^2) = 2 \times 10^{-6}$

eq. (38) numerically. The result of such computations has been plotted in fig. 9. They were made for an assumed value $k/(1 - \nu^2) = 2 \times 10^{-6}$, corresponding to $t/a = 4.76 \times 10^{-3}$ for $\nu = 0.3$ and to $t/a = 4.88 \times 10^{-3}$ for $\nu = 0$, ratios which may easily be encountered in shell design.

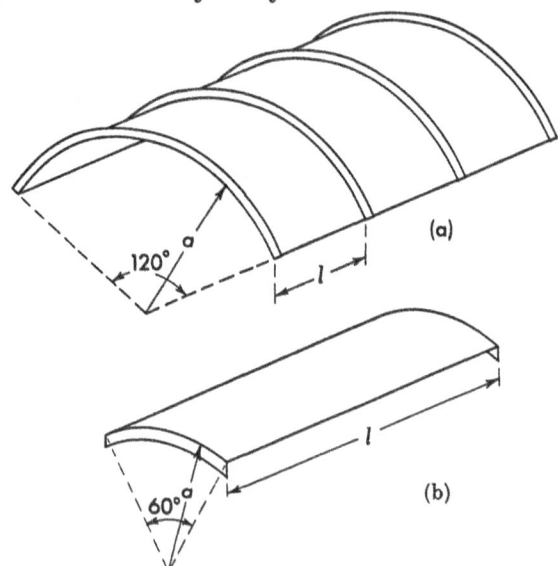

Fig. 10. Two typical shell roofs, (a) short cylinder, (b) long cylinder

Now let us consider two reinforced concrete roofs, figs. 10a, b. The first one consists of three shells between stiff ribs, and l/a may be, say, 0.6. For $n = 1$, this yields $\lambda = 5.24$. The smaller one of the two damping exponents is then $\varkappa_1 = 5.1$, and a disturbance which begins with the value 1 at the edge $\phi = 0$ will have decayed to $\exp(-5.1 \times 2.094) = 0.23 \times 10^{-4}$ at $\phi = 120°$. For higher harmonics, $n > 1$, the decay will be even greater. If one requires a decay to only 0.01, one may allow l/a to be as great as 2.6.

Quite to the contrary, the barrel vault, fig. 10b, may have $l/a = 2.5$, hence $\lambda = 1.26$ for the first harmonic. Fig. 9 yields $\varkappa_1 = 2.23$, and over an angle of $60° = 1.047$ an edge disturbance will decay from 1 to $\exp(-2.23 \times 1.047) = 0.097 \approx 10\%$. This, of course, is still very much and will have to be considered when the boundary conditions at the other edge are formulated. There will be even less decay if the shell is thicker, but, still, for the higher harmonics, $n = 4, 5, \ldots$, the values of λ and hence of $\varkappa_1$ are high enough to make the corresponding edge disturbances more or less local.

5.4.1.2 One Boundary Only

In those cases where the disturbance is localized at one edge, the general solution may be considerably simplified. If we measure the angular coordinate from the edge under consideration, then the terms $j = 5, 6, 7, 8$ in the preceding formulas evidently are unsuitable, since they describe just the contrary of a local disturbance: stresses and displacements that increase exponentially the farther away we go from the edge $\phi = 0$. Therefore, these terms must be dropped, and we are left with the following formulas:

$$u_n = e^{-\varkappa_1 \phi}[(A_1 + A_2)\cos\mu_1\phi + i(A_1 - A_2)\sin\mu_1\phi]$$
$$+ e^{-\varkappa_2 \phi}[(A_3 + A_4)\cos\mu_2\phi + i(A_3 - A_4)\sin\mu_2\phi],$$
$$v_n = e^{-\varkappa_1 \phi}[(B_1 + B_2)\cos\mu_1\phi + i(B_1 - B_2)\sin\mu_1\phi]$$
$$+ e^{-\varkappa_2 \phi}[(B_3 + B_4)\cos\mu_2\phi + i(B_3 - B_4)\sin\mu_2\phi],$$
$$w_n = e^{-\varkappa_1 \phi}[(C_1 + C_2)\cos\mu_1\phi + i(C_1 - C_2)\sin\mu_1\phi]$$
$$+ e^{-\varkappa_2 \phi}[(C_3 + C_4)\cos\mu_2\phi + i(C_3 - C_4)\sin\mu_2\phi].$$

Obviously, the coefficients appearing in these formulas must all have real values, and they may be expressed by eqs. (28), (29) on pp. 229-230, substituting, of course, for the α's and β's the quantities defined on p. 241.

When we put the expressions for u_n, v_n, w_n into the formulas (34) (omitting the summation) and then go back to the elastic law (9) and the equations (1d, e) we see that the n-th harmonics of all displacements and stress resultants assume the form:

$$f = c\left[e^{-\varkappa_1 \phi}((a_1 \bar{C}_1 + a_2 \bar{C}_2)\cos\mu_1\phi + (a_1 \bar{C}_2 - a_2 \bar{C}_1)\sin\mu_1\phi)\right. \tag{40}$$
$$\left. + e^{-\varkappa_2 \phi}((a_3 \bar{C}_3 + a_4 \bar{C}_4)\cos\mu_2\phi + (a_3 \bar{C}_4 - a_4 \bar{C}_3)\sin\mu_2\phi)\right]_{\sin}^{\cos} \frac{\lambda x}{a}.$$

Table 2. *Cylinder Loaded along a Generator*

f	c	a_1
u	1	$\bar{\alpha}_1$
v	1	$\bar{\beta}_1$
w	1	1
$w^{\cdot}$	1	$-\varkappa_1$
N_ϕ	D/a	$1 + k - (\varkappa_1 \bar{\beta}_1 + \mu_1 \bar{\beta}_2) - \nu \lambda \bar{\alpha}_1 + k(\varkappa_1^2 - \mu_1^2)$
N_x	D/a	$-\lambda \bar{\alpha}_1 + \nu - \nu(\varkappa_1 \bar{\beta}_1 + \mu_1 \bar{\beta}_2) + k \lambda^2$
$N_{\phi x}$	$\dfrac{D(1-\nu)}{2a}$	$-(1+k)(\varkappa_1 \bar{\alpha}_1 + \mu_1 \bar{\alpha}_2) + \lambda \bar{\beta}_1 - k \lambda \varkappa_1$
$N_{x\phi}$	$\dfrac{D(1-\nu)}{2a}$	$-(\varkappa_1 \bar{\alpha}_1 + \mu_1 \bar{\alpha}_2) + (1+k) \lambda \bar{\beta}_1 + k \lambda \varkappa_1$
M_ϕ	K/a^2	$1 + (\varkappa_1^2 - \mu_1^2) - \nu \lambda^2$
M_x	K/a^2	$-\lambda^2 + \lambda \bar{\alpha}_1 + \nu(\varkappa_1^2 - \mu_1^2) + \nu(\varkappa_1 \bar{\beta}_1 + \mu_1 \bar{\beta}_2)$
$M_{\phi x}$	$\dfrac{K(1-\nu)}{2a^2}$	$-2\lambda \varkappa_1 - (\varkappa_1 \bar{\alpha}_1 + \mu_1 \bar{\alpha}_2) - \lambda \bar{\beta}_1$
$M_{x\phi}$	$\dfrac{K(1-\nu)}{a^2}$	$-\lambda(\varkappa_1 + \bar{\beta}_1)$
Q_ϕ	K/a^3	$\varkappa_1(\lambda^2 - 1 - \varkappa_1^2 + 3\mu_1^2) + (1-\nu) \lambda^2 \bar{\beta}_1$
Q_x	$\dfrac{K}{2a^3}$	$\begin{aligned} &-2\lambda^3 + 2\lambda(\varkappa_1^2 - \mu_1^2) + 2\lambda^2 \bar{\alpha}_1 \\ &+ (1-\nu)[(\varkappa_1^2 - \mu_1^2) \bar{\alpha}_1 + 2\varkappa_1 \mu_1 \bar{\alpha}_2] + (1+\nu) \lambda(\varkappa_1 \bar{\beta}_1 + \mu_1 \bar{\beta}_2) \end{aligned}$
S_ϕ	$\dfrac{K}{2a^3}$	$\begin{aligned} &2\varkappa_1[-1 + (2-\nu)\lambda^2 - \varkappa_1^2 + 3\mu_1^2] + (1-\nu) \lambda(\varkappa_1 \bar{\alpha}_1 + \mu_1 \bar{\alpha}_2) \\ &+ 3(1-\nu) \lambda^2 \bar{\beta}_1 \end{aligned}$

The factor c and the coefficients a_1, a_2 are given in Table 2. To find a_3 and a_4, we use the same formulas, only changing the subscript 1 to 2 for $\varkappa$ and μ and the subscripts 1, 2 to 3, 4 for α, $\bar{\beta}$.

To solve a specific problem, we have to proceed in the following way: We choose the order n of the harmonic which we want to investigate and compute from the dimensions of the shell (a, t, l) the parameters k and λ. We then find $\varkappa_1$, μ_1, $\varkappa_2$, μ_2 from eq. (38) and α_j, β_j $(j = 1 \ldots 4)$ by solving the first two of eq. (37) for A_j, B_j with $C_j = 1$. We are then prepared to establish the boundary conditions. From the table we find the boundary values of those forces and displacements which appear in the boundary conditions, and using the numerical values of the $\bar{\alpha}_j$, $\bar{\beta}_j$, we obtain four linear equations for $\bar{C}_1$,

Table 2. (Continued)

a_2	symmetry	x factor
$\bar{\alpha}_2$	sym.	cos
$\bar{\beta}_2$	anti.	sin
0	sym.	sin
μ_1	anti.	sin
$-(\varkappa_1\bar{\beta}_2 - \mu_1\bar{\beta}_1) - \nu\lambda\bar{\alpha}_2 - 2k\varkappa_1\mu_1$	sym.	sin
$-\lambda\bar{\alpha}_2 - \nu(\varkappa_1\bar{\beta}_2 - \mu_1\bar{\beta}_1)$	sym.	sin
$-(1+k)(\varkappa_1\bar{\alpha}_2 - \mu_1\bar{\alpha}_1) + \lambda\bar{\beta}_2 + k\lambda\mu_1$	anti.	cos
$-(\varkappa_1\bar{\alpha}_2 - \mu_1\bar{\alpha}_1) + (1+k)\lambda\bar{\beta}_2 - k\lambda\mu_1$	anti.	cos
$-2\varkappa_1\mu_1$	sym.	sin
$\lambda\bar{\alpha}_2 - 2\nu\varkappa_1\mu_1 + \nu(\varkappa_1\bar{\beta}_2 - \mu_1\bar{\beta}_1)$	sym.	sin
$2\lambda\mu_1 - (\varkappa_1\bar{\alpha}_2 - \mu_1\bar{\alpha}_1) - \lambda\bar{\beta}_2$	anti.	cos
$\lambda(\mu_1 - \bar{\beta}_2)$	anti.	cos
$\mu_1(1 - \lambda^2 + 3\varkappa_1^2 - \mu_1^2) + (1-\nu)\lambda^2\bar{\beta}_2$	anti.	sin
$-4\lambda\varkappa_1\mu_1 + 2\lambda^2\bar{\alpha}_2 + (1-\nu)[(\varkappa_1^2 - \mu_1^2)\bar{\alpha}_2 - 2\varkappa_1\mu_1\bar{\alpha}_1] + (1+\nu)\lambda(\varkappa_1\bar{\beta}_2 - \mu_1\bar{\beta}_1)$	sym.	cos
$2\mu_1[1 - (2-\nu)\lambda^2 + 3\varkappa_1^2 - \mu_1^2] + (1-\nu)\lambda(\varkappa_1\bar{\alpha}_2 - \mu_1\bar{\alpha}_1) + 3(1-\nu)\lambda^2\bar{\beta}_2$	anti.	sin

$\bar{C}_2$, $\bar{C}_3$, $\bar{C}_4$. When these equations have been solved, the table will yield numerical values for all the stress resultants which we may want to know.

Along the edges $\phi = $ const we find forces N_ϕ, $N_{\phi x}$, Q_ϕ and moments M_ϕ, $M_{\phi x}$. To them considerations may be applied similar to those which lead to eqs. (32). Since the edge is straight, there is no force F_t (fig. 4b), and the effective shear T_ϕ is identical with the shear $N_{\phi x}$. But there is an effective transverse force

$$S_\phi = Q_\phi + \frac{M'_{\phi x}}{a}. \tag{41}$$

Its coefficients may also be found in Table 2.

5.4.1.3 Symmetric Stress System

The second special case of practical importance is that of a shell having boundaries at $\phi = \pm\phi_0$ and having there such conditions that the stress system will be symmetric with respect to the generator $\phi = 0$. In this case we cannot discard the solutions $j = 5 \ldots 8$, but we may combine the real exponential functions to hyperbolic sines and cosines, writing

$$
\begin{aligned}
w_n = {} & (C_1 + C_2 + C_5 + C_6)\, \mathrm{Cosh}\varkappa_1\phi \cos\mu_1\phi \\
& +i(-C_1 + C_2 + C_5 - C_6)\, \mathrm{Sinh}\varkappa_1\phi \sin\mu_1\phi \\
& +(-C_1 - C_2 + C_5 + C_6)\, \mathrm{Sinh}\varkappa_1\phi \cos\mu_1\phi \\
& +i(C_1 - C_2 + C_5 - C_6)\, \mathrm{Cosh}\varkappa_1\phi \sin\mu_1\phi .
\end{aligned}
\tag{42}
$$

From symmetry we now conclude that the coefficients of the antimetric solutions $\mathrm{Cosh}\varkappa_1\phi \sin\mu_1\phi$ and $\mathrm{Sinh}\varkappa_1\phi \cos\mu_1\phi$ must be zero. This leads to the relations $C_1 = C_6$ and $C_2 = C_5$. Dropping everywhere a common factor 2, we may write the solution in the following form:

$$
\begin{aligned}
w_n = {} & (C_1 + C_2)\, \mathrm{Cosh}\,\varkappa_1\phi \cos\mu_1\phi - i(C_1 - C_2)\, \mathrm{Sinh}\varkappa_1\phi \sin\mu_1\phi \\
& + (C_3 + C_4)\, \mathrm{Cosh}\varkappa_2\phi \cos\mu_2\phi - i(C_3 - C_4)\, \mathrm{Sinh}\varkappa_2\phi \sin\mu_2\phi
\end{aligned}
$$

or, with the notation (28) from p. 229:

$$
\begin{aligned}
w_n = {} & \overline{C}_1 \mathrm{Cosh}\varkappa_1\phi \cos\mu_1\phi - \overline{C}_2 \mathrm{Sinh}\varkappa_1\phi \sin\mu_1\phi \\
& + \overline{C}_3 \mathrm{Cosh}\varkappa_2\phi \cos\mu_2\phi - \overline{C}_4 \mathrm{Sinh}\varkappa_2\phi \sin\mu_2\phi .
\end{aligned}
$$

To find u_n and v_n, we have to replace C_j in eq. (42) by A_j or B_j, respectively. Again dropping the factor 2 and using the notation (29), we find these expressions:

$$
\begin{aligned}
u_n = {} & (\overline{\alpha}_1 \overline{C}_1 + \overline{\alpha}_2 \overline{C}_2)\, \mathrm{Cosh}\varkappa_1\phi \cos\mu_1\phi - (\overline{\alpha}_1 \overline{C}_2 - \overline{\alpha}_2 \overline{C}_1)\, \mathrm{Sinh}\varkappa_1\phi \sin\mu_1\phi \\
& + (\overline{\alpha}_3 \overline{C}_3 + \overline{\alpha}_4 \overline{C}_4)\, \mathrm{Cosh}\varkappa_2\phi \cos\mu_2\phi - (\overline{\alpha}_3 \overline{C}_4 - \overline{\alpha}_4 \overline{C}_3)\, \mathrm{Sinh}\varkappa_2\phi \sin\mu_2\phi ,
\end{aligned}
$$

$$
\begin{aligned}
v_n = {} & -(\overline{\beta}_1 \overline{C}_1 + \overline{\beta}_2 \overline{C}_2)\, \mathrm{Sinh}\varkappa_1\phi \cos\mu_1\phi + (\overline{\beta}_1 \overline{C}_2 - \overline{\beta}_2 \overline{C}_1)\, \mathrm{Cosh}\varkappa_1\phi \sin\mu_1\phi \\
& - (\overline{\beta}_3 \overline{C}_3 + \overline{\beta}_4 \overline{C}_4)\, \mathrm{Sinh}\varkappa_2\phi \cos\mu_2\phi + (\overline{\beta}_3 \overline{C}_4 - \overline{\beta}_4 \overline{C}_3)\, \mathrm{Cosh}\varkappa_2\phi \sin\mu_2\phi ,
\end{aligned}
$$

When we go back to eqs. (9) and (1d, e) we find that some of the stress resultants (the symmetric group) are expressed by formulas similar to those for u and w, i. e.

$$
\begin{aligned}
f = c\big[& (a_1 \overline{C}_1 + a_2 \overline{C}_2)\, \mathrm{Cosh}\varkappa_1\phi \cos\mu_1\phi - (a_1 \overline{C}_2 - a_2 \overline{C}_1)\, \mathrm{Sinh}\varkappa_1\phi \sin\mu_1\phi \\
& + (a_3 \overline{C}_3 + a_4 \overline{C}_4)\, \mathrm{Cosh}\varkappa_2\phi \cos\mu_2\phi \\
& - (a_3 \overline{C}_4 - a_4 \overline{C}_3)\, \mathrm{Sinh}\varkappa_2\phi \sin\mu_2\phi \big]{\textstyle\frac{\cos}{\sin}}\frac{\lambda x}{a} ,
\end{aligned}
\tag{43a}
$$

while the rest (the antimetric group) looks like

$$f = -c[(a_1 \bar{C}_1 + a_2 \bar{C}_2) \operatorname{Sinh} \varkappa_1 \phi \cos \mu_1 \phi - (a_1 \bar{C}_2 - a_2 \bar{C}_1) \operatorname{Cosh} \varkappa_1 \phi \sin \mu_1 \phi$$
$$+ (a_3 \bar{C}_3 + a_4 \bar{C}_4) \operatorname{Sinh} \varkappa_2 \phi \cos \mu_2 \phi \qquad\qquad (43\,\mathrm{b})$$
$$- (a_3 \bar{C}_4 - a_4 \bar{C}_3) \operatorname{Cosh} \varkappa_2 \phi \sin \mu_2 \phi] {\cos \atop \sin} \frac{\lambda x}{a}.$$

The coefficients c, a_1, a_2 are those already presented in Table 2, and the column marked "symmetry" in this Table indicates whether the quantity belongs to the symmetric or to the antimetric group. The coefficients a_3, a_4 are again found by simply changing subscripts.

5.4.2 Barrel Vaults

5.4.2.1 The Differential Equation and its Solution

The theory explained on the preceding pages may be simplified considerably in the case of barrel vaults, in which l is much greater than the radius a of the shell. In Chapter 3 we saw that an edge load applied to the straight edges $\phi = $ const of such a shell cannot be carried by membrane forces N_ϕ. Bending moments M_ϕ and transverse forces Q_ϕ are needed to carry the load away from the edge, and forces N_x and $N_{x\phi}$ are needed to transmit the load to the ends of the span. However, the deformation produced, in particular the lengthwise curvature $\partial^2 w/\partial x^2$, will not be sufficient to allow for bending moments M_x of any importance.

We may, therefore, simplify the theory by dropping most of the moments and keeping, besides the membrane forces, only the moment M_ϕ and the force Q_ϕ. One of the conditions of equilibrium, eq. (1e), then becomes trivial, and another one, eq. (1f), yields $N_{x\phi} = N_{\phi x}$. In the remaining 4 equations, (1a–d), there are only 5 unknowns left and we may eliminate all but one of them. We choose to retain M_ϕ.

Eq. (1d) yields

$$Q_\phi = \frac{1}{a} M_\phi^{\boldsymbol\cdot}, \qquad\qquad (44\,\mathrm{a})$$

and with $p_x = p_\phi = p_r = 0$ we find from eqs. (1c, b, a):

$$N_\phi = -Q_\phi^{\boldsymbol\cdot} = -\frac{1}{a} M_\phi^{\boldsymbol{\cdot\cdot}},$$

$$N_{x\phi}' = Q_\phi - N_\phi^{\boldsymbol\cdot} = \frac{1}{a}(M_\phi^{\boldsymbol\cdot} + M_\phi^{\boldsymbol{\cdot\cdot\cdot}}), \qquad\qquad (44\,\mathrm{b–d})$$

$$N_x'' = -N_{x\phi}^{\boldsymbol{\cdot\cdot}} = -\frac{1}{a}(M_\phi^{\boldsymbol{\cdot\cdot}} + M_\phi^{\boldsymbol{\cdot\cdot\cdot\cdot}}).$$

To obtain a differential equation for M_ϕ, we have to consider the deformation of the shell. The formulas (9f–h) concern the neglected moments and are of no interest here. In eqs. (9a–d) we drop, of course, the terms with K, which are of no numerical importance. Eqs. (9c) and (9d) are then identical, and we have 4 equations left, just enough for the 4 unknowns M_ϕ, u, v, w.

We next eliminate the displacements. From eqs. (9a, b) we find

$$u' = \frac{a}{D(1 - v^2)} (N_x - v N_\phi). \tag{45a}$$

After differentiating this with respect to ϕ and eq. (9c) with respect to x/a, we find from both

$$v'' = \frac{a}{D(1 - v^2)} [2(1 + v) N'_{x\phi} - N^{\cdot\cdot}_x + v N^{\cdot\cdot}_\phi]. \tag{45b}$$

Lastly, we get from eqs. (9a, b) an expression for $v^{\cdot} + w$, and combining this with the preceding formula for v'', we obtain

$$w'' = \frac{a}{D(1 - v^2)} [N''_\phi - v N^{\cdot\cdot}_\phi + N^{\cdot\cdot}_x - v N''_x - 2(1 + v) N'_{x\phi}]. \tag{45c}$$

All this may now be introduced into eq. (9e) which, for this purpose, must be differentiated twice with respect to x/a:

$$M''_\phi = \frac{K}{D a(1 - v^2)} [N''_\phi - v N^{\cdot\cdot}_\phi + (1 - v^2) N''_\phi{}^{\cdot\cdot} - v N^{\cdot\cdot}_\phi + v N^{IV}_\phi$$
$$+ N^{\cdot\cdot}_x - v N''_x + N^{\cdot\cdot}_x - v^2 N^{IV}_x - 2(1 + v)(N^{\cdot\cdot}_{x\phi} + N^{\cdot\cdot\cdot}_{x\phi})$$
$$- 2v(1 + v) N''_{x\phi}].$$

This equation yields the final equation for M_ϕ when we eliminate the forces on its right-hand side with the help of the conditions of equilibrium (44). Again using the abbreviation $k = K/Da^2$, we obtain

$$M^{\cdot\cdot}_\phi{}^{\cdot\cdot} + (2 + v) M''_\phi{}^{\cdot\cdot\cdot} + 2 M^{\cdot\cdot}_\phi{}^{\cdot} + (1 + 2v) M^{IV}_\phi{}^{\cdot\cdot} + 2(2 + v) M''_\phi{}^{\cdot\cdot} \tag{46}$$
$$+ M^{\cdot\cdot}_\phi + v M^{VI}_\phi{}^{\cdot\cdot} + (1 + v)^2 M^{IV}_\phi{}^{\cdot\cdot} + (2 + v) M''_\phi{}^{\cdot\cdot} + \frac{1 - v^2}{k} M^{IV}_\phi = 0.$$

This is a partial differential equation of the 8th order, and it replaces the set (13) or eq. (17). We may solve it by the same means as applied in Section 5.4.1, putting

$$M_\phi = C e^{m\phi} \sin\frac{\lambda x}{a} \quad \text{with} \quad \lambda = \frac{n \pi a}{l}.$$

Introducing this into eq. $(46)_l$ we find an equation of the 8th degree for m:

$$m^8 + [-(2 + v) \lambda^2 + 2] m^6 + [(1 + 2v) \lambda^4 - 2(2 + v) \lambda^2 + 1] m^4$$
$$+ [-v \lambda^6 + (1 + v)^2 \lambda^4 - (2 + v) \lambda^2] m^2 + \frac{1 - v^2}{k} \lambda^4 = 0.$$

When we compare this equation with (38), we see that the coefficients of certain terms differ considerably. If our approximation shall be admissible, these terms must be unimportant and we had best drop them completely. What remains, is very simple:

$$m^8 + 2m^6 + m^4 + \frac{1-\nu^2}{k}\lambda^4 = 0.$$

The eight solutions of this equation are

$$m = \pm\sqrt{-\frac{1}{2} \pm \sqrt{\frac{1}{4} \pm i\lambda^2\sqrt{\frac{1-\nu^2}{k}}}}.\qquad(47)$$

Since they depend on only one parameter,

$$\lambda^2\sqrt{\frac{1-\nu^2}{k}} = 2\pi^2\sqrt{3(1-\nu^2)}\,\frac{a^3}{l^3 t}\,n^3 = \zeta^4 n^2,$$

their real and imaginary parts as defined by eqs. (39) may be tabulated as its functions. The result is shown in fig. 11. We may now proceed in

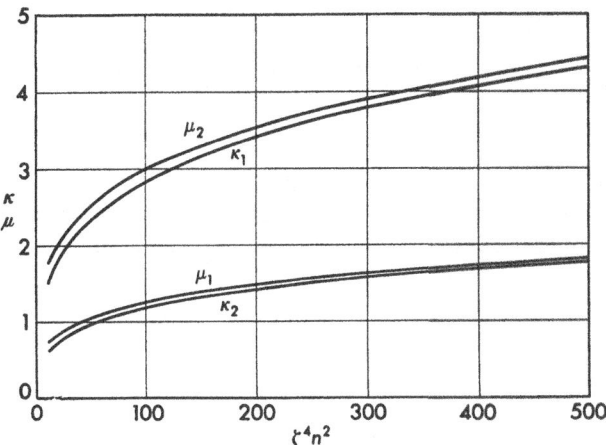

Fig. 11. Real and imaginary parts of $m = \varkappa + i\mu$ from eq. (47)

the same way as in the exact theory and treat in full detail the two cases that either the conditions at opposite boundaries do not interfere (isolated boundary) or that the stress system is symmetrical.

5.4.2.2 Isolated Boundary

We assume one boundary to be located at $\phi = 0$ and the other one to be so far away that its influence is negligible for the displacements and stress resultants in the vicinity of $\phi = 0$. For the study of this boundary zone we may then drop all those solutions where m has a positive real part. The remaining four may be grouped in pairs and written in real form, using A and B for the linear combinations of the C:

$$M_\phi = a[e^{-\varkappa_1\phi}(A_1\cos\mu_1\phi + B_1\sin\mu_1\phi)$$
$$+ e^{-\varkappa_2\phi}(A_2\cos\mu_2\phi + B_2\sin\mu_2\phi)]\sin\frac{\lambda x}{a}. \tag{48}$$

We do not have to bother here with complex constants α_j, β_j, because we have only one differential equation (46) instead of the set (13) and therefore no such thing as the three linear equations (37). This simplifies considerably the formulas for the stress resultants and displacements. To find them from eqs. (44) and (45), we need the derivatives of M_ϕ. We have:

$$M_\phi^{\cdot} = a[e^{-\varkappa_1\phi}((-\varkappa_1 A_1 + \mu_1 B_1)\cos\mu_1\phi + (-\mu_1 A_1 - \varkappa_1 B_1)\sin\mu_2\phi)$$
$$+ e^{-\varkappa_2\phi}((-\varkappa_2 A_2 + \mu_2 B_2)\cos\mu_2\phi$$
$$+ (-\mu_2 A_2 - \varkappa_2 B_2)\sin\mu_2\phi)]\sin\frac{\lambda x}{a}$$

which may be written

$$M_\phi^{\cdot} = a[e^{-\varkappa_1\phi}(A_1^{(1)}\cos\mu_1\phi + B_1^{(1)}\sin\mu_1\phi)$$
$$+ e^{-\varkappa_2\phi}(A_2^{(1)}\cos\mu_2\phi + B_2^{(1)}\sin\mu_2\phi)]\sin\frac{\lambda x}{a}.$$

Table 3. Barrel Vault

f	c	a_1
M_ϕ	a	A_1
Q_ϕ	1	$A_1^{(1)}$
N_ϕ	-1	$A_1^{(2)}$
$N_{x\phi}$	$-1/\lambda$	$A_1^{(3)} + A_1^{(1)}$
N_x	$1/\lambda^2$	$A_1^{(4)} + A_1^{(2)}$
u	$-\dfrac{a}{D(1-\nu^2)\,\lambda^3}$	$A_1^{(4)} + (1+\nu\lambda^2)\,A_1^{(2)}$
v	$+\dfrac{a}{D(1-\nu^2)\,\lambda^4}$	$A_1^{(5)} - [(2+\nu)\,\lambda^2 - 1]\,A_1^{(3)} - 2(1+\nu)\,\lambda^2\,A_1^{(1)}$
w	$-\dfrac{a}{D(1-\nu^2)\,\lambda^4}$	$A_1^{(6)} - (2\lambda^2 - 1)\,A_1^{(4)} + \lambda^2(\lambda^2 - 2 - \nu)\,A_1^{(2)}$
$w^{\cdot}$	$-\dfrac{a}{D(1-\nu^2)\,\lambda^4}$	$A_1^{(7)} - (2\lambda^2 - 1)\,A_1^{(5)} + \lambda^2(\lambda^2 - 2 - \nu)\,A_1^{(3)}$

The same linear relations which lead from the A, B to the $A^{(1)}$, $B^{(1)}$ will lead from these to the coefficients $A^{(2)}$, $B^{(2)}$ of $M_\phi^{..}$ and so on. Writing $A^{(0)}$ and $B^{(0)}$ for A and B, we compute the "iterated coefficients" from the following recurrence formulas:

$$A_{\frac{1}{2}}^{(\iota+1)} = -\varkappa_1 A_{\frac{1}{2}}^{(\iota)} + \mu_1 B_{\frac{1}{2}}^{(\iota)}, \qquad B_{\frac{1}{2}}^{(\iota+1)} = -\mu_1 A_{\frac{1}{2}}^{(\iota)} - \varkappa_1 B_{\frac{1}{2}}^{(\iota)}. \quad (49)$$

In numerical work it frequently happens that only the iterated coefficients for even ι or only those for odd ι are needed. The work will then be speeded up by using the following double-step formulas:

$$A_{\frac{1}{2}}^{(\iota+2)} = (\varkappa_1^2 - \mu_1^2) A_{\frac{1}{2}}^{(\iota)} - 2\varkappa_1 \mu_1 B_{\frac{1}{2}}^{(\iota)},$$

$$B_{\frac{1}{2}}^{(\iota+2)} = 2\varkappa_1 \mu_1 A^{(\iota)} + (\varkappa_1^2 - \mu_1^2) B_{\frac{1}{2}}^{(\iota)}. \quad (49')$$

In eqs. (44) all those stress resultants which are not of negligible magnitude have been expressed in terms of the derivatives of M_ϕ. When the solution (48) is introduced here, they all assume the form

$$f = c[e^{-\varkappa_1\phi}(a_1 \cos\mu_1\phi + b_1 \sin\mu_1\phi$$
$$+ e^{-\varkappa_2\phi}(a_2 \cos\mu_2\phi + b_2 \sin\mu_2\phi]^{\cos}_{\sin} \frac{\lambda x}{a}, \quad (50)$$

and the coefficients a_1, b_1, a_2, b_2 may be expressed by the iterated coefficients. The factor c and these expressions for a_1, b_1 have been

Table 3. (Continued)

b_1	x factor
B_1	sin
$B_1^{(1)}$	sin
$B_1^{(2)}$	sin
$B_1^{(3)} + B_1^{(1)}$	cos
$B_1^{(4)} + B_1^{(2)}$	sin
$B_1^{(4)} + (1 + \nu\lambda^2) B_1^{(2)}$	cos
$B_1^{(5)} - [(2 + \nu)\lambda^2 - 1] B_1^{(3)} - 2(1 + \nu)\lambda^2 B_1^{(1)}$	sin
$B_1^{(6)} - (2\lambda^2 - 1) B_1^{(4)} + \lambda^2(\lambda^2 - 2 - \nu) B_1^{(2)}$	sin
$B_1^{(7)} - (2\lambda^2 - 1) B_1^{(5)} + \lambda^2(\lambda^2 - 2 - \nu) B_1^{(3)}$	sin

collected in Table 3; a_2 and b_2 are obtained simply by using A_2 and B_2 instead of A_1 and B_1. As soon as these coefficients have been established, eqs. (45) may be used to obtain similar formulas for the displacements. They can also be found in Table 3.

With the help of this Table we may easily solve any boundary problem which lies within reach of this theory. For an example, let us consider an unsupported boundary $\phi = 0$, where the external forces and moments are prescribed as

$$M_\phi = \overline{M}_{\phi n} \sin \frac{n \pi x}{l}, \qquad Q_\phi = \overline{Q}_{\phi n} \sin \frac{n \pi x}{l},$$

$$N_\phi = \overline{N}_{\phi n} \sin \frac{n \pi x}{l}, \qquad N_{x \phi} = \overline{N}_{x \phi n} \cos \frac{n \pi x}{l}.$$

Formula (50) yields for $\phi = 0$:

$$f = c[a_1 + a_2] \frac{\sin}{\cos} \frac{\lambda x}{a},$$

and from the Table of coefficients we read easily the following equations:

$$\begin{aligned}
A_1 + A_2 &= \overline{M}_{\phi n}/a, \\
A_1^{(1)} + A_2^{(1)} &= \overline{Q}_{\phi n}, \\
A_1^{(2)} + A_2^{(2)} &= -\overline{N}_{\phi n}, \\
A_1^{(3)} + A_2^{(3)} &= -\lambda \overline{N}_{x \phi n} - \overline{Q}_{\phi n}.
\end{aligned} \qquad (51)$$

When we have found the numerical values of $\varkappa_1$, $\varkappa_2$, μ_1, μ_2, from eq. (47) we may use the recurrence (49) to express all the $A^{(i)}$ in terms of A_1, B_1, A_2, B_2 and have then four linear equations for these constants. After they have been solved, the recurrence formula (49) and the Table will yield the coefficients for all stress resultants and displacements and hence the complete solution of the problem.

5.4.2.3 Symmetric Case

When we have two edges arranged symmetrically about $\phi = 0$ and not far enough apart to make their mutual influence negligible, we write our solution in the following form:

$$\begin{aligned}
M_\phi = a[&A_1 \operatorname{Cosh} \varkappa_1 \phi \cos \mu_1 \phi + B_1 \operatorname{Sinh} \varkappa_1 \phi \sin \mu_1 \phi \\
&+ A_2 \operatorname{Cosh} \varkappa_2 \phi \cos \mu_2 \phi + B_2 \operatorname{Sinh} \varkappa_2 \phi \sin \mu_2 \phi] \sin \frac{\lambda x}{a}. \quad (52)
\end{aligned}$$

The derivatives with respect to ϕ have the same form, if they are of even order:

$$\begin{aligned}
\frac{\partial^i M_\phi}{\partial \phi} = a[&A_1^{(i)} \operatorname{Cosh} \varkappa_1 \phi \cos \mu_1 \phi + B_1^{(i)} \operatorname{Sinh} \varkappa_1 \phi \sin \mu_1 \phi \\
&+ A_2^{(i)} \operatorname{Cosh} \varkappa_2 \phi \cos \mu_2 \phi + B_2^{(i)} \operatorname{Sinh} \varkappa_2 \phi \sin \mu_2 \phi] \sin \frac{\lambda x}{a},
\end{aligned}$$

but the form

$$\frac{\partial^{\iota} M_{\phi}}{\partial \phi^{\iota}} = a[A_1^{(\iota)} \, \text{Sinh} \varkappa_1 \phi \cos \mu_1 \phi + B_1^{(\iota)} \, \text{Cosh} \varkappa_1 \phi \sin \mu_1 \phi$$

$$+ A_2^{(\iota)} \, \text{Sinh} \varkappa_2 \phi \cos \mu_2 \phi + B_2^{(\iota)} \, \text{Cosh} \varkappa_2 \phi \sin \mu_2 \phi] \sin \frac{\lambda x}{a}$$

if their order ι is an odd number. For all coefficients the recurrence formulas are

$$A_{\frac{1}{2}}^{(\iota+1)} = \varkappa_1 A_{\frac{1}{2}}^{(\iota)} + \mu_1 B_{\frac{1}{2}}^{(\iota)}, \qquad B_{\frac{1}{2}}^{(\iota+1)} = -\mu_1 A_{\frac{1}{2}}^{(\iota)} + \varkappa_1 B_{\frac{1}{2}}^{(\iota)}. \qquad (53)$$

When these expressions for the derivatives of M_{ϕ} are introduced into eqs. (44) and (45), stress resultants and displacements will assume one of the two following forms:

$$f = c[a_1 \, \text{Cosh} \varkappa_1 \phi \cos \mu_1 \phi + b_1 \, \text{Sinh} \varkappa_1 \phi \sin \mu_1 \phi$$

$$+ a_2 \, \text{Cosh} \varkappa_2 \phi \cos \mu_2 \phi + b_2 \, \text{Sinh} \varkappa_2 \phi \sin \mu_2 \phi] \frac{\cos}{\sin} \frac{\lambda x}{a},$$

$$f = c[a_1 \, \text{Sinh} \varkappa_1 \phi \cos \mu_1 \phi + b_1 \, \text{Cosh} \varkappa_1 \phi \sin \mu_1 \phi$$

$$+ a_2 \, \text{Sinh} \varkappa_2 \phi \cos \mu_2 \phi + b_2 \, \text{Cosh} \varkappa_2 \phi \sin \mu_2 \phi] \frac{\cos}{\sin} \frac{\lambda x}{a},$$

(54a, b)

and the coefficients c, a_1, b_1, a_2, b_2 will be the same as in the preceding case, listed in Table 3. Eq. (54a) applies if the superscripts are all even, eq. (54b) applies if they are all odd.

5.4.3 Simplified Barrel-vault Theory

5.4.3.1 Isolated Boundary

When doing the numerical work which leads to fig 11, one finds that the term $\frac{1}{4}$ under the second radical in eq. (47) is without importance in the range of values $\zeta^2 n$ which is of practical interest. The term $\frac{1}{2}$ under the first radical, however, is responsible for the differences within each pair of curves shown in the figure. This indicates that we may safely neglect 1 compared with $\zeta^4 n^2$, but not compared with $\zeta^2 n$. But if we do even that and drop the term $\frac{1}{2}$ in eq. (47), our formulas become extremely simple, though not all too accurate. We have then

$$\left. \begin{aligned} m &= \pm \sqrt{\pm \sqrt{\pm i \, n^2}} \cdot \zeta, \\ \varkappa_1 &= \mu_2 = \zeta \sqrt{n} \cos \frac{\pi}{8} = \frac{1}{2} \zeta \sqrt{n} \sqrt{2 + \sqrt{2}}, \\ \varkappa_2 &= \mu_1 = \zeta \sqrt{n} \sin \frac{\pi}{8} = \frac{1}{2} \zeta \sqrt{n} \sqrt{2 - \sqrt{2}}, \end{aligned} \right\} \qquad (55)$$

and when we introduce this into the recurrence formulas (49) and in the two-step formulas (49′), they assume the following form:

$$
\left.\begin{aligned}
A_1^{(\iota+1)} &= \frac{1}{2}\,\zeta\sqrt{n}\left(-A_1^{(\iota)}\sqrt{2+\sqrt{2}} + B_1^{(\iota)}\sqrt{2-\sqrt{2}}\right), \\[4pt]
B_1^{(\iota+1)} &= \frac{1}{2}\,\zeta\sqrt{n}\left(-A_1^{(\iota)}\sqrt{2-\sqrt{2}} - B_1^{(\iota)}\sqrt{2+\sqrt{2}}\right), \\[4pt]
A_2^{(\iota+1)} &= \frac{1}{2}\,\zeta\sqrt{n}\left(-A_2^{(\iota)}\sqrt{2-\sqrt{2}} + B_2^{(\iota)}\sqrt{2+\sqrt{2}}\right), \\[4pt]
B_2^{(\iota+1)} &= \frac{1}{2}\,\zeta\sqrt{n}\left(-A_2^{(\iota)}\sqrt{2+\sqrt{2}} - B_2^{(\iota)}\sqrt{2-\sqrt{2}}\right),
\end{aligned}\right\} \tag{56}
$$

and

$$
\left.\begin{aligned}
A_1^{(\iota+2)} &= \frac{1}{\sqrt{2}}\,\zeta^2 n\,(A_1^{(\iota)} - B_1^{(\iota)}), &\qquad B_1^{(\iota+2)} &= \frac{1}{\sqrt{2}}\,\zeta^2 n\,(A_1^{(\iota)} + B_1^{(\iota)}), \\[4pt]
A_2^{(\iota+2)} &= -\frac{1}{\sqrt{2}}\,\zeta^2 n\,(A_2^{(\iota)} + B_2^{(\iota)}), &\qquad B_2^{(\iota+2)} &= \frac{1}{\sqrt{2}}\,\zeta^2 n\,(A_2^{(\iota)} - B_2^{(\iota)}).
\end{aligned}\right\} \tag{56′}
$$

The simplicity of these expressions makes it possible to proceed very far in general terms. We see from eqs. (56′) that the ratio of the coefficients $A^{(\iota+2)}$, $B^{(\iota+2)}$ to $A^{(\iota)}$, $B^{(\iota)}$ is of the order $\zeta^2 n$ and that consequently in any line of Table 3 the coefficients of lower order may now be neglected compared with the highest one. Furthermore, eqs. (56), (56′) may be used to express all the iterated coefficients still left in this table in terms of the original constants A_1, B_1, A_2, B_2. In this way Table 4 has been obtained, which must be used together with eq. (50).

As an application of Table 4 we may write the equations which, on the level of this simplified theory, replace the set (51). We need only to read the boundary values of $M_{\phi n}$, $Q_{\phi n}$, $N_{\phi n}$, $N_{x\phi n}$ from the columns c, a_1, a_2 of the Table. This yields:

$$
\left.\begin{aligned}
A_1 \quad\quad\quad &+ A_2 &&= \frac{\overline{M}_{\phi n}}{a}, \\[6pt]
A_1 - (\sqrt{2}-1)\,B_1 + (\sqrt{2}-1)\,A_2 \quad &- B_2 &&= -\frac{2}{\sqrt{2+\sqrt{2}}}\,\frac{\overline{Q}_{\phi n}}{\zeta\sqrt{n}}, \\[6pt]
A_1 \quad - B_1 \quad - A_2 \quad &- B_2 &&= -\sqrt{2}\,\frac{\overline{N}_{\phi n}}{\zeta^2 n}, \\[6pt]
(\sqrt{2}-1)\,A_1 \quad - B_1 \quad - A_2 + (\sqrt{2}-1)\,B_2 &&&= \frac{2}{\sqrt{2+\sqrt{2}}}\,\frac{\lambda \overline{N}_{\phi n}}{\zeta^3 n\sqrt{n}}.
\end{aligned}\right\} \tag{57}
$$

When only one of the right-hand members of these equations is different from zero, we may obtain handy formulas for the solution which will now be given. In these formulas, of course, we systematically neglect 1 compared with $\zeta^2 n$ and λ^2 compared with 1.

When, at a free edge $\phi = 0$, moments

$$M_\phi = \overline{M}_{\phi n} \sin \frac{n \pi x}{l}$$

are applied, the bending moment in the shell is

$$M_\phi = \frac{\overline{M}_{\phi n}}{\sqrt{2}} \left[(1 + \sqrt{2}) e^{-\varkappa_1 \phi} (\cos \varkappa_2 \phi + \sin \varkappa_2 \phi) \right.$$
$$\left. - e^{-\varkappa_2 \phi} (\cos \varkappa_1 \phi - \sin \varkappa_1 \phi) \right] \sin \frac{n \pi x}{l} . \tag{58}$$

By comparison with eq. (48) we may read from this formula that

$$A_1 = B_1 = \frac{\overline{M}_{\phi n}}{a} \cdot \frac{1 + \sqrt{2}}{\sqrt{2}}, \qquad -A_2 = B_2 = \frac{\overline{M}_{\phi n}}{a} \cdot \frac{1}{\sqrt{2}},$$

and then we may find all the stress resultants and all the displacements from eq. (50) and Table 4. In particular we may find the dis-

Table 4. Barrel Vault, Isolated Boundary

f	c	a_1	b_1	a_2	b_2	x factor
M_ϕ	a	A_1	B_1	A_2	B_2	sin
Q_ϕ	$\zeta \sqrt{n} \, \varrho$	$-A_1 + \psi B_1$	$-\psi A_1 - B_1$	$-\psi A_2 + B_2$	$-A_2 - \psi B_2$	sin
N_ϕ	$-\zeta^2 n/\sqrt{2}$	$A_1 - B_1$	$A_1 + B_1$	$-A_2 - B_2$	$A_2 - B_2$	sin
$N_{x\phi}$	$-\dfrac{\zeta^3 n^{3/2}}{\lambda} \varrho$	$-\psi A_1 + B_1$	$-A_1 - \psi B_1$	$A_2 - \psi B_2$	$\psi A_2 + B_2$	cos
N_x	$\zeta^4 n^2/\lambda^2$	$-B_1$	A_1	B_2	$-A_2$	sin
u	$-\dfrac{\lambda a^3}{K \zeta^4 n^2}$	$-B_1$	A_1	B_2	$-A_2$	cos
v	$\dfrac{a^3}{K \zeta^3 n^{3/2}} \varrho$	$\psi A_1 + B_1$	$-A_1 + \psi B_1$	$-A_2 - \psi B_2$	$\psi A_2 - B_2$	sin
w	$\dfrac{a^3}{\sqrt{2} \, K \zeta^2 n}$	$A_1 + B_1$	$-A_1 + B_1$	$-A_2 + B_2$	$-A_2 - B_2$	sin
$w^{\textbf{·}}$	$-\dfrac{a^3}{K \zeta \sqrt{n}} \varrho$	$A_1 + \psi B_1$	$-\psi A_1 + B_1$	$\psi A_2 + B_2$	$-A_2 + \psi B_2$	sin

Abbreviations: $\varrho = \dfrac{1}{2} \sqrt{2 + \sqrt{2}} = 0.925$,

$\psi = \sqrt{2} - 1 = 0.414$.

placements $\bar{u}, \bar{v}, \bar{w}, \bar{w}^{\cdot}$ at the edge $\phi = 0$:

$$\bar{u}_n = \frac{\overline{M}_{\phi n}}{D(1-\nu^2)} \frac{\zeta^4 n^2}{\lambda^3},$$

$$\bar{v}_n = \frac{\overline{M}_{\phi n}}{D(1-\nu^2)} \frac{\zeta^5 n^{5/2}}{\lambda^4} \sqrt{2} \sqrt{2+\sqrt{2}},$$

$$\bar{w}_n = \frac{\overline{M}_{\phi n}}{D(1-\nu^2)} \frac{\zeta^6 n^3}{\lambda^4} (2+\sqrt{2}),$$

$$\bar{w}_n^{\cdot} = -\frac{\overline{M}_{\phi n}}{D(1-\nu^2)} \frac{\zeta^7 n^{7/2}}{\lambda^4} \sqrt{2} \sqrt{2+\sqrt{2}}.$$

These values are needed when the moment load is used as a redundant force system in a statically indeterminate shell structure.

When the free edge $\phi = 0$ is loaded by normal forces

$$N_\phi = \overline{N}_{\phi n} \sin \frac{n \pi x}{l},$$

we find in the same way the following formula for the bending moment

$$M_\phi = -(\sqrt{2}+1) \frac{\overline{N}_{\phi n} a}{\zeta^2 n} \left[e^{-\varkappa_1 \phi}(\cos\varkappa_2 \phi + \sqrt{2} \sin\varkappa_2 \phi) \right.$$
$$\left. - e^{-\varkappa_2 \phi} \cos\varkappa_1 \phi \right] \sin \frac{n \pi x}{l}, \tag{59}$$

and consequently the following expressions for the edge displacements:

$$\bar{u}_n = -\frac{\overline{N}_{\phi n} a}{D(1-\nu^2) \lambda^3} \zeta^2 n (2+\sqrt{2}),$$

$$\bar{v}_n = -\frac{\overline{N}_{\phi n} a}{D(1-\nu^2) \lambda^4} \zeta^3 n^{3/2} (2+\sqrt{2})^{3/2},$$

$$\bar{w}_n = -\frac{\overline{N}_{\phi n} a}{D(1-\nu^2) \lambda^4} \zeta^4 n^2 (3+2\sqrt{2}),$$

$$\bar{w}_n^{\cdot} = \frac{\overline{N}_{\phi n} a}{D(1-\nu^2) \lambda^4} \zeta^5 n^{5/2} \sqrt{2} \sqrt{2+\sqrt{2}}.$$

When the free edge $\phi = 0$ is loaded by tangential shearing forces

$$N_{x\phi} = \overline{N}_{x\phi n} \cos \frac{n \pi x}{l},$$

the bending moment in the shell is given by

$$M_\phi = -\frac{\overline{N}_{x\phi n} a}{2\zeta^3 n^{3/2}} \sqrt{2+\sqrt{2}} \left[e^{-\varkappa_1 \phi}(\cos\varkappa_2 \phi + (\sqrt{2}+1) \sin\varkappa_2 \phi) \right.$$
$$\left. - e^{-\varkappa_2 \phi}(\cos\varkappa_1 \phi + (\sqrt{2}-1) \sin\varkappa_1 \phi) \right] \sin \frac{n \pi x}{l}, \tag{60}$$

and the displacements at the edge are

$$\bar{u}_n = -\frac{\bar{N}_{x\phi n}\,a}{D(1-\nu^2)\,\lambda^2}\,\zeta\,\sqrt{n}\,\sqrt{2}\,\sqrt{2+\sqrt{2}}\,,$$

$$\bar{v}_n = -\frac{\bar{N}_{x\phi n}\,a}{D(1-\nu^2)\,\lambda^3}\,\zeta^2\,n\big(2+\sqrt{2}\big)\,,$$

$$\bar{w}_n = -\frac{\bar{N}_{x\phi n}\,a}{D(1-\nu^2)\,\lambda^3}\,\zeta^3\,n^{3/2}\,\sqrt{2}\,\sqrt{2+\sqrt{2}}\,,$$

$$\overline{\dot{w}}_n = \frac{\bar{N}_{x\phi n}\,a}{D(1-\nu^2)\,\lambda^3}\,\zeta^4\,n^2.$$

The fourth kind of an edge load which we have to investigate, consists of a transverse shearing force

$$Q_\phi = \bar{Q}_{\phi n}\sin\frac{n\pi x}{l}.$$

The bending moment is then

$$M_\phi = \frac{\bar{Q}_{\phi n}\,a}{2\zeta\,\sqrt{n}}\,\sqrt{2+\sqrt{2}}\,\Big[e^{-\varkappa_1\phi}\big((1+\sqrt{2})\cos\varkappa_2\phi + (1+2\sqrt{2})\sin\varkappa_2\phi\big)$$
$$-e^{-\varkappa_2\phi}\big((1+\sqrt{2})\cos\varkappa_1\phi - \sin\varkappa_1\phi\big)\Big]\sin\frac{n\pi x}{l}\,,\tag{61}$$

and the displacements at the edge are

$$\bar{u}_n = \frac{\bar{Q}_{\phi n}\,a}{D(1-\nu^2)\,\lambda^3}\,\zeta^3\,n^{3/2}\,\sqrt{2}\,\sqrt{2+\sqrt{2}}\,,$$

$$\bar{v}_n = \frac{\bar{Q}_{\phi n}\,a}{D(1-\nu^2)\,\lambda^4}\,\zeta^4\,n^2\big(3+2\sqrt{2}\big)\,,$$

$$\bar{w}_n = \frac{\bar{Q}_{\phi n}\,a}{D(1-\nu^2)\,\lambda^4}\,\zeta^5\,n^{5/2}\,\big(2+\sqrt{2}\big)^{3/2},$$

$$\overline{\dot{w}}_n = -\frac{\bar{Q}_{\phi n}\,a}{D(1-\nu^2)\,\lambda^4}\,\zeta^6\,n^3\big(2+\sqrt{2}\big).$$

In all the preceding cases the displacements at the edge were supposed not to be restricted. We may, of course, also consider cases where one or more of the displacement components are given, either because the edge is supported or as a condition of symmetry or antimetry. We shall here mention three such cases which will be useful in practical applications.

Suppose a radial line load

$$P = P_n\sin\frac{n\pi x}{l}$$

to be applied along the generator $\phi = 0$ of the shell (fig. 12). When the edges of the shell are far enough away to have no influence or when their influence is to be evaluated and added later, we may assume that stresses and deformations are distributed symmetrically to both

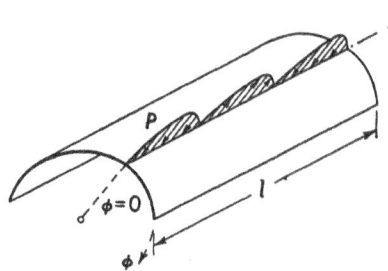

sides of the loaded generator. We have then at $\phi = 0$ the boundary conditions

$$N_{x\phi} \overset{!}{=} 0, \quad v = 0, \quad w^{\cdot} = 0,$$

and for the right half of the shell ($\phi \geqq 0$) we have the fourth condition that

$$Q_{\phi n} = -\frac{1}{2} P_n.$$

Fig. 12. Radial line load applied to the generator $\phi = 0$

From these four boundary conditions we may derive four linear equations similar to eqs. (57), and they yield the following expressions for the four constants:

$$A_1 = -B_2 = P_n \frac{\sqrt{2 + \sqrt{2}}}{8\zeta \sqrt{n}},$$

$$A_2 = -B_1 = P_n \frac{\sqrt{2 - \sqrt{2}}}{8\zeta \sqrt{n}}. \tag{62}$$

In a quite similar way a tangential line load (fig. 13)

$$P = P_n \sin \frac{n \pi x}{l}$$

may be treated. This is an antimetric case, and we have therefore at $\phi = 0$:

$$M_\phi = 0, \quad u = 0, \quad w = 0, \quad N_{\phi n} = -\frac{1}{2} P_n.$$

The result is this:

$$A_1 = -A_2 = -B_1 = -B_2 = \frac{P_n}{4\sqrt{2}\,\zeta^3 n}. \tag{63}$$

As the third case in this group we consider a shear load

$$T = T_n \cos \frac{n \pi x}{l}$$

applied along the generator $\phi = 0$ as shown in fig. 14. The resulting stress system is symmetrical to both sides of the loaded line, and at $\phi = 0$ the boundary conditions

$$Q_\phi = 0, \quad v = 0, \quad w^{\cdot} = 0, \quad N_{x\phi n} = \frac{1}{2} T_n$$

hold for the side $\phi \geqq 0$. They lead to the following expressions for the constants in eq. (48):

$$A_1 = B_2 = T_n \frac{\lambda \sqrt{2 - \sqrt{2}}}{8 \zeta^3 n^{3/2}},$$

$$A_2 = B_1 = - T_n \frac{\lambda \sqrt{2 + \sqrt{2}}}{8 \zeta^3 n^{3/2}}.$$

(64)

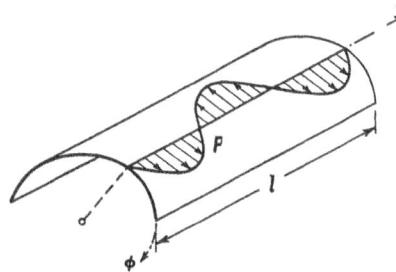 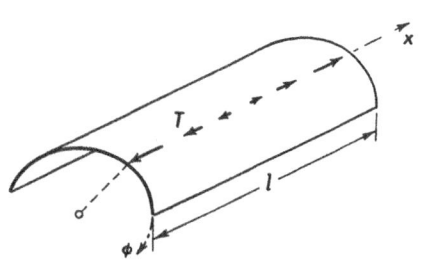

Fig. 13. Circumferential line load applied to the generator $\phi = 0$

Fig. 14. Line load applied to the generator $\phi = 0$ in the direction of this generator

5.4.3.2 Symmetric Case

When we want to apply the simplified theory to the symmetric case, we have to introduce eqs. (55) into eqs. (53). This yields the following set of recurrence formulas:

$$A_1^{(i+1)} = \frac{1}{2} \zeta \sqrt{n} \left[A_1^{(i)} \sqrt{2 + \sqrt{2}} + B_1^{(i)} \sqrt{2 - \sqrt{2}} \right],$$

$$B_1^{(i+1)} = \frac{1}{2} \zeta \sqrt{n} \left[- A_1^{(i)} \sqrt{2 - \sqrt{2}} + B_1^{(i)} \sqrt{2 + \sqrt{2}} \right],$$

$$A_2^{(i+1)} = \frac{1}{2} \zeta \sqrt{n} \left[A_2^{(i)} \sqrt{2 - \sqrt{2}} + B_2^{(i)} \sqrt{2 + \sqrt{2}} \right],$$

$$B_2^{(i+1)} = \frac{1}{2} \zeta \sqrt{n} \left[- A_2^{(i)} \sqrt{2 + \sqrt{2}} + B_2^{(i)} \sqrt{2 - \sqrt{2}} \right].$$

(65)

When they are introduced into Table 3 and when again 1 is neglected compared with $\zeta^2 n$, Table 5 results. This table must be used in connection with eqs. (54), and there is again a column "Symmetry" which indicates whether a quantity belongs to the symmetric group and hence to eq. (54a) or to the antimetric group and hence to eq. (54b).

In the symmetrical shell the boundary conditions must be written for some finite value $\phi = \phi_0$ and are less simple than those for an isolated boundary at $\phi = 0$. It is, therefore, not advisable to ask for ready-to-use formulas similar to eqs. (58) to (61).

Table 5. Barrel Vault, Symmetric Case

t	c	a_1	b_1	a_2	b_2	sym-metry	x-factor
M_ϕ	a	A_1	B_1	A_2	B_2	sym.	sin
Q_ϕ	$\zeta \sqrt{n}\,\varrho$	$A_1 + \psi B_1$	$-\psi A_1 + B_1$	$\psi A_2 + B_2$	$-A_2 + \psi B_2$	anti.	sin
N_ϕ	$-\zeta^2 n/\sqrt{2}$	$A_1 + B_1$	$-A_1 + B_1$	$-A_2 + B_2$	$-A_2 - B_2$	sym.	sin
$N_{x\phi}$	$-\dfrac{\zeta^3 n^{3/2}}{\lambda}\varrho$	$\psi A_1 + B_1$	$-A_1 + \psi B_1$	$-A_2 - \psi B_2$	$\psi A_2 - B_2$	anti.	cos
N_x	$\zeta^4 n^2/\lambda^2$	B_1	$-A_1$	$-B_2$	A_2	sym.	sin
u	$-\dfrac{\lambda a^3}{K \zeta^4 n^2}$	B_1	$-A_1$	$-B_2$	A_2	sym.	cos
v	$\dfrac{a^3}{K \zeta^3 n^{3/2}}\varrho$	$-\psi A_1 + B_1$	$-A_1 - \psi B_1$	$A_2 - \psi B_2$	$\psi A_2 + B_2$	anti.	sin
w	$-\dfrac{a^3}{\sqrt{2}\,K \zeta n^2}$	$-A_1 + B_1$	$-A_1 - B_1$	$A_2 + B_2$	$-A_2 + B_2$	sym.	sin
w^*	$\dfrac{a^3}{K \zeta \sqrt{n}}\varrho$	$A_1 - \psi B_1$	$\psi A_1 + B_1$	$\psi A_2 - B_2$	$A_2 + \psi B_2$	anti.	sin

For quantities marked "sym."
use eq. (54a),
for quantities marked "anti."
use eq. (54b).

Abbreviations: $\varrho = \dfrac{1}{2}\sqrt{2 + \sqrt{2}} = 0.925$,

$$\psi = \sqrt{2} - 1 = 0.414.$$

5.4.4 Examples

5.4.4.1 Half-filled Pipe

After these preparations, we may treat some examples which will illustrate the practical application of the formulas and the results which may be obtained with their help.

Fig. 15a shows a pipe which is only half filled with water. In order to make the problem as simple as possible, we assume that both ends of the pipe are supported by rings and that there are expansion joints so that we have $N_x = 0$ as a boundary condition for $x = 0$ and for $x = l$.

The upper half of the shell does not carry any load and therefore all membrane forces in it are zero. In the lower half we have

$$p_x = p_\phi = 0, \qquad p_r = -\gamma a \cos\phi,$$

using the notation of Chapter 3. The corresponding membrane forces are given by eqs. (III-8) with $p_0 = 0$. However, in these formulas x was measured from the midspan section of the pipe, while we are now counting it from one end. We have, therefore, in the present notation

$$N_\phi = -\gamma a^2 \cos\phi, \qquad N_{x\phi} = \gamma a\left(\frac{l}{2} - x\right)\sin\phi,$$

$$N_x = -\frac{1}{2}\gamma x(l - x)\cos\phi.$$

At the limits of validity of this formula, $\phi = 90°$ and $\phi = 270°$, both normal forces are zero, but there is a shear which does not find its

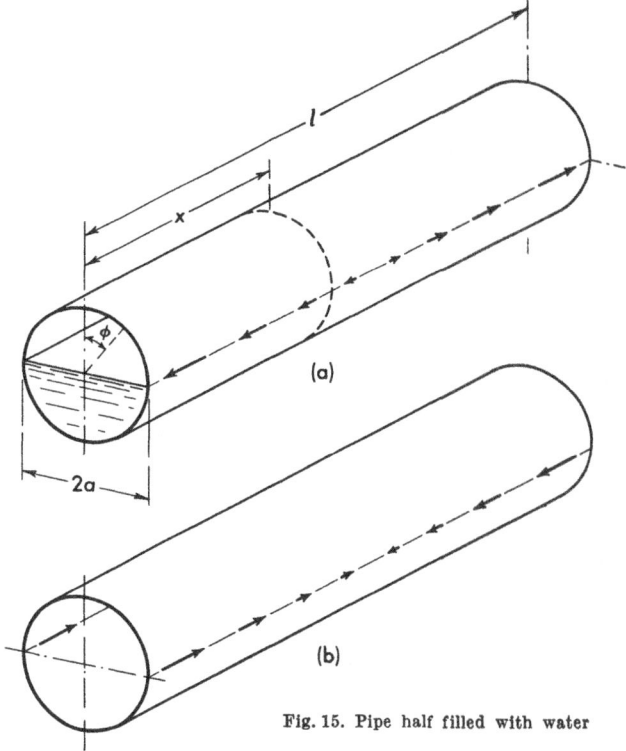

(a)

(b)

Fig. 15. Pipe half filled with water

counterpart among the stress resultants in the upper half of the shell. It must therefore be applied as an external load as shown in fig. 15a. The real pipe has, of course, no such load, and we must compensate it by adding the load shown in fig. 15b. It consists of tangential forces

$$T = -\gamma a\left(\frac{l}{2} - x\right)$$

similar to those shown in fig. 14 and applied along two generators. This load produces bending stresses, and we may use eq. (64) to find them and all the stress resultants connected with them, if only we expand the load T in a FOURIER series:

$$T = -\frac{4\gamma a l}{\pi^2}\left(\cos\frac{\pi x}{l} + \frac{1}{9}\cos\frac{3\pi x}{l} + \frac{1}{25}\cos\frac{5\pi x}{l} + \cdots\right).$$

The general term of this series,

$$T_n = -\frac{4\gamma a l}{\pi^2 n^2}, \qquad n = \text{odd},$$

may be introduced into eqs. (64), and then Table 4 and eq. (50) may be applied.

There are four equal stress systems in the shell which are all described by these equations, but with a different meaning of the variable ϕ appearing there. The first of these stress systems emanates from the load at our edge $\phi = 90°$ and extends in damped oscillations around the lower half of the shell. We find its stress resultants when we replace ϕ in eqs. (48) and (50) by $\phi - 90°$, e. g.:

$$M_{\phi n} = a\left[e^{-\varkappa_1(\phi - 90°)}\left(A_1 \cos\varkappa_2(\phi - 90°) + B_1 \sin\varkappa_2(\phi - 90°)\right)\right.$$

$$\left. + e^{-\varkappa_2(\phi - 90°)}\left(A_2 \cos\varkappa_1(\phi - 90°) + B_2 \sin\varkappa_1(\phi - 90°)\right)\right]\sin\frac{\lambda x}{a}.$$

This formula is valid for $\phi \geqq 90°$, without an upper limit for ϕ. In such shells to which the simplified barrel vault theory is applicable, it cannot be expected that at a short distance from the loaded generator the stresses will already have dropped to insignificant magnitude. They may die out somewhere on the lower half of the circumference, but they may as well spread much farther, and there is no reason why it should not happen that they are perceptible for more than 360° and even several times around the whole circumference.

There is a second stress system which emanates from the same load, but which spreads first over the upper half of the shell:

$$M_{\phi n} = a\left[e^{-\varkappa_1(90° - \phi)}\left(A_1 \cos\varkappa_2(90° - \phi) + \cdots\right)\right]\sin\frac{\lambda x}{a}$$

$$\text{for}\quad \phi \leqq 90°,$$

and in a similar way the load at $\phi = -90°$ (or $+270°$) yields two more stress systems:

$$M_{\phi n} = a\left[e^{-\varkappa_1(\phi + 90°)}\left(A_1 \cos\varkappa_2(\phi + 90°) + \cdots\right)\right]\sin\frac{\lambda x}{a}$$

$$\text{for}\quad \phi \geqq -90°,$$

$$M_{\phi n} = a\left[\left(e^{-\varkappa_1(270° - \phi)}\left(A_1 \cos\varkappa_2(270° - \phi) + \cdots\right)\right]\sin\frac{\lambda x}{a}$$

$$\text{for}\quad \phi \leqq 270°.$$

All four stress systems must be superposed to obtain the complete result.

Some figures have been computed from these formulas for the following data:

$$l = 40.0 \text{ ft}, \quad a = 4.0 \text{ ft}, \quad t = 0.5 \text{ in.}, \quad \nu = 0.3, \quad \gamma = 62.4 \text{ lb/ft}^3.$$

For the first harmonic, $n = 1$, one has then

$$\varkappa_1 = 2.185, \quad \varkappa_2 = 0.904,$$

and each of the four parts of $M_{\phi 1}$ extends over a little more than half the circumference of the cylinder until it becomes negligibly small.

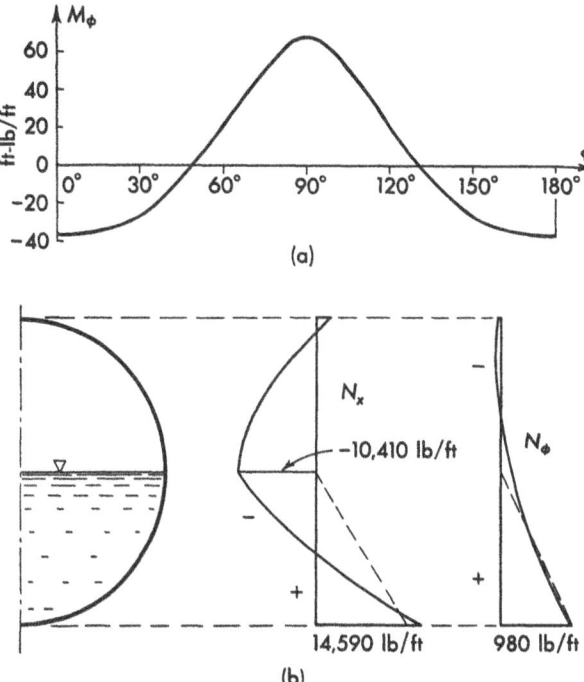

Fig. 16. Stress resultants in a half-filled pipe, (a) hoop moment M_ϕ at midspan, (b) normal forces N_x and N_ϕ at midspan. The broken lines give the membrane forces corresponding to Fig. 15 a

The next harmonic, $n = 3$, dies out much faster, and it contributes only 5% to the total, so that it does not seem worthwhile to compute higher harmonics.

Some results have been plotted in fig. 16 for the cross section at $x = l/2$. The bending moments are by no means localized, and their magnitude is such that the maximum circumferential fiber stress $\sigma_x = 1632$ lb/in.^{2} comes close to the maximum axial stress $\sigma_x = 2080$ lb/in.^{2} in the completely filled pipe. The distribution of the longitudinal force N_x is quite different from that in a full pipe. A compression zone develops in the middle third of the cross section, and the top part is almost unstressed. In the lower half of the shell the hoop force N_ϕ shows almost the same distribution as the water pressure, but it extends upward beyond the water level.

These diagrams give a good qualitative idea of the magnitude and distribution of the stresses in the shell, but they are far from being quantitatively reliable. There are two causes of errors involved in the solution given. One lies in the basic assumptions of what we call the simplified barrel vault theory, and the other one in the fact that we removed only the worst discrepancy of the membrane solution, the unbalanced shear, but did not bother about the discontinuity of the membrane deformations of the upper and lower halves of the pipe. We shall now eliminate these two deficiencies one after the other, restricting our attention to the first harmonic $n = 1$.

When we cut the shell in two parts along the water level, the upper half, having no membrane forces, will not be deformed. The deformation of the lower half is described by eqs. (III–23) if we put $p_0 = 0$ there. On the boundaries $\phi = \pm 90°$ they yield $u = w = 0$, and there are discrepancies only in v and $w\dot{}$. We are interested in their first harmonic. When we remember that, in eqs. (III–23), x is measured from a midspan point, we find for $\phi = 90°$

$$E\,t\,v_1 = \frac{4\gamma\,l^3}{\pi^3}\left[\frac{l}{\pi^2 a} + (2 + \nu)\frac{a}{l}\right],$$

$$E\,t\,w_1\dot{} = \frac{4\gamma\,l^3}{\pi^3}\left[\frac{l}{\pi^2 a} + \frac{2a}{l} + \frac{\pi^2 a^3}{l^3}\right].$$

These displacements represent a gap between the upper and lower halves of the shell. We may remove it by applying bending moments $\overline{M}_{\phi 1}\sin\pi x/l$ and hoop forces $\overline{N}_{\phi 1}\sin\pi x/l$ along the edges $\phi = \pm 90°$ of the half cylinders. Their magnitudes must be chosen so, that each shell comes half the way, and this condition yields two equations for $\overline{M}_{\phi 1}$ and $\overline{N}_{\phi 1}$.

To obtain these equations, we may either start from the formulas for the symmetric case (p. 252), or we may apply twice the formulas for an isolated boundary (p. 249). In the latter case, which we shall adopt here, only one edge of each half cylinder is loaded at a time, and the resulting stress resultants not only are considered in the upper or lower shell but, disregarding the opposite gap, are followed all around the shell and even farther, if their magnitude requires it.

The edge displacements produced by the redundant forces and moments $\overline{N}_{\phi 1}$ and $\overline{M}_{\phi 1}$ are given by eqs. (58) and (59). Their sum must be equal to one half of the membrane displacements just given, and in those formulas we may, within the limits of correctness of the simplified theory, neglect all but the first term. Thus we arrive at the following equations:

$$(2 + \sqrt{2})\, a\, \overline{N}_{\phi 1} - \zeta^2\, \sqrt{2}\, \overline{M}_{\phi 1} = \frac{\gamma\, a^3}{\pi\, \zeta^3}\, \sqrt{2}\, \sqrt{2 - \sqrt{2}}\ ,$$

$$a\, \overline{N}_{\phi 1} - \zeta^2\, \overline{M}_{\phi 1} = -\frac{\gamma\, a^3}{\pi\, \zeta^5}\, \sqrt{2 - \sqrt{2}}\ .$$

When the edge $\phi = +90°$ of both half shells is loaded with the redundant forces and moments following from these equations and when the corresponding stress resultants at the opposite edge ($\phi = 270°$ or $\phi = -90°$) are transferred from the upper to the lower half and vice versa and so on as often as need be, then a stress system will be obtained which closes the gap at $\phi = 90°$ without changing the situation at $\phi = -90°$. We have then simply to apply an identical load at the edges $\phi = -90°$ which will close the gap there. The sum of the two stress systems thus found must be added as a correction to the stresses which we determined previously and which are represented in fig. 16.

In fig. 17 the first harmonic of M_ϕ and of N_x is compared for the two computations. The solid lines give the results of the simplest approach; the broken lines take care of the corrections just explained. For the bending moment, the difference is considerable, although the ‘order of magnitude and the general distribution are well represented by the simplest computation. The longitudinal force N_x is not much changed by the correction, except that the break in the middle of the curve has been rounded.

We shall now compare these results with those of the next better theory, which uses eq. (47) as it stands for the determination of $\varkappa_1$, $\varkappa_2$, μ_1, μ_2. The procedure is essentially the same as before, except that the ready-to-use formula (64) is not available and one must at once set up a system of four linear equations for the constants A_1, B_1, A_2, B_2, expressing the conditions that at $\phi = 90°$ we have $Q_\phi = 0$ and v, w, $N_{x\phi}$ equal to certain values following from the discrepancies of the membrane solution. This has been done for the first harmonic, $n = 1$, and the results have been entered as dotted lines in fig. 17. The force N_x is again almost the same, and the bending moment is very different, but this time the correction goes toward the other side, yielding lower peak values. This outcome suggests the following general policy for treating similar problems: One may use the simplified barrel vault theory to remove the worst discrepancies of the membrane theory (i. e. unbalanced forces). If the stresses found in this way are of such importance that it is necessary to have correct values, then the complete theory described in Section 5.4.2 must be applied, and it is then not worthwhile to do anything less than satisfy all reasonable conditions at once.

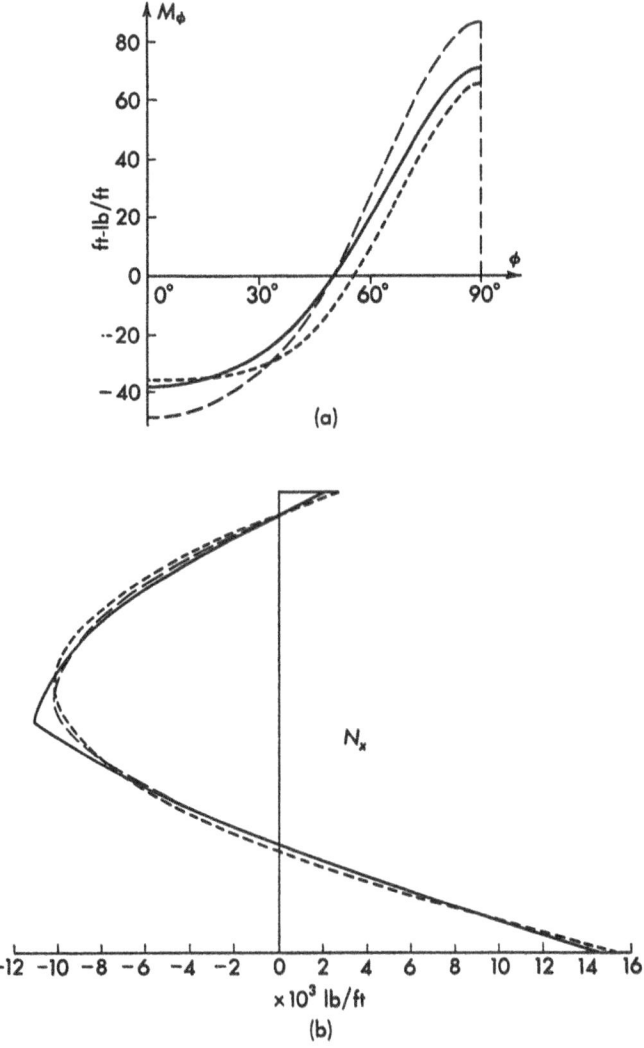

Fig. 17. Stress resultants in a half-filled pipe, (a) hoop moment at midspan, plotted over one quarter of the circumference, (b) N_x plotted over the vertical diameter

5.4.4.2 Barrel-vault Roof

Fig. 18 shows a barrel-vault roof such as is frequently used to cover a large rectangular area without intermediate supports. The roof is a continuous structure consisting of a number of cylindrical shells and edge beams. There is always one beam more than there are shells. For greater clarity, only these shells and beams are shown in the figure. There are, of course, diaphragms at the ends of the shells and columns at the ends of the beams.

The stress analysis of this structure is rather tedious and it is avoided by considering two limiting cases: (a) a single barrel vault with two edge beams, (b) a structure consisting of an infinite number of shells and edge beams. Case (a) comes close to the situation in the outer half of the outer shells, while case (b) approaches the stresses in the inner part of the shell roof. Since usually the stresses in the two

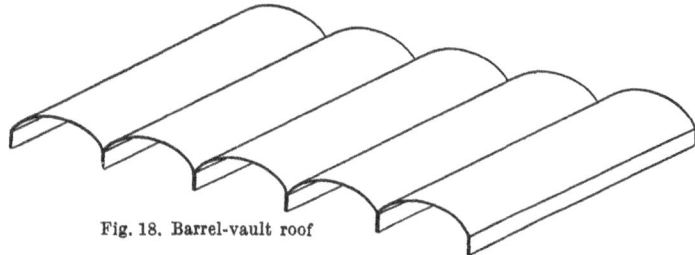

Fig. 18. Barrel-vault roof

cases are not all too different, they yield a sufficient basis from which to judge the adequacy of the construction and the dimensions of the necessary reinforcement rods.

We use case (a) here as an example and consider the shell structure whose cross section is shown in figs. 19a and 20. The span in the x direction is 75 ft. The load on the shell is assumed to be $p = 55$ lb/ft².

The analysis starts from the membrane forces and displacements. There are two kinds of formulas available; the explicit formulas (III–16) and (III–24), and the FOURIER series representations (III–15) and (III–28). We choose the latter, because they fit the formulas of the bending theory which we intend to use. We have then only to deal with the amplitudes $N_{\phi n}$ etc. which still depend on ϕ but not on x. Each number which we write then stands for a sine or cosine distribution in the spanwise direction. The computations must be made for each n separately; it will be enough to explain them for $n = 1$.

According to eqs. (III–15) the first harmonic of the hoop force is $N_{\phi 1} = -2310$ [lb/ft] $\cos\phi$. At the edges $\phi = \pm\phi_0 = \pm 35.15°$ it assumes the value of -1889 lb/ft. The vertical component of this edge force may be transmitted to the edge beam and is an addition to the first harmonic of its weight of 155 lb/ft. The horizontal component of 1542 lb/ft, however, must be applied as an external thrust to the shell (fig. 19b), if a membrane stress system is to be at all possible. Since this thrust does not act on the real shell, we must at once superpose an outward pull of the same magnitude (fig. 19c). This is the first of several edge loads which produce bending stresses in the shell. Other such edge loads are acting between the shell and the beam. They are a vertical force (having sine distribution) and a shear (with cosine distribution like the first harmonic of the membrane shear $N_{x\phi1}$ from eq. (III-15), but additional to it). For the shell, all these forces may be expressed as boundary values of the hoop force, the transverse force, and the shearing force. We shall denote them by $\overline{N}_{\phi1}, \overline{Q}_{\phi1}, \overline{N}_{x\phi1}$. They must be so chosen that for the complete stress system there is

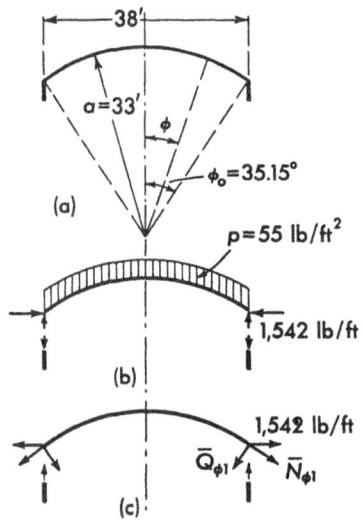

(a)

(b)

(c)

Fig. 19. Cross section of an isolated barrel vault, (a) dimensions, (b) load and corresponding edge forces in a membrane-stress system, (c) additional edge forces needed for continuity of deformation

no discrepancy between the vertical deflections of shell and beam nor between the strains ϵ_x of both along the edge where they are connected and no need for the external thrust shown in fig. 19b.

Additionally, there may be a clamping moment $\overline{M}_{\phi1}$ at the edge of the shell, whose counterpart produces torsion in the beam. It is often assumed to be zero on the ground that the torsional rigidity of the edge beam is rather small. This reason is not always convincing, and the assumption should be used with caution. If one decides to take the moment $\overline{M}_{\phi1}$ as another redundant quantity in the computation, it becomes a logical necessity to consider also the lateral deflection of the edge beam. This makes the stress analysis a good deal lengthier, and only for this reason shall we here drop the clamping moment from our considerations.

We begin the computation of the redundant quantities by applying to the edge $\phi = 35.15°$ of the shell, one after the other, the unit loads $\overline{N}_{\phi1} = 1$ lb/ft, $\overline{Q}_{\phi1} = 1$ lb/ft, $\overline{N}_{x\phi1} = 1$ lb/ft. In each case $\overline{M}_{\phi1} = 0$. The corresponding stress resultants and displacements may be computed with any of the three theories explained on the preceding twenty pages. We choose the simplest one, the Simplified Barrel Vault Theory. Since the shell is symmetric with respect to the vertical plane $\phi = 0$, all unit loads are also applied symmetrically at the edge $\phi = -35.15°$, and we have to use Table 5 with eqs. (54). This requires the computation of $\lambda = \pi a/l = 1.382$ and of $\zeta = 5.71$. Eqs. (55) yield then

$$\varkappa_1 = \mu_2 = 5.28, \qquad \varkappa_2 = \mu_1 = 2.185,$$

and we are now ready to find all the trigonometric and hyperbolic functions occurring in eqs. (54).

With the help of Table 5 we may now write for each unit load case a set of four equations, similar to the set (57), but slightly more complicated, and from it we find the values of the constants A_1, B_1, A_2, B_2 which belong to this case. Entering Table 5 with these values, we find the deflections v_1, w_1 of the edge and the force N_{x1}. From the latter we calculate $\sigma_{x1} = N_{x1}/t$ or, if we prefer, $\epsilon_{x1} = \sigma_{x1}/E$. When we assume $\nu = 0$, continuity of ϵ_x is identical with continuity of σ_x, and we may save the division by E.

In this way the following numerical results have been obtained:

for $\overline{N}_{\phi 1} = 1$ lb/ft:
$E\,t\,v_1 = 1.036 \times 10^4$ lb, $E\,t\,w_1 = -3.433 \times 10^5$ lb, $\sigma_{x1} = 252.5$ lb/ft^2;

for $\overline{Q}_{\phi 1} = 1$ lb/ft:
$E\,t\,v_1 = 6.02 \times 10^4$ lb, $E\,t\,w_1 = -2.295 \times 10^6$ lb, $\sigma_{x1} = 1155$ lb/ft^2;

for $\overline{N}_{x\phi 1} = 1$ lb/ft:
$E\,t\,v_1 = -1.191 \times 10^3$ lb, $E\,t\,w_1 = 3.295 \times 10^4$ lb, $\sigma_{x1} = -47.4$ lb/ft^2.

Of the deflections only the combination $v_1 \sin\phi_0 - w_1 \cos\phi_0$, i. e. the vertical deflection, is of interest.

We must now apply the same unit loads in the opposite direction to the upper edge of the beams, keeping in mind that the figures always represent the amplitudes of loads whose spanwise distribution follows a sine or cosine law. There is no need for reproducing here the details of the elementary calculations which yield the first harmonic of the vertical deflection δ_1 of the beam and of the stress σ_1 in its top fiber.

We may now write the final set of three equations which state the following facts:

1) There is no horizontal thrust applied from outside to the springing line of the vault, i. e. the sum of $\overline{N}_{\phi 1} \cos\phi_0 - \overline{Q}_{\phi 1} \sin\phi_0$ and of the membrane contribution of 1542 lb/ft must vanish.

2) The vertical deflection $v_1 \sin\phi_0 - w_1 \cos\phi_0$ of the edge of the shell equals the deflection δ_1 of the beam. For the shell it consists of the membrane deflection from eqs. (III–28), i. e.

$$E\,t\,(v_1 \sin\phi_0 - w_1 \cos\phi_0) = 1.457 \times 10^5 \text{ lb,}$$

and of the contributions of the redundant forces:

$$E\,t\,(v_1 \sin\phi_0 - w_1 \cos\phi_0) = 2.867 \times 10^5 \text{ [ft] } \overline{N}_{\phi 1}$$
$$+ 1.912 \times 10^6 \text{ [ft] } \overline{Q}_{\phi 1} - 2.762 \times 10^4 \text{ [ft] } \overline{N}_{x\phi 1}.$$

For the beam the deflection is similarly composed of four terms.

3) The stress σ_{x1} (i. e. $E\,\epsilon_{x1}$) along the edge of the shell must equal the stress σ_x in the top fiber of the beam. This equation has the same kinds of terms as the preceding one.

Two of these equations are exactly the type which is used in the theory of statically indeterminate systems. The first one seems to be different. It is, however, the degenerated form of the equation expressing continuity of the horizontal deflection, degenerated by the assumption that the bending rigidity of the beam is zero.

When these equations are formulated and solved, the following results are found:

$$\bar{N}_{\phi 1} = 1736 \text{ lb/ft}, \quad \bar{Q}_{\phi 1} = -214.5 \text{ lb/ft}, \quad \bar{N}_{x\phi 1} = 3013 \text{ lb/ft}.$$

The final values of the constants of integration may now be calculated by superposition, and then Table 5 may be used a last time to write numerical expressions for all the quantities that may be of interest. Some of them have been plotted in fig. 20.

From this figure the following conclusions of general validity may be drawn: The σ_x diagram shows that the shell and the edge beams

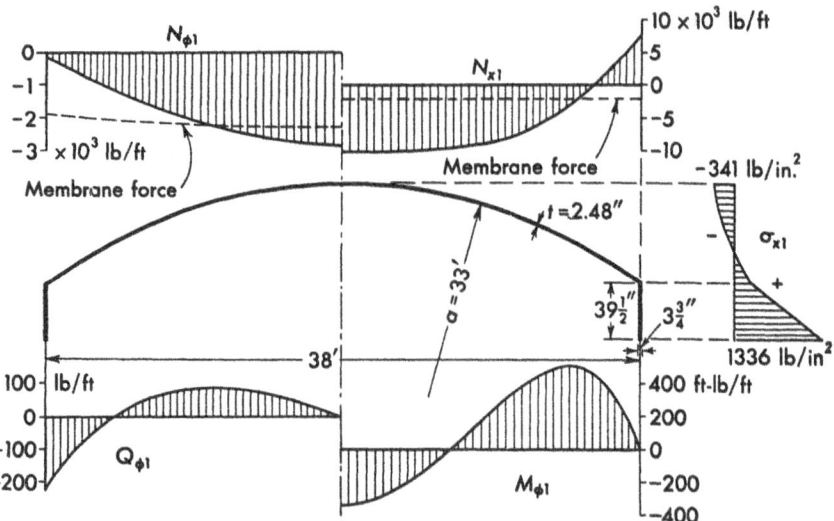

Fig. 20. Stress resultants in the barrel vault of Fig. 19

cooperate in such a way that the shell is essentially the compression zone and the edge beams are the tension zone of a huge composite beam. The dimensions of the shell and the edge beams determine whether the zero of stress lies in the one or in the other. The bending moments are not restricted to a small zone near the edge but are spread over the entire area. Since we assumed with more or less justification that

$M_{\phi 1} = 0$ at the edge, the moment diagram is shaped accordingly. A more realistic assumption would lead to a finite clamping moment. The hoop force $N_{\phi 1}$ drops to almost zero at the edge. The small value there is needed to compensate the horizontal component of Q_1. The longitudinal force $N_{x 1}$ is on the whole considerably larger than its membrane value.

The computation just described is simplified if the constants A_1, B_1, A_2, B_2 are used as redundant quantities. In this case the condition $M_{\phi 1} = 0$ must be added as a fourth equation to the final set of three, but the investigation of the unit loads becomes superfluous. It depends much on personal preferences whether one chooses this way or the other. The use of the constants A, B as key unknowns is rather abstract and more subject to the danger of undiscovered errors. On the other hand, the gain in numerical simplicity is not as large as might appear at the first glance, mainly because all the unit load cases use the same equations, only with different right-hand sides, so that the elimination may be done in common.

Since the Simplified Barrel Vault Theory is based upon rather far-reaching approximations, its results are not very reliable. If more exact figures are needed, the theory of Section 5.4.2 must be used. The computation runs along the same lines but is lengthier because it is based on Table 3 and requires the use of the recurrence formulas (53).

The first harmonic, $n = 1$, of course, does not represent the complete solution of the problem. Since the membrane forces, eqs. (III–15), contain only odd harmonics, $n = 1, 3, 5, \ldots$, there will be no bending stresses of even order. It may be left to the reader to work out the figures for $n = 3$ and $n = 5$ and to see how much they add to the stresses in the shell. He will find that the general conclusions drawn from fig. 20 remain unaltered.

5.5 Cylindrical Tanks and Related Problems

5.5.1 Differential Equation

A cylindrical water tank with a vertical axis and filled to its rim (fig. 21) is subjected to a load

$$p_r = \gamma (h - x). \tag{66}$$

Its membrane forces are only hoop forces

$$N_\phi = \gamma a (h - x), \tag{67}$$

increasing linearly with the depth below the water level. They lead to a hoop strain ϵ_ϕ and hence to a radial displacement, independent of the coordinate ϕ,

$$w = \frac{\gamma a^2}{D(1 - \nu^2)} (h - x),$$ (68)

which may be derived from the formulas (III–20) or found immediately from elementary considerations.

If the wall thickness t is a constant, w increases linearly with the depth below the water level. But at the lower edge the tank wall is

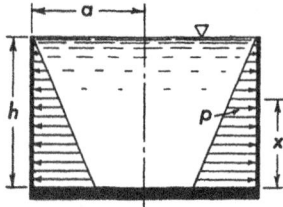

Fig. 21. Cylindrical water tank

connected to a flat or curved bottom, and there the displacement w cannot develop freely. The restraint requires a transverse force Q_x and a bending moment M_x transmitted from the bottom to the cylindrical shell and leads to the particular problem we have to consider here.

For a shell of constant thickness t the solution is a special case of the one described in Section 5.3. Because of the axial symmetry of the stress system, it is considerably simpler than the general solution, and therefore we shall not derive it from that by specialization but shall give the reader an independent approach. In doing so, we take advantage of the opportunity to extend the theory to those shells where t is a function of x, a case often met in tank design.

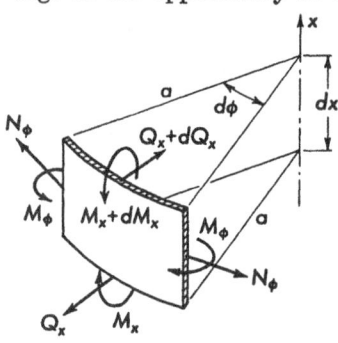

Fig. 22. Shell element

Fig. 22 shows the shell element. Of all the forces shown in figs. 1a, b, only N_ϕ, Q_x, M_x, and M_ϕ appear. Most of the rest are zero because of symmetry, and N_x has been omitted since the vertical stresses due to the weight of the shell may be found easily without using shell theory.

There are only two conditions of equilibrium which are not trivial: those for the forces in a radial direction and those for the moments with respect to a horizontal tangent to the cylinder. They are

$$Q'_x + N_\phi = p_r a,$$
$$M'_x - a Q_x = 0.$$ (69)

The elastic law may be found from eqs. (9) by dropping v and all derivatives with respect to ϕ. We may also safely neglect the small terms with the factor K in the normal force and the term u' in M_x. The resulting relations are so simple that they are immediately evident:

$$N_\phi = \frac{D}{a}\,(w + \nu\,u'),$$

$$N_x = \frac{D}{a}\,(u' + \nu\,w),$$

$$M_x = \frac{K}{a^2}\,w''. \tag{70a}$$

Since, in our particular problem, $N_x = 0$, we may use the second equation to eliminate u from the first one. We get:

$$N_\phi = \frac{D(1 - \nu^2)}{a}\,w. \tag{70b}$$

We now have 4 equations, (69) and (70), for the 4 unknowns N_ϕ, Q_x, M_x, w. The elimination follows the same lines as in the general case. We first eliminate Q_x from the two conditions of equilibrium. This yields:

$$M_x'' + a\,N_\phi = p_r\,a^2.$$

Then we use the elastic law (70) to express the remaining stress resultants in terms of the displacement w, and here we shall not forget that t and hence D and K may depend on x. Thus we arrive at the differential equation

$$(K w'')'' + D a^2 (1 - \nu^2)\,w = p_r\,a^4. \tag{71}$$

It is of the fourth order, allowing for two boundary conditions at each edge of the shell. This is half of the number we had for the higher harmonics. The reduction is due to the fact that two conditions, refering to u or N_x, have become trivial, and that two more, concerning v or $N_{x\phi}$, do not fit into the particular kind of axial symmetry to which we have confined our theory.

5.5.2 Solution for Constant Thickness

5.5.2.1 Homogeneous Solution

When the wall thickness t does not vary with x, eq. (71) reads simply

$$K w^{\mathrm{IV}} + D a^2 (1 - \nu^2)\,w = p_r\,a^4. \tag{72}$$

We shall first consider the homogeneous equation, putting $p_r \equiv 0$. It has constant coefficients, and its solution must, therefore, consist of four terms of the type

$$w = C\,e^{\lambda x/a}.$$

Introducing this into eq. (72) and putting

$$\varkappa^4 = \frac{D(1 - \nu^2)\, a^2}{4K} = 3(1 - \nu^2)\frac{a^2}{t^2},$$

we find for λ the equation

$$\lambda^4 + 4\varkappa^4 = 0.$$

It has 4 solutions, $\lambda = \pm(1 \pm i)\varkappa$, each yielding one independent solution w. These are two pairs of conjugate complex functions. Sum and difference of the functions of each pair are purely real or purely imaginary and constitute another set of four independent homogeneous solutions. Using them as elements, we may write the general solution in the form:

$$w = e^{-\varkappa x/a}\left[C_1 \cos\frac{\varkappa x}{a} + C_2 \sin\frac{\varkappa x}{a}\right] + e^{+\varkappa x/a}\left[C_3 \cos\frac{\varkappa x}{a} + C_4 \sin\frac{\varkappa x}{a}\right].$$

$$(73\,\mathrm{a})$$

For boundary conditions we need the slope w'/a. We have

$$w' = -\varkappa\, e^{-\varkappa x/a}\left[(C_1 - C_2) \cos\frac{\varkappa x}{a} + (C_1 + C_2) \sin\frac{\varkappa x}{a}\right]$$

$$+ \varkappa\, e^{+\varkappa x/a}\left[(C_3 + C_4) \cos\frac{\varkappa x}{a} - (C_3 - C_4) \sin\frac{\varkappa x}{a}\right].$$

$$(73\,\mathrm{b})$$

Introducing w into eq. (70a), we find the bending moment and then from eq. (69) the transverse shear:

$$M_x = \frac{2K\varkappa^2}{a^2}\left[e^{-\varkappa x/a}\left(C_1 \sin\frac{\varkappa x}{a} - C_2 \cos\frac{\varkappa x}{a}\right)\right.$$

$$\left. - e^{+\varkappa x/a}\left(C_3 \sin\frac{\varkappa x}{a} - C_4 \cos\frac{\varkappa x}{a}\right)\right],$$

$$(73\,\mathrm{c, d})$$

$$Q_x = \frac{2K\varkappa^3}{a^3}\left[e^{-\varkappa x/a}\left((C_1 + C_2) \cos\frac{\varkappa x}{a} - (C_1 - C_2) \sin\frac{\varkappa x}{a}\right)\right.$$

$$\left. - e^{+\varkappa x/a}\left((C_3 - C_4) \cos\frac{\varkappa x}{a} + (C_3 + C_4) \sin\frac{\varkappa x}{a}\right)\right].$$

In these formulas the terms with C_1 and C_2 are oscillating functions of x which decrease exponentially when x increases. The other two terms show the opposite tendency; they are also damped oscillations, but they decrease with decreasing x. In many applications the cylinder is long enough to make $e^{-\varkappa l/a}$ a rather small quantity, negligible compared with unity. Then the values of w and of the stress resultants at one end, say $x = 0$, depend almost exclusively on the constants C_1 and C_2; their values at the other end, $x = l$, depend on C_3 and C_4. In such cases, it is useful to write the solution in the following form:

$$w = e^{-\varkappa x/a}\left(A_1 \cos\frac{\varkappa x}{a} + A_2 \sin\frac{\varkappa x}{a}\right)$$
$$+ e^{-\varkappa(l-x)/a}\left(B_1 \cos\frac{\varkappa(l-x)}{a} + B_2 \sin\frac{\varkappa(l-x)}{a}\right) \tag{74}$$

and to determine, independent of each other, A_1, A_2 from two boundary conditions at $x = 0$ and B_1 and B_2 from two conditions at the end $x = l$ of the cylinder.

There is still another way of writing the solution, combining the cosine and the sine in each parenthesis into a sine with a phase angle ψ:

$$w = A\,e^{-\varkappa x/a}\sin\left(\frac{\varkappa x}{a} + \psi_1\right) + B\,e^{-\varkappa(l-x)/a}\sin\left(\frac{\varkappa(l-x)}{a} + \psi_2\right). \tag{75a}$$

The derivative is

$$w' = \varkappa\,\sqrt{2}\left[-A\,e^{-\varkappa x/a}\sin\left(\frac{\varkappa x}{a} + \psi_1 - \frac{\pi}{4}\right)\right.$$
$$\left.+ B\,e^{-\varkappa(l-x)/a}\sin\left(\frac{\varkappa(l-x)}{a} + \psi_2 - \frac{\pi}{4}\right)\right], \tag{75b}$$

and the stress resultants are

$$M_x = -\frac{2K\varkappa^2}{a^2}\left[A\,e^{-\varkappa x/a}\cos\left(\frac{\varkappa x}{a} + \psi_1\right)\right.$$
$$\left.+ B\,e^{-\varkappa(l-x)/a}\cos\left(\frac{\varkappa(l-x)}{a} + \psi_2\right)\right],$$
$$Q_x = \frac{2\sqrt{2}\,K\varkappa^3}{a^3}\left[A\,e^{-\varkappa x/a}\sin\left(\frac{\varkappa x}{a} + \psi_1 + \frac{\pi}{4}\right)\right. \tag{75c, d}$$
$$\left.- B\,e^{-\varkappa(l-x)/a}\sin\left(\frac{\varkappa(l-x)}{a} + \psi_2 + \frac{\pi}{4}\right)\right].$$

Here the four constants are A, B, ψ_1, ψ_2. Since only two of them appear as linear factors, the expression (75a) is not suitable for cases where a set of four equations for the constants must be established. But this form of the solution is particularly useful for simple cases, where ψ_1 or ψ_2 can be determined at a glance from a homogeneous boundary condition. How this may be done, we shall see in some examples.

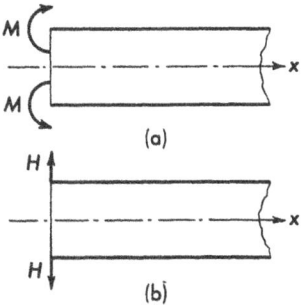

Fig. 23. Edge load on a semi-infinite cylinder

We consider a cylinder (fig. 23a) which begins at $x = 0$ and extends far enough in the direction of positive x that we may forget about the conditions at the far end while determining A and ψ_1. At the edge $x = 0$ bending moments

$M_x = M$ are applied as a load. Since Q_x is supposed to be zero, we see from eq. (75d) that $\psi_1 = -\pi/4$. The constant A then follows from eq. (75c):

$$-\frac{2K\varkappa^2}{a^2}\,A\cos(-\pi/4) = M,$$

and when we introduce the result into eqs. (75), we find

$$\left.\begin{aligned}
M_x &= M\,\sqrt{2}\,e^{-\varkappa x/a}\cos\left(\frac{\varkappa x}{a} - \frac{\pi}{4}\right), \\
Q_x &= -\frac{2M\varkappa}{a}\,e^{-\varkappa x/a}\sin\frac{\varkappa x}{a}, \\
N_\phi &= -\frac{2\sqrt{2}\,M\varkappa^2}{a}\,e^{-\varkappa x/a}\sin\left(\frac{\varkappa x}{a} - \frac{\pi}{4}\right).
\end{aligned}\right\} \tag{76}$$

When these results are used for the solution of statically indeterminate structures, the deflection w and the slope w'/a at the end $x = 0$ are needed. They are

$$w = \frac{M a^2}{2K\varkappa^2}, \qquad \frac{w'}{a} = -\frac{M a}{K\varkappa}. \tag{76'}$$

When the end of the cylinder is acted upon by radial forces H (fig. 23b), we have the boundary conditions $M_x = 0$, $Q_x = H$, and we see at once from eq. (75c) that in this case $\psi_1 = \pm\pi/2$. We arbitrarily choose the plus sign and find then from eq. (75d) and the second boundary condition that

$$A = \frac{H a^3}{2K\varkappa^3}.$$

The stress resultants are then

$$\left.\begin{aligned}
M_x &= \frac{H a}{\varkappa}\,e^{-\varkappa x/a}\sin\frac{\varkappa x}{a}, \\
Q_x &= \sqrt{2}\,H\,e^{-\varkappa x/a}\cos\left(\frac{\varkappa x}{a} + \frac{\pi}{4}\right), \\
N_\phi &= 2H\varkappa\,e^{-\varkappa x/a}\cos\frac{\varkappa x}{a},
\end{aligned}\right\} \tag{77}$$

and at the edge $x = 0$ we have

$$w = \frac{H a^3}{2K\varkappa^3}, \qquad \frac{w'}{a} = -\frac{H a^3}{2K\varkappa^2}. \tag{77'}$$

When we put $H = -M\varkappa/a$ and add the formulas for the two cases, we obtain the stress resultants for a shell loaded by moments as in fig. 23a, but with a rigid bulkhead at the end, so that there $w = 0$. The slope at the end is then

$$\frac{w'}{a} = -\frac{M a}{2K\varkappa}, \tag{78'}$$

and the stress resultants are

$$M_x = M\,e^{-\varkappa x/a}\cos\frac{\varkappa\,x}{a},$$

$$Q_x = -\frac{\sqrt{2}\,M\,\varkappa}{a}\,e^{-\varkappa x/a}\sin\left(\frac{\varkappa\,x}{a}+\frac{\pi}{4}\right),$$ (78)

$$N_\phi = -\frac{2\,M\,\varkappa^2}{a}\,e^{-\varkappa x/a}\sin\frac{\varkappa\,x}{a}.$$

If we replace H by $-2H$ in eqs. (77) and add to them a multiple of eqs. (76) such that $w' = 0$ at the end, then we have the stress result-

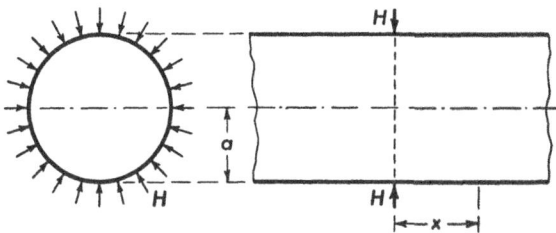

Fig. 24. Infinite cylinder carrying a uniform line load H

ants for the right half of the cylinder shown in fig. 24. They are

$$M_x = -\frac{H\,a}{2\varkappa\sqrt{2}}e^{-\varkappa x/a}\sin\left(\frac{\varkappa\,x}{a}-\frac{\pi}{4}\right),$$

$$Q_x = -\tfrac{1}{2}H\,e^{-\varkappa x/a}\cos\frac{\varkappa\,x}{a},$$ (79)

$$N_\phi = -\tfrac{1}{2}H\sqrt{2}\,e^{-\varkappa x/a}\sin\left(\frac{\varkappa\,x}{a}+\frac{\pi}{4}\right).$$

The deflection under the load is

$$w = -\frac{H\,a^3}{2\,K\,\varkappa^3}.$$

We shall come back to this case on p. 281.

5.5.2.2 Water Tanks

We may now resume the tank problem explained on p. 272. If a cylindrical water tank is filled to the rim (fig. 21), the membrane deflection w as given by eq. (68) happens to be an exact solution of the differential equation (72) of the bending theory. Furthermore, it fulfils at the upper edge $x = h$ the conditions $w = 0$, $w'' = 0$, $w''' = 0$ and hence all boundary conditions if the edge is either free or simply supported. On the lower edge, however, the membrane deflection is considerable and incompatible with the presence of a tank bottom. Therefore, it is necessary to apply there external forces H and moments M as shown in fig. 23, which will produce bending stresses, but we may drop the B terms from the solution (74) unless the tank is rather shallow.

Under these circumstances, the complete expression for the deflection is

$$w = \frac{\gamma a^2}{D(1-\nu^2)}(h-x) + e^{-\varkappa x/a}\left(A_1 \cos\frac{\varkappa x}{a} + A_2 \sin\frac{\varkappa x}{a}\right),$$

and its first derivative is

$$w' = -\frac{\gamma a^3}{D(1-\nu^2)} - \varkappa e^{-\varkappa x/a}\left[(A_1 - A_2)\cos\frac{\varkappa x}{a} + (A_1 + A_2)\sin\frac{\varkappa x}{a}\right].$$

In the simplest case the bottom plate is so thick that it may be assumed to be rigid. The boundary conditions at $x = 0$ are then $w = 0$,

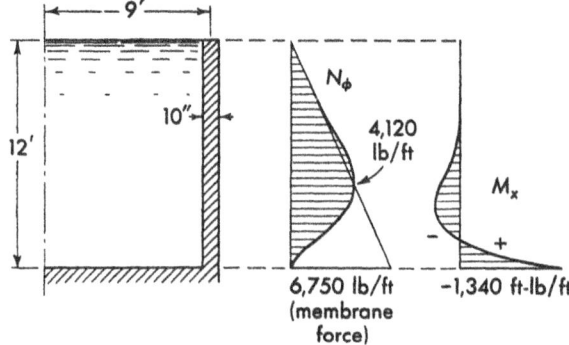

Fig. 25. Stress resultants in a cylindrical tank wall with clamped base. The straight line in the N_ϕ diagram represents the membrane force, and the shaded diagram the total hoop force

$w' = 0$. They yield two equations for A_1, A_2. Solving them and introducing the result into w, we find

$$w = \frac{\gamma a^2}{E t}\left[h - x - h e^{-\varkappa x/a}\cos\frac{\varkappa x}{a} + \left(\frac{a}{\varkappa} - h\right)e^{-\varkappa x/a}\sin\frac{\varkappa x}{a}\right].$$

The equations (70 b, a) and (69) yield now the stress resultants as follows:

$$N_\phi = \gamma a\left[h - x - h e^{-\varkappa x/a}\cos\frac{\varkappa x}{a} + \left(\frac{a}{\varkappa} - h\right)e^{-x/\varkappa a}\sin\frac{\varkappa x}{a}\right],$$

$$M_x = -\frac{\gamma a t}{\sqrt{12(1-\nu^2)}}\left[\left(\frac{a}{\varkappa} - h\right)e^{-\varkappa x/a}\cos\frac{\varkappa x}{a} + h e^{-\varkappa x/a}\sin\frac{\varkappa x}{a}\right],$$

$$Q_x = \frac{\gamma t \varkappa}{\sqrt{12(1-\nu^2)}}\left[\left(\frac{a}{\varkappa} - 2h\right)e^{-\varkappa x/a}\cos\frac{\varkappa x}{a} + \frac{a}{\varkappa}e^{-\varkappa x/a}\sin\frac{\varkappa x}{a}\right].$$

The results are presented in fig. 25. The ordinates of the N_ϕ diagram may also be interpreted as representing the deflection w. One may recognize in these diagrams the clamping of the lower edge, the ensuing moments, and the dying out of the disturbance far below the upper edge.

When the shell is connected to other structural elements which are not rigid enough to be considered as undeformable, we had best use the notations of the theory of statically indeterminate structures. As an example of this let us consider the connection of shell and ceiling in the tank shown in fig. 26.

When we use the coordinate x as indicated in this figure, eq. (68) reads:

$$w = \frac{\gamma\, a^2}{D(1 - \nu^2)}\, x.$$

Additionally we have the deflections produced by the transverse forces Q_x and by the clamping moments M_x transmitted from the ceiling slab.

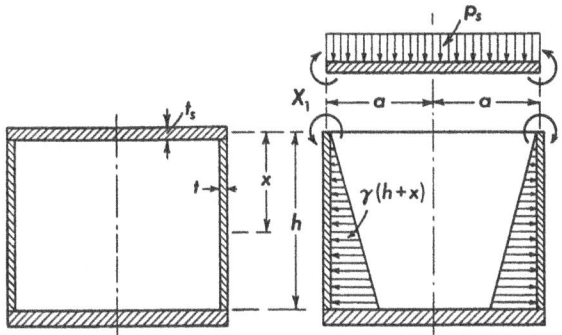

Fig. 26. Cylindrical tank with elastic roof and bottom, (a) meridional section, (b) roof slab and shell cut apart to show the redundant moment X_1

In this slab, the forces Q_x of the shell will produce a plane stress system which certainly will not lead to any appreciable deformation. We therefore have $w = 0$ at the edge, and the effect of the clamping moment is then represented by eqs. (78). This clamping moment M_x is our redundant quantity and will now be called X_1 (fig. 26b). The rotation of a line element dx of the shell, situated at the edge $x = 0$, is

$$\frac{w'}{a} = \frac{\gamma\, a^2}{D(1 - \nu^2)} - \frac{a}{2K\varkappa}\, X_1,$$

the first term being due to the water pressure on the tank wall, the second to the redundant moment.

On the other hand, the slab of thickness t_s, carrying a load p_s and subjected to the action of the redundant moment X_1 coming from the shell (positive as indicated in fig. 26), has at its edge the slope[1]

$$\omega = \frac{p_s\, a^3}{8K_s(1 + \nu)} + \frac{a}{K_s(1 + \nu)}\, X_1.$$

---

[1] The formulas for circular plates needed here may easily be found from those given in TIMOSHENKO, S. and S. Woinowsky-Krieger: Theory of Plates and Shells., 2nd ed., New York 1959, pp. 51–69. – They are found ready for use in BEYER, K.: Die Statik im Eisenbetonbau. 2nd ed., vol. 2, Berlin 1934, pp. 652–660.

Here,

$$K_s = \frac{E\,t_s^3}{12(1 - \nu^2)}$$

is the bending rigidity of the slab, and the two terms show the influence of the load p_s and of the redundant X_1.

Using the notations of the theory of statically indeterminate structures, we denote the relative rotation between shell and slab, $\omega - w'/a$, by δ_{10} if produced by the external loads (γ and p_s), and by δ_{11} if produced by $X_1 = 1$. From the preceding formulas we read:

$$\delta_{10} = -\frac{\gamma a^2}{D(1 - \nu^2)} + \frac{p_s a^3}{8K_s(1 + \nu)},$$

$$\delta_{11} = \frac{a}{2K\varkappa} + \frac{a}{K_s(1 + \nu)}.$$

The condition of continuity of deformation is then

$$\delta_{11} X_1 + \delta_{10} = 0.$$

From it we find X_1 and then w, N_ϕ, M_x, Q_x.

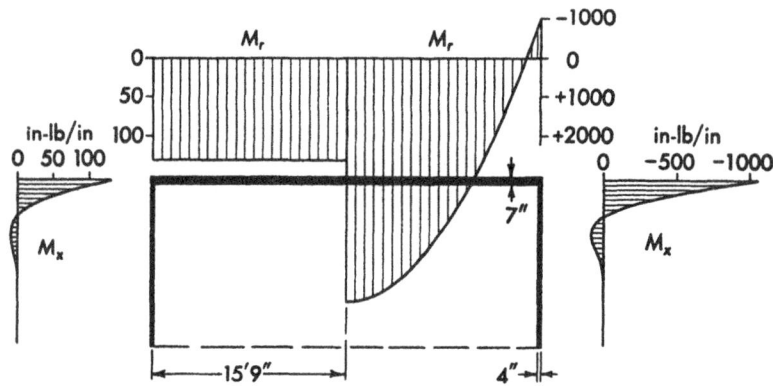

Fig. 27. Bending moments in the roof slab and the cylindrical wall of the tank in Fig. 26.
Left side: water pressure, right side: weight of the slab

When we separate the two external loads, the solution may easily be written in general terms. If there is only the water pressure and no load p_s on the slab, we have

$$w = \frac{\gamma a^3}{D(1 - \nu^2)}\left[\frac{x}{a} - \frac{K_s(1 + \nu)}{2K\varkappa + K_s(1 + \nu)}\frac{1}{\varkappa}e^{-\varkappa x/a}\sin\frac{\varkappa x}{a}\right],$$

$$M_x = \frac{2\gamma a K K_s \varkappa}{D(1 - \nu)[2K\varkappa + K_s(1 + \nu)]}e^{-\varkappa x/a}\cos\frac{\varkappa x}{a}.$$

If there is a uniform load p_s on the ceiling, but no water pressure in the tank, the deflection and the bending moment in the shell are

$$w = \frac{p_s a^4}{8\varkappa[2K\varkappa + K_s(1 + \nu)]} e^{-\varkappa x/a} \sin\frac{\varkappa x}{a},$$

$$M_x = -\frac{p_s a^2}{4} \frac{K\varkappa}{2K\varkappa + K_s(1 + \nu)} e^{-\varkappa x/a} \cos\frac{\varkappa x}{a}.$$

The results for both cases are illustrated by the diagrams, fig. 27.

In a quite similar way many other problems may be treated, e. g. the tank with a non-rigid bottom plate resting on the ground and the tank with a concentric partition wall (fig. 28). This latter structure, which is often used in water towers, has two redundant quantities, the clamping moments of the two concentric shells. In many cases a conical or spherical tank bottom is preferred to a flat plate. In this case also a bending problem arises. It will be explained in connection with the bending theory of such shells in the next chapter (pp. 340, 374).

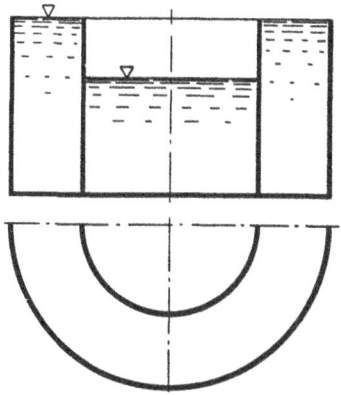

Fig. 28. Cylindrical tank with concentric partition wall

5.5.2.3 Cylinder Subjected to a Ring of Radial Forces

A problem of general interest closely connected to the tank problem is shown in fig. 29. To a cylindrical shell of length l a radial load is applied in its plane of symmetry $x = 0$. It is uniformly distributed

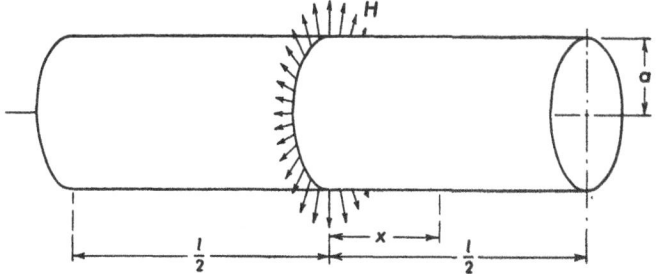

Fig. 29. Finite cylinder carrying a uniform line load

over the circumference of the circular cross-section and has the intensity H per unit length of this circle. We ask for the stress resultants in the shell and in particular for the radial displacement $w = w_0$ at $x = 0$.

The plane $x = 0$ is a plane of symmetry for stress and deformation. It is therefore sufficient to consider only the right half of the shell, $x > 0$, and to formulate four boundary conditions for its edges. At $x = 0$, we have no slope, $w' = 0$, and the transverse shear Q_x must be in equilibrium with one half of the applied load: $Q_x = H/2$. At the far end, $x = l/2$, we have no load and no support, hence $M_x = 0$, $Q_x = 0$.

Since there is no distributed surface load, eq. (73) or eq. (74) is the complete solution. When the cylinder is long enough, we may use the A terms of eq. (74) and forget about the boundary conditions at $x = l/2$. This is exactly what we did on p. 277; but now we want to cover cylinders of any length and in particular the very short ones. In this case we had better use eq. (73) and find from the four boundary conditions four equations for C_1, C_2, C_3, C_4. Their solution is

$$C_{1,3} = \frac{H a \varkappa}{4 E t} \left[\pm 1 + \frac{2(\text{Cosh}^2 \lambda + \cos^2 \lambda)}{\text{Sinh} 2\lambda + \sin 2\lambda} \right],$$

$$C_{2,4} = \frac{H a \varkappa}{4 E t} \left[1 \pm \frac{2(\text{Cosh}^2 \lambda - \cos^2 \lambda)}{\text{Sinh} 2\lambda + \sin 2\lambda} \right], \qquad \lambda = \frac{\varkappa l}{2a}.$$

Introducing this into eq. (73) and putting $x = 0$, we find the displacement w_0 (the greatest one occurring):

$$w_0 = C_1 + C_3 = \frac{H a \varkappa}{E t} \frac{\text{Cosh}^2 \lambda + \cos^2 \lambda}{\text{Sinh} 2\lambda + \sin 2\lambda}. \tag{80}$$

The greatest hoop force is found on the same circle. It is

$$N_{\phi 0} = \frac{E t}{a} w_0 = H \varkappa \frac{\text{Cosh}^2 \lambda + \cos^2 \lambda}{\text{Sinh} 2\lambda + \sin 2\lambda}.$$

With increasing x the hoop force N_ϕ is distributed in the form of damped waves which we may find from eq. (73), and for negative values of x the distribution is symmetric. Cutting the shell in half lengthwise we may see easily that the integral of the hoop force over the total length of the cylinder must equal $-H a$.

In cases like this one, where only the peak value of distributed forces is of interest, the result may be represented in terms of an effective width. This is the length b of a fictitious shell in which a uniform distribution of the force $H a$ would yield the correct peak value N_ϕ of the hoop force. From this definition it follows that

$$b = \frac{H a}{N_{\phi 0}} = \frac{a}{\varkappa} \frac{\text{Sinh} 2\lambda + \sin 2\lambda}{\text{Cosh}^2 \lambda + \cos^2 \lambda}.$$

In the limiting case of a very short cylinder, $l \to 0$, both sines may be replaced by their argument, and both cosines replaced by 1, and then we have $b \to l$. When l is finite, b is always smaller, and for $l \to \infty$ the effective width has a limiting value

$$b_{\max} = \frac{2a}{\varkappa} \approx 1.57 \sqrt{at}.$$

It appears that this is always much smaller than the radius of the cylinder. If the cylinder is made for the sole purpose of carrying the load H, it is scarcely worthwhile to make it longer than $4a/\varkappa$, the effective width then being 92.5% of $b_{\max}$.

We shall now apply our results to two problems. The first one concerns a plane plate of thickness t_p which is subjected to a uniform biaxial stress. In this plate a hole of radius a is drilled and its edge reinforced by a cylinder as shown in fig. 30. The reinforcement is perfect, if there is no stress concentration in the plate. This happens, if a radial load H applied to the cylinder according to fig. 29 produces there the same radial displacement w_0 as it would do when applied to the edge of the missing piece of plate of radius a. Now, this latter displacement is

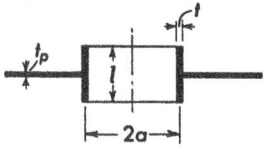

$$w = \frac{H a(1-\nu)}{E t_p},$$

and equating this to w_0 as we just found it, we get a relation for the thickness t_p:

$$\frac{t_p}{t} = \frac{1-\nu}{\sqrt[4]{3(1-\nu^2)}} \sqrt{\frac{t}{a}} \frac{\operatorname{Sinh}2\lambda + \sin 2\lambda}{\operatorname{Cosh}^2\lambda + \cos^2\lambda}.$$

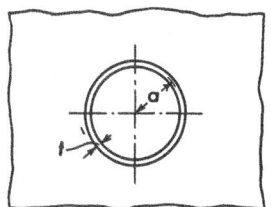

Fig. 30. Hole in a flat plate reinforced by a short cylindrical shell

It is evident that the plate must be much thinner than the cylinder. Of course, this relation is valid only when the stress in the plate is the same in all directions. In any other case we have to apply the theory for the higher harmonics in the shell, as it has been developed in Section 5.3 of this chapter.

The second problem to which we shall apply our results is that of a cylinder under constant internal pressure p and stiffened by a

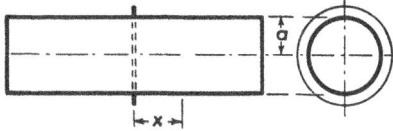

Fig. 31
Cylindrical gas tank with one stiffening ring

ring in the plane $x = 0$ (fig. 31). If there were no ring, the pressure would simply produce a constant hoop force $N_\phi = p\,a$, and from the pressure acting on the bulkheads an axial force $N_x = p\,a/2$ will result. For all points sufficiently far away from the bulkheads the hoop strain is then according to eq. (III–17):

$$\epsilon_\phi = \frac{p\,a}{D} \frac{2-\nu}{2(1-\nu^2)}$$

and hence the radial displacement

$$w = \frac{p a^2}{E t} \frac{2 - \nu}{2}.$$

Now this displacement is not possible in the plane $x = 0$ because of the ring. Between the ring and the shell radial forces H will be transmitted, just as shown in fig. 29 but of opposite direction. We call them $P = -H$. In the ring they produce a hoop force $F = P a$ and hence a radial displacement $w_r = P a^2/EA$, when A is the cross-sectional area of the ring. The displacement of the shell, which is additional to the value w due to p, is found from eq. (80). If the cylinder is long enough, we may replace the transcendental factor by $\frac{1}{2}$. The total displacement of the cylinder opposite the ring is then

$$w_s = \frac{p a^2}{E t} \frac{2 - \nu}{2} - \frac{P a \varkappa}{2 E t}.$$

From the fact that the ring is connected to the cylinder, we conclude that $w_r = w_s$ and thus find P and the force F in the ring:

$$F = \frac{p a^2 A (2 - \nu)}{2 a t + A \varkappa}.$$

The ring participates with this force in carrying the load p and thus relieves the shell of some of its hoop stresses. We may say that the ring carries the total load acting on a strip of the shell of width

$$b^* = \frac{P}{p} = \frac{a A (2 - \nu)}{2 a t + A \varkappa}.$$

This width depends on the cross section of the ring and tends toward a maximum $(2 - \nu) a/\varkappa$, when $A \to \infty$.

5.5.2.4 Cylinder with many Rings

Often cylindrical shells are built which have many stiffening rings as shown in fig. 32 (pipe lines, submarines, airplane fuselages). When the spacing of the rings is more than, say, $3a/\varkappa$, they do not much in-

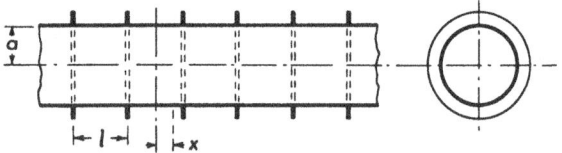

Fig. 32. Cylindrical shell with many equidistant rings

fluence each other, and then the preceding formulas may be applied. If the spacing is closer, the mutual influence becomes important. To treat this stress problem, we use a special set of elementary solutions which may be obtained from eq. (73) by combining the two cos terms and the two sin terms. It is:

$$w = A_1 \operatorname{Cosh}\frac{\varkappa\, x}{a}\cos\frac{\varkappa\, x}{a} + A_2 \operatorname{Sinh}\frac{\varkappa\, x}{a}\sin\frac{\varkappa\, x}{a}$$
$$+ B_1 \operatorname{Cosh}\frac{\varkappa\, x}{a}\sin\frac{\varkappa\, x}{a} + B_2 \operatorname{Sinh}\frac{\varkappa\, x}{a}\cos\frac{\varkappa\, x}{a}. \tag{81}$$

The meaning of the constants A, B is not the same here as in eq. (74). A characteristic of this solution is that the A terms describe a deflection which is symmetric with respect to the plane $x = 0$, and the B terms one which is antimetric. Now, in our case of a cylinder with many rings, we choose the plane $x = 0$ halfway between two rings and then need only the symmetric part of w. The constants A_1 and A_2 may then be determined from two boundary conditions at $x = l/2$, and identical conditions will then automatically be fulfilled at $x = -l/2$. The conditions in our case are that the slope must be zero, and that the deflection w of the shell is the same as that of the ring loaded radially by twice the transverse force Q_x coming from the shell.

When A is the cross section of each ring, these conditions yield the following equations:

$$x = \frac{l}{2} \quad : \quad w' = 0,$$
$$x = \frac{l}{2} \quad : \quad w = \frac{2 Q_x a^2}{E A}.$$

In these equations we introduce a solution composed of the inhomogeneous part

$$w = \frac{2-\nu}{2}\,\frac{p a^2}{E t}$$

and the A terms of eq. (81). Solving for A_1 and A_2 and introducing the result into w, we find the final solution. To write it in a concise form, we define two dimensionless parameters:

$$\eta = \frac{8 K \varkappa^3}{E A a} = \frac{2 t \sqrt{a\,t}}{A\sqrt[4]{3(1-\nu^2)}}, \qquad \zeta = \frac{\varkappa l}{2a} = \frac{\sqrt[4]{3(1-\nu^2)}\;l}{2\sqrt{a\,t}}.$$

With these abbreviations we have

$$w = \frac{2-\nu}{2}\,\frac{p a^2}{E t}\Big\{1 - \Big[(\operatorname{Cosh}\zeta\sin\zeta + \operatorname{Sinh}\zeta\cos\zeta)\operatorname{Cosh}\frac{\varkappa\, x}{a}\cos\frac{\varkappa\, x}{a}$$
$$+ (\operatorname{Cosh}\zeta\sin\zeta - \operatorname{Sinh}\zeta\cos\zeta)\operatorname{Sinh}\frac{\varkappa\, x}{a}\sin\frac{\varkappa\, x}{a}\Big]$$
$$\times [\operatorname{Cosh}\zeta\,(\operatorname{Sinh}\zeta + \eta\operatorname{Cosh}\zeta) + \cos\zeta\,(\sin\zeta - \eta\cos\zeta)]^{-1}\Big\}$$

Now we may find the bending moment M_x from eq. (70a), but we cannot indiscriminately use eq. (70b) for N_ϕ, because that equation has been derived under the assumption $N_x = 0$. This is true for the homogeneous part of the solution, but not for the membrane forces, and the term 1 in the braces represents the combined influence of the membrane part of N_ϕ and of the force N_x due to the pressure on the closing bulkheads at the ends of the shell. For this part of the solution the corresponding hoop force is simply $p\,a$. We have therefore

$$N_\phi = p\,a\left\{1 - \frac{2-\nu}{2}[\;]\times[\;]^{-1}\right\},$$

the brackets being the same as in the expression for w, and

$$M_x = \frac{(2-\nu)\,p\,a\,t}{4\sqrt{3(1-\nu^2)}}\left[(\mathrm{Cosh}\,\zeta\,\sin\zeta + \mathrm{Sinh}\,\zeta\,\cos\zeta)\,\mathrm{Sinh}\frac{\varkappa x}{a}\sin\frac{\varkappa x}{a}\right.$$
$$\left. - (\mathrm{Cosh}\,\zeta\,\sin\zeta - \mathrm{Sinh}\,\zeta\,\cos\zeta)\,\mathrm{Cosh}\frac{\varkappa x}{a}\cos\frac{\varkappa x}{a}\right]$$
$$\times\,[\mathrm{Cosh}\,\zeta\,(\mathrm{Sinh}\,\zeta + \eta\,\mathrm{Cosh}\,\zeta) + \cos\zeta\,(\sin\zeta - \eta\cos\zeta)]^{-1}.$$

The force F in the ring is best found from the deflection at $x = l/2$:

$$F = \frac{E\,A\,w}{a}$$
$$= \frac{(2-\nu)\,p\,a\sqrt{a\,t}}{\sqrt[4]{3(1-\nu^2)}}\;\frac{\mathrm{Cosh}^2\zeta - \cos^2\zeta}{\mathrm{Cosh}\,\zeta\,(\mathrm{Sinh}\,\zeta + \eta\,\mathrm{Cosh}\,\zeta) + \cos\zeta\,(\sin\zeta - \eta\cos\zeta)}\,.$$

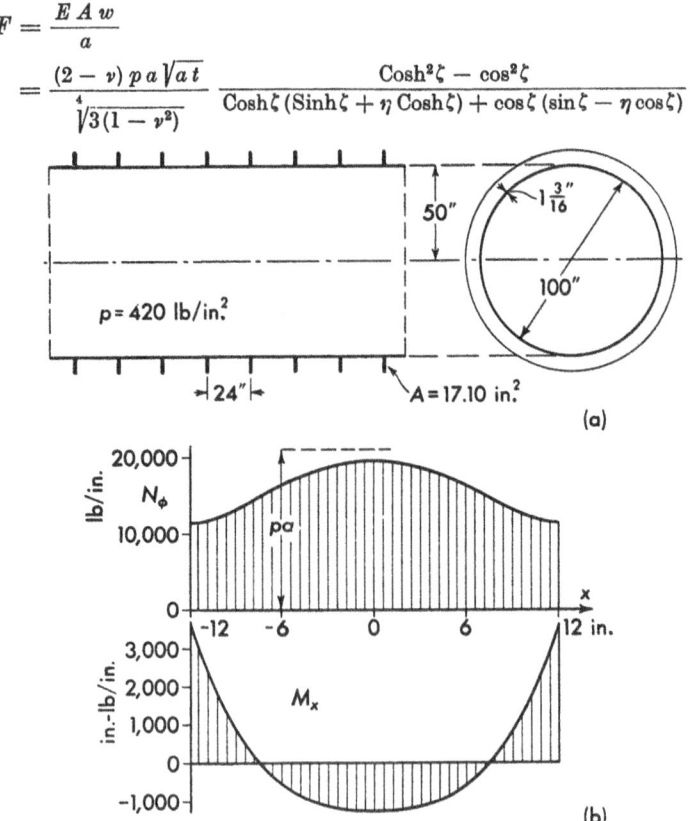

Fig. 33. Cylinder with many rings, (a) dimensions. (b) stress resultants in a 24-in. length between two rings

When the rings are far enough apart, these formulas must be used to find F and N_ϕ, M_x for different points along the shell. The maximum of w and N_ϕ may be found at $x = 0$, as illustrated by fig. 33, but it may also occur that it is found elsewhere.

It is of interest to see what happens when the rings are rather closely spaced. Let us first consider the limiting case $l \to 0$. In order to arrive at a reasonable result, we must assume that with increasing number of rings, the individual ring becomes weaker so that the ratio A/l remains a constant. Then $\zeta \to 0$, the product $\eta\,\zeta = \text{const}$, and N_ϕ and F/l approach finite limits:

$$\lim N_\phi = p\,a\,\frac{t + \nu A/2l}{t + A/l}, \qquad \lim \frac{F}{l} = p\,a\,\frac{2-\nu}{2}\,\frac{A/l}{t + A/l}.$$

This corresponds to an anisotropic shell which we shall study in detail in Section 5.6. If l is small, but finite, F/l is smaller than the limiting value, the shell taking a larger share of the total load. The ratio

$$\frac{F/l}{\lim F/l} = \frac{t + A/l}{A/l}\,\frac{1}{\zeta}\,\frac{\text{Cosh}^2\zeta - \cos^2\zeta}{\text{Cosh}\zeta\,(\text{Sinh}\zeta + \eta\,\text{Cosh}\zeta) + \cos\zeta\,(\sin\zeta - \eta\,\cos\zeta)}$$

may then give an idea about how far away we are from the ideal case, if we treat the shell as a homogeneous, but anisotropic structure.

5.5.3 Shell of Variable Thickness

As shown in fig. 25, the hoop force N_ϕ in a cylindrical tank increases from zero at the water level to rather considerable values at greater depths. This suggests making the wall thickness increase from top to bottom. In concrete tanks it is the rule to choose a linear variation, say

$$t = \alpha\,x,$$

where x is measured downward from that level where t would be zero if extrapolated beyond the upper edge of the tank (fig. 34). The water pressure is then $p = \gamma(x - h_0)$, and the rigidities are functions of x:

$$D = \frac{E\,\alpha}{1 - \nu^2}\,x, \qquad K = \frac{E\,\alpha^3}{12(1 - \nu^2)}\,x^3.$$

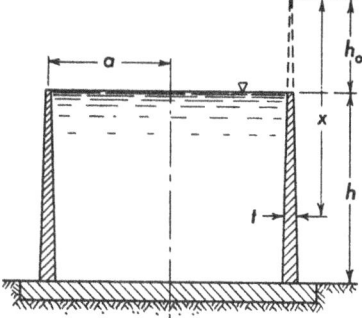

Fig. 34. Cylindrical water tank with variable wall thickness

Introducing all this into eq. (71), we arrive at the differential equation of the bending problem presented by fig. 34:

$$(x^3 w'')'' + \frac{12(1 - \nu^2)\,a^2}{\alpha^2}\,x\,w = \frac{12\,(1 - \nu^2)\,\gamma\,a^4}{E\,\alpha^3}\,(x - h_0). \qquad (82)$$

A particular solution may easily be found:

$$w = \frac{\gamma a^2}{E \alpha} \frac{x - h_0}{x}. \tag{83}$$

It corresponds to the membrane forces $N_\phi = \gamma a (x - h_0)$, but since the deflection is no longer a linear function of x, we have a curvature of the generators and hence a bending moment

$$M_x = - \frac{\gamma a^2 \alpha^2}{6 (1 - \nu^2)} h_0. \tag{84}$$

It is constant and very small.

To satisfy the boundary conditions, we still need the complete solution of the homogeneous equation. When we introduce

$$\varrho^4 = \frac{12 (1 - \nu^2)}{\alpha^2},$$

this equation may be written in the following form:

$$\frac{a}{x} \left(\frac{x^3}{a^3} w'' \right)'' + \varrho^4 w = 0.$$

As one may easily verify, this is identical with the following equation:

$$\frac{a}{x} \left[\frac{x^2}{a^2} \left(\frac{a}{x} \left[\frac{x^2}{a^2} w' \right]' \right)' \right]' + \varrho^4 w = 0.$$

Here we see that the first term is the result of the repeated application to w of the operator

$$L(\ ) = \frac{a}{x} \left[\frac{x^2}{a^2} (\ )' \right]'.$$

Through the use of this symbol, the differential equation becomes simpler in appearance and easier to handle:

$$LL(w) + \varrho^4 w = 0. \tag{85}$$

It may be written in the following alternative forms:

$$L\big(L(w) + i \varrho^2 w\big) - i \varrho^2 (L(w) + i \varrho^2 w) = 0,$$
$$L\big(L(w) - i \varrho^2 w\big) + i \varrho^2 (L(w) - i \varrho^2 w) = 0.$$

From these we recognize that the solutions of the second order equations

$$L(w) \pm i \varrho^2 w = 0 \tag{86a, b}$$

must be solutions of eq. (85). Since the new equations have an imaginary coefficient, their solutions are complex-valued functions of x, and those of (86b) are conjugate complex to those of (86a). From this fact it follows that they are linearly independent of each other, and hence that two independent solutions of either equation (86) together will form a complete system of four independent solutions of eq. (85).

And furthermore, it follows that their real and imaginary parts, being each a linear combination of two such solutions, will also satisfy eq. (85), although they are not solutions to (86a) or (86b).

To find these functions, it is enough to solve eq. (86a). When the differential operator is written in full, this equation reads:

$$\frac{x}{a}\,w'' + 2\,w' + i\,\varrho^2 w = 0.$$

It may be transformed into a BESSEL equation by introducing a complex variable, putting

$$\eta = y\,\sqrt{i} = 2\varrho\,\sqrt{\frac{ix}{a}}, \quad \zeta = w\,\sqrt{\frac{x}{a}}.$$

Performing the transformation, we find

$$\eta^2\frac{d^2\zeta}{d\eta^2} + \eta\,\frac{d\zeta}{d\eta} + (\eta^2 - 1)\zeta = 0.$$

The solutions of this equation are the BESSEL functions of the first order of the complex argument η:

$$\zeta = A\,J_1(\eta) + B\,H_1^{(1)}(\eta).$$

They have complex values. They are connected with the functions of zero order by the relations[1]

$$J_1(\eta) = -\,\frac{dJ_0(\eta)}{d\eta}, \qquad H_1^{(1)}(\eta) = -\,\frac{dH_0^{(1)}(\eta)}{d\eta}.$$

The real and imaginary parts of J_0 and $H_0^{(1)}$ may be considered as real functions of the real variable y. The functions so defined are the THOMSON functions. They are introduced by the following formulas:

$$J_0(\eta) = J_0(y\,\sqrt{i}) = \text{ber}\,y - i\,\text{bei}\,y,$$

$$H_0^{(1)}(\eta) = H_0^{(1)}(y\,\sqrt{i}) = -\,\frac{2}{\pi}\,(\text{kei}\,y + i\,\text{ker}\,y).$$

Differentiating with respect to η and then separating real and imaginary parts, we find a corresponding set of relations for the first order functions:

$$J_1(\eta) = \frac{1}{\sqrt{2}}\,[(\text{bei}\,'y - \text{ber}\,'y) + i(\text{bei}\,'y + \text{ber}\,'y)],$$

$$H_1^{(1)}(\eta) = \frac{\sqrt{2}}{\pi}\,[(\text{ker}\,'y + \text{kei}\,'y) + i(\text{ker}\,'y - \text{kei}\,'y)],$$

where the prime indicates the derivatives of these functions with respect to their argument y. Since real and imaginary parts of $J_1(\eta)$ and $H_1^{(1)}(\eta)$ and any constant multiples and linear combinations thereof

---

[1] For the relations concerning BESSEL functions see e. g. WATSON, G. N.: Theory of BESSEL Functions. 2nd ed. Cambridge 1944, pp. 45, 74, or any other book on BESSEL functions.

are solutions of eq. (85), we may choose the derivatives of the THOMSON functions as elementary solutions and write

$$\zeta = C_1 \operatorname{ber}'y + C_2 \operatorname{bei}'y + C_3 \operatorname{ker}'y + C_4 \operatorname{kei}'y. \qquad (87)$$

Before we proceed in the discussion of the shell problem, some remarks must be made on the subject of the THOMSON functions. They are useful in various problems (for another one see p. 345) and have been tabulated by many authors under almost as many different notations[1]. Any one of these tables will help in handling the numerical side of the problem as easily as if we had to deal with trigonometric or hyperbolic functions. But in order to find from eq. (87) formulas for the slope and the stress resultants, we still need some simple formulas concerning the higher derivatives of the THOMSON functions. They may be derived from the fact that $J_0(\eta)$ and $H_0^{(1)}(\eta)$ satisfy a differential equation of the BESSEL type. We give them here without proof:

$$\left.\begin{aligned}
\frac{d^2}{dy^2} \operatorname{ber} y &= -\operatorname{bei} y - \frac{1}{y} \operatorname{ber}'y, \\
\frac{d^2}{dy^2} \operatorname{bei} y &= \operatorname{ber} y - \frac{1}{y} \operatorname{bei}'y, \\
\frac{d^2}{dy^2} \operatorname{ker} y &= -\operatorname{kei} y - \frac{1}{y} \operatorname{ker}'y, \\
\frac{d^2}{dy^2} \operatorname{kei} y &= \operatorname{ker} y - \frac{1}{y} \operatorname{kei}'y.
\end{aligned}\right\} \qquad (88)$$

With the help of these formulas we find from eq. (87):

$$w = \frac{1}{\sqrt{x}}[C_1 \operatorname{ber}'y + C_2 \operatorname{bei}'y + C_3 \operatorname{ker}'y + C_4 \operatorname{kei}'y],$$

$$\begin{aligned}
\frac{dw}{dx} = -\frac{1}{2x\sqrt{x}}[&C_1(2\operatorname{ber}'y + y\operatorname{bei} y) + C_2(2\operatorname{bei}'y - y\operatorname{ber} y) \\
&+ C_3(2\operatorname{ker}'y + y\operatorname{kei} y) + C_4(2\operatorname{kei}'y - y\operatorname{ker} y)],
\end{aligned}$$

$$N_\phi = \frac{E\,a}{a}\sqrt{x}\,[C_1 \operatorname{ber}'y + C_2 \operatorname{bei}'y + C_3 \operatorname{ker}'y + C_4 \operatorname{kei}'y],$$

$$\begin{aligned}
M_x = \frac{E\,a^3}{48(1-\nu^2)}\sqrt{x}\,[&C_1(-y^2\operatorname{bei}'y + 4y\operatorname{bei} y + 8\operatorname{ber}'y) \qquad (89)\\
&+ C_2(y^2\operatorname{ber}'y - 4y\operatorname{ber} y + 8\operatorname{bei}'y) \\
&+ C_3(-y^2\operatorname{kei}'y + 4y\operatorname{kei} y + 8\operatorname{ker}'y) \\
&+ C_4(y^2\operatorname{ker}'y - 4y\operatorname{ker} y + 8\operatorname{kei}'y)],
\end{aligned}$$

$$\begin{aligned}
Q_x = -\frac{E\,a^2}{4\sqrt{3(1-\nu^2)}\,a}\sqrt{x}\,[&C_1(-y\operatorname{ber} y + 2\operatorname{bei}'y) - C_2(y\operatorname{bei} y + 2\operatorname{ber}'y) \\
&+ C_3(-y\operatorname{ker} y + 2\operatorname{kei}'y) - C_4(y\operatorname{kei} y + 2\operatorname{ker}'y)].
\end{aligned}$$

---

[1] See FLÜGGE, W.: Four-Place Tables of Transcendental Functions. London 1954, pp. 62–83 and the literature given there on pp. 66–67.

In fig. 35 a, b the derivatives of the four THOMSON functions are shown. From these graphs one may see that the solutions C_1 and C_2 are such that the deflection and all stress resultants decrease in damped oscillations when y (and hence x) decreases. They describe disturbances originating from the lower edge of the cylinder. In the same way the solutions C_3 and C_4 belong to a disturbance at the upper edge. In many practical cases the decrease of the elementary solutions from one end of the generator to the other is strong enough to make the boundary conditions at both ends independent of each other, as we have seen on p. 278 for constant wall thickness.

Most tables of the THOMSON functions go up only to $y = 10$. It is useful therefore to know that for large arguments these functions may be approximated by asymptotic series. For our purposes it is enough to use the first term of each of these series:

$$
\begin{aligned}
\operatorname{ber} y &\approx (2\pi y)^{-1/2} \exp \frac{y}{\sqrt{2}} \cos\left(\frac{y}{\sqrt{2}} - \frac{\pi}{8}\right), \\
\operatorname{ber}' y &\approx (2\pi y)^{-1/2} \exp \frac{y}{\sqrt{2}} \cos\left(\frac{y}{\sqrt{2}} + \frac{\pi}{8}\right), \\
\operatorname{bei} y &\approx (2\pi y)^{-1/2} \exp \frac{y}{\sqrt{2}} \sin\left(\frac{y}{\sqrt{2}} - \frac{\pi}{8}\right), \\
\operatorname{bei}' y &\approx (2\pi y)^{-1/2} \exp \frac{y}{\sqrt{2}} \sin\left(\frac{y}{\sqrt{2}} + \frac{\pi}{8}\right), \\
\operatorname{ker} y &\approx \left(\frac{\pi}{2y}\right)^{1/2} \exp\left(-\frac{y}{\sqrt{2}}\right) \cos\left(\frac{y}{\sqrt{2}} + \frac{\pi}{8}\right), \\
\operatorname{ker}' y &\approx -\left(\frac{\pi}{2y}\right)^{1/2} \exp\left(-\frac{y}{\sqrt{2}}\right) \cos\left(\frac{y}{\sqrt{2}} - \frac{\pi}{8}\right), \\
\operatorname{kei} y &\approx -\left(\frac{\pi}{2y}\right)^{1/2} \exp\left(-\frac{y}{\sqrt{2}}\right) \sin\left(\frac{y}{\sqrt{2}} + \frac{\pi}{8}\right), \\
\operatorname{kei}' y &\approx \left(\frac{\pi}{2y}\right)^{1/2} \exp\left(-\frac{y}{\sqrt{2}}\right) \sin\left(\frac{y}{\sqrt{2}} - \frac{\pi}{8}\right).
\end{aligned}
\tag{90}
$$

For $y = 10$ the error made in using these formulas is still several percent, sometimes more, and slowly decreases with increasing y.

The application of the formulas (89) to a reinforced concrete tank is shown in fig. 36. From the particular solution (83) we find for the lower edge the deflection and the slope

$$
w = \frac{\gamma a^2}{E\alpha} \frac{h}{h_0 + h}, \qquad \frac{dw}{dx} = \frac{\gamma a^2}{E\alpha} \frac{h_0}{(h_0 + h)^2}.
$$

In order to fulfill the boundary conditions of a clamped edge, $w = 0$, $dw/dx = 0$, we must superpose the first two terms of the homogeneous solution (87). From the dimensions of the tank we find with $\nu = 0$:

$$
\alpha = 0.0556, \qquad \varrho = 7.90
$$

and for $x_{max} = 16.5$ ft we have $y_{max} = 21.40$. We may easily calculate the corresponding values of the THOMSON functions and their first

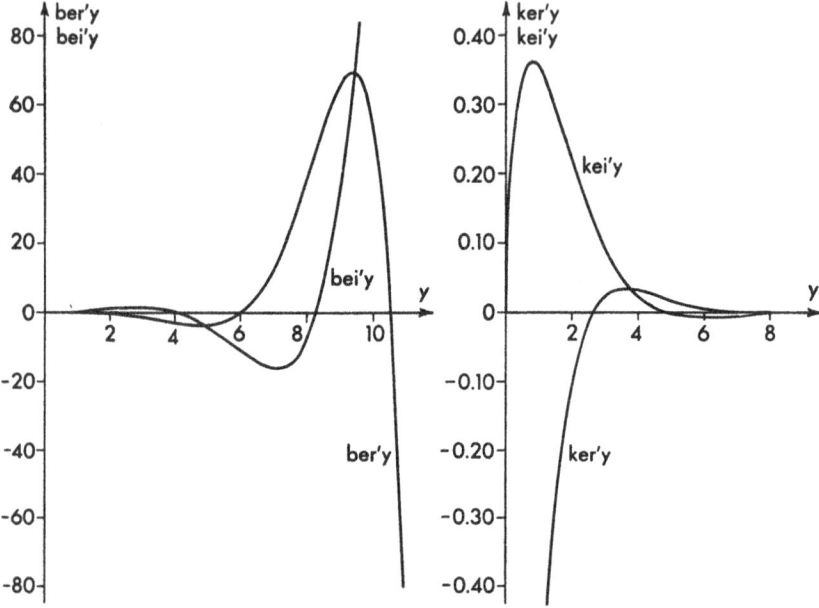

Fig. 35. Derivatives of the THOMSON functions

derivatives from the asymptotic expressions (90), and then we may write two linear equations for C_1 and C_2, expressing that the deflection

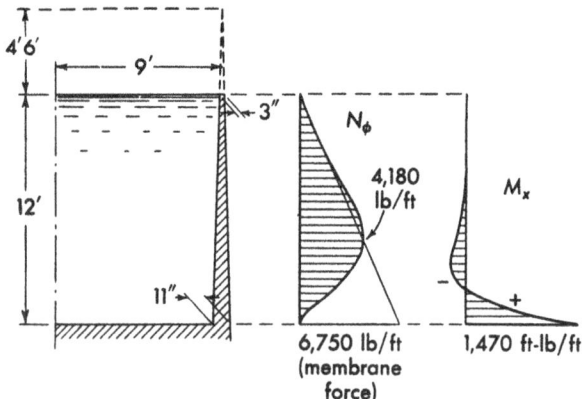

Fig. 36. Stress resultants in a cylindrical tank with variable wall thickness

and the slope following from (89) with $y = y_{max}$ are equal and opposite to those following from (83). Having determined C_1 and C_2 we easily find w, N_ϕ, M_x. The results are shown in fig. 36. The diagrams are rather similar to those shown in fig. 25, the two tanks being the same but for the wall thickness. Upon closer inspection one recognizes that in fig. 36 the clamping moment is higher, while the negative maximum of M_x in the thinner part of the wall is considerably lower than in the tank with $t = $ const. The maximum of the hoop force is 1.5% higher in the thinner wall.

At the upper edge, $x = 4.5$ ft, the inhomogeneous solution satisfies the conditions $w = 0$, $M_x = 0$, $Q_x = 0$, if we neglect the small moment (84), and the contribution of the terms with C_1 and C_2 is negligibly small. It is therefore not necessary to use terms with C_3 and C_4 if the edge is free or simply supported.

5.6 Anisotropic Shells

5.6.1 Elastic Law

5.6.1.1 Plywood Shell

So far, we have based our formulas, in particular the elastic law (9), (12), on the assumption that the shell is a wall of thickness t, made of an isotropic and homogeneous material. We shall now consider some typical cases of anisotropic shells.

The simplest among them is the plywood shell. We shall still simplify the problem by restricting our attention to the most commonly used type of plywood, the symmetric, three-layer material (fig. 37). The grain of the two outer layers runs at right angles to that of the middle layer.

Before we can write the elastic law of the shell element, we must know HOOKE's law for the individual layer. Since wood displays much more rigidity in the direction of the grain than across, this law is not symmetric with respect to x and ϕ. For the inner layer we assume it in the following form:

$$\left.\begin{aligned}
\sigma_x &= E_1 \epsilon_x + E_\nu \epsilon_\phi, \\
\sigma_\phi &= E_\nu \epsilon_x + E_2 \epsilon_\phi, \\
\tau_{x\phi} &= G \gamma_{x\phi}.
\end{aligned}\right\} \tag{91}$$

Here the strains ϵ_x, ϵ_ϕ, $\gamma_{x\phi}$ are defined as usual, and the four moduli E_1, E_2, E_ν, G are all independent of each other. In particular, there is no relation connecting the shear modulus G with the other moduli, since the well-known relation between E, ν, G holding for isotropic bodies is derived from the fact of their isotropy.

If the outer layers are made of the same kind of wood – and this we shall assume – then their elastic law is the same, except that the moduli E_1 and E_2 change places.

In fig. 37 it has been assumed that the grain of the inner layer is running in the x direction. If this is true, E_1 is the common modulus of elasticity of the wood, i. e. the one for stresses in the direction of the grain, while E_2 is the much smaller cross-grain modulus. When in a shell the grain runs circumferentially in the middle layer and lengthwise in the other two, we must identify E_2 with the common modulus and E_1 with the cross-grain modulus.

When we want to establish the relations which are to take the place of eqs. (9), we may use without change the kinematic relations (5) and the definitions (7) of the stress resultants. But when we introduce

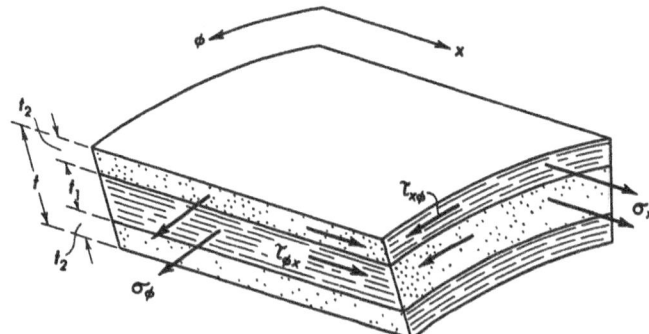

Fig. 37. Element of a plywood shell

HOOKE's law, we must use it in the form (91) for the middle layer only and exchange E_1 and E_2 while integrating over the outer layers. This leads to the definition of the following rigidities:
extensional rigidities:

$$D_x = E_1\, t_1 + 2 E_2\, t_2, \quad D_\phi = E_2\, t_1 + 2 E_1\, t_2, \quad D_\nu = E_\nu\, t; \quad \text{(92a–c)}$$

shear rigidity:

$$D_{x\phi} = G\, t; \tag{92d}$$

bending rigidities:

$$K_x = \frac{1}{12}\left[E_2(t^3 - t_1^3) + E_1\, t_1^3\right],$$

$$K_\phi = \frac{1}{12}\left[E_1(t^3 - t_1^3) + E_2\, t_1^3\right], \tag{92e–g}$$

$$K_\nu = \frac{1}{12}\, E\, t^3;$$

twisting rigidity:

$$K_{x\phi} = \frac{1}{12}\, G\, t^3. \tag{92h}$$

The elastic law of the plywood shell appears then in the following form:

$$N_\phi = \frac{D_\phi}{a}(v^\cdot + w) + \frac{D_\nu}{a}u' + \frac{K_\phi}{a^3}(w + w^{\cdot\cdot}),$$

$$N_x = \frac{D_x}{a}u' + \frac{D_\nu}{a}(v^\cdot + w) - \frac{K_x}{a^3}w'',$$

$$N_{\phi x} = \frac{D_{x\phi}}{a}(u^\cdot + v') + \frac{K_{x\phi}}{a^3}(u^\cdot + w'^\cdot),$$

$$N_{x\phi} = \frac{D_{x\phi}}{a}(u^\cdot + v') + \frac{K_{x\phi}}{a^3}(v' - w'^\cdot),$$

$$M_\phi = \frac{K_\phi}{a^2}(w + w^{\cdot\cdot}) + \frac{K_\nu}{a^2}w'',$$

$$M_x = \frac{K_x}{a^2}(w'' - u') + \frac{K_\nu}{a^2}(w^{\cdot\cdot} - v^\cdot),$$

$$M_{\phi x} = \frac{K_{x\phi}}{a^2}(2w'^\cdot + u^\cdot - v'),$$

$$M_{x\phi} = \frac{2K_{x\phi}}{a^2}(w'^\cdot - v').$$

(93 a–h)

These formulas contain as a special case eqs. (9) for the isotropic shell. We need only replace in eqs. (91) the moduli E_1 and E_2 by $E/(1 - \nu^2)$, E_ν by $E\,\nu/(1 - \nu^2)$ and G by $E/2(1 + \nu)$ and make the necessary changes in the definitions (92) of the rigidities.

Another special case is obtained when we put $t_2 = 0$, $t_1 = t$ in eqs. (92). We have then the rigidities and the elastic law of a shell which is made of one solid board of wood or of such crystalline materials which have the elastic anisotropy described by eqs. (91).

When $N_{x\phi}$, $N_{\phi x}$, and $M_{\phi x}$ from eqs. (93) are introduced in the sixth condition of equilibrium, eq. (1f), it is identically satisfied. We may eliminate Q_ϕ and Q_x from the other five equilibrium conditions as we did before, and thus arrive again at eqs. (2). The introduction of the new elastic law into these equations is postponed until p. 307.

5.6.1.2 Double-walled Shell

Occasionally it is desired to build a shell of exceptionally high bending stiffness. This may be necessary to increase its buckling strength or to make it capable of carrying very concentrated local loads. A substantial increase of stiffness without an uneconomic increase of dead weight can be achieved by making the wall hollow, i. e. by composing it of two concentric cylindrical slabs and a connecting gridwork of thin ribs. Sections $\phi = $ const and $x = $ const of such a shell are shown in fig. 38. The rib system consists of a set of circumferential ribs, the rings, and a set of longitudinal ribs, the stringers.

When the ribs are few and far between, we have to deal with a structure composed of shell panels and of ribs, and we have to analyze it as such. But when the ribs are closely and evenly spaced, it is worthwhile to consider the limiting case of very closely spaced and correspondingly weak ribs. In this case we have to deal with an anisotropic shell.

Before we can go into any details of stress and strain, we must define a middle surface. Contrary to the smooth shells considered thus far, there is no cylinder which halves the thickness and which would be equally acceptable for both sections, fig. 38a and b. Now,

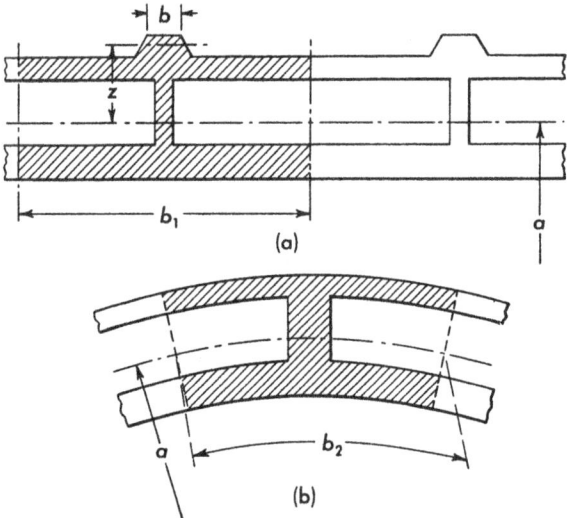

Fig. 38. Sections $\phi = $ const and $x = $ const through a double-walled shell

when we look back, we may see that also for the smooth shell our choice of the middle surface as the one which halves the thickness t, was lastly arbitrary. The faces $x = $ const of the shell elements were trapezoidal, and their centroids did not lie on the middle surface. It was exactly this fact which gave rise to some of the queer terms in eqs. (9), as discussed on p. 216. Therefrom we conclude that the word "middle" in the term "middle surface" must not be taken all too literally and that any reference surface is welcome which lies somewhere in the middle of the thickness. In some cases it is convenient to choose it so that one of the moments S_ϕ, S_x, or S defined by eqs. (96d–f) becomes zero, but there is no need to satisfy this requirement.

We consider now a section $\phi = $ const of the shell (fig. 38a). It consists of a periodic repetition of the shaded part, whose length is equal to the distance b_1 of the rings. The width of the rib may be a function of z. We denote it by b. For those values of z which belong to the slab areas, the width of the rib is not defined and will not be needed.

Since there are normal stresses in both the slabs and the ribs, N_ϕ and M_ϕ are integrals of these stresses over the cross sections of the slabs and the rib. We shall distinguish these two parts of the integrals by attaching the letter s or r to the lower end of the integral sign. We have then

$$N_\phi = \int_s \sigma_\phi \, dz + \int_r \sigma_\phi \left(\frac{b}{b_1}\right) dz,$$

$$M_\phi = \int_s \sigma_\phi z \, dz + \int_r \sigma_\phi \left(\frac{b}{b_1}\right) z \, dz.$$

(94a, b)

We now turn our attention to the section $x = \text{const}$, fig. 38b. Let b_2 be the distance of adjacent stringers, measured on the middle surface. The width of the rib may again be denoted by b, although this is, of course, a different function of z from the one just used in eqs. (94a, b). It is also understood that the subscript r at the integral signs now denotes the stringer section. We have then

$$N_x = \int_s \sigma_x \left(1 + \frac{z}{a}\right) dz + \int_r \sigma_x \frac{b}{b_2} \, dz,$$

$$M_x = \int_s \sigma_x \left(1 + \frac{z}{a}\right) z \, dz + \int_r \sigma_x \frac{b}{b_2} z \, dz.$$

(94c, d)

The factor $(1 + z/a)$ in the slab integrals expresses the slightly trapezoidal shape of the slab section.

There are, of course, no shearing stresses on the lateral surfaces of the ribs. Consequently, there is also no shearing stress $\tau_{x\phi}$ or $\tau_{\phi x}$ in the cross section of the ribs, and all the tangential shear is carried by the slabs alone. The shearing forces and twisting moments are therefore

$$N_{\phi x} = \int_s \tau_{\phi x} \, dz, \qquad N_{x\phi} = \int_s \tau_{\phi x} \left(1 + \frac{z}{a}\right) dz,$$

$$M_{\phi x} = \int_s \tau_{\phi x} z \, dz, \qquad M_{x\phi} = \int_s \tau_{\phi x} \left(1 + \frac{z}{a}\right) z \, dz.$$

(94e–h)

In order to obtain the elastic law of the shell, we must by means of HOOKE's law express the stresses in terms of the strains and then use the kinematic relations (5) to express the strains by u, v, w. For the slabs HOOKE's law is represented by eqs. (6), but in the ribs we have simply

$$\sigma_x = E \, \epsilon_x, \qquad \sigma_\phi = E \, \epsilon_\phi.$$

This difference in the elastic laws for the one-dimensional and the two-dimensional parts of the shell has the remarkable consequence that a normal stress, e. g. σ_x, along two adjacent fibers of the stringer and the slab is not the same, although ϵ_x is, because one stress is influenced by ϵ_ϕ while the other is not.

When we go in detail through the procedure described, we find the following relations:

$$N_\phi = \frac{D_\phi}{a}(v^{\cdot}+w) - \frac{S_\phi}{a^2}(w+w^{\cdot\cdot}) + \frac{K_\phi}{a^3}(w+w^{\cdot\cdot}) + \frac{\nu D}{a}u' - \frac{\nu S}{a^2}w'',$$

$$N_x = \frac{D_x}{a}u' - \frac{S_x}{a^2}w'' + \frac{\nu D}{a}(v^{\cdot}+w) + \frac{\nu S}{a^2}(v^{\cdot}-w^{\cdot\cdot}),$$

$$N_{\phi x} = \frac{(1-\nu)D}{2a}(u^{\cdot}+v') - \frac{(1-\nu)S}{2a^2}(u^{\cdot}-v'+2w'^{\cdot})$$
$$+ \frac{(1-\nu)K}{2a^3}(u^{\cdot}+w'^{\cdot}),$$

$$N_{x\phi} = \frac{(1-\nu)D}{2a}(u^{\cdot}+v') + \frac{(1-\nu)S}{a^2}(v'-w'^{\cdot}) + \frac{(1-\nu)K}{2a^3}(v'-w'^{\cdot}),$$

$$M_\phi = -\frac{S_\phi}{a}(v^{\cdot}+w) + \frac{K_\phi}{a^2}(w+w^{\cdot\cdot}) - \frac{\nu S}{a}u' + \frac{\nu K}{a^2}w'', \qquad (95\,\text{a–h})$$

$$M_x = -\frac{S_x}{a}u'^{\cdot} + \frac{K_x}{a^2}w'' - \frac{\nu S}{a}(v^{\cdot}+w) - \frac{\nu K}{a^2}(v^{\cdot}-w^{\cdot\cdot}),$$

$$M_{\phi x} = -\frac{(1-\nu)S}{2a}(u^{\cdot}+v') + \frac{(1-\nu)K}{2a^2}(u^{\cdot}-v'+2w'^{\cdot}),$$

$$M_{x\phi} = -\frac{(1-\nu)S}{2a}(u^{\cdot}+v') - \frac{(1-\nu)K}{a^2}(v'-w'^{\cdot}).$$

These formulas, which constitute the elastic law of the double-walled shell, contain a large number of rigidities. They are defined as follows:
extensional rigidities:

$$D_\phi = \frac{E}{1-\nu^2}\int_s dz + E\int_r \frac{b}{b_1}\,dz,$$

$$D_x = \frac{E}{1-\nu^2}\int_s\left(1+\frac{z}{a}\right)dz + E\int_r \frac{b}{b_2}\,dz, \qquad D = \frac{E}{1-\nu^2}\int_s dz; \qquad (96\,\text{a–c})$$

rigiditiy moments:

$$S_\phi = \frac{E}{1-\nu^2}\int_s z\,dz + E\int_r \frac{b}{b_1}z\,dz,$$

$$S_x = \frac{E}{1-\nu^2}\int_s\left(1+\frac{z}{a}\right)z\,dz + E\int_r \frac{b}{b_2}z\,dz, \qquad S = \frac{E}{1-\nu^2}\int_s z\,dz; \qquad (96\,\text{d–f})$$

bending rigidities:

$$K_\phi = \frac{E}{1-\nu^2}\int_s z^2\,dz + E\int_r \frac{b}{b_1}z^2\,dz,$$

$$K_x = \frac{E}{1-\nu^2}\int_s z^2\,dz + E\int_r \frac{b}{b_2}z^2\,dz, \qquad K = \frac{E}{1-\nu^2}\int_s z^2\,dz. \qquad (96\,\text{g–i})$$

When we compare eqs. (95) with the elastic law for the plywood shell, eqs. (93), we find a strong similarity. We may formally obtain eqs. (93) from eqs. (95) by dropping the terms with S_ϕ and with S and making the following substitutions which may easily be understood from a comparison of the corresponding definitions:

$$S_x \to K_x, \quad \nu D \to D_\nu, \quad \nu K \to K_\nu,$$

$$\frac{1}{2}(1-\nu)D \to D_{x\phi}, \quad \frac{1}{2}(1-\nu)K \to K_{x\phi}.$$

5.6.1.3 Gridwork Shell

Before we study the most common type of an anisotropic shell, the single slab reinforced by ribs, we shall have a look at a shell which consists of ribs only. There must be ribs in two directions, which we again call rings and stringers. We assume that these ribs are connected rigidly at their intersections. If their centroidal axes happen to lie on the same cylinder, we choose it as the middle surface of the shell. In general, however, the axes of the rings do not intersect those of the stringers, and then again the choice of the reference cylinder is arbitrary. It may or may not be convenient to choose it so that it passes through the axes of at least one sort of ribs. For the purpose of our

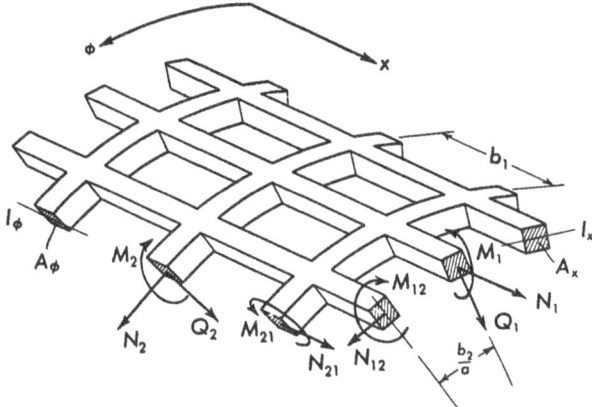

Fig. 39. Element of a gridwork shell

analysis we choose it so that the axes of all the ribs are on its outside, those of the rings at a distance c_ϕ and those of the stringers at a distance c_x.

Fig. 39 shows an element of the shell. At the centroids of the ring sections (which we suppose to coincide with their shear centers) we apply axial forces N_2, bending moments M_2, and transverse shearing forces Q_2. The stress resultants of the shell are the forces per unit length of a section $\phi = $ const, viz.

$$N_\phi = \frac{N_2}{b_1}, \qquad Q_\phi = \frac{Q_2}{b_1}. \tag{97a, b}$$

The bending moments of the shell must be referred to a generator of
the middle surface as its axis:

$$M_\phi = \frac{M_2 - N_2 c_\phi}{b_1}. \tag{97c}$$

Similarly we define for a section $x = $ const the stress resultants

$$N_x = \frac{N_1}{b_2}, \qquad Q_x = \frac{Q_1}{b_2}, \qquad M_x = \frac{M_1 - N_1 c_x}{b_2}. \tag{97d–f}$$

A particular problem arises when it comes to shearing forces and
twisting moments. We may apply torques M_{12} of any magnitude at
the ends of the stringers without an imme-
diate obligation of applying anything simultaneously to the rings. Speaking in terms
of a shell element, these torques correspond to a twisting moment

$$M_{x\phi} = \frac{M_{12}}{b_2}. \tag{97g}$$

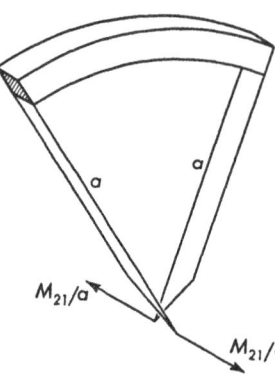

Fig. 40. Torsion of a ring element

When we try to apply to similar torques
M_{21} to the rings, we see that these
moments at opposite sides of the shell
element have slightly divergent axes and,
therefore, are not in equilibrium with each
other. There is only one way to apply a
self-equilibrating system of torques: We
attach to each end of each ring a rigid lever ending on the axis of
the cylinder and there we apply forces M_{21}/a as shown in fig. 40 for
one ring. This is equivalent to applying simultaneously at the center
of each cross section a twisting moment M_{21} and a shearing force M_{21}/a.
Under the combined action of these forces and moments the ring element has the same stresses as an element of a helical spring of zero
pitch, i. e., shear and torsion but no bending.

Considering fig. 39 again as a shell element, we find the torques M_{21}
equivalent to a twisting moment

$$M_{\phi x} = \frac{M_{21}}{b_1}, \tag{97h}$$

and the simultaneous forces M_{21}/a make a contribution $M_{21}/a\, b_1$ to
the shear $N_{\phi x}$.

In addition to this shear M_{21}/a we may still apply forces N_{12} and N_{21} as shown in fig. 39. For the shell element they yield a shearing force

$$N_{x\phi} = \frac{N_{12}}{b_2} \qquad (97\,\text{i})$$

and a contribution N_{21}/b_1 to $N_{\phi x}$. The moment equilibrium for a radial axis requires that these forces per unit length be equal, hence

$$\frac{N_{12}}{b_2} = \frac{N_{21}}{b_1}.$$

The total shearing force in the section $\phi = \text{const}$ is then

$$N_{\phi x} = \frac{N_{21}}{b_1} + \frac{M_{21}}{a\,b_1} = N_{x\phi} + \frac{M_{\phi x}}{a}. \qquad (97\,\text{j})$$

This equation is identical with eq. (1f), the sixth condition of equilibrium, which hence is automatically satisfied.

After having studied the forces and moments acting on the grid element, we may attempt to write the elastic law. When doing this, we shall make use of the fact that the rings are thin in order to prevent the appearance of too many terms of minor importance.

Let the cross section of a ring be A_ϕ (fig. 39). The strain at its centroid may be found from eq. (5b) with $z = c_\phi$, neglecting z compared with a in the denominator. When we multiply this strain by $E\,A_\phi$, we have the axial force N_2 and hence

$$N_\phi = \frac{E\,A_\phi}{b_1} \left(\frac{v^{\cdot}}{a} - \frac{c_\phi\,w^{\cdot\cdot}}{a^2} + \frac{w}{a} \right). \qquad (98\,\text{a})$$

In the same way we find the other normal force, using eq. (5a) with $z = c_x$:

$$N_x = \frac{E\,A_x}{b_2} \left(\frac{u'}{a} - \frac{c_x\,w''}{a^2} \right). \qquad (98\,\text{b})$$

The moment M_2 depends on the elastic change of curvature of the ring which, for a thin ring, is $w^{\cdot\cdot}/a^2$. When I_ϕ is the moment of inertia for the centroidal axis of the section, we have

$$M_\phi = \frac{E\,I_\phi}{b_1} \frac{w^{\cdot\cdot}}{a^2} - \frac{E\,A_\phi c_\phi}{b_1} \left(\frac{v^{\cdot}}{a} + \frac{w}{a} \right) + \frac{E\,A_\phi c_\phi^2}{b_1} \frac{w^{\cdot\cdot}}{a^2}. \qquad (98\,\text{c})$$

Similarly we find

$$M_x = \frac{E\,I_x}{b_2} \frac{w''}{a^2} - \frac{E\,A_x c_x}{b_2} \frac{u'}{a} + \frac{E\,A_x c_x^2}{b_2} \frac{w''}{a^2}. \qquad (98\,\text{d})$$

The forces N_{12} and N_{21} produce bending deformations in the ribs, as shown in fig. 41a. Each bar between two joints becomes S-shaped, and the straight lines connecting adjacent joints are no longer at right angles to each other. For the analysis of this deformation we isolate the elementary period of the gridwork, as shown on a larger scale in fig. 41 b. It ends at the center points of four grid bars, and one easily recognizes that these are inflection points. The right angle between the tangents at the joint is conserved, and since a rigid-body rotation

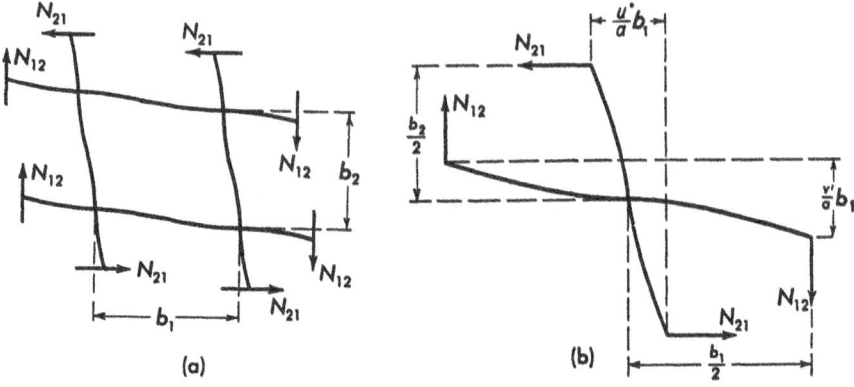

Fig. 41. Bending of the ribs in the tangential plane

does not matter for the deformation we seek, we assume that the joint does not rotate at all. Using twice the well-known formula for the end deflection of a cantilever beam, we find that

$$\frac{u^{\cdot}}{a}\, b_2 = 2\,\frac{N_{21}\, b_2^3}{24\, E\, I_r}, \qquad \frac{v'}{a}\, b_1 = 2\,\frac{N_{12}\, b_1^3}{24\, E\, I_s}.$$

In these equations I_r and I_s are the relevant moments of inertia of rings and stringers, respectively. Using eq. (97i) and the unnumbered relation following it, we find the shear deformation

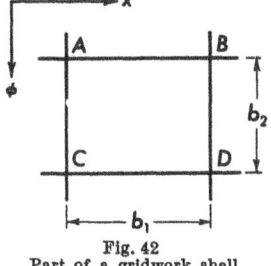

Fig. 42
Part of a gridwork shell

$$\frac{u^{\cdot} + v'}{a} = N_{x\phi} \cdot \frac{1}{12 a}\left(\frac{a\, b_1\, b_2^2}{E\, I_r} + \frac{a\, b_1^2\, b_2}{E\, I_s}\right). \quad (98e)$$

This result is independent of the initial assumption that the joint does not rotate. Indeed, when the whole configuration of fig. 41 b is rotated in its plane, $u^{\cdot}$ gains as much as v' loses or vice versa.

The last deformation we have to consider, is a twisting of the shell element. We consider the element shown in fig. 42. The bars AB and CD are parts of stringers, AC and BD parts of rings.

When the torque M_{12} is applied to the bar AB, this bar is twisted, and the torsion theory says that the twist θ_1 is

$$\theta_1 = \frac{M_{12}}{G\, J_x}\,,$$

where J_x depends on the shape of the stringer section. The bars AC and BD, which originally were parallel, are then rotated with respect to each other by the angle

$$\theta_1 b_1 = \frac{M_{12} b_1}{G J_x} = M_{x\phi} \frac{b_1 b_2}{G J_x},$$

which is also equal to $b_1 w''/a^2$; hence

$$M_{x\phi} = \frac{G J_x}{a^2 b_2} w''. \tag{98f}$$

When this deformation occurs, the bars AB and CD also rotate with respect to each other, corresponding to a twisting of the stringers. As we have seen, this requires the combined action of a torque M_{21} and a shear M_{21}/a. In the curved bars they produce the twist

$$\theta_2 = \frac{M_{21}}{G J_\phi},$$

where J_ϕ is the torsional stiffness factor of the rings. The relative rotation of the bars AB and CD is then

$$\theta_2 b_2 = \frac{M_{21} b_2}{G J_\phi} = M_{\phi x} \cdot \frac{b_1 b_2}{G J_\phi} = \frac{w'' b_2}{a^2},$$

whence

$$M_{\phi x} = \frac{G J_\phi}{a^2 b_1} w''. \tag{98g}$$

When we eliminate w'' from eqs. (98f, g) we find the relation

$$M_{\phi x} = M_{x\phi} \cdot \frac{J_\phi b_2}{J_x b_1}, \tag{99}$$

which the two twisting moments must satisfy. We see from it that the two twisting moments may be equal or very different, depending on the dimensions and the spacing of the ribs.

Last of all, we might still write a relation for the shear $N_{\phi x}$, using eqs. (97j) and (98e, g).

We may bring all these equations into a form similar to eqs. (93) and (95), when we introduce the following rigidities

$$\left.\begin{aligned}
D_\phi &= \frac{E A_\phi}{b_1}, \quad D_x = \frac{E A_x}{b_2}, \\[4pt]
D_{x\phi} &= \frac{12}{b_1 b_2} \left(\frac{b_2}{E I_r} + \frac{b_1}{E I_s} \right)^{-1}, \\[4pt]
S_\phi &= \frac{E A_\phi c_\phi}{b_1}, \quad S_x = \frac{E A_x c_x}{b_2}, \\[4pt]
K_\phi &= \frac{E(I_\phi + A_\phi c_\phi^2)}{b_1}, \quad K_x = \frac{E(I_x + A_x c_x^2)}{b_2}, \\[4pt]
K_{\phi x} &= \frac{G J_\phi}{b_1}, \quad K_{x\phi} = \frac{G J_x}{b_2}.
\end{aligned}\right\} \tag{100}$$

There are no moments $S_{x\phi}$ and $S_{\phi x}$ since we neglected the influence of the eccentricities c_ϕ and c_x in our equations for twisting and shear where it is of minor importance.

With the help of the rigidities just defined, eqs. (98) may be rewritten in the following, more convenient form:

$$N_\phi = \frac{D_\phi}{a}(v^\cdot + w) - \frac{S_\phi}{a^2} w^{\cdot\cdot}, \qquad N_x = \frac{D_x}{a} u' - \frac{S_x}{a^2} w'',$$

$$N_{\phi x} = \frac{D_{x\phi}}{a}(u^\cdot + v') + \frac{K_{\phi x}}{a^3} w'^\cdot, \quad N_{x\phi} = \frac{D_{x\phi}}{a}(u^\cdot + v'),$$

$$\hspace{9cm} \text{(101 a–h)}$$

$$M_\phi = \frac{K_\phi}{a^2} w^{\cdot\cdot} - \frac{S_\phi}{a}(v^\cdot + w), \qquad M_x = \frac{K_x}{a^2} w'' - \frac{S_x}{a} u',$$

$$M_{\phi x} = \frac{K_{\phi x}}{a^2} w'^\cdot, \qquad\qquad M_{x\phi} = \frac{K_{x\phi}}{a^2} w'^\cdot.$$

These equations constitute the elastic law of the gridwork shell. The differences from the equations (95) for the double-walled shell may be traced back to two causes. The first one is the absence of all terms containing POISSON's ratio. It is clear that the lateral contraction of a rib does not affect the deformation of the grid element. This is in agreement with the fact, that the ν terms in eqs. (95) all have a factor D, S, or K, i. e. one of those rigidities whose definitions do not contain a rib integral.

The second cause of differences between eqs. (95) and (101) is the fact that for the gridwork shell we did not strictly adhere to the assumption of plane cross sections. This decision appears to be reasonable, since we neglected also the influence of possible local deformations of the bars between the joints as well as the influence of the warping constraint on the torsion terms. Since this latter would increase the order of the differential equations, it should not be introduced without evidence of a need to do so.

5.6.1.4 Shell with Rings and Stringers

The last type of anisotropic shell which we discuss here is the most important one: the shell of uniform thickness reinforced by closely spaced rings or stringers or both (fig. 43). We may handle this case in two ways: either we superpose the stress resultants of an isotropic shell and those of a gridwork, or we use eqs. (95) and (96) with the understanding that the slab integrals are now to be extended over one slab only. This second way is to be recommended for concrete shells and similar structures, for which it well represents the facts. All that is to be said about it has already been said in Section 5.6.1.2.

For the thin shells of airplane fuselages eqs. (95) have a serious drawback which excludes their use. In a double-walled shell the twisting moments are carried by shearing stresses $\tau_{x\phi}$ or $\tau_{\phi x}$ having opposite directions in the two slabs. The contribution of the ribs is practically nil and has been neglected in eqs. (95g, h). It is quite different when the shell consists of only one very thin wall and a set of sturdy stiffeners, particularly when these have tubular cross sections. Then the twisting rigidity of the wall is next to nothing, and almost all the twisting stiffness of the shell comes from the torsional rigidity

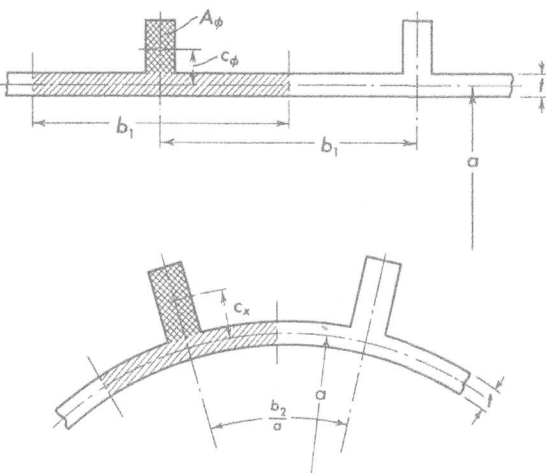

Fig. 43. Sections $\phi = $ const and $x = $ const through a shell with rings and stringers

of the ribs. Therefore, we must introduce this torsional rigidity GJ as we did in Section 5.6.1.3 and must superpose the grid formulas (101) and the elastic law of the wall.

Since in eqs. (101) quantities of the order z/a have been neglected compared with unity, it would be useless to combine them with eqs. (9), but rather we use the simplified relations (12). When we choose the middle surface of the wall as the middle surface of the entire shell, we arrive at the following elastic law:

$$
\left.
\begin{aligned}
N_\phi &= \frac{D_\phi}{a}\,(v^{\cdot} + w) + \frac{D_\nu}{a}\,u' - \frac{S_\phi}{a^2}\,w^{\cdot\cdot}, \\[4pt]
N_x &= \frac{D_x}{a}\,u' + \frac{D_\nu}{a}\,(v^{\cdot} + w) - \frac{S_x}{a^2}\,w'', \\[4pt]
N_{\phi x}^{\bar{}} &= \frac{D_{x\phi}}{a}\,(u^{\cdot} + v') + \frac{K_{\phi x}}{a^3}\,w'^{\cdot}, \\[4pt]
N_{x\phi} &= \frac{D_{x\phi}}{a}\,(u^{\cdot} + v'), \\[4pt]
M_\phi &= \frac{K_\phi}{a^2}\,w^{\cdot\cdot} + \frac{K_\nu}{a^2}\,w'' - \frac{S_\phi}{a}\,(v^{\cdot} + w), \\[4pt]
M_x &= \frac{K_x}{a^2}\,w'' + \frac{K_\nu}{a^2}\,w^{\cdot\cdot} - \frac{S_x}{a}\,u', \\[4pt]
M_{\phi x} &= \frac{K_{\phi x}}{a^2}\,w'^{\cdot}, \qquad M_{x\phi} = \frac{K_{x\phi}}{a^2}\,w'^{\cdot}.
\end{aligned}
\right\}
\tag{102}
$$

When we eliminate w'' from the last two equations, we obtain the relation which the elasticity of the shell imposes on $M_{\phi x}$ and $M_{x\phi}$.

The rigidity constants in eqs. (102) are the sums of the corresponding constants occurring in eqs. (12) and (101). There is, however, one exception to this simple superposition of the rigidities of the wall and the ribs. In the gridwork shell shearing forces $N_{x\phi}$ and $N_{\phi x}$ are transmitted through bending stresses in the ribs as shown in fig 41 a, and the shear stiffness $D_{x\phi}$ from eq. (100) is usually rather small. When the ribs are connected with a coherent wall, this bending deformation is not possible, and all the shear is carried by the wall alone. We must therefore omit the term with $D_{x\phi}$ in eqs. (101 c, d) when we superpose the grid and the wall.

When we keep this in mind, we find the following expressions for the rigidity parameters in eqs. (102):

$$
\begin{aligned}
D_\phi &= \frac{E\,t}{1-\nu^2} + \frac{E\,A_\phi}{b_1}, & D_x &= \frac{E\,t}{1-\nu^2} + \frac{E\,A_x}{b_2}, \\[2mm]
D_\nu &= \frac{E\,t\,\nu}{1-\nu^2}, & D_{x\phi} &= \frac{E\,t}{2(1+\nu)}, \\[2mm]
S_\phi &= \frac{E\,A_\phi\,c_\phi}{b_1}, & S_x &= \frac{E\,A_x\,c_x}{b_2}, \\[2mm]
K_\phi &= \frac{E\,t^3}{12(1-\nu^2)} + \frac{E\,(I_\phi + A_\phi\,c_\phi^2)}{b_1}, \\[2mm]
K_x &= \frac{E\,t^3}{12(1-\nu^2)} + \frac{E\,(I_x + A_x\,c_x^2)}{b_2}, \\[2mm]
K_\nu &= \frac{E\,t^3\,\nu}{12(1-\nu^2)}, \\[2mm]
K_{\phi x} &= \frac{E\,t^3}{12(1+\nu)} + \frac{G\,J_\phi}{b_1}, & K_{x\phi} &= \frac{E\,t^3}{12(1+\nu)} + \frac{G\,J_x}{b_2}.
\end{aligned}
\tag{103}
$$

Here the notations of fig. 43 and of the preceding section have been used.

The rigidity moments S_ϕ, S_x are positive when the ribs lie on the outside of the wall and are negative in the opposite case. Their presence should not be overlooked when dealing with actual shell problems. In the rare case that $c_\phi = c_x$, they may be eliminated from the formulas by choosing another middle surface.

A gridwork shell must always have both kinds of ribs: rings and stringers. Rings alone or stringers alone would not make a coherent structure. Quite differently, the reinforced shell may have only rings or only stringers, and in fact these are the shells which occur most frequently. They have the same elastic law (102), if only we drop from eqs. (103) the terms with A_x, I_x, J_x or those with A_ϕ, I_ϕ, J_ϕ. The two twisting rigidities, $K_{x\phi}$ and $K_{\phi x}$, are then of very different magnitude, and so are the twisting moments $M_{x\phi}$ and $M_{\phi x}$.

5.6.2 Differential Equations for the Shell with Ribs

We obtain three linear differential equations similar to eqs. (13) or (15), when we introduce any one of the preceding elastic laws into eqs. (2a–c). It is enough to show the result for eqs. (102):

$$
\begin{aligned}
a^2 D_x u'' + a^2 D_{x\phi} u^{\cdot\cdot} + a^2 (D_\nu + D_{x\phi}) v'^{\cdot} + a^2 D_\nu w' - a S_x w''' & \\
+ K_{\phi x} w''^{\cdot} &= -p_x a^4, \\[4pt]
a^2 (D_\nu + D_{x\phi}) u'^{\cdot} + a(a D_\phi + S_\phi) v^{\cdot\cdot} + a^2 D_{x\phi} v'' - (K_\phi + a S_\phi) w^{\cdot\cdot} & \\
- (K_\nu + K_{x\phi}) w''^{\cdot} + a(a D_\phi + S_\phi) w^{\cdot} &= -p_\phi a^4, \\[4pt]
-a S_x u''' + a^2 D_\nu u' - a S_\phi v^{\cdot\cdot} + a^2 D_\phi v^{\cdot} + K_x w^{\mathrm{IV}} & \\
+ (2K_\nu + K_{\phi x} + K_{x\phi}) w''^{\cdot\cdot} + K_\phi w^{\cdot\cdot} - 2a S_\phi w^{\cdot\cdot} + a^2 D_\phi w & \\
&= p_r a^4.
\end{aligned}
\tag{104}
$$

Comparing these differential equations with eqs. (13) for the isotropic cylinder, we see that the theory of the anisotropic shell is not more involved than that of the isotropic shell. There are only a few new terms, and many of the old ones are missing; even compared with the simplified equations (15) the difference is not great. The only additional complication lies in the fact that the coefficients are no longer as simple as before but depend on many rigidities D, S, K, which are independent of each other. However, in most practical applications some of them will be zero or negligibly small.

5.7 Folded Structures

On p. 164 we have seen how imperfect an instrument the membrane theory of folded structures is. Therefore, a bending theory is absolutely necessary for a realistic stress analysis, and an outline of this theory will be given on the following pages.

It is natural to start from the membrane solution, to find its imperfections, and to amend them. This procedure is identical with the usual analysis of statically indeterminate structures. The principal system is characterized by the fact that the rigid connections of the plate strips at the edges have been replaced by piano hinges, which can transmit the shear T_m, but which cannot transmit plate bending moments M_y from one strip to the next. Bending moments of this kind will later be chosen as the redundant quantities. Because of the edge shears the principal system itself is statically indeterminate. We shall use here the superscript (o) to indicate the load action in this piano hinge system and not in the principal system used for the membrane theory.

We now study the deformation of the hinged system. Fig. 44 shows part of a cross section $x = \text{const}$ of the prism with the plate strips $m - 1, m, m + 1$. Each of these strips carries in its own plane a load S_m and acts as a beam of large depth. Consequently, there are deflections v_{m-1}, v_m etc. as shown in the figure, and these deflections are subject to the differential equation of beam bending

$$\frac{d^2 v_m}{dx^2} = \frac{12}{E \, t_m \, h_m^3} \, M_m.$$

(105)

The beam moment M_m in this equation is the sum of the moments which were called $M_m^{(0)}$ and $M_m^{(1)}$ in eqs. (III–45) and (III–46).

We shall here use the FOURIER series form given in Section 3.5.2 and consequently write

$$v_m = v_{m,n} \sin\frac{n \pi x}{l}.$$

We have then from eqs. (105) and (III–53), (III–54):

$$v_{m,n} = \frac{6}{E \, t_m \, h_m^2} \, \frac{l^3}{n^3 \, \pi^3} \left(- \frac{2l}{h_m \, n \, \pi} S_{m,n} + T_{m-1,n} + T_{m,n} \right).$$

(106)

The corner m in fig. 44 is at the same time a point of two strips. With the strip m it has to undergo the displacement $v_m = v_{m,n} \sin n \pi x/l$ as shown, while the strip $m + 1$ requires that there be a displace-

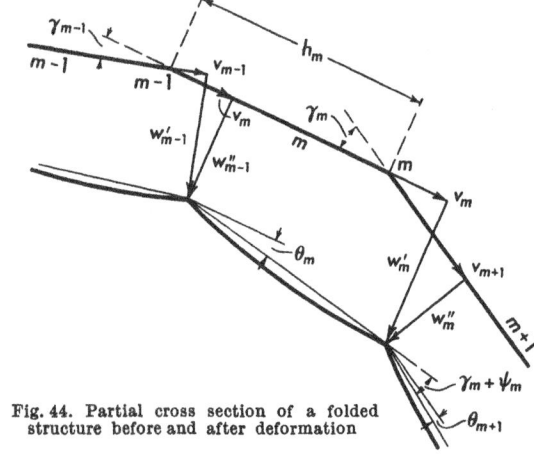

Fig. 44. Partial cross section of a folded structure before and after deformation

ment v_{m+1}. To reconcile these two requirements, we remember that the plate strips are thin and long. They do not offer any substantial resistance to lengthwise bending and twisting, and therefore we are allowed to add arbitrary displacements normal to each plate strip, as long as we keep its cross section straight. Making use of this possibility, we add to v_m the normal displacement w_m' at the corner m (but another displacement w_{m-1}'' at the corner $m - 1$), and we combine v_{m+1} with a normal displacement w_m''. By simple trigonometry we find the following relations, which we may write for the local quantities $(v_m, \ldots)$ or for their amplitudes $(v_{m,n}, \ldots)$:

$$w'_m \sin\gamma_m = v_{m+1} - v_m \cos\gamma_m,$$
$$w''_m \sin\gamma_m = v_{m+1} \cos\gamma_m - v_m. \tag{107}$$

This additional deformation is, of course, accompanied by bending moments M_x and twisting moments M_{xy}. Neglecting them, as we do, is equivalent to neglecting the moments M_x and $M_{x\phi}$ in the barrel-vault theory (see p. 247). However, we are not allowed to neglect bending moments M_y, and to find their magnitude is the essential objective of the theory we are about to develop.

Because of the displacements w''_{m-1} and w'_m of its ends, the straight line $m-1, m$ rotates clockwise by the angle θ_m with the amplitude

$$\theta_{m,n} = \frac{w'_{m,n} - w''_{m-1,n}}{h_m}, \tag{108}$$

and the line $m, m+1$ rotates by a similar angle θ_{m+1}. The increase of the angle γ_m is their difference and has the amplitude

$$\psi_{m,n} = \theta_{m+1,n} - \theta_{m,n} = \frac{w'_{m+1,n} - w''_{m,n}}{h_{m+1}} - \frac{w'_{m,n} - w''_{m-1,n}}{h_m}. \tag{109}$$

We may use eqs. (107) to express w', w'' in terms of v and then eq. (106) to express v in terms of the loads S and the edge shears T that go with them in the hinged structure. It may be left to the reader to work out this somewhat lengthy formula.

The preceding equations apply to any vertical edge load brought upon the hinged structure. We may apply them in particular to the actual load (more exactly to the n-th term of its FOURIER expansion). The forces $S_{m,n}$ in eq. (106) are then those computed from eqs. (III–52), and the edge shears $T_{m,n}$ are those obtained from solving a set of equations (III–54) with those $S_{m,n}$ on the right-hand sides. We shall designate the ensuing deformation in eqs. (106), (108), (109) by the superscript (o), i. e. $v_{m,n}^{(o)}$, $\theta_{m,n}^{(o)}$, $\psi_{m,n}^{(o)}$. They are the deformations of the principal system under the given load.

In the actual structure the strips are not connected by piano hinges but are so fixed that a relative rotation ψ_m cannot take place. It is prevented by bending moments, which deform the straight cross sections shown in fig. 44 into gentle curves whose tangents meet at the same angles γ_m as do the strips in the unstressed structure.

The moment M_y transmitted across the edge $m = r$ from the strip r to the strip $r+1$ is denoted by M_r. It depends on x as

$$M_r = M_{r,n} \sin\frac{n\pi x}{l}.$$

We now have to study the internal force system set up by applying this moment with $M_{r,n} = 1$ as an external unit load to the hinged system. We replace the moment as shown in fig. 45a by the two loads of figs. 45b, c. The forces and moments shown in fig. 45b are in local equilibrium, if we give them a sinusoidal distribution with the amplitudes

$$P_{r-1,n} = \frac{1}{h_r \cos\phi_r}, \qquad P_{r+1,n} = \frac{1}{h_{r+1}\cos\phi_{r+1}}.$$

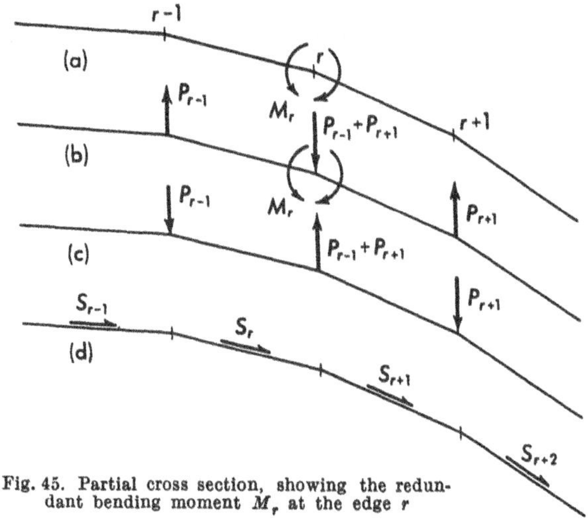

Fig. 45. Partial cross section, showing the redundant bending moment M_r at the edge r

The forces of fig. 45c are applied as loads to the entire structure. They are equivalent to sinusoidal tangential loads in the planes of four strips, having the amplitudes

$$
\begin{aligned}
S^{(r)}_{r-1,n} &= -\frac{1}{h_r \sin\gamma_{r-1}}, \\[2mm]
S^{(r)}_{r,n} &= \frac{1}{h_{r+1}\sin\gamma_r} + \frac{1}{h_r\cos\phi_r}\left(\frac{\cos\phi_{r-1}}{\sin\gamma_{r-1}} + \frac{\cos\phi_{r+1}}{\sin\gamma_r}\right), \\[2mm]
S^{(r)}_{r+1,n} &= -\frac{1}{h_r\sin\gamma_r} - \frac{1}{h_{r+1}\cos\phi_{r+1}}\left(\frac{\cos\phi_r}{\sin\gamma_r} + \frac{\cos\phi_{r+2}}{\sin\gamma_{r+1}}\right), \\[2mm]
S^{(r)}_{r+2,n} &= \frac{1}{h_{r+1}\sin\gamma_{r+1}}.
\end{aligned}
\qquad (110)
$$

These quantities must be introduced on the right-hand sides of eqs. (III–55), which then will yield the set of edge shears $T_{m,n} = T^{(r)}_{m,n}$ that goes with the unit load $M_{r,n} = 1$.

The loads $S^{(r)}$ and the edge shears $T^{(r)}$ must now be introduced into eqs. (106) to find the displacements $v^{(r)}_{m,n}$, and the subsequent use of eqs. (107) and (109) yields the angles $\psi^{(r)}_{m,n}$ by which the strips rotate with respect to each other in the hinges. All these computations must be made separately for every r from $r = 1$ to $r = k - 1$. The results of these computations are distinguished by the corresponding super-scripts.

The angle $\psi_{m,n}^{(r)}$ represents only the deformation pertaining to the external forces shown in fig. 45c. In addition we have still some move-ment in the hinges due to the forces and moments of fig. 45b. Since this load is in local equilibrium it only causes bending of the strips r and $r + 1$, leading to the rotations of the end tangents shown in fig. 46. Instead of first writing the rotations caused by M_r at different hinges and then adding the effects of different such moments on the rotation at the hinge m, we may at once write the angle $\omega_m = \omega_m' + \omega_m''$ caused by the action of moments M_{m-1}, M_m, M_{m+1}. Since all these quantities are distributed sinusoidally, we write the relation in terms of their

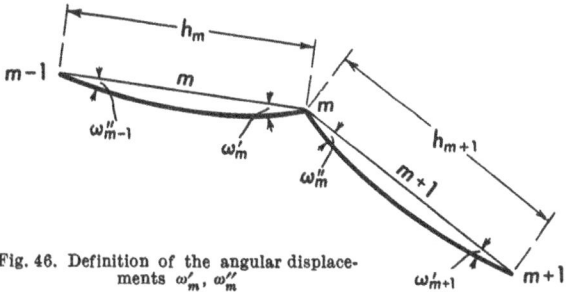

Fig. 46. Definition of the angular displacements ω_m', ω_m''

amplitudes. From a well-known beam formula we find that

$$\omega_{m,n} = \frac{h_m}{6 K_m} (M_{m-1,n} + 2 M_{m,n}) + \frac{h_{m+1}}{6 K_{m+1}} (2 M_{m,n} + M_{m+1,n}), \quad (111)$$

where K_m is the bending stiffness of the m-th strip, calculated from eq. (8b) with $t = t_m$.

Now it is time to collect all the contributions to the relative rotation taking place at the hinge m. The given loads yield the value $\psi_{m,n}^{(o)}$ obtained from eq. (109) in the way already described. The sinusoidal distribution of moments M_r at an arbitrary hinge r makes a contri-bution of amplitude $\psi_{m,n}^{(r)} M_{r\,n}$, and we have to write the sum of all these contributions from $r = 1$ to $r = k - 1$ (see fig. III–33). Lastly, there is the contribution $\omega_{m,n}$ of eq. (111). The sum of all these is the relative rotation of the strips m and $m + 1$ in the hinge m, and since there is no hinge in the actual structure, this rotation must equal zero:

$$\sum_{r=1}^{k-1} \psi_{m,n}^{(r)} M_{r,n} + \frac{h_m}{6 K_m} M_{m-1,n} + \frac{1}{3} \left(\frac{h_m}{K_m} + \frac{h_{m+1}}{K_{m+1}} \right) M_{m,n}$$
$$+ \frac{h_{m+1}}{6 K_{m+1}} M_{m+1,n} = - \psi_{m,n}^{(o)}. \quad (112)$$

There are $k - 1$ such equations for the edges $m = 1, 2, \ldots, k - 1$, and these are just enough linear equations for the $k - 1$ unknown moments $M_{r,n}$. When these equations have been solved, we may easily retrace our steps and calculate all the stress resultants and displacements we desire.

Thus far we have restricted the analysis to the case that all loads were applied at the edges only, as shown in fig. III–34. That these loads were assumed to be vertical is not a serious restriction of generality. When they are inclined or horizontal, only eqs. (III–43) have to be rewritten. But most of the important loads, such as the weight of the structure, are distributed over the surface of the strips, and we still have to extend the theory to cover this case.

The first thing to do is to consider each strip m as a plate strip supported along the edges $m - 1$ and m and carrying its load by bending moments M_y and the inevitable shearing forces Q_y. Of course, we again neglect the plate bending moments M_x and the twisting moments M_{xy}, and then each element of width dx in the x direction is a beam of span h_m in y direction carrying its own load. If the load is vertical, the reactions at the ends of the span h_m may and should also be assumed as vertical, and the inverse of these reactions are the loads P_m to be used when applying the theory already described. In addition to the moments M_y caused by the redundant edge moments M_r there are now the moments M_y caused by the distributed load in these beams. They may be of considerable magnitude, especially when the folded structure consists of only a few plate strips, and this is the essential economic disadvantage of folded structures when they have to compete with cylindrical shells.

The moments M_y in the beams h_m produce, of course, contributions to the angles ω, which must be added to eq. (111) and which will be carried over into eqs. (112). Since they do not depend on the redundant quantities $M_{r,n}$, they ultimately are an addition to the right-hand side of these equations.

Chapter 6

BENDING STRESSES IN SHELLS OF REVOLUTION

6.1 Differential Equations

6.1.1 Conditions of Equilibrium

When dealing with the bending stresses of shells of revolution, we use the same coordinates as in Chapter 2 where the direct stresses in the same kind of shells were treated: The angle θ between the meridian and an arbitrary datum meridian, and the colatitude ϕ (figs. II–1 and II–2). For the derivatives with respect to these coordinates we shall use the dash-and-dot notation explained on p. 88.

Fig. 1 shows the shell element which is cut out by two pairs of adjacent coordinate lines. In fig. 1 a the forces which act on this element are shown. We find there all those which have already been used in the membrane theory (fig. II–2), but additionally there are the transverse forces Q_ϕ and Q_θ which are peculiar to the bending theory. In fig. 1 b the bending and twisting moments are shown, represented by vectors along the axes of these moments. This system of forces

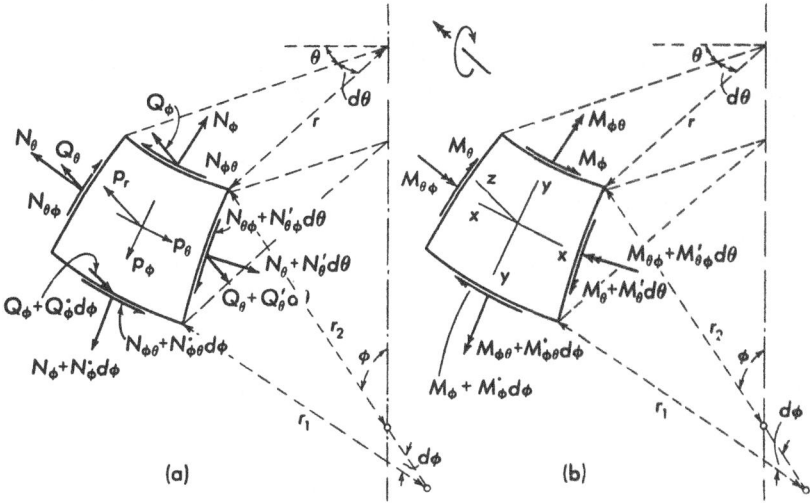

Fig. 1. Shell element

and moments must satisfy six conditions of equilibrium, three for the force components in the direction of the loads p_θ, p_ϕ, p_r and three for the moments with respect to the reference axes x, y, z in fig. 1 b, which are two tangents and the normal to the middle surface.

The first three of these equations are substantially the same as eqs. (II–6) of the membrane theory, but they contain, in addition, the contributions of the transverse forces.

The two forces $Q_\phi \cdot r \, d\theta$ include the small angle $d\phi$ and, therefore, have the resultant $Q_\phi \cdot r \, d\theta \cdot d\phi$ parallel to a tangent to the meridian and pointing in the direction of decreasing ϕ. It must be included in eq. (II–6a). Similarly, the two forces $Q_\theta \cdot r_1 \, d\phi$ have a resultant parallel to a tangent to a parallel circle. It may be found from the horizontal components of these forces, since their vertical components are exactly parallel and therefore cancel each other. The horizontal components make an angle $d\theta$ and yield the resultant $Q_\theta \cdot r_1 \, d\phi \cdot \sin\phi \cdot d\theta$ which is opposite in direction to the load $p_\theta \, dA$ and must be included in eq. (II–6b).

If the transverse forces on opposite sides of the shell element do not have the same magnitude, their difference is a radial force which enters eq. (II–6c). Both Q_ϕ and Q_θ make such contributions, pointing in the direction of the load $p_r \, dA$. Their sum is

$$(Q_\phi \, r \, d\theta)^{\cdot} \, d\phi + (Q_\theta \, r_1 \, d\phi)' \, d\theta = (Q_\phi \, r)^{\cdot} \, d\phi \, d\theta + r_1 \, Q_\theta' \, d\phi \, d\theta.$$

When we introduce all these new terms in eqs. (II–6), we must remember that a common factor $d\phi \, d\theta$ has already been dropped when these equations were derived and that the same factor must also be dropped from the new terms. Then the following equations will be found:

$$(r \, N_\phi)^{\cdot} + r_1 \, N_{\theta\phi}' - r_1 \, N_\theta \cos\phi - r \, Q_\phi = -r \, r_1 \, p_\phi,$$

$$(r \, N_{\phi\theta})^{\cdot} + r_1 \, N_\theta' + r_1 \, N_{\theta\phi} \cos\phi - r_1 \, Q_\theta \sin\phi = -r \, r_1 \, p_\theta, \quad (1\,\mathrm{a\text{–}c})$$

$$r_1 \, N_\theta \sin\phi + r \, N_\phi + r_1 \, Q_\theta' + (r \, Q_\phi)^{\cdot} = r \, r_1 \, p_r.$$

We may now turn to the moment equilibrium and begin with the moments with respect to the axis x in fig. 1b. There we have: the difference between the two bending moments $M_\phi \cdot r \, d\theta$ on opposite sides of the shell element, the difference between the twisting moments $M_{\theta\phi} \cdot r_1 \, d\phi$ on the other two sides, and the couple made by the two transverse forces $Q_\phi \cdot r \, d\theta$. But there is still another term: The two moment vectors $M_\theta \cdot r_1 \, d\phi$ are not strictly parallel, and we may apply to them the same reasoning which we applied on page 21 to the forces $N_{\theta\phi} \cdot r_1 \, d\phi$. We find a resultant moment $M_\theta \cdot r_1 \, d\phi \cdot \cos\phi \cdot d\theta$ about the axis x, in the same direction as the couple of the transverse forces. We have, therefore, the following condition of equilibrium:

$$(M_\phi \cdot r \, d\theta)^{\cdot} \, d\phi + (M_{\theta\phi} \cdot r_1 \, d\phi)' \, d\theta - Q_\phi \cdot r \, d\theta \cdot r_1 \, d\phi$$

$$-M_\theta \cdot r_1 \, d\phi \cdot \cos\phi \cdot d\theta = 0$$

which may be simplified to

$$(r \, M_\phi)^{\cdot} + r_1 \, M_{\theta\phi}' - r_1 \, M_\theta \cos\phi = r \, r_1 \, Q_\phi. \qquad (1\,\mathrm{d})$$

A similar equation may be found for the y axis of fig. 1b. Besides the derivatives of the moments $M_\theta \cdot r_1 \, d\phi$ and $M_{\phi\theta} \cdot r \, d\theta$ and the couple formed by the forces $Q_\theta \cdot r_1 \, d\phi$, it contains a contribution of the twisting moments $M_{\theta\phi} \cdot r_1 \, d\phi$. Just as we found for the forces $N_\theta \cdot r_1 \, d\phi$ on p. 20, these moments have a resultant $M_{\theta\phi} \cdot r_1 \, d\phi \cdot d\theta$ which points in the direction of a radius of a parallel circle. A component of this resultant moment enters our equation. When we drop the factor $d\phi \, d\theta$ from all terms, this equation will read as follows:

$$(r \, M_{\phi\theta})^{\cdot} + r_1 \, M_\theta' + r_1 \, M_{\theta\phi} \cos\phi = r \, r_1 \, Q_\theta. \tag{1e}$$

The last of the six equations of equilibrium contains all the moments about a normal to the middle surface. They are the two couples made up by the shearing forces $N_{\theta\phi} \, r_1 \, d\phi$ and $N_{\phi\theta} \, r \, d\theta$, the other component of the resultant of the moments $M_{\theta\phi} \cdot r_1 \, d\phi$, and the resultant of the moments $M_{\phi\theta} \cdot r \, d\theta$:

$$r \, r_1 \, N_{\theta\phi} - r \, r_1 \, N_{\phi\theta} - r_1 \, M_{\theta\phi} \sin\phi + r \, M_{\phi\theta} = 0.$$

Because $r = r_2 \sin\phi$ this equation may also be written as

$$\frac{M_{\phi\theta}}{r_1} - \frac{M_{\theta\phi}}{r_2} = N_{\phi\theta} - N_{\theta\phi}. \tag{1f}$$

The equations (1a–f) describe the equilibrium of a group of forces and couples in space. They contain as special cases eqs. (II–6a–c) for the membrane forces in a shell of revolution and eqs. (V–1a–f) of the bending theory of the circular cylinder.

6.1.2 Deformations

The conditions of equilibrium (1a–f) are 6 equations for 10 stress resultants. They are therefore not sufficient to determine these unknowns. The additional equations needed are, of course, found in the elastic law of the shell, i. e. in the relations between the stress resultants and the displacements of the middle surface. These relations are a counterpart to the elastic law (V–9) of the cylindrical shell.

To establish these equations, we may proceed in the same way as we did when deriving eqs. (V–9), and thus arrive at one version of the bending theory of shells of revolution. As we have seen (p. 211), this theory is exact (except for an occasional dropping of terms with t^5, t^7 etc.) for a certain anisotropic material. Some of the elastic moduli of this material are infinite so that $\epsilon_z \equiv \gamma_{z\phi} \equiv \gamma_{z\theta} \equiv 0$. Since these strains cannot be of more than local importance in a thin shell, the theory so obtained cannot be very wrong when applied to real shells, and since it is exact at least for the special material, it is free from internal contradictions. On the following pages we shall derive and use the elastic equations pertaining to this theory.

Later on (p. 355) we shall develop another, competing set of elastic relations, which is less consistent in its assumptions but which has the merit of leading to simple equations for the axisymmetric theory.

We start from eqs. (II–59). As they stand, they relate the strains ϵ_ϕ, ϵ_θ, $\gamma_{\phi\theta}$ at a point of the middle surface to the displacements u, v, w of this point and their derivatives. However, we may use them for an arbitrary point A at a distance z from the middle surface, if we write its displacements u_A, v_A, w_A instead of u, v, w, and $r_1 + z$, $r_2 + z$ instead of r_1 and r_2:

$$\epsilon_\phi = \frac{v_A' + w_A}{r_1 + z}, \qquad \epsilon_\theta = \frac{u_A' + v_A \cos\phi + w_A \sin\phi}{(r_2 + z)\sin\phi},$$

$$\gamma_{\phi\theta} = \frac{u_A'}{r_1 + z} - \frac{u_A \cos\phi - v_A'}{(r_2 + z)\sin\phi}. \tag{2}$$

In these formulas u_A, v_A, w_A must now be expressed in terms of the displacements u, v, w of the corresponding point A_0 on the middle surface. To do this, we use the fundamental assumption that normals to the middle surface remain normals when the shell is deformed.

Fig. 2 shows a section through the shell in the plane of a parallel circle. The heavy line is part of the parallel circle on the middle sur-

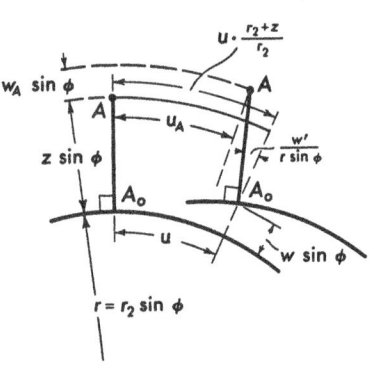

Fig. 2
Displacements of the points A_0 and A, projected on the plane of a parallel circle

face, and the line marked A_0A is the projection of the real line $A_0A = z$ which is normal to the middle surface. When the shell is deformed, these two lines come into new positions, but the right angle between them is conserved. When the point A_0 undergoes a displacement u along the parallel circle, the point A will move through a distance $u\,\frac{r_2 + z}{r_2}$ and this is a contribution to its displacement u_A. The radial displacement w and the meridional displacement v of A have no immediate influence on u_A. But when w depends on θ, the element $r\,d\theta$ of the parallel circle rotates by the angle w'/r, and the normal A_0A rotates through the same angle, thus moving the point A backward by $w'\,z/r$. In all, we have

$$u_A = u\,\frac{r_2 + z}{r_2} - w'\,\frac{z}{r}. \tag{3a}$$

A similar formula for v_A may be found from a meridional section of the shell. As such, we may use fig. V–2 which was originally designed for a circular cylinder, but we must now read r_1 instead of-a for the radius of curvature of the middle surface. With the figure we may also use eq. (V–3b) which was derived from it, applying the same change of notation:

$$v_A = v \frac{r_1 + z}{r_1} - w \cdot \frac{z}{r_1}. \tag{3b}$$

Lastly, we have again the simple relation

$$w_A = w \tag{3c}$$

which indicates that the length of the line $A_0 A$ does not change during deformation, at least not enough to affect the kinematics of the deformation.

Combining eqs. (2) and (3), we find the strains at a distance z from the middle surface in terms of the displacements u, v, w at the middle surface and of their derivatives:

$$\left.\begin{aligned}
\epsilon_\phi &= \frac{v^{\boldsymbol{\cdot}}}{r_1} - \frac{v}{r_1} \frac{r_1^{\boldsymbol{\cdot}}}{r_1} \frac{z}{r_1 + z} - \frac{w^{\boldsymbol{\cdot\cdot}}}{r_1} \frac{z}{r_1 + z} + \frac{w^{\boldsymbol{\cdot}}}{r_1} \frac{r_1^{\boldsymbol{\cdot}}}{r_1} \frac{z}{r_1 + z} + \frac{w}{r_1 + z}, \\[2mm]
\epsilon_\theta &= \frac{u'}{r} + \frac{v}{r_1} \cot\phi \cdot \frac{r_1 + z}{r_2 + z} - \frac{w''}{r \sin\phi} \frac{z}{r_2 + z} \\[2mm]
&\quad - \frac{w^{\boldsymbol{\cdot}}}{r_1} \cot\phi \frac{z}{r_2 + z} + \frac{w}{r_2 + z}, \\[2mm]
\gamma_{\phi\theta} &= \frac{u^{\boldsymbol{\cdot}}}{r_2} \frac{r_2 + z}{r_1 + z} - \frac{u r_1}{r_2^2} \frac{r_2 + z}{r_1 + z} \cot\phi + \frac{v'}{r_1 \sin\phi} \frac{r_1 + z}{r_2 + z} \\[2mm]
&\quad - \frac{w^{\boldsymbol{\cdot}}z}{r_1 \sin\phi} \left(\frac{1}{r_2 + z} + \frac{r_1}{r_2} \frac{1}{r_1 + z} \right) \\[2mm]
&\quad + \frac{w'}{r_2} \frac{\cot\phi}{\sin\phi} \left(\frac{z}{r_2 + z} + \frac{r_1}{r_2} \frac{z}{r_1 + z} \right).
\end{aligned}\right\} \tag{4}$$

These expressions may now be entered on the right-hand side of HOOKE's law

$$\sigma_\phi = \frac{E}{1 - \nu^2} (\epsilon_\phi + \nu \epsilon_\theta), \qquad \sigma_\theta = \frac{E}{1 - \nu^2} (\epsilon_\theta + \nu \epsilon_\phi),$$

$$\tau_{\phi\theta} = \frac{E}{2(1 + \nu)} \gamma_{\phi\theta}$$

and the expressions so obtained may be introduced into the integrals (I–1a, b, d, e, g–j) which define the stress resultants (replacing, of course, the subscripts x, y by ϕ, θ). The result is the elastic law for a shell of revolution. With the notation (V–8) for the rigidities of the shell, this law assumes the following form:

$$N_\phi = D \left[\frac{v^{\boldsymbol{\cdot}} + w}{r_1} + \nu \, \frac{u' + v\cos\phi + w\sin\phi}{r} \right]$$
$$+ \frac{K}{r_1^2} \frac{\dot{r}_2 - r_1}{r_2} \left[\frac{v - w^{\boldsymbol{\cdot}}}{r_1} \cdot \frac{r_1^{\boldsymbol{\cdot}}}{r_1} + \frac{w^{\boldsymbol{\cdot\cdot}} + w}{r_1} \right],$$

$$N_\theta = D \left[\frac{u' + v\cos\phi + w\sin\phi}{r} + \nu \, \frac{v^{\boldsymbol{\cdot}} + w}{r_1} \right]$$
$$- \frac{K}{r\,r_1} \frac{r_2 - r_1}{r_2} \left[-\frac{v}{r_1} \frac{r_2 - r_1}{r_2} \cos\phi + \frac{w\sin\phi}{r_2} + \frac{w''}{r} + \frac{w^{\boldsymbol{\cdot}}\cos\phi}{r_1} \right],$$

$$N_{\phi\theta} = D \, \frac{1-\nu}{2} \left[\frac{u^{\boldsymbol{\cdot}}}{r_1} + \frac{v' - u\cos\phi}{r} \right] + \frac{K}{r_1^2} \frac{1-\nu}{2} \frac{r_2 - r_1}{r_2} \left[\frac{u^{\boldsymbol{\cdot}}}{r_1} \frac{r_2 - r_1}{r_2} \right.$$
$$\left. + \frac{u}{r_2} \frac{r_1 - r_2}{r_2} \cot\phi + \frac{w^{\boldsymbol{\cdot}\prime}}{r} - \frac{w'}{r} \frac{r_1}{r} \cos\phi \right],$$

$$N_{\theta\phi} = D \, \frac{1-\nu}{2} \left[\frac{u^{\boldsymbol{\cdot}}}{r_1} + \frac{v' - u\cos\phi}{r} \right]$$
$$+ \frac{K}{r\,r_1} \frac{1-\nu}{2} \frac{r_2 - r_1}{r_2} \left[\frac{v'}{r_1} \frac{r_2 - r_1}{r_2} - \frac{w^{\boldsymbol{\cdot}\prime}}{r_1} + \frac{w'\cos\phi}{r} \right], \qquad (5\,\text{a–h})$$

$$M_\phi = K \left[\frac{1}{r_1^2} \left(w^{\boldsymbol{\cdot\cdot}} - w^{\boldsymbol{\cdot}} \frac{r_1^{\boldsymbol{\cdot}}}{r_1} - w \frac{r_1 - r_2}{r_2} \right) - \frac{v^{\boldsymbol{\cdot}}}{r_1 r_2} + \frac{v}{r_1^2} \frac{r_1^{\boldsymbol{\cdot}}}{r_1} \right.$$
$$\left. + \nu \frac{w''}{r^2} + \nu \frac{w^{\boldsymbol{\cdot}}\cos\phi}{r\,r_1} - \nu \frac{u'}{r\,r_2} - \nu \frac{v\cos\phi}{r\,r_1} \right],$$

$$M_\theta = K \left[\frac{w''}{r^2} + \frac{w^{\boldsymbol{\cdot}}\cos\phi}{r\,r_1} - \frac{w}{r_2^2} \frac{r_2 - r_1}{r_1} - \frac{u'}{r\,r_1} - \frac{v\cos\phi}{r\,r_1} \frac{2r_2 - r_1}{r_2} \right.$$
$$\left. + \frac{v}{r_1^2} \left(w^{\boldsymbol{\cdot\cdot}} - w^{\boldsymbol{\cdot}} \frac{r_1^{\boldsymbol{\cdot}}}{r_1} \right) - \frac{v}{r_1^2} \left(v^{\boldsymbol{\cdot}} - v \frac{r_1^{\boldsymbol{\cdot}}}{r_1} \right) \right],$$

$$M_{\phi\theta} = K \, \frac{1-\nu}{2} \left[2 \frac{w^{\boldsymbol{\cdot}\prime}}{r\,r_1} - 2 \frac{w'}{r^2} \cos\phi - \frac{u^{\boldsymbol{\cdot}}}{r_1 r_2} \frac{2r_1 - r_2}{r_1} \right.$$
$$\left. + \frac{u}{r_2^2} \frac{2r_1 - r_2}{r_1} \cot\phi - \frac{v'}{r\,r_1} \right],$$

$$M_{\theta\phi} = K \, \frac{1-\nu}{2} \left[2 \frac{w^{\boldsymbol{\cdot}\prime}}{r\,r_1} - 2 \frac{w'}{r^2} \cos\phi - \frac{u^{\boldsymbol{\cdot}}}{r_1 r_2} + \frac{u}{r_2^2} \cot\phi - \frac{v'}{r\,r_1} \frac{2r_2 - r_1}{r_2} \right].$$

These formulas need some comment. They are based on two assumptions – that the displacements are small, and that the normals to the middle surface are conserved as such during deformation. These assumptions have their origin in the theory of plane plates, and there they lead to a very simple and appealing elastic law, stating that the normal and shearing forces depend on the strains in the middle surface and the bending and twisting moments depend on the change of curvature and twist of this surface. In the elastic law for shells of revolution there are many additional terms. We have already met with some of them in the theory of cylindrical shells (p. 214), but in eqs. (5) there are still more. Through them the elastic change of curvature influences the normal and shearing forces, and the strains in the middle surface influence the moments.

As in the case of the cylindrical shell, these terms are of minor importance, and it is well worthwhile to consider a simpler elastic law in which they are missing. We arrive at a set of consistent formulas when we trace the undesired terms back to their sources and there apply some reasonable simplifications.

There are two such sources, the factors $(r_1 + z)$, $(r_2 + z)$ in eqs. (I–1) and the same quantities in the kinematic relations (2) and (3). Realizing that z is very small compared with the radii, we put $r_1 + z \approx r_1$ and $r_2 + z \approx r_2$ in all these equations and then repeat all that we have done before. This leads to the following elastic law, which takes the place of eqs. (5):

$$N_\phi = D\left[\frac{v^\cdot + w}{r_1} + v\,\frac{u' + v\cos\phi + w\sin\phi}{r}\right],$$

$$N_\theta = D\left[\frac{u' + v\cos\phi + w\sin\phi}{r} + v\,\frac{v^\cdot + w}{r_1}\right],$$

$$N_{\phi\theta} = N_{\theta\phi} = D\frac{1-v}{2}\left[\frac{u^\cdot}{r_1} + \frac{v' - u\cos\phi}{r}\right],$$

$$M_\phi = K\left[\frac{1}{r_1}\left(\frac{w^\cdot}{r_1}\right)^\cdot + \frac{v}{r}\left(\frac{w''}{r} + \frac{w^\cdot}{r_1}\cos\phi\right)\right],$$ (6a–f)

$$M_\theta = K\left[\frac{1}{r}\left(\frac{w''}{r} + \frac{w^\cdot}{r_1}\cos\phi\right) + \frac{v}{r_1}\left(\frac{w^\cdot}{r_1}\right)^\cdot\right],$$

$$M_{\phi\theta} = M_{\theta\phi} = K(1-v)\left[\frac{w''}{r_1 r} - \frac{w'}{r^2}\cos\phi\right].$$

In this simplified version of the elastic law there is no difference between the two shearing forces and none between the two twisting moments. It is therefore incompatible with eq. (1f) unless $r_1 = r_2$. The violation of a condition of equilibrium seems to be a serious matter, but in this particular case it may easily be justified. Both sides of eq. (1f) are small differences between quantities which are almost equal. If these two differences are not exactly equal, as eq. (1f) demands, only a slight adjustment of the values of the shearing forces and of the twisting moments is needed to satisfy this equation, and since we noted earlier that we have one equation too many, we may simply disregard the sixth equation of equilibrium.

We have now the following balance of unknowns and equations: If we use the exact elastic law, we have 6 equations (1) (among them one identity) and 8 equations (5), altogether 13 independent equations for 13 unknowns (N_ϕ, N_θ, $N_{\phi\theta}$, $N_{\theta\phi}$, M_ϕ, M_θ, $M_{\phi\theta}$, $M_{\theta\phi}$, Q_ϕ, Q_θ, u, v, w). If we use the simplified form of the elastic law, we have 6 equations (1) (among them one which is unimportant) and 6 equations (6), together 11 useful equations for 11 unknowns (the two shears and the two twisting moments being equal). In both cases a sufficient number of equations is available, and the general procedure would be to eliminate all but u, v, w and thus to arrive at a set of three differential equations for the displacements, as we did for the cylindrical shell. This will be done in this book for two particular cases, the sphere and the cone. When the loads are symmetric, a simpler procedure is possible which is shown in the next Section.

6.1.3 Axially Symmetric Case

Frequently we have to deal with stress systems that have the same axial symmetry as the shell itself. In this case the preceding equations simplify considerably. First of all, we must drop all derivatives with respect to θ. Second, many of the stress resultants vanish identically: the shearing forces $N_{\phi\theta}$, $N_{\theta\phi}$, the twisting moments $M_{\phi\theta}$, $M_{\theta\phi}$, and the transverse shear Q_{θ}, but not Q_{ϕ}. Also, the load component p_{θ} is zero. Then nothing is left in eqs. (1 b, e, f), and the other three conditions of equilibrium simplify to

$$(r\,N_\phi)^{\textbf{.}} - r_1\,N_\theta\cos\phi - r\,Q_\phi = -r\,r_1\,p_\phi\,,$$

$$(r\,Q_\phi)^{\textbf{.}} + r_1\,N_\theta\sin\phi + r\,N_\phi = \quad r\,r_1\,p_r\,, \qquad (7\,\text{a–c})$$

$$(r\,M_\phi)^{\textbf{.}} - r_1\,M_\theta\cos\phi \qquad\quad = \quad r\,r_1\,Q_\phi\,.$$

In the elastic law (5) we drop the displacement u, which vanishes identically, and again all prime derivatives. We are then left with the following relations:

$$N_\phi = D\left[\frac{v^{\textbf{.}}+w}{r_1} + \frac{v}{r_2}\,(v\cot\phi + w)\right]$$

$$\qquad + \frac{K}{r_1^3}\,\frac{r_2-r_1}{r_2}\left[(v-w^{\textbf{.}})\frac{r_1}{r_1} + w^{\textbf{..}} + w\right],$$

$$N_\theta = D\left[\frac{1}{r_2}\,(v\cot\phi + w) + \frac{v}{r_1}\,(v^{\textbf{.}}+w)\right]$$

$$\qquad - \frac{K}{r_1 r_2}\,\frac{r_2-r_1}{r_2}\left[-\frac{v}{r_1}\,\frac{r_2-r_1}{r_2}\cot\phi + \frac{w^{\textbf{.}}}{r_1}\cot\phi + \frac{w}{r_2}\right], \qquad (8\,\text{a–d})$$

$$M_\phi = \frac{K}{r_1}\left[\left(\frac{w^{\textbf{.}}-v}{r_1}\right)^{\textbf{.}} + (v^{\textbf{.}}+w)\frac{r_2-r_1}{r_1 r_2} + \frac{v}{r_2}\,(w^{\textbf{.}}-v)\cot\phi\right],$$

$$M_\theta = \frac{K}{r_1}\left[\frac{w^{\textbf{.}}-v}{r_2}\cot\phi - \frac{v\cot\phi + w}{r_2}\,\frac{r_2-r_1}{r_2} + v\left(\frac{w^{\textbf{.}}-v}{r_1}\right)^{\textbf{.}}\right].$$

The simplified elastic law (6) reduces to the following equations:

$$N_\phi = D\left[\frac{v^\cdot + w}{r_1} + \nu\,\frac{v\cos\phi + w\sin\phi}{r}\right],$$

$$N_\theta = D\left[\frac{v\cos\phi + w\sin\phi}{r} + \nu\,\frac{v^\cdot + w}{r_1}\right],$$

$$M_\phi = \frac{K}{r_1}\left[\left(\frac{w^\cdot}{r_1}\right)^\cdot + \nu\,\frac{w^\cdot\cos\phi}{r}\right],$$

$$M_\theta = \frac{K}{r_1}\left[\frac{w^\cdot\cos\phi}{r} + \nu\left(\frac{w^\cdot}{r_1}\right)^\cdot\right].$$

(9a–d)

On this level of accuracy the meridional displacement v no longer influences the bending moments.

6.2 Axially Symmetric Loads

Since the differential equations are so much simpler when the stress system has axial symmetry, we now assume that such symmetry is present. Additionally, we shall assume that the wall thickness t is a constant. In this case the theory has to start from eqs. (7) and (8).

As we saw on p. 218 for a cylinder and as we shall see on p. 355 for a sphere, the membrane theory describes the stresses in a shell satisfactorily, if we are able to provide those boundary conditions which the membrane shell requires. But we have also seen (pp. 25, 28, 35, 38) that the desirable boundary conditions are not always those which the membrane forces could fulfill. It then becomes necessary to apply to the edge of the shell additional forces N_ϕ, Q_ϕ and moments M_ϕ and thus to bridge the gap between the desired deformation and the membrane deformation. Since these edge loads are beyond the possibilities of the membrane theory, they must produce bending stresses, and to deal with these stress systems is the principal purpose of the bending theory. Therefore, we now drop the load terms from the conditions of equilibrium (7) and put

$$p_\phi \equiv p_r \equiv 0.$$

Later, on p. 353 we shall take up the load terms and shall see that we did not lose anything essential.

6.2.1 Spherical Shell

6.2.1.1 Differential Equations

After the circular cylinder, the sphere is one of the simplest surfaces of revolution. Therefore, we shall now develop the bending theory of a spherical shell.

When the radius of the middle surface is a, we have

$$r_1 = r_2 = a, \quad r = a \sin\phi.$$

The conditions of equilibrium (7) are then

$$
\begin{aligned}
(N_\phi \sin\phi)^{\cdot} - N_\theta \cos\phi - Q_\phi \sin\phi &= 0, \\
(Q_\phi \sin\phi)^{\cdot} + N_\theta \sin\phi + N_\phi \sin\phi &= 0, \\
(M_\phi \sin\phi)^{\cdot} - M_\theta \cos\phi &= a\, Q_\phi \sin\phi,
\end{aligned}
\qquad (10\,\text{a–c})
$$

and the elastic law (8) simplifies to

$$
\left.
\begin{aligned}
N_\phi &= \frac{D}{a}\,[(v^{\cdot} + w) + \nu(v \cot\phi + w)], \\
N_\theta &= \frac{D}{a}\,[(v \cot\phi + w) + \nu(v^{\cdot} + w)],
\end{aligned}
\right\}
\qquad (11\,\text{a, b})
$$

$$
M_\phi = \frac{K}{a^2}\,[(w^{\cdot} - v)^{\cdot} + \nu(w^{\cdot} - v)\cot\phi],
$$

$$
M_\theta = \frac{K}{a^2}\,[(w^{\cdot} - v)\cot\phi + \nu(w^{\cdot} - v)^{\cdot}].
$$

The last two equations suggest the introduction of an auxiliary variable

$$\chi = \frac{w^{\cdot} - v}{a} \qquad (12)$$

which represents the angle by which an element $a\,d\phi$ of the meridian rotates during deformation. Using χ, we obtain the following form for the moment relations:

$$
\begin{aligned}
M_\phi &= \frac{K}{a}\,[\chi^{\cdot} + \nu \chi \cot\phi], \\
M_\theta &= \frac{K}{a}\,[\chi \cot\phi + \nu \chi^{\cdot}].
\end{aligned}
\qquad (11\,\text{c, d})
$$

The conditions of equilibrium (10), the elastic law (11), and the definition (12) are together a set of 8 equations for as many unknowns, viz. the stress resultants N_ϕ, N_θ, Q_ϕ, M_ϕ, M_θ, and the displacements v, w, χ. This set may be reduced to a pair of equations for Q_ϕ and χ.

One of these is easily found. We only have to introduce the bending moments from eqs. (11 c, d) into eq. (10 c):

$$\chi^{\cdot\cdot} + \chi^{\cdot} \cot\phi - \chi(\cot^2\phi + \nu) = \frac{a^2}{K}\, Q_\phi. \qquad (13\,\text{a})$$

The second equation necessarily contains eqs. (10a, b) and (11a, b). From the latter two we have

$$v^{\cdot} + w = \frac{a}{D(1 - \nu^2)}\,(N_\phi - \nu N_\theta),$$

$$v \cot\phi + w = \frac{a}{D(1 - \nu^2)}\,(N_\theta - \nu N_\phi)$$

and by differentiating the second of these equations, we obtain for a shell of constant wall thickness

$$v^{\cdot} \cot\phi - \frac{v}{\sin^2\phi} + w^{\cdot} = \frac{a}{D(1-v^2)}(N_\theta^{\cdot} - v N_\phi^{\cdot}).$$

When we eliminate $v^{\cdot}$ and w from these three relations, we arrive at an expression for $w^{\cdot} - v$ which, according to eq. (12), equals $a \chi$. It is

$$a \chi = \frac{a}{D(1-v^2)}[N_\theta^{\cdot} - v N_\phi^{\cdot} + (1+v)(N_\theta - N_\phi)\cot\phi]. \quad (14)$$

We now use the conditions of equilibrium (10a, b) to express N_ϕ and N_θ in terms of Q_ϕ. When we eliminate N_θ we find that

$$(N_\phi \sin^2\phi)^{\cdot} + (Q_\phi \sin\phi \cos\phi)^{\cdot} = 0.$$

This equation evidently expresses the fact that the (vertical) resultant of all forces transmitted through a parallel circle of radius $a \sin\phi$ does not depend on ϕ. Since we dropped the surface loads p_ϕ, p_r, this is the condition of equilibrium for a zone of the shell limited by two adjacent parallel circles. When we integrate the equation, writing

$$N_\phi \sin\phi + Q_\phi \cos\phi = - \frac{P}{2\pi a \sin\phi}, \quad (14')$$

the constant of integration P is this vertical resultant which, of course, must be known. We have then

$$N_\phi = -Q_\phi \cot\phi - \frac{P}{2\pi a} \frac{1}{\sin^2\phi} \quad (15a)$$

and from eq. (10b) we find

$$N_\theta = -Q_\phi^{\cdot} + \frac{P}{2\pi a} \frac{1}{\sin^2\phi}. \quad (15b)$$

These expressions may now be introduced on the right-hand side of eq. (14). When we arrange the terms according to the derivatives of Q_ϕ, we arrive at the following differential equation:

$$Q_\phi^{\cdot\cdot} + Q_\phi^{\cdot} \cot\phi - (\cot^2\phi - v) Q_\phi = -D(1-v^2) \chi. \quad (13b)$$

It is remarkable that the terms with P have dropped out.

Eqs. (13a, b) are a pair of second-order differential equations for the variables χ and Q_ϕ. Since we dropped the surface loads p_ϕ, p_r, they describe the stresses in a spherical shell which is loaded at its edges and, possibly, by a concentrated force P at the top.

The left-hand sides of eqs. (13a, b) are very similar to each other. This similarity suggests defining a linear differential operator

$$L(\ldots) = (\ldots)^{\cdot\cdot} + (\ldots)^{\cdot} \cot\phi - (\ldots) \cot^2\phi. \quad (16)$$

With this operator they assume the following form

$$L(\chi) - \nu\,\chi = \frac{a^2}{K}\,Q_\phi,$$

$$L(Q_\phi) + \nu\,Q_\phi = -D(1 - \nu^2)\,\chi. \tag{17a, b}$$

We may now easily separate the unknowns by substituting either Q_ϕ from (17a) into (17b) or χ from (17b) into (17a):

$$LL(\chi) - \nu^2\,\chi = -\frac{D(1 - \nu^2)\,a^2}{K}\,\chi,$$

$$LL(Q_\phi) - \nu^2\,Q_\phi = -\frac{D(1 - \nu^2)\,a^2}{K}\,Q_\phi. \tag{18a, b}$$

Either one of these equations may be used to solve the problem. When we have found, say, Q_ϕ from eq. (18b), we may find χ from eq. (17b) by simple differentiations, and then all other quantities may be obtained from preceding formulas.

We rewrite eq. (18b) in the form

$$L\,L(Q_\phi) + 4\varkappa^4\,Q_\phi = 0 \tag{19}$$

with

$$\varkappa^4 = \frac{D(1 - \nu^2)\,a^2}{4\,K} - \frac{\nu^2}{4} = 3\,(1 - \nu^2)\,\frac{a^2}{t^2} - \frac{\nu^2}{4}. \tag{20}$$

Eq. (19) may be written in either of the following forms:

$$L(L(Q_\phi) + 2\,i\,\varkappa^2\,Q_\phi) - 2\,i\,\varkappa^2(L(Q_\phi) + 2\,i\,\varkappa^2\,Q_\phi) = 0,$$

$$L(L(Q_\phi) - 2\,i\,\varkappa^2\,Q_\phi) + 2\,i\,\varkappa^2(L(Q_\phi) - 2\,i\,\varkappa^2\,Q_\phi) = 0,$$

which show that the solutions of the two second-order equations

$$L(Q_\phi) \pm 2\,i\,\varkappa^2\,Q_\phi = 0 \tag{21a, b}$$

satisfy eq. (18b). Because of the factor i in the second term these solutions have complex values, and those of eq. (21b) are conjugate complex to those of eq. (21a). The two pairs of solutions are therefore linearly independent and constitute together a set of four solutions of eq. (18b). Since the real and imaginary parts of these functions are each a linear combination of two of them, they are also solutions of eq. (18b), and since they have real values (the imaginary parts after dropping the constant factor i), they are more suitable for practical purposes. To find them, we need only to solve one second-order differential equation, say (21a). How this may be done we shall see in the next sections.

6.2.1.2 Solution Using Hypergeometric Series

When we write eq. (21a) in full, we have

$$\ddot{Q}_\phi + \dot{Q}_\phi \cot\phi - Q_\phi \cot^2\phi + 2\,i\,\varkappa^2\,Q_\phi = 0. \tag{22}$$

This is a second-order differential equation with variable coefficients. By introducing new variables, putting

$$\cos^2\phi = x, \quad Q_\phi = z\sin\phi, \tag{23}$$

we may transform it into a standard type:

$$\frac{d^2z}{dx^2} + \frac{1-5x}{2x(1-x)}\cdot\frac{dz}{dx} - \frac{1-2i\varkappa^2}{4x(1-x)}\cdot z = 0. \tag{24}$$

This is a hypergeometric equation, and from the general theory of this type of differential equations[1] we may obtain the following information:

The equation

$$\frac{d^2z}{dx^2} + \frac{\gamma-(1+\alpha+\beta)x}{x(1-x)}\cdot\frac{dz}{dx} - \frac{\alpha\beta}{x(1-x)}\cdot z = 0$$

has the solutions

$$z_a = F(\alpha,\beta,\gamma;x) \equiv 1 + \frac{\alpha\beta}{1!\gamma}x + \frac{\alpha(\alpha+1)\cdot\beta(\beta+1)}{2!\cdot\gamma(\gamma+1)}x^2$$
$$+ \frac{\alpha(\alpha+1)(\alpha+2)\cdot\beta(\beta+1)(\beta+2)}{3!\cdot\gamma(\gamma+1)(\gamma+2)}x^3 + \cdots$$

and

$$z_b = x^{1-\gamma}F(\alpha-\gamma+1,\ \beta-\gamma+1,\ 2-\gamma;x),$$

and the two power series converge for $0 \leqq x < 1$.

When we compare the general form of the hypergeometric equation with eq. (24), we find that in our case

$$\alpha = \frac{1}{4}(3+\sqrt{5+8i\varkappa^2}), \quad \beta = \frac{1}{4}(3-\sqrt{5+8i\varkappa^2}), \quad \gamma = \frac{1}{2}.$$

The two elementary solutions are then these:

$$z_a = 1 + \frac{1-2i\varkappa^2}{2!}x + \frac{(1-2i\varkappa^2)(11-2i\varkappa^2)}{4!}x^2$$
$$+ \frac{(1-2i\varkappa^2)(11-2i\varkappa^2)(29-2i\varkappa^2)}{6!}x^3$$
$$+ \frac{(1-2i\varkappa^2)(11-2i\varkappa^2)(29-2i\varkappa^2)(55-2i\varkappa^2)}{8!}x^4 + \cdots,$$

$$z_b = x^{1/2} + \frac{5-2i\varkappa^2}{3!}x^{3/2} + \frac{(5-2i\varkappa^2)(19-2i\varkappa^2)}{5!}x^{5/2}$$
$$+ \frac{(5-2i\varkappa^2)(19-2i\varkappa^2)(41-2i\varkappa^2)}{7!}x^{7/2}$$
$$+ \frac{(5-2i\varkappa^2)(19-2i\varkappa^2)(41-2i\varkappa^2)(71-2i\varkappa^2)}{9!}x^{9/2} + \cdots.$$

As explained on p. 324 for the solutions Q_ϕ of eq. (21), the real and imaginary parts of z_a and z_b are a set of four independent solutions

---

[1] See e. g. INCE, E. L.: Ordinary Differential Equations, p. 161. New York 1944.

of the shell problem. They are:

$$z = z_1 = \frac{i}{2}(z_a + \bar{z}_a) = 1 + \frac{1}{2!}\cos^2\phi + \frac{1}{4!}(11 - 4\varkappa^4)\cos^4\phi$$

$$+ \frac{1}{6!}(319 - 164\varkappa^4)\cos^6\phi$$

$$+ \frac{1}{8!}(17545 - 10456\varkappa^4 + 16\varkappa^8)\cos^8\phi + \cdots,$$

$$z = z_2 = \frac{1}{2}(z_b + \bar{z}_b) = \cos\phi + \frac{5}{3!}\cos^3\phi + \frac{1}{5!}(95 - 4\varkappa^4)\cos^5\phi$$

$$+ \frac{1}{7!}(3895 - 260\varkappa^4)\cos^7\phi + \cdots, \qquad (25)$$

$$z = z_3 = \frac{i}{2}(z_a - \bar{z}_a) = \varkappa^2\left[\cos^2\phi + \cos^4\phi + \frac{1}{360}(359 - 4\varkappa^4)\cos^6\phi\right.$$

$$\left. + \frac{1}{210}(209 - 4\varkappa^4)\cos^8\phi + \cdots\right],$$

$$z = z_4 = \frac{i}{2}(z_b - \bar{z}_b) = \frac{\varkappa^2}{3}\left[\cos^3\phi + \frac{6}{5}\cos^5\phi\right.$$

$$+ \frac{1}{840}(1079 - 4\varkappa^4)\cos^7\phi + \frac{1}{7560}(10063 - 68\varkappa^4)\cos^9\phi + \cdots\right].$$

A linear combination of these four elementary solutions, having four arbitrary constants C_1, C_2, C_3, C_4, will be the general solution z of our problem. For the transverse force Q_ϕ it yields the expression

$$Q_\phi = z\sin\phi = (C_1 z_1 + C_2 z_2 + C_3 z_3 + C_4 z_4)\sin\phi. \qquad (26a)$$

Up to this point we admitted the possibility that the loads applied to the edge of the shell have a vertical resultant P. In order to keep our formulas simpler, we now shall restrict ourselves to the most important case that P = 0. We find then from eqs. (15):

$$N_\phi = -(C_1 z_1 + C_2 z_2 + C_3 z_3 + C_4 z_4)\cos\phi,$$

$$N_\theta = -Q_\phi' = -(C_1 z_1' + C_2 z_2' + C_3 z_3' + C_4 z_4')\sin\phi \qquad (26b, c)$$

$$- (C_1 z_1 + C_2 z_2 + C_3 z_3 + C_4 z_4)\cos\phi.$$

Here $z_1' \ldots z_4'$ are another four transcendental functions, linearly independent of $z_1 \ldots z_4$. The series from which they are computed, may easily be found by differentiating those just given for $z_1 \ldots z_4$ with respect to ϕ.

Before we can write formulas for the other stress resultants, we need $L(Q_\phi)$. When we introduce z_a, z_b into (21a), $\bar{z}_a$, $\bar{z}_b$ into (21b), we find

$$L(z_a\sin\phi) = -2i\varkappa^2 z_a\sin\phi, \qquad L(z_b\sin\phi) = -2i\varkappa^2 z_b\sin\phi,$$

$$L(\bar{z}_a\sin\phi) = +2i\varkappa^2 \bar{z}_a\sin\phi, \qquad L(\bar{z}_b\sin\phi) = +2i\varkappa^2 \bar{z}_b\sin\phi$$

and when we express here $z_a \ldots \bar{z}_b$ in terms of $z_1 \ldots z_4$, we arrive at the following relations:

$$L(z_1 \sin\phi) = -2\varkappa^2 z_3 \sin\phi, \quad L(z_2 \sin\phi) = -2\varkappa^2 z_4 \sin\phi,$$

$$L(z_3 \sin\phi) = +2\varkappa^2 z_1 \sin\phi, \quad L(z_4 \sin\phi) = +2\varkappa^2 z_2 \sin\phi,$$

and hence

$$L(Q_\phi) = 2\varkappa^2(-C_1 z_3 - C_2 z_4 + C_3 z_1 + C_4 z_2) \sin\phi.$$

Now we may use eq. (17 b), which yields

$$D(1 - \nu^2)\,\chi = [C_1(2\varkappa^2 z_3 - \nu z_1) + C_2(2\varkappa^2 z_4 - \nu z_2)$$

$$-C_3(2\varkappa^2 z_1 + \nu z_3) - C_4(2\varkappa^2 z_2 + \nu z_4)] \sin\phi \qquad (26\text{d})$$

and then eqs. (11 c, d) which yield the moments

$$M_\phi = \frac{K}{D(1 - \nu^2)\,a} [C_1(2\varkappa^2 \dot{z}_3 - \nu \dot{z}_1) + C_2(2\varkappa^2 \dot{z}_4 - \nu \dot{z}_2)$$

$$- C_3(2\varkappa^2 \dot{z}_1 + \nu \dot{z}_3) - C_4(2\varkappa^2 \dot{z}_2 + \nu \dot{z}_4)] \sin\phi$$

$$+ \frac{K}{D(1 - \nu)\,a} [C_1(2\varkappa^2 z_3 - \nu z_1) + C_2(2\varkappa^2 z_4 - \nu z_2)$$

$$- C_3(2\varkappa^2 z_1 + \nu z_3) - C_4(2\varkappa^2 z_2 + \nu z_4)] \cos\phi,$$

$$\qquad (26\text{e, f})$$

$$M_\theta = \frac{K\,\nu}{D(1 - \nu^2)\,a} \cdot [C_1(2\varkappa^2 \dot{z}_3 - \nu \dot{z}_1) + C_2(2\varkappa^2 \dot{z}_4 - \nu \dot{z}_2)$$

$$- C_3(2\varkappa^2 \dot{z}_1 + \nu \dot{z}_3) - C_4(2\varkappa^2 \dot{z}_2 - \nu \dot{z}_4)] \sin\phi$$

$$+ \frac{K}{D(1 - \nu)\,a} [C_1(2\varkappa^2 z_3 - \nu z_1) + C_2(2\varkappa^2 z_4 - \nu z_2)$$

$$- C_3(2\varkappa^2 z_1 + \nu z_3) - C_4(2\varkappa^2 z_2 + \nu z_4)] \cos\phi.$$

These formulas are the general solution for the bending problem of a spherical shell, subjected to forces and moments applied to its edges.

The four constants $C_1 \ldots C_4$ must be determined from four boundary conditions, two at each edge of a spherical zone. In the simplest cases, these conditions may refer to the bending moment M_ϕ or the rotation χ, to the transverse force Q_ϕ or the horizontal displacement $r\,\epsilon_\theta$.

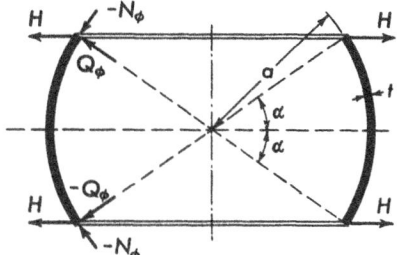

Fig. 3
Axial section through a spherical shell

Before we enter into a discussion of the limitations to which this solution is subject, we shall consider an example. Both edges of a spherical zone (fig. 3) are subjected to uniformly distributed radial loads, H being the force per unit length of the circumference. This force may be resolved into a transverse force $Q_\phi = \pm H \cos\alpha$ and a normal force $N_\phi = -H \sin\alpha$, as indicated on the left-hand side of the figure. These components are, of course, in the relation required by eq. (15a), and the boundary conditions for our example are these:

$$\text{at } \phi = 90° - \alpha: \ Q_\phi = \ \ H \cos\alpha, \quad M_\phi = 0,$$

$$\text{at } \phi = 90° + \alpha: \ Q_\phi = -H \cos\alpha, \quad M_\phi = 0.$$

Because of the symmetry of the shell and of the load with respect to the parallel circle $\phi = 90°$ we shall need only the functions $z_2(\phi)$ and $z_4(\phi)$ which have the same symmetry, and it will be enough to determine their constants C_2 and C_4 from the conditions at one edge.

For numerical work we choose $t = 1$ in., $a = 15$ in., $\alpha = 10°$, and $\nu = 0.3$. This yields $\varkappa = 4.98$. When we introduce this into the series (25) for z_2, we find

$$z_2 = \cos\phi + 0.833 \cos^3\phi - 19.67 \cos^5\phi$$

$$- 30.93 \cos^7\phi - 27.23 \cos^9\phi \pm \ldots .$$

The derivative z_2' looks still worse:

$$z_2' = -\sin\phi(1 + 2.500 \cos^2\phi - 98.35 \cos^4\phi - 216.5 \cos^6\phi$$

$$-245.1 \cos^8\phi \pm \ldots).$$

These figures suggest that the coefficients of the series increase more, the farther we go. This, however, is not true. The increase is essentially due to the powers of $\varkappa^4$ in the numerators of these coefficients, and if we go far enough, the factorial in the denominator will increase faster and will make the coefficients decrease. But the thinner the shell, the larger $\varkappa$, and the farther we have to go in these series before convergence becomes apparent and before a numerical result of even moderate accuracy can be obtained.

The remaining details of the calculation are not worth while recording here. We first determine z_2, z_4 and their first derivatives for $\phi = 80°$. From these we formulate the boundary conditions in terms of C_2 and C_4, and we shall find from them $C_2 = 4.73 H$, $C_4 = 3.873 H$. Then the formulas (26) may be used to compute the bending moment M_ϕ, the transverse force Q_ϕ, and the hoop force N_θ represented in fig. 4. The diagrams give an idea of the non-uniformity of the hoop stress in the cross section of the ring, and they show how the edge load is distributed across the width of the spherical zone by bending and shear.

One may easily understand that it is practically impossible to apply the solution (26) to shells whose $\varkappa$ is substantially greater than it is in our example. Even for very moderate values of $\varkappa$, the numerical work becomes considerable, if we are interested in colatitudes much different from $90°$. What can be done in such a case, may be explained for a shell which extends from, say $\phi = 50°$ to $\phi = 90°$. At and near the lower border the solution (26) may be used; with some effort it may be feasible to go with the series (25) as far up as $\phi = 70°$. For the upper half of the meridian, the transformation (23) is replaced by another one, which brings the zero of the auxiliary variable x into the vicinity of $\phi = 50°$. This will again lead to a hypergeometric equation, but its parameters will be different, and all our formulas have to be re-

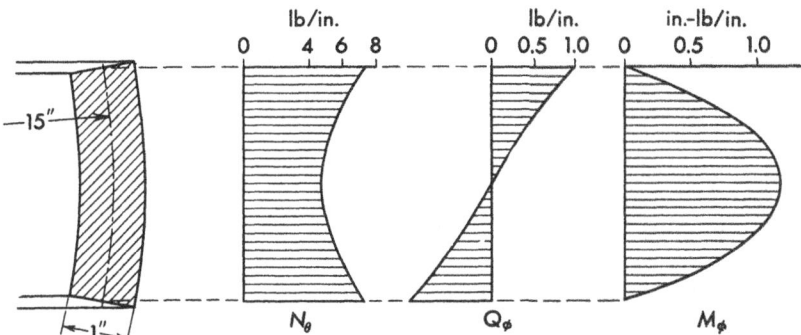

Fig. 4. Spherical shell as in Fig. 3, axial section and stress resultants

made. Again, four independent solutions will be found which may be multiplied by constants, say $C_1^* \ldots C_4^*$, but these constants are not arbitrary. They depend on the set $C_1 \ldots C_4$ through the fact that both solutions must represent the same function of ϕ. This will be assured, if somewhere halfways down on the meridian $Q_\phi, \dot{Q}_\phi, \ddot{Q}_\phi, \dddot{Q}_\phi$ as computed from both solutions are the same. These are four conditions which will yield four linear relations between $C_1^* \ldots C_4^*$ and $C_1 \ldots C_4$. Together with the boundary conditions at both edges they suffice to determine all eight constants.

When the shell is closed at the top, two of the boundary conditions are lost and must be replaced by the statement that at $\phi = 0$ all stress resultants and all displacements must be finite. This cannot be checked on the series (25), since $\phi = 0$ is the limit of their domain of convergence, and only the procedure of analytical prolongation just described can help. We then use instead of (23) the transformation

$$\sin^2\phi = x, \quad Q_\phi = z \sin\phi$$

and we shall be led to a hypergometric equation which has one regular solution and one with a singularity at $\phi = 0$.

6.2.1.3 Asymptotic Solution for Thin-walled Shells

When we are particularly interested in large values of the parameter $\varkappa$, we may think of expanding the solution of eq. (22) into a series in negative powers of $\varkappa$. For this purpose it is convenient to employ another transformation, which has the advantage that the first derivative of the unknown disappears entirely and that the variability of the remaining coefficient is not all too large. We put

$$Q_\phi = \frac{y}{\sqrt{\sin\phi}} \tag{27}$$

and obtain from (22) the differential equation

$$y'' + y\left(\frac{2 - 3\cot^2\phi}{4} - \lambda^2\right) = 0. \tag{28}$$

Here

$$\lambda^2 = -2i\,\varkappa^2$$

is an imaginary constant of large absolute value.

If we were to neglect the first term in the parenthesis as being small compared with λ^2, eq. (28) would have the simple solution

$$y = e^{\lambda\phi}.$$

Since λ is complex, this function describes oscillations with exponentially increasing or decreasing amplitude. As an approximate solution of eq. (28) it is better, the larger $|\lambda|$ is, and we may therefore assume the solution as the product of this approximation and a series of descending powers of λ,

$$y = e^{\lambda\varphi} \sum_{n=0}^{\infty} \lambda^{-n} y_n(\phi) \tag{29}$$

with $y_0(\phi) \equiv 1$. When (29) is introduced into the differential equation (28), the following relation results:

$$e^{\lambda\varphi} \sum \left(\lambda^{-n} y_n'' + 2\lambda^{-n+1} y_n' + \frac{1}{4}\lambda^{-n}(2 - 3\cot^2\phi)\, y_n\right) = 0,$$

and from it we obtain a recurrence formula for the y_n:

$$y_{n+1}' = -\frac{1}{2}\, y_n'' - \frac{1}{8}(2 - 3\cot^2\phi)\, y_n. \tag{30}$$

It yields y_{n+1} from y_n by an integration, and therefore a new constant enters with every step. It is easily seen that we may choose these constants quite arbitrarily. If we add, for instance, such a constant c_n to y_n, then y_{n+1} will be increased by $c_n\, y_1$, y_{n+2} by $c_n\, y_2$ etc. The whole sum in (29) will then be increased by a factor $(1 + c_n\, \lambda^{-n})$ which is without interest, since at a later stage we shall multiply the solution y with an appropriate constant to meet the boundary conditions.

When we really perform the integrations indicated by the recurrence formula (30), we obtain the following functions:

$$y_0(\phi) = 1,$$
$$y_1(\phi) = -\frac{1}{8}(5\phi + 3\cot\phi),$$
$$y_2(\phi) = \frac{5}{128}(5\phi^2 + 6\phi\cot\phi - 3\cot^2\phi).$$

For practical computation purposes it is often preferable and never objectionable to tabulate these functions with the complement angle

$$\psi = 90° - \phi$$

as the argument and to choose the constants of integration so that the functions are either even or odd in ψ. When this is done, the following set is obtained:

$$y_0(\psi) = 1,$$
$$y_1(\psi) = \frac{1}{8}(5\psi - 3\tan\psi),$$
$$y_2(\psi) = \frac{5}{128}(5\psi^2 - 6\psi\tan\psi - 3\tan^2\psi),$$
$$y_3(\psi) = \frac{5}{3072}(120\psi + 25\psi^3 - 216\tan\psi - 45\psi^2\tan\psi$$
$$- 45\psi\tan^2\psi - 63\tan^3\psi),$$
$$y_4(\psi) = \frac{5}{98304}(2400\psi^2 + 125\psi^4 - 5760\psi\tan\psi - 300\psi^3\tan\psi$$
$$- 6624\tan^2\psi - 450\psi^2\tan^2\psi - 1260\psi\tan^3\psi - 2835\tan^4\psi).$$

If need should be, more of them may be found from the recurrence formula (30).

In the formulas for the stress resultants we shall need the derivatives

$$\dot{y}_n = dy_n/d\phi = -dy_n/d\psi.$$

The first four are these:

$$\dot{y}_1 = -\frac{1}{8}(2 - 3\tan^2\psi),$$
$$\dot{y}_2 = -\frac{5}{64}(2\psi - 6\tan\psi - 3\tan^2\psi - 3\tan^3\psi),$$
$$\dot{y}_3 = \frac{5}{1024}(32 - 10\psi^2 + 60\psi\tan\psi + 150\tan^2\psi + 15\psi^2\tan^2\psi$$
$$+ 30\psi\tan^3\psi + 63\tan^4\psi),$$
$$\dot{y}_4 = \frac{5}{24576}(240\psi - 50\psi^3 + 4752\tan\psi + 450\psi^2\tan\psi + 2610\psi\tan^2\psi$$
$$+ 75\psi^3\tan^2\psi + 6462\tan^3\psi + 225\psi^2\tan^3\psi + 945\psi\tan^4\psi + 2835\tan^5\psi).$$

A short table of these functions is given on p. 333. It may not be sufficient for all practical problems, but it will be useful as initial information for the general layout of numerical work.

The functions y_n may now be introduced into eq. (29). Since λ, is double-valued,

$$\lambda = \sqrt{-2i\varkappa} = \pm (1-i)\varkappa$$

this will yield two linearly independent solutions which may be multiplied by arbitrary constants A and B:

$$y = A\, e^{\varkappa\phi}\, e^{-i\varkappa\phi} \sum_{n=0}^{\infty} \frac{y_n}{(1-i)^n \varkappa^n} + B\, e^{-\varkappa\phi}\, e^{i\varkappa\phi} \sum_{n=0}^{\infty} \frac{(-1)^n\, y_n}{(1-i)^n\, \varkappa^n}.$$

The first one of these functions decreases exponentially when we proceed from the base to the top of the shell, the second one does so when we proceed in the opposite direction. The A solution will therefore describe a stress system caused by loads applied at the base $\phi = \phi_1$ (fig. 5), and the B solution will correspond to loads at an upper edge $\phi = \phi_2$, hence will not appear at all in a shell which is closed at the top. Since in thin shells both stress systems are only of local importance, it is useful to introduce local coordinates in the border zones, putting $\phi = \phi_1 - \omega_1$ in the A solution and $\phi = \phi_2 + \omega_2$ in the B solution. We may then absorb a constant factor $e^{\varkappa(1-i)\phi_1}$ or $e^{\varkappa(1-i)\phi_2}$ in the constants A and B respectively and may write

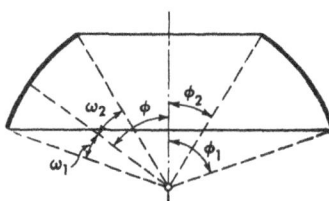

Fig. 5. Spherical shell with two edges

$$y = A\, e^{-\varkappa\omega_1}(\cos\varkappa\omega_1 + i\sin\varkappa\omega_1) \sum_{n=0}^{\infty} \frac{y_n}{(1-i)^n \varkappa^n}$$

$$+ B\, e^{-\varkappa\omega_2}(\cos\varkappa\omega_2 + i\sin\varkappa\omega_2) \sum_{n=0}^{\infty} \frac{(-1)^n\, y_n}{(1-i)^n \varkappa^n}.$$

After separating real and imaginary parts in these expressions, we may write y in the following form:

$$y = A\, e^{-\varkappa\omega_1}[(Y_1 \cos\varkappa\omega_1 - Y_2 \sin\varkappa\omega_1) + i(Y_2 \cos\varkappa\omega_1 + Y_1 \sin\varkappa\omega_1)] \quad (31)$$
$$+ B\, e^{-\varkappa\omega_2}[(Y_3 \cos\varkappa\omega_2 - Y_4 \sin\varkappa\omega_2) + i(Y_4 \cos\varkappa\omega_2 + Y_3 \sin\varkappa\omega_2)],$$

where $Y_1 \ldots Y_4$ are four series in descending powers of $\varkappa$:

$$\left.\begin{aligned}
Y_1 &= 1 + \frac{y_1}{2\varkappa} \qquad\quad - \frac{y_3}{4\varkappa^3} - \frac{y_4}{4\varkappa^4} - \frac{y_5}{8\varkappa^5} \qquad\quad + \frac{y_7}{16\varkappa^7} + \frac{y_8}{16\varkappa^8} + \cdots, \\
Y_2 &= \frac{y_1}{2\varkappa} + \frac{y_2}{2\varkappa^2} + \frac{y_3}{4\varkappa^3} \qquad\quad - \frac{y_5}{8\varkappa^5} - \frac{y_6}{8\varkappa^6} - \frac{y_7}{16\varkappa^7} \qquad\quad + \cdots, \\
Y_3 &= 1 - \frac{y_1}{2\varkappa} \qquad\quad + \frac{y_3}{4\varkappa^3} - \frac{y_4}{4\varkappa^4} + \frac{y_5}{8\varkappa^5} \qquad\quad - \frac{y_7}{16\varkappa^7} + \frac{y_8}{16\varkappa^8} - \cdots, \\
Y_4 &= -\frac{y_1}{2\varkappa} + \frac{y_2}{2\varkappa^2} - \frac{y_3}{4\varkappa^3} \qquad\quad + \frac{y_5}{8\varkappa^5} - \frac{y_6}{8\varkappa^6} + \frac{y_7}{16\varkappa^7} \qquad\quad - \cdots.
\end{aligned}\right\} \quad (32)$$

We may then use eqs. (17b) and (21a) to express the second principal variable, χ, in terms of Q_ϕ:

$$D(1 - \nu^2)\chi = (2i\varkappa^2 - \nu)Q_\phi = (2i\varkappa^2 - \nu)\frac{y}{\sqrt{\sin\phi}},$$

and introducing y from (31) here, we find

$$\begin{aligned}
\chi = \frac{1}{D(1-\nu^2)\sqrt{\sin\phi}} \{&A e^{-\varkappa\omega_1}[(-(2\varkappa^2 Y_2 + \nu Y_1)\cos\varkappa\,\omega_1 \\
&- (2\varkappa^2 Y_1 - \nu Y_2)\sin\varkappa\,\omega_1) \\
&+ i((2\varkappa^2 Y_1 - \nu Y_2)\cos\varkappa\,\omega_1 - (2\varkappa^2 Y_2 + \nu Y_1)\sin\varkappa\,\omega_1)] \\
&+ B e^{-\varkappa\omega_2}[(-(2\varkappa^2 Y_4 + \nu Y_3)\cos\varkappa\,\omega_2 - (2\varkappa^2 Y_3 - \nu Y_4)\sin\varkappa\,\omega_2) \\
&+ i((2\varkappa^2 Y_3 - \nu Y_4)\cos\varkappa\,\omega_2 - (2\varkappa^2 Y_4 + \nu Y_3)\sin\varkappa\,\omega_2)]\}.
\end{aligned}$$

Table 6. Functions y_n *and* y_n'

ψ	y_1	y_2	y_3	y_4	y_1'	y_2'	y_3'	y_4'
$0°$	0	0	0	0	$-.2500$	0	.1562	0
$5°$	.0217	$-.0012$	$-.0139$	$-.0039$	$-.2471$	.0277	.1638	.0901
$10°$	.0430	$-.0049$	$-.0291$	$-.0160$	$-.2383$	.0579	.1871	.1897
$15°$	.0631	$-.0115$	$-.0473$	$-.0376$	$-.2231$	.0936	.2287	.3097
$20°$	.0817	$-.0215$	$-.0703$	$-.0711$	$-.2003$	.1382	.2936	.466
$25°$	.0978	$-.0360$	$-.1005$	$-.1206$	$-.1685$	.1964	.3899	.681
$30°$	.1107	$-.0564$	$-.1413$	$-.1928$	$-.1250$	.2748	.531	.994
$35°$	.1192	$-.0848$	$-.1976$	$-.2988$	$-.0661$	.3834	.741	1.472
$40°$	.1217	$-.1246$	$-.2769$	$-.458$	.0140	.538	1.058	2.240
$45°$	.1159	$-.1808$	$-.3916$	$-.705$	.1250	.764	1.556	3.552
$50°$	.0985	$-.2614$	$-.563$	-1.109	.2826	1.109	2.380	5.95
$55°$	.0644	$-.3804$	$-.831$	-1.812	.515	1.661	3.834	10.75
$60°$	.0050	$-.562$	-1.278	-3.153	.875	2.602	6.64	21.50
$65°$	$-.0952$	$-.858$	-2.092	-6.045	1.475	4.362	12.75	49.6
$70°$	$-.2667$	-1.380	-3.785	-13.51	2.581	8.119	28.66	140.5

Both Q_ϕ and χ are complex-valued functions of ϕ. As explained on p. 324, their real and imaginary parts represent independent solutions of the fourth-order shell problem, whose general solution with four free constants A_1, A_2, B_1, B_2 is therefore:

$$\begin{aligned}
Q_\phi = \frac{1}{\sqrt{\sin\phi}} \{&e^{-\varkappa\omega_1}[A_1(Y_1\cos\varkappa\,\omega_1 - Y_2\sin\varkappa\,\omega_1) \\
&+ A_2(Y_2\cos\varkappa\,\omega_1 + Y_1\sin\varkappa\,\omega_1)] \\
&+ e^{-\varkappa\omega_2}[B_1(Y_3\cos\varkappa\,\omega_2 - Y_4\sin\varkappa\,\omega_2) \\
&+ B_2(Y_4\cos\varkappa\,\omega_2 + Y_3\sin\varkappa\,\omega_2)]\}
\end{aligned} \qquad (33)$$

and a similar expression for χ. The normal forces N_ϕ and N_θ may be found from eqs. (15) and the bending moments M_ϕ and M_θ from eqs. (11c, d). For greater simplicity we shall again assume that $P = 0$. In this case the formulas for all the stress resultants and deformations

Table 7. Coefficients for Spherical Shells

f	c	f_1
Q_ϕ	1	Y_1
χ	$\dfrac{1}{D(1-\nu^2)}$	$-(2\varkappa^2 Y_2 + \nu Y_1)$
N_ϕ	1	$-Y_1 \cot\phi$
N_θ	1	$\dfrac{1}{2} Y_1 \cot\phi - [Y_1 + \varkappa(Y_1 + Y_2)]$
χ	$\dfrac{1}{D(1-\nu^2)}$	$\dfrac{1}{2}(2\varkappa^2 Y_2 + \nu Y_1)\cot\phi - (2\varkappa^2 Y_2' + \nu Y_1')$ $+ 2\varkappa^3(Y_1 - Y_2) - \nu\varkappa(Y_1 + Y_2)$
M_ϕ	$\dfrac{K}{Da(1-\nu^2)}$	$\dfrac{1}{2}(1-2\nu)(2\varkappa^2 Y_2 + \nu Y_1)\cot\varphi - (2\varkappa^2 Y_2' + \nu Y_1')$ $+ 2\varkappa^3(Y_1 - Y_2) - \nu\varkappa(Y_1 + Y_2)$
M_θ	$\dfrac{K}{Da(1-\nu^2)}$	$-\dfrac{1}{2}(2-\nu)(2\varkappa^2 Y_2 + \nu Y_1)\cot\varphi - \nu(2\varkappa^2 Y_2' + \nu Y_1')$ $+ 2\nu\varkappa^3(Y_1 - Y_2) - \nu^2\varkappa(Y_1 + Y_2)$

may be written in the form

$$f = \frac{c}{\sqrt{\sin\phi}} \{e^{-\varkappa\omega_1}[(A_1 f_1 + A_2 f_2)\cos\varkappa\,\omega_1 + (A_2 f_1 - A_1 f_2)\sin\varkappa\,\omega_1]$$
$$+ e^{-\varkappa\omega_2}[(B_1 g_1 + B_2 g_2)\cos\varkappa\,\omega_2 + (B_2 g_1 - B_1 g_2)\sin\varkappa\,\omega_2]\}. \tag{34}$$

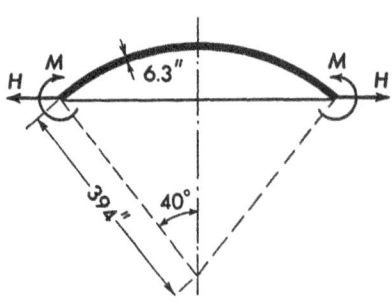

Fig. 6. Spherical tank bottom

The values of the constant c and the functions f_1, f_2 are given in Table 7. Similar expressions for g_1, g_2 are obtained as indicated at the bottom of the table.

The formulas of this table represent the solution of the problem, but this solution still needs some discussion. The series employed here are not truly convergent. One may see this at once in the vicinity of the point $\phi = 0$. There $\cot\phi \to \infty$ and the higher n, the more strongly y_n approaches infinity. The solution has the character of an asymptotic solution. For a fixed number of terms in each of the series (32) our formulas approach the true solution better, the larger $\varkappa$ is; but for a given $\varkappa$

Table 7. (Continued)

f_2
Y_2
$2\varkappa^2 Y_1 - \nu Y_2$
$-Y_2 \cot\phi$
$\frac{1}{2}Y_2 \cot\phi - [Y_2' + \varkappa(Y_2 - Y_1)]$
$-\frac{1}{2}(2\varkappa^2 Y_1 - \nu Y_2)\cot\phi + (2\varkappa^2 Y_1' - \nu Y_2') + 2\varkappa^3(Y_1 + Y_2) + \nu\varkappa(Y_1 - Y_2)$
$-\frac{1}{2}(1 - 2\nu)(2\varkappa^2 Y_1 - \nu Y_2)\cot\phi + (2\varkappa^2 Y_1' - \nu Y_2') + 2\varkappa^3(Y_1 + Y_2) + \nu\varkappa(Y_1 - Y_2)$
$\frac{1}{2}(2 - \nu)(2\varkappa^2 Y_1 - \nu Y_2)\cot\phi + \nu(2\varkappa^2 Y_1' - \nu Y_2') + 2\nu\varkappa^3(Y_1 + Y_2) + \nu^2\varkappa(Y_1 - Y_2)$

To obtain g_1 and g_2 from f_1 and f_2: Replace Y_1 by Y_3, Y_2 by Y_4, $\varkappa$ by $-\varkappa$.

they always keep from it a distance that cannot be decreased indefinitely by using more terms of the series.

The use of the method may be illustrated by analyzing the spherical bottom of a reinforced concrete water tank (fig. 6). This shell has been investigated by J. E. EKSTRÖM[1] for the edge load $H = 1095$ lb/in, $M = -9715$ in·lb/in. When we assume with EKSTRÖM $\nu = 0$, we find $\varkappa = 10.42$. This value is rather high for the first method and will give considerable trouble with the power series (25). On the other hand, $\varkappa$ is not large enough for the simple solution explained under 6.2.1.4, but it is just right for the asymptotic solution under consideration.

We pick from Table 6 the numerical values y_n and y_n' for $\phi = 40°$ and find from eqs. (32) with slide-rule accuracy:

$$Y_1 = 0.995, \quad Y_2 = 0.00340, \quad Y_1' = 0.01289, \quad Y_2' = 0.01919.$$

These series converge well, e. g.

$$Y_1' = 0.01355 - 0.00053 - 0.00013.$$

Using eq. (34) and Table 7, we may now easily formulate two equations for the constants A_1 and A_2, expressing the boundary conditions that

---

[1] Ing. Vetensk. Akad. Stockholm, Hdl. 121 (1933), p. 126.

$Q_\phi = -H \sin 40° = -704$ lb/in and $M_\phi = -9715$ in·lb/in. They yield $A_1 = -581$ lb/in, $A_2 = +194.0$ lb/in. Introducing these figures into eq. (34) and making further use of the table, we may compute all the stress resultants. Two of them, Q_ϕ and M_ϕ, are shown in fig. 7. As far

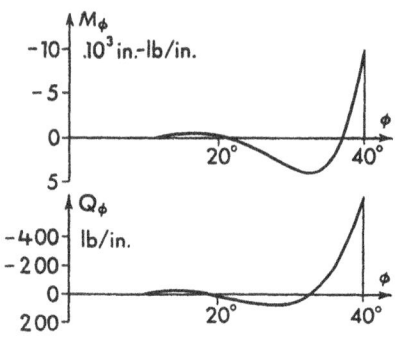

up the meridian as $\phi = 20°$ the series (32) are easy to handle. Beyond that the convergence becomes unsatisfactory, but the stress resultants are already so small that they have no practical interest.

Fig. 7. Stress resultants in the shell of Fig. 6

6.2.1.4 Simplified Asymptotic Solution

If one tried to apply the asymptotic theory just presented to a reinforced concrete dome with $\varkappa = 30$, he would find that each of the series $Y_1 \ldots Y_4$ is practically reduced to its first term. This simplifies appreciably the numerical work, but there still remains a chance for a more drastic simplification. All displacements, deformations, and stress resultants have the form

$$e^{\pm \varkappa \phi} f(\phi) \; {\cos \atop \sin} \varkappa \phi,$$

where $f(\phi)$ does not vary much in the interesting range of the coordinate ϕ. Every derivative of such a product has the same form but an additional factor $\varkappa$. When $\varkappa$ is sufficiently large, it is possible to neglect the lower derivatives of a variable compared with the highest one, unless a coefficient is extremely large.

When we apply this idea to the operator L, eq. (16), we recognize that the lower derivatives are multiplied by $\cot \phi$ and $\cot^2 \phi$. Now, for $\phi > 30°$ or even $>25°$, $\cot \phi$ is still of rather moderate size, and then the operator L simplifies to the second derivative:

$$L(\ldots) = (\ldots)^{\cdot\cdot}.$$

Eqs. (21 a, b) then become differential equations with constant coefficients:

$$Q_\phi^{\cdot\cdot} \pm 2 i \varkappa^2 Q_\phi = 0,$$

and their combined solution is

$$Q_\phi = c_1 e^{(1+i) \varkappa \phi} + c_2 e^{(1-i) \varkappa \phi} + c_3 e^{-(1+i) \varkappa \phi} + c_4 e^{-(1-i) \varkappa \phi}.$$

There are different ways of writing this in a real form. They all consist in using as a fundamental system four linear combinations of the complex exponentials, e. g.

$$Q_\phi = e^{\varkappa\phi}(A_1\cos\varkappa\phi + A_2\sin\varkappa\phi) + e^{-\varkappa\phi}(B_1\cos\varkappa\phi + B_2\sin\varkappa\phi) \quad (35)$$

or, with another set of constants,

$$Q_\phi = C_1\operatorname{Cosh}\varkappa\phi\,\cos\varkappa\phi + C_2\operatorname{Sinh}\varkappa\phi\,\sin\varkappa\phi + C_3\operatorname{Cosh}\varkappa\phi\,\sin\varkappa\phi \\ + C_4\operatorname{Sinh}\varkappa\phi\,\cos\varkappa\phi. \quad (36)$$

The constants A and B or C are determined by the boundary conditions at the edges $\phi = \phi_1$ and $\phi = \phi_2$ of, the shell. If these edges are not too close together, the A terms will predominate at the lower and the B terms at the upper edge and, because of the rapid increase or decrease of the exponential factors, the predominance of one pair at each edge is often so strong that for all practical purposes the two pairs of boundary conditions are independent of each other. In such cases it is useful to introduce again (see p. 332) the coordinates $\omega_1 = \phi_1 - \phi$ and $\omega_2 = \phi - \phi_2$ and, with a different meaning of the notation for the constants, to write

$$Q_\phi = e^{-\varkappa\omega_1}(A_1\cos\varkappa\,\omega_1 + A_2\sin\varkappa\,\omega_1) + e^{-\varkappa\omega_2}(B_1\cos\varkappa\,\omega_2 + B_2\sin\varkappa\,\omega_2). \quad (37\text{a})$$

Formulas for the normal forces may easily be established by introducing (37a) into eqs. (15) from which, of course, the P terms must be dropped:

$$N_\phi = -Q_\phi\cot\phi,$$

$$N_\theta = -Q_\phi^{\cdot} = -\varkappa e^{-\varkappa\omega_1}[(A_1 - A_2)\cos\varkappa\,\omega_1 + (A_1 + A_2)\sin\varkappa\,\omega_1] \\ + \varkappa e^{-\varkappa\omega_2}[(B_1 - B_2)\cos\varkappa\,\omega_2 + (B_1 + B_2)\sin\varkappa\,\omega_2]. \quad (37\text{b, c})$$

The rotation χ of the shell element is found by introducing (37a) into eq. (17b), using there the simplified form of the operator L and neglecting the term with Q_ϕ against $L(Q_\phi)$:

$$D(1 - \nu^2)\,\chi = -Q_\phi^{\cdot\cdot} = 2\varkappa^2 e^{-\varkappa\omega_1}[A_2\cos\varkappa\,\omega_1 - A_1\sin\varkappa\,\omega_1] \\ - 2\varkappa^2 e^{-\varkappa\omega_2}[B_2\cos\varkappa\,\omega_2 - B_1\sin\varkappa\,\omega_2]. \quad (37\text{d})$$

When we introduce this into the elastic law (11c, d), we do not only neglect χ compared with $\chi^{\cdot}$, but we also drop the term $\nu^2/4$ in eq. (20) and arrive at the following formulas for the bending moments:

$$M_\phi = -\frac{a}{4\varkappa^4}Q_\phi^{\cdots} = \frac{a}{2\varkappa}e^{-\varkappa\omega_1}[(A_1 + A_2)\cos\varkappa\,\omega_1 - (A_1 - A_2)\sin\varkappa\,\omega_1] \\ - \frac{a}{2\varkappa}e^{-\varkappa\omega_2}[(B_1 + B_2)\cos\varkappa\,\omega_2 - (B_1 - B_2)\sin\varkappa\,\omega_2],$$

$$M_\theta = \nu\,M_\phi. \quad (37\text{e, f})$$

In many practical cases, in particular in the simpler ones, it is useful to write the solution (37a) in still another form which has as free constants two amplitudes C_1, C_2, and two phase angles ψ_1, ψ_2:

$$Q_\phi = C_1 e^{-\varkappa \omega_1} \sin(\varkappa \omega_1 + \psi_1) + C_2 e^{-\varkappa \omega_2} \sin(\varkappa \omega_2 + \psi_2). \quad (38\text{a})$$

Compared with eq. (37a), this version has the disadvantage that it will not be possible to obtain *linear* equations for the constants ψ_1, ψ_2; but when it is possible to see at a glance which values the phase angles will have, then eq. (38a) has the advantage that only two linear equations will be needed instead of four. Going through the same procedure as before, we again find the relations (37b, f) for N_ϕ and M_θ and for the other quantities the following formulas:

$$N_\theta = \varkappa \sqrt{2} \left[- C_1 e^{-\varkappa \omega_1} \sin\left(\varkappa \omega_1 + \psi_1 - \frac{\pi}{4}\right) \right.$$
$$\left. + C_2 e^{-\varkappa \omega_2} \sin\left(\varkappa \omega_2 + \psi_2 - \frac{\pi}{4}\right) \right],$$

$$D(1 - \nu^2) \chi = 2\varkappa^2 [C_1 e^{-\varkappa \omega_1} \cos(\varkappa \omega_1 + \psi_1) \qquad (38\text{c-e})$$
$$+ C_2 e^{-\varkappa \omega_2} \cos(\varkappa \omega_2 + \psi_2)],$$

$$M_\phi = \frac{a}{\varkappa \sqrt{2}} \left[C_1 e^{-\varkappa \omega_1} \sin\left(\varkappa \omega_1 + \psi_1 + \frac{\pi}{4}\right) \right.$$
$$\left. - C_2 e^{-\varkappa \omega_2} \sin\left(\varkappa \omega_2 + \psi_2 + \frac{\pi}{4}\right) \right].$$

Eqs. (38) represent the form of the solution which is most frequently used when dealing with practical problems. We may derive from it ready-to-use formulas for two important cases, described by figs. 8a, b.

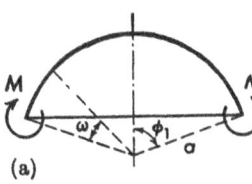

(a)

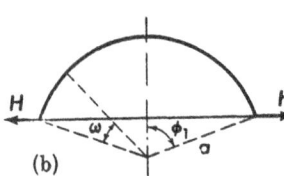

(b)

Fig. 8
Edge loads at a spherical cap

In both cases we must drop the C_2 solution and we may then write ω instead of ω_1. In the first case, fig. 8a, we have at $\phi = \phi_1$, $\omega = 0$ the conditions $Q_\phi = 0$ and $M_\phi = M$, which lead to

$$\psi_1 = 0, \qquad C_1 = \frac{2 \varkappa M}{a}. \qquad (39)$$

For the application to statically indeterminate structures, it is good to know for $\omega = 0$ the rotation χ and the hoop-strain. They are

$$\chi = \frac{4 \varkappa^3 M}{D(1 - \nu^2) a}, \qquad \epsilon_\theta = \frac{2 \varkappa^2 M}{D(1 - \nu^2) a}. \qquad (39')$$

In the second case, fig. 8b, the boundary conditions are $M_\phi = 0$ and $Q_\phi = -H \sin\phi_1$. They yield

$$\psi_1 = -\frac{\pi}{4}, \qquad C_1 = H \sqrt{2} \sin\phi_1 \qquad (40)$$

and

$$\chi = \frac{2 \varkappa^2 H \sin\phi_1}{D(1 - \nu^2)}, \qquad \epsilon_\theta = \frac{H}{D(1 - \nu^2)} (2\varkappa \sin\phi_1 - \nu \cos\phi_1). \qquad (40')$$

The application of these formulas may be seen in two examples.

When a spherical container (fig. 9) is partially filled with a hot liquid, not only the weight of the content will lead to stresses in the wall but also the difference in temperature between the hot and the cold parts of the shell. We intend to find these thermal stresses, assuming that the sphere is just half filled.

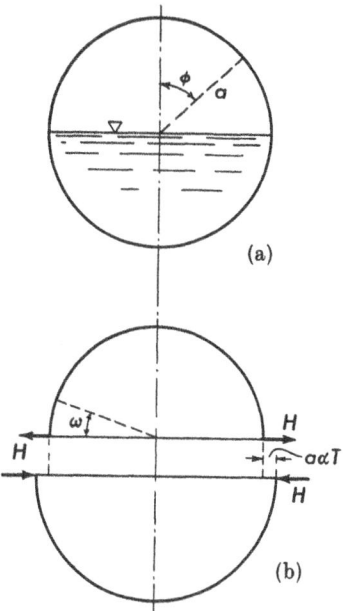

(a)

(b)

Fig. 9. Spherical tank half filled with a hot liquid

When we cut the cold and the hot hemispheres apart along the equator, there will be no thermal stress, but a gap of width $a \alpha T$ between the edges, α being the coefficient of thermal expansion and T the difference in temperature (fig. 9b). To close this gap, radial forces H must be applied which will bend the upper shell outward by $\frac{1}{2} a \alpha T$ and the lower shell inward by the same amount, thus producing hoop strains $\epsilon_\theta = \pm\frac{1}{2}\alpha T$. From symmetry it follows that the rotation χ will then be the same at both edges so that the tangent to the meridian will be continuous without the application of moments M. The upper hemisphere is then exactly in the situation described by fig. 8b and from eq. (40') we find with $\phi_1 = 90°$:

$$H = \frac{D(1 - \nu^2) \alpha T}{4 \varkappa}, \qquad C_1 = H \sqrt{2}.$$

Now all the formulas (38) may be used to find anything we want. In fig. 10 the hoop force N_θ and the meridional bending moment M_ϕ are plotted for a rather thin steel shell, having $\varkappa = 28.73$. The diagrams show that the thermal stresses are limited to a zone of about 6° on each side of the discontinuity. In such cases the real forces and moments will be smaller because the local change of temperature is never as sudden as we assumed it to be. If the shell is thicker, the zone of thermal stresses will be wider, and then a slight smoothing out of the discontinuity of the temperature will not greatly influence the result of the stress analysis.

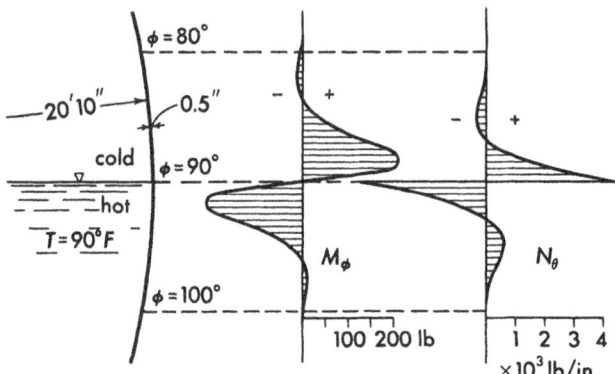

Fig. 10. Thermal stresses in the tank of Fig. 9 a

A second example may explain the co-operation of different shells. Fig. 11 shows a cylindrical boiler drum closed by a hemispherical end.

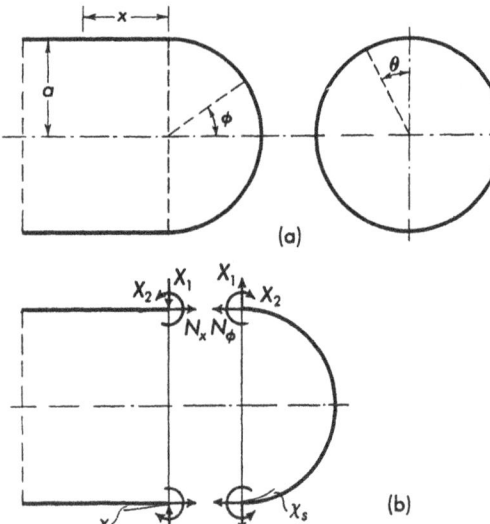

(a)

(b)

Fig. 11. Hemispherical boiler end, (a) axial and transverse sections, (b) cylinder and hemisphere cut apart to show the redundant stress resultants X_1, X_2

When an internal pressure p is applied, the membrane theory will yield the following normal forces:

in the cylinder:

$$N_x = \frac{1}{2} p a,$$

$$N_\theta = p a,$$

in the sphere:

$$N_\phi = N_\theta = \frac{1}{2} p a.$$

At the juncture of the two shells ($x = 0$ and $\phi = 90°$), the meridional forces N_x and N_ϕ are in equilibrium with each other, but there is a discrepancy in the hoop strains. In the cylinder we have

$$\epsilon_\theta = \epsilon_{\theta c} = \frac{1}{D(1 - \nu^2)} (N_\theta - \nu N_x) = \frac{p a(2 - \nu)}{2 D(1 - \nu^2)}$$

and in the sphere

$$\epsilon_\theta = \epsilon_{\theta s} = \frac{1}{D(1 - \nu^2)} (N_\theta - \nu N_\phi) = \frac{p\,a\,(1 - \nu)}{2D(1 - \nu^2)}.$$

The membrane forces will therefore only be possible if we separate the two shells (fig. 11 b), and then the deformation will be such that the edges do not fit together. We may make them match by applying radial forces X_1 and, perhaps, moments X_2 in order to make the tangents of the deformed meridians also coincide. We consider these two quantities as redundant in a statically indeterminate structure and have already introduced the pertinent notation. The deformations corresponding to the redundant quantities are the radial gaping

$$D(1 - \nu^2)\,a(\epsilon_{\theta s} - \epsilon_{\theta c}) = D(1 - \nu^2)\,(w_s - w_c) = \delta_1$$

and the angular gaping

$$D(1 - \nu^2)\,(\chi_s - \chi_c) = \delta_2.$$

They may be written as

$$\delta_1 = \delta_{10} + X_1\,\delta_{11} + X_2\,\delta_{12},$$

$$\delta_2 = \delta_{20} + X_1\,\delta_{21} + X_2\,\delta_{22},$$

thus separating the term due to the membrane forces:

$$\delta_{10} = -\frac{p\,a^2}{2}, \qquad \delta_{20} = 0,$$

and the terms due to a unit of X_1 or of X_2, respectively. These latter deformations are connected with bending stresses in both shells. For the cylinder, we use the formulas of 5.5.2.1, in particular eqs. (V–76′) and (V–77′), and for the sphere eqs. (39′) and (40′).

We assume that both shells have the same wall thickness t. Then the constants $\varkappa$ used for cylinder and sphere are the same, at least if we decide to neglect the small term $\nu^2/4$ in eq. (20). From eqs. (40′) and (V–77′) we find

$$\delta_{11} = 2\varkappa a + \frac{D(1 - \nu^2)\,a^3}{2K\varkappa^3} = 4\varkappa a,$$

$$\delta_{21} = 2\varkappa^2 - \frac{D(1 - \nu^2)\,a^2}{2K\varkappa^2} = 0,$$

and from (39′) and (V–76′):

$$\delta_{12} = 2\varkappa^2 - \frac{D(1 - \nu^2)\,a^2}{2K\varkappa^2} = 0,$$

$$\delta_{22} = \frac{4\varkappa^3}{a} + \frac{D(1 - \nu^2)\,a}{K\varkappa} = \frac{8\varkappa^3}{a}.$$

The conditions for the compatibility of deformations,

$$\delta_1 = 0, \qquad \delta_2 = 0,$$

read therefore

$$-\frac{p\,a^2}{2} + X_1 \cdot 4\varkappa\,a = 0, \qquad X_2 \cdot \frac{8\varkappa^3}{a} = 0,$$

and their solution is obviously

$$X_1 = \frac{p\,a}{8\varkappa}, \qquad X_2 = 0.$$

We may now find all stress resultants for the sphere from eqs. (38) with

$$C_1 = X_1\,\sqrt{2} = \frac{p\,a}{4\,\sqrt{2}\,\varkappa}, \qquad \psi_1 = -\frac{\pi}{4}$$

and all those for the cylinder from eqs. (V–77) with $H = -X_1$, in particular the bending moment

$$M_\phi = \frac{p\,a^2}{8\,\varkappa^2}\,e^{-\varkappa\omega}\sin\varkappa\,\omega$$

in the sphere and

$$M_x = -\frac{p\,a^2}{8\,\varkappa^2}\,e^{-\varkappa x/a}\sin\frac{\varkappa\,x}{a}$$

in the cylinder, as well as the hoop force (including the membrane force)

$$N_\theta = p\,a\left(\frac{1}{2} + \frac{1}{4}\,e^{-\varkappa\omega}\cos\varkappa\,\omega\right)$$

in the sphere and

$$N_\theta = p\,a\left(1 - \frac{1}{4}\,e^{-\varkappa x/a}\cos\frac{\varkappa\,x}{a}\right)$$

in the cylinder. These stress resultants are shown in fig. 12 for $t/a = 0.010$ and $\nu = 0.3$. In the N_θ diagram one recognizes the continuous transition of the hoop force (shaded diagram) replacing the

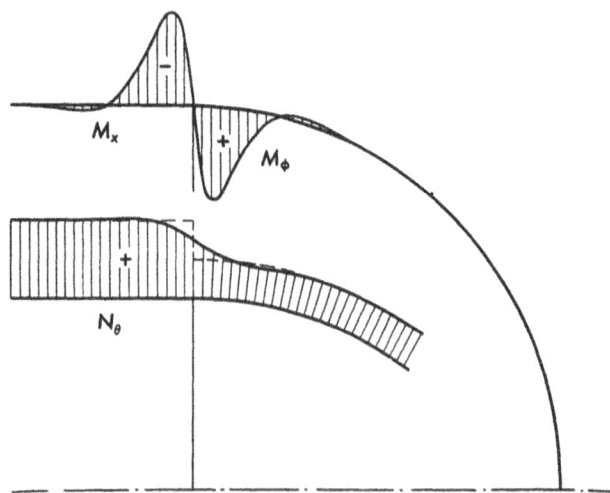

Fig. 12. Stress resultants (discontinuity stresses) at the juncture of a cylindrical boiler drum and a hemispherical boiler end

discontinuity of the membrane theory (broken line). In the bending moment diagram both maxima are equal, $\pm 0.244 \times 10^{-3}\, p\, a^2$, and the zero between them coincides exactly with the boundary between cylinder and sphere. This fact indicates that a welding seam should be placed right along this line and not at a short distance away from it where it might easily get into a region of maximum bending stress.

It is interesting to compare fig. 12 with the stresses in a vessel like fig. 13a. This pressure vessel is closed by a shallow spherical cap of radius $b = a/\sin\phi_1$. The membrane forces in this cap are

$$N_\phi = N_\theta = \frac{p\,b}{2} = \frac{p\,a}{2\sin\phi_1},$$

whereas the cylinder has the same membrane forces as in the preceding example. The particular difficulty of the problem consists in the fact

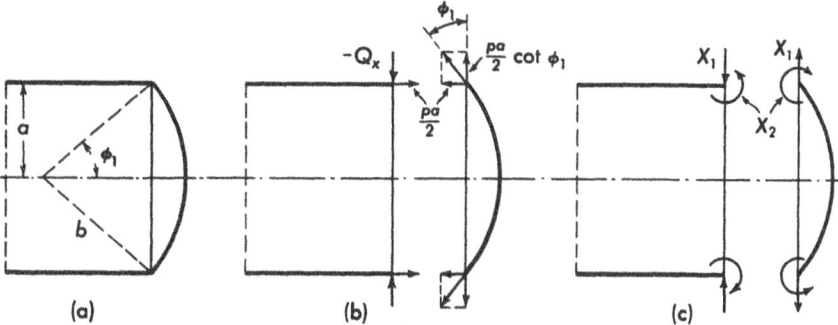

Fig. 13. Shallow spherical boiler end, (a) axial section, (b) forces in the principal system, (c) redundant quantities

that these membrane forces cannot be used as the internal forces of a principal system, because they are not in equilibrium with each other and the pressure p. We see this at once when we cut the two shells apart (fig. 13b). The axial component $pa/2$ of the force N_ϕ in the sphere is transmitted to the cylinder as a force N_x, but the radial component of N_ϕ has, so far, no counterpart on the left-hand side of the cut. To procure it, we must still apply a transverse force

$$Q_x = -\frac{p\,a}{2}\cot\phi_1$$

at the edge of the cylinder, and the bending stresses which it causes are a part of the stresses of the principal system.

From this basis we may proceed in the usual way and apply radial forces X_1 and moments X_2 along the edges of both shells (fig. 13c), choosing their magnitudes so as to restore the continuity of deformation. There is no difficulty in formulating the equations which express the continuity of ϵ_θ and χ, but the computation soon gets rather bulky and we leave it to the reader to work out the details.

A result which has been obtained in this way is represented in fig. 14. In this example it has been assumed that the walls of the cylinder and the sphere have the same thickness t with $t/a = 0.010$ and $\phi_1 = 45°$. The N_θ diagram shows that a zone with high compressive stress develops on both sides of the edge, indicating the usefulness of a reinforcing ring along this line. The distribution of bending moments

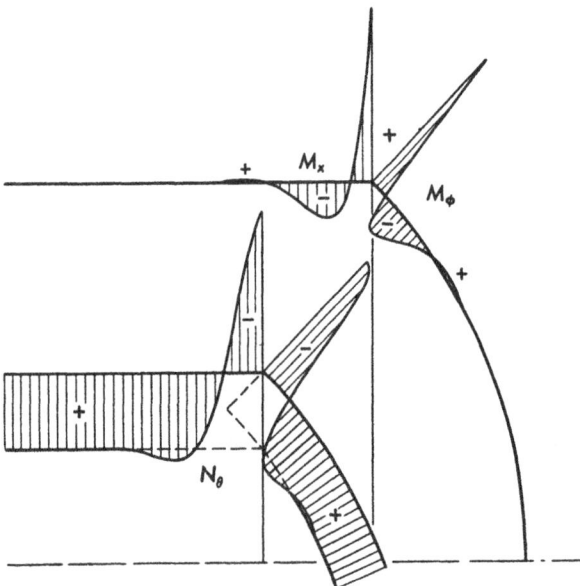

Fig. 14. Stress resultants at the juncture of a cylindrical boiler drum and a shallow spherical boiler end

is also entirely different from fig. 12. Instead of having a zero at the joint between the shells, the moment has a sharp peak there, and this peak is so high ($8.92 \times 10^{-3} \, pa^2$) that it was impossible to draw the ordinates of the moments in figs. 12 and 14 to the same scale.

These results indicate clearly that it is not a good practice to have a sharp edge between the boiler end and the boiler drum, and one may easily conclude that an almost sharp edge in the meridian, rounded by an arc of great curvature, is almost as bad. If for some reason or other the edge cannot be avoided, one should at least provide a strong stiffening ring there. The stress analysis will then be still more involved, but it will pay because of the partial reduction of the bending stresses in the shell.

6.2.1.5 Bending Stresses in the Vicinity of the Apex

Of the three solutions explained in the preceding sections, only the first one is applicable, after some modifications, to the top of the shell and its immediate vicinity. It is, however, subject to the difficulties resulting from the slow convergence of the power series employed.

In some cases another approach is possible. In the vicinity of $\phi = 0$, $\cot\phi$ may be expanded into a LAURENT series:

$$\cot\phi = \frac{1}{\phi} - \frac{\phi}{3} - \frac{\phi^3}{45} - \cdots,$$

and if ϕ is small enough, we may approximate $\cot\phi$ by ϕ^{-1}. When this is done in the coefficients of eq. (22), it reads

$$Q_\phi'' + \phi^{-1} Q_\phi' - \phi^{-2} Q_\phi + 2i\varkappa^2 Q_\phi = 0.$$

This is almost BESSEL's equation. When we introduce a new independent variable

$$\xi = \varkappa\sqrt{2i}\,\phi,$$

it will assume the standard form

$$\frac{d^2 Q_\phi}{d\xi^2} + \frac{1}{\xi}\frac{dQ_\phi}{d\xi} + \left(1 - \frac{1}{\xi^2}\right)Q_\phi = 0. \tag{41}$$

On p. 289 we met this equation, and we saw that its solutions are the BESSEL functions of the first order of the complex argument ξ. It was explained there, how these functions may be written in terms of the first derivatives of the THOMSON functions of the real variable

$$x = \varkappa\phi\sqrt{2}$$

so that we may write

$$Q_\phi' = C_1[(\text{bei}\,'x - \text{ber}\,'x) + i(\text{bei}\,'x + \text{ber}\,'x)] \\ + C_2[(\text{ker}\,'x + \text{kei}\,'x) + i(\text{ker}\,'x - \text{kei}\,'x)].$$

We know that real and imaginary parts of this expression will separately satisfy eq. (19), and so will any linear combination of these four functions. In this way we finally arrive at the following form of the general solution:

$$Q_\phi = A_1\,\text{ber}\,'x + A_2\,\text{bei}\,'x + B_1\,\text{ker}\,'x + B_2\,\text{kei}\,'x. \tag{42a}$$

When we introduce this into (17) to find χ, we make use of eqs. (V–88) which allow the elimination of all derivatives higher than the first of the THOMSON functions, and we obtain

$$D(1 - \nu^2)\chi = A_1(2\varkappa^2\,\text{bei}\,'x - \nu\,\text{ber}\,'x) - A_2(2\varkappa^2\,\text{ber}\,'x + \nu\,\text{bei}\,'x) \\ + B_1(2\varkappa^2\,\text{kei}\,'x - \nu\,\text{ker}\,'x) - B_2(2\varkappa^2\,\text{ker}\,'x + \nu\,\text{kei}\,'x). \tag{42b}$$

We may now use eqs. (15) with $P = 0$ to find the normal forces and eqs. (11c, d) for the bending moments:

$$N_\phi = -\phi^{-1}Q_\phi,$$

$$N_\theta = \varkappa\sqrt{2}[A_1(\text{bei } x + x^{-1}\text{ber }'x) - A_2(\text{ber } x - x^{-1}\text{bei }'x)$$
$$+ B_1(\text{kei } x + x^{-1}\text{ker }'x) - B_2(\text{ker } x - x^{-1}\text{kei }'x)],$$

$$M_\phi = \frac{K\varkappa\sqrt{2}}{Da(1-\nu^2)}\left[A_1\left[2\varkappa^2\left(\text{ber } x - \frac{1-\nu}{x}\text{bei}'x\right) + \nu\left(\text{bei } x + \frac{1-\nu}{x}\text{ber}'x\right)\right]\right.$$
$$+ A_2\left[2\varkappa^2\left(\text{bei } x + \frac{1-\nu}{x}\text{ber}'x\right) - \nu\left(\text{ber } x - \frac{1-\nu}{x}\text{bei}'x\right)\right]$$
$$+ B_1\left[2\varkappa^2\left(\text{ker } x - \frac{1-\nu}{x}\text{kei}'x\right) + \nu\left(\text{kei } x + \frac{1-\nu}{x}\text{ker}'x\right)\right]$$
$$\left.+ B_2\left[2\varkappa^2\left(\text{kei } x + \frac{1-\nu}{x}\text{ker}'x\right) - \nu\left(\text{ker } x - \frac{1-\nu}{x}\text{kei}'x\right)\right]\right],$$

(42c)

$$M_\theta = \frac{K\varkappa\sqrt{2}}{Da(1-\nu^2)}\left[A_1\left[2\varkappa^2\left(\nu\,\text{ber } x + \frac{1-\nu}{x}\text{bei}'x\right) + \nu\left(\nu\,\text{bei } x - \frac{1-\nu}{x}\text{ber}'x\right)\right]\right.$$
$$+ A_2\left[2\varkappa^2\left(\nu\,\text{bei } x - \frac{1-\nu}{x}\text{ber}'x\right) - \nu\left(\nu\,\text{ber } x + \frac{1-\nu}{x}\text{bei}'x\right)\right]$$
$$+ B_1\left[2\varkappa^2\left(\nu\,\text{ker } x + \frac{1-\nu}{x}\text{kei}'x\right) + \nu\left(\nu\,\text{kei } x - \frac{1-\nu}{x}\text{ker}'x\right)\right]$$
$$\left.+ B_2\left[2\varkappa^2\left(\nu\,\text{kei } x - \frac{1-\nu}{x}\text{ker}'x\right) - \nu\left(\nu\,\text{ker } x + \frac{1-\nu}{x}\text{kei}'x\right)\right]\right].$$

In the preceding Section we have seen that the bending stress system consists of two parts: one which assumes large values near the lower edge of the shell and decreases in damped oscillations as we go up the meridian, and another one which is in the same way related to the upper edge. The same is true here. The A terms in eqs. (42) are regular functions of x and hence of ϕ which increase as ϕ increases. Consequently, they are associated with the lower (or outer) edge of the shallow shell. The B terms show the opposite behavior. The functions ker x and ker $'x$ are infinite for $x = 0$, and they as well as kei x and kei $'x$ decrease in damped oscillations as x increases. They describe the stresses caused by loads acting at the edge of a hole or by a concentrated force applied at the top of the shell. If there is neither such a force nor a hole, we must ask that the solution be regular at $x = 0$, and this requires that $B_1 = B_2 = 0$.

We may use the solution (42) to study the effect of a light dishing of a circular plate. When the plate is plane and carries a uniformly distributed load p (fig. 15a), the radial bending moment is

$$M_r = \frac{p}{16}(3 + \nu)(b^2 - r^2)$$

and the tangential moment

$$M_\theta = \frac{p}{16}[(3 + \nu)b^2 - (1 + 3\nu)r^2].$$

On a sphere (fig. 15b), a vertical load p, constant per unit of projected area, has the components

$$p_\phi = p\cos\phi\sin\phi, \qquad p_r = -p\cos^2\phi$$

and leads to the membrane forces

$$N_\phi = -\frac{1}{2}pa, \qquad N_\theta = \frac{1}{2}pa(1 - 2\cos^2\phi),$$

as one may easily verify from eqs. (II–10) and (II–6c). In order to have only vertical reactions at the edge of the shell, we have to superpose the horizontal load $H = \frac{1}{2}pa\cos\alpha$ indicated in fig. 15b. It may be resolved into a transverse force $Q_\phi = -H\sin\alpha$ and a normal force $N_\phi = H\cos\alpha$. With this information, we go into formulas (42). Since there is no hole at the center of the shell, we drop the B terms and find A_1, A_2 from the conditions that for $\phi = \alpha$ there is

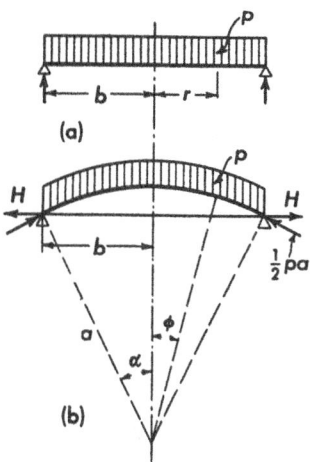

(a)

(b)

Fig. 15. Circular plate and shallow spherical shell

$$Q_\phi = -\frac{1}{2}pa\cos\alpha\sin\alpha, \qquad M_\phi = 0.$$

In the general case this will lead to rather clumsy formulas for A_1 and A_2, and it is advisable to introduce in time the particular numerical data of the shell under consideration. However, if we assume $\nu = 0$, simple expressions are obtained:

$$A_1 = -\frac{pa\cos\alpha\sin\alpha\,(x_0\,\mathrm{bei}\,x_0 + \mathrm{ber}'x_0)}{x_0(\mathrm{ber}'x_0\,\mathrm{bei}\,x_0 - \mathrm{bei}'x_0\,\mathrm{ber}\,x_0) + \mathrm{ber}'^2x_0 + \mathrm{bei}'^2x_0},$$

$$A_2 = +\frac{pa\cos\alpha\sin\alpha\,(x_0\,\mathrm{ber}\,x_0 - \mathrm{bei}'x_0)}{x_0(\mathrm{ber}'x_0\,\mathrm{bei}\,x_0 - \mathrm{bei}'x_0\,\mathrm{ber}\,x_0) + \mathrm{ber}'^2x_0 + \mathrm{bei}'^2x_0},$$

$$x_0 = \varkappa\sqrt{2}\,\alpha$$

From these, the figs. 16 and 17 have been computed which may illustrate the behavior of such slightly dished circular plates.

Fig. 16 shows the meridional bending moment M_ϕ and the hoop force N_θ for two shells both having $\alpha = 10°$, but $b/t = 5$ and $= 25$. In the thicker shell the bending moment has the same distribution as in a flat plate (broken line) and is only slightly smaller. The hoop forces are not much compared with the almost constant membrane value, and their distribution over the radius is distinctly different from that of the membrane forces. In the central part of the plate we have compression, and at least a part of the load is carried by vault action, but in a wide border zone the hoop stress is positive. This zone replaces the missing foot ring; it resists by its hoop force the radial thrust of the inner part.

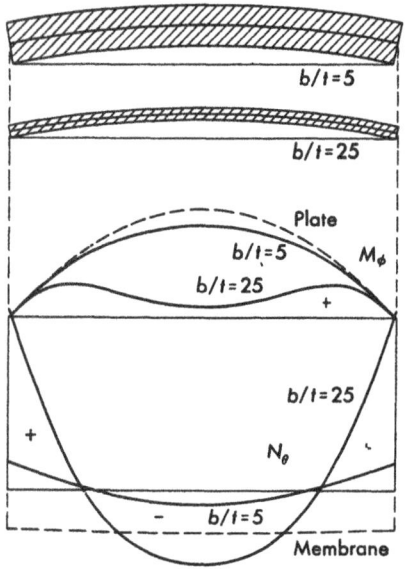

Fig. 16. Stress resultants in two slightly dished circular plates

The thinner shell represents a transitional case between the platelike thicker shell and a typical thin shell. The bending moments show clearly the tendency to concentrate in a border zone, although this zone is still rather wide, and the inner part is by no means free from bending. In the N_θ diagram the zero has moved outward and in the central part N_θ not only has approached the membrane value but has overshot it considerably. When we make the shell still thinner this will be remedied, the border zone will become still narrower and the positive peak value of N_θ will become still higher.

This tendency becomes clearer from fig. 17. Here M_ϕ for the center and N_θ for the edge of the shell have been plotted against b/t. One recognizes the rapid decline of the bending moment which may even become negative and finally will approach zero in rapidly damped oscillations. On the other hand, N_θ increases more and more since the membrane thrust of the shell requires a certain total amount of hoop stresses which are concentrated in a zone which becomes narrower as the shell is made thinner. This indicates clearly the necessity of a strong reinforcing ring at the edge of a thin shell, while a thick shell (say $b/t = 5$) may well take care of itself, behaving essentially like a plate.

The second and more important application of the formulas is made when studying the stresses near a small hole at the top of the shell, such as a manhole in a boiler end or a skylight in a dome. The roof structure shown in fig. II–5 may serve as an example. The shell has two edges, and each of them must be treated in a different way. The upper edge lies at $\phi = 9.96°$. In this region the solutions given in Sections 6.2.1.3 and .4 cannot be used, and the exact solution of Section 6.2.1.2 is too laborious; but the basic assumptions underlying eqs. (42) are fulfilled. The B terms of these equations are suitable for describing the bending stresses near this edge, but the A terms would assume their largest values far down the meridian, i. e. in a region where eqs. (42) are no longer applicable. They cannot be used to treat the bending problem at the lower edge, but there the solutions of the

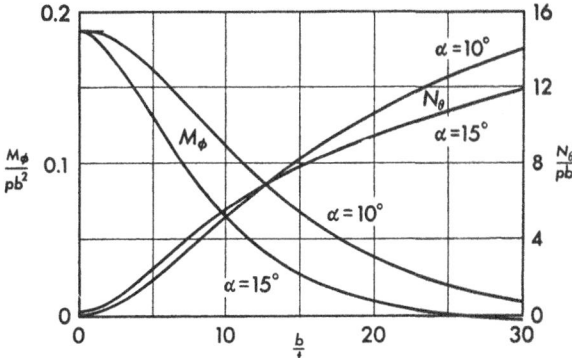

Fig. 17. Stress resultants (N_θ at the edge, M_ϕ at the center) for a slightly dished circular plate

preceding sections may be applied, e. g. eqs. (37), from which the B terms must then be dropped.

Under the load specified on p. 26 there is a membrane force $N_\phi = -2580$ lb/ft at the edge of the shell, and this force acts as an eccentric thrust on the ring. It produces there not only a compressive hoop force but also a bending moment which leads to a uniform rotation of all cross sections of the ring (see Appendix). This deformation does not match the membrane deformation of the shell, which consists only of a decrease of the diameter of the opening without an appreciable rotation of the tangent to the meridian. Therefore, horizontal forces H and moments M must act between the ring and the shell, which can be determined as redundant quantities in a statically indeterminate system. For the analysis one needs the deformations of ring and shell under unit loads, $H = 1$ lb/ft and $M = 1$ ft·lb/ft. For the shell they are described by the B terms of eqs. (42). The calculations have been made, and some of the results have been plotted in fig. 18. The diagrams show that the bending moments are restricted to a rather narrow zone; at 2 ft from the edge not much of the disturbance is left. The diagrams show also how the elastic yielding of the ring makes the meridional force N_ϕ decrease and the hoop force N_θ increase.

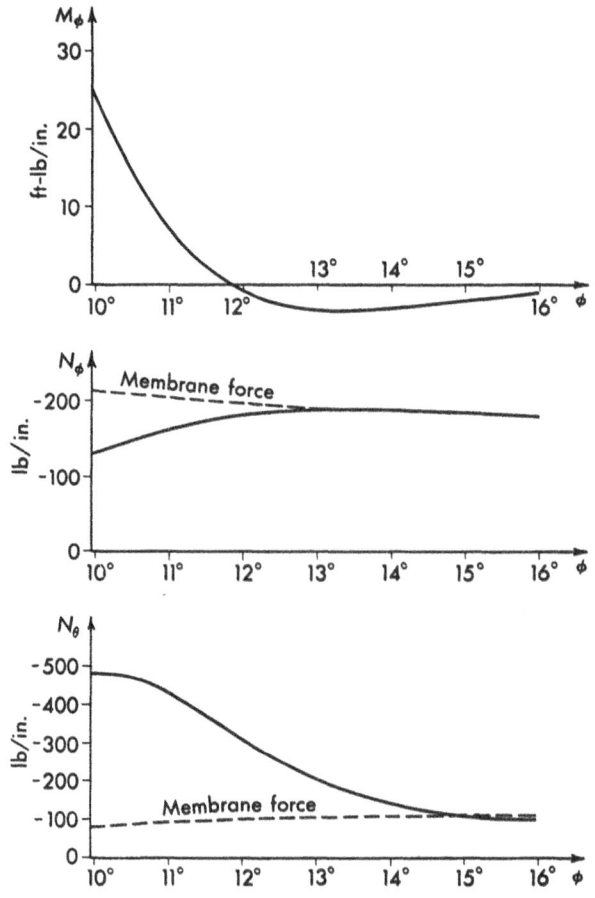

Fig. 18. Stress resultants near the lantern ring of the shell roof of Fig. II.5

6.2.1.6 Concentrated Load at the Apex

On p. 24 we have seen that a shell of revolution may carry a concentrated force at the top but that in this case the membrane forces must have a singularity of the order ϕ^{-2}. It may be expected that in the vicinity of the singular point the deformation will be such that bending moments of appreciable magnitude appear, and they will, of course, modify but not remove the singularity. The solution based on the differential equation (41) is the instrument for a detailed study of this problem.

This equation has been derived from eqs. (17), in which we permitted the stress resultants acting across any parallel circle to have a vertical resultant P. This force P may now be identified with the load applied at the apex of the shell, positive when downward.

Since the force P does not appear in eqs. (17), the solution (42a) is still valid, but since we are interested only in the stresses in the vicinity of the apex, we must now drop the A terms. The same is true for eq. (42b), but we must now use eqs. (15) as they stand and find from them

$$N_\phi = -\phi^{-1}(B_1 \operatorname{ker}'x + B_2 \operatorname{kei}'x) - \frac{P}{2\pi a}\phi^{-2}, \qquad (42\,\mathrm{c}',\,\mathrm{d}')$$

$$N_\theta = \varkappa\sqrt{2}\,[B_1(\operatorname{kei}x + x^{-1}\operatorname{ker}'x) - B_2(\operatorname{ker}x - x^{-1}\operatorname{kei}'x)] + \frac{P}{2\pi a}\phi^{-2}.$$

These formulas take the place of eqs. (42c, d), while eqs. (42e, f) again remain unchanged except for the dropping of the A terms.

We now have to determine the constants B_1 and B_2. Obviously, this cannot be done from a condition of equilibrium as we did for the membrane forces on p. 24, since eq. (14′) already assures the equilibrium of a shell element at the apex. We may go a step further and ask for finite displacements or even for finite deformations. The first thing to be done in this regard is to make $\chi = 0$ for $x = 0$. Now $\operatorname{kei}' 0 = 0$, $\operatorname{ker}' 0 = \infty$, and therefore the coefficient of $\operatorname{ker}'x$ in ·eq. (42b) must vanish. This coefficient is $-\nu B_1 - 2\varkappa^2 B_2$ and hence

$$B_2 = -\frac{\nu}{2\varkappa^2}B_1.$$

Now we look at the normal forces. Using the relation between the two constants, we may write eq. (42c′) as:

$$N_\phi = -\varkappa\sqrt{2}\,B_1\,x^{-1}\left(\operatorname{ker}'x - \frac{\nu}{2\varkappa^2}\operatorname{kei}'x\right) - \frac{P\varkappa^2}{\pi a}x^{-2}.$$

To proceed further, we need the series expansions of the THOMSON functions involved. Each of them consists of a power series and of one or more singular terms:

$$\operatorname{ker}'x = -x^{-1} + (\tfrac{1}{16}x^3 + \ldots)\ln x + \text{power series in } x,$$

$$\operatorname{kei}'x = -(\tfrac{1}{2}x + \ldots)\ln x + \text{power series in } x,$$

and the two power series have no constant term. Since $x^2\ln x \to 0$, N_ϕ has two singular terms, one with x^{-2} and another with $\ln x$:

$$N_\phi = \left(\varkappa\sqrt{2}\,B_1 - \frac{P\varkappa^2}{\pi a}\right)x^{-2} - \frac{\nu B_1}{2\sqrt{2}\,\varkappa}\ln x + \cdots.$$

It is not possible to make both of them vanish, but we may rid ourselves of the stronger one by choosing

$$B_1 = \frac{P\varkappa}{\sqrt{2}\,\pi\,a}.$$

The meridional force has then a logarithmic singularity

$$N_\phi = -\frac{\nu\,P}{4\pi\,a}\ln x + \cdots$$

which disappears entirely for $\nu = 0$.

When we collect the singular terms in eq. (42d'), we find that N_θ has the same singularity as N_ϕ. The strains ϵ_ϕ and ϵ_θ are then of the same type, and the displacements which are integrals of the strains, will be finite.

With the particular values for the constants, eqs. (42) yield the following formulas:

$$\left.\begin{aligned}
Q_\phi &= \frac{P\varkappa}{\sqrt{2}\,\pi\,a}\left[\ker'x - \frac{\nu}{2\varkappa^2}\,\kei'x\right], \\[2mm]
\chi &= \frac{P\,a}{2\sqrt{2}\,\pi\,K\,\varkappa}\,\kei'x, \\[2mm]
N_\phi &= -\frac{P\varkappa^2}{\pi\,a\,x}\left[\ker'x - \frac{\nu}{2\varkappa^2}\,\kei'x + x^{-1}\right], \\[2mm]
N_\theta &= \frac{P\varkappa^2}{\pi\,a}\left[\kei x + x^{-1}\ker'x + x^{-2} + \frac{\nu}{2\varkappa^2}(\ker x - x^{-1}\kei'x)\right]. \\[2mm]
M_\phi &= \frac{P}{2\pi}\left[\ker x - (1-\nu)\,x^{-1}\kei'x\right], \\[2mm]
M_\theta &= \frac{P}{2\pi}\left[\nu\ker x + (1-\nu)\,x^{-1}\kei'x\right].
\end{aligned}\right\} \quad (43)$$

Numerical results obtained from these formulas are shown in fig. 19 for $t/a = 0.01$ and $\nu = 0$. The transverse forces and the bending moments are localized in the vicinity of the singular point. Both normal forces assume finite values at the apex, and with increasing ϕ they approach the values given by the membrane theory.

The singularities of Q_ϕ and of M_ϕ are identical with those which occur in a flat circular plate. Indeed, when we replace in the preceding formulas $\ker'x$ and $\kei'x$ by their singular parts and put in the same degree of approximation

$$\ker x \approx -\ln x,$$

then we obtain

$$Q_\phi = -\frac{P\varkappa}{\sqrt{2}\,\pi\,a}\,x^{-1} = -\frac{P}{2\pi}\,\frac{1}{a\phi},$$

$$M_\phi = M_\theta = -\frac{P}{4\pi}(1+\nu)\ln x = -\frac{P}{4\pi}(1+\nu)(\ln a\phi - \text{const}),$$

and these are exactly the singularities known from the theory of circular plates where the plane polar coordinate r takes the place of $a\,\phi$.

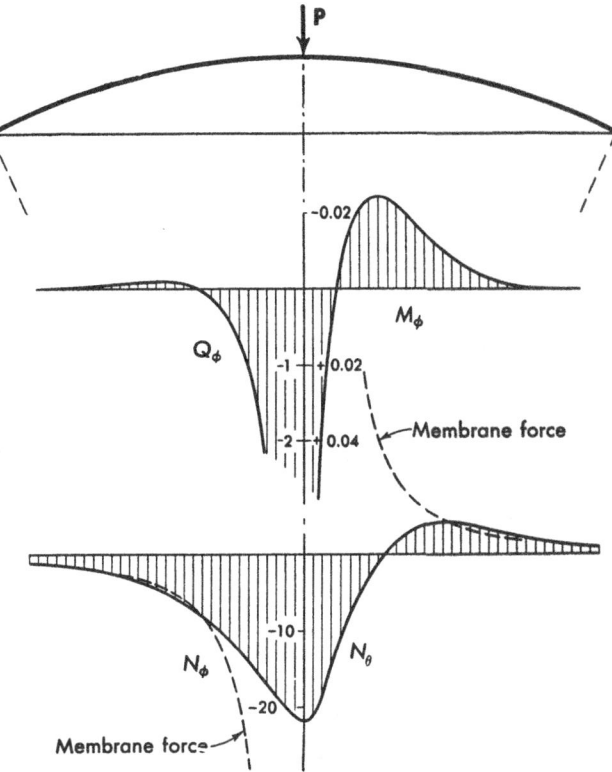

Fig. 19. Stress resultants near a concentrated force applied to a spherical shell. Units of the scales: for Q_ϕ, N_ϕ, N_θ: P/a; for M_ϕ: P

6.2.1.7 Surface Loads

When writing the differential equations (10), we dropped the surface loads. Consequently, all differential equations turned out to be homogeneous, and we worked out their solution in great detail. When we now want to keep p_ϕ, p_r in the equations, we only need to look for one particular solution of the differential equations.

It may be left to the reader to repeat the process of elimination described in Section 6.2.1.1. After writing $-p_\phi\,a\,\sin\phi$ and $+p_r\,a\,\sin\phi$ on the right-hand sides of eqs. (10a, b), respectively, he will find that no changes occur up to eq. (14). The equations (15) must be replaced by the following:

$$N_\phi = -Q_\phi \cot\phi - F(\phi),$$
$$N_\theta = -Q_\phi' + F(\phi) + p_r\,a, \qquad\qquad (44\text{a, b})$$

where
$$F(\phi) = \frac{a}{\sin^2\phi} \int (p_\phi \sin\phi - p_r \cos\phi) \sin\phi \, d\phi.$$

When this is introduced into eq. (14), a load term appears on the right-hand side, and eqs. (17) become these:

$$L(\chi) - \nu\chi = \frac{a^2}{K} Q_\phi,$$

$$L(Q_\phi) + \nu Q_\phi = -D(1 - \nu^2)\chi + (1 + \nu) p_\phi a + \dot{p}_r a.$$

(45 a, b)

Elimination of χ yields at last the equation

$$LL(Q_\phi) + 4\varkappa^4 Q_\phi = \Phi(\phi) \equiv (1 + \nu) a L(p_\phi) + a L(\dot{p}_r) \\ - \nu(1 + \nu) p_\phi a - \nu \dot{p}_r a.$$

(46)

We want to find a particular solution of this equation for a few important cases of loading.

6.2.1.7.1 Internal Pressure. We see at once that with $p_\phi = 0$, $p_r = p$ the right-hand side of eq. (46) vanishes. We are then back to the homogeneous problem with the only difference that with $Q_\phi = 0$ eq. (44a) is identical with the membrane equation (II–10) and yields for the particular solution a non-vanishing value of N_ϕ.

6.2.1.7.2 Weight of the Shell. With

$$p_\phi = p \sin\phi, \qquad p_r = -p \cos\phi$$

we find
$$\Phi(\phi) = -(1 + \nu)(2 + \nu) p a \sin\phi.$$

Since $L(\sin\phi) = -\sin\phi$, there exists a particular solution of eq. (46) in the form
$$Q_\phi = C \sin\phi,$$

and one verifies easily that it is
$$Q_\phi = -(1 + \nu)(2 + \nu) \frac{p a}{1 + 4\varkappa^4} \sin\phi.$$

6.2.1.7.3 Hydrostatic Pressure. Any hydrostatic pressure may be replaced by the sum of a constant pressure and another hydrostatic pressure with a different position of the zero-pressure level. Since we have dealt already with the case of constant pressure, we may assume the variable part in its most convenient form

$$p_\phi = 0, \qquad p_r = \gamma a \cos\phi$$

where γ is the specific weight of the liquid inside the shell.

For this load we have

$$\Phi(\phi) = (1 + \nu) \gamma a^2 \sin\phi$$

and consequently
$$Q_\phi = \frac{1+\nu}{1+4\varkappa^4} \gamma a^2 \sin\phi.$$

6.2.1.7.4 Centrifugal Force. We denote by μ the mass of the shell per unit area of its middle surface. Whe the shell rotates with the angular velocity ω about its axis, the centrifugal force acting on each element has the components

$$p_\phi = \mu\, a\, \omega^2 \sin\phi\, \cos\phi, \qquad p_r = \mu\, a\, \omega^2 \sin^2\phi.$$

We find easily that in this case

$$\varPhi(\phi) = -(3+\nu)(5+\nu)\,\mu\, a^2\, \omega^2 \sin\phi\, \cos\phi$$

and

$$Q_\phi = -\frac{(3+\nu)(5+\nu)}{25+4\varkappa^4}\,\mu\, a^2\, \omega^2 \sin\phi\, \cos\phi.$$

In all cases except the trivial first one, there appears a denominator containing $\varkappa^4$. Since this is a rather large number, for thin shells in particular, the transverse shear Q_ϕ is very small as compared with the normal forces N_ϕ and N_θ. We conclude that the membrane solution, which assumes $Q_\phi \equiv 0$, is almost identical with a particular solution of the bending equations. This justifies once more the general use of the membrane solutions in lieu of particular solutions of the bending theory. In the case of an exceptionally thick shell, of course, there always exists the possibility of using one of the preceding particular solutions or a similar one and thus improving the accuracy of the stress analysis.

6.2.2 Shells Having a Meridian of Arbitrary Shape

6.2.2.1 Elastic Law

When we adapted the general elastic law (8) to the special case of a spherical shell, we obtained the simple equations (11). The subsequent processing of these equations is contingent upon their simplicity. When we want to proceed in a similar way in a more general case, we first must simplify the elastic law. But, strangely enough, the simpler law (9) is too simple and does not lend itself to the method used for the sphere. What we need is an intermediate form, and this we shall now derive.

The terms $(v\dot{} + w)/a$ and $(v\cot\phi + w)/a$ in eqs. (11a, b) may easily be recognized as the strains ϵ_ϕ and ϵ_θ of line elements on the middle surface. Therefore, eqs. (11a, b) may be written as

$$N_\phi = D(\epsilon_\phi + \nu\,\epsilon_\theta), \qquad N_\theta = D(\epsilon_\theta + \nu\,\epsilon_\phi), \tag{47a, b}$$

and the simplified equations (9) for the general shell of revolution have the same form with

$$\varepsilon_\phi = \frac{v^\cdot + w}{r_1}, \qquad \varepsilon_\theta = \frac{v \cot\phi + w}{r_2}. \qquad (48\,a,\,b)$$

The general equations (8) contain additional terms with the factor K. Since $K/D = t^2/12a^2$, these terms are rather small, and it is not serious to sacrifice them, if they stand in the way to the solution of the problem.

The moment equations are less simple. The term $\chi^\cdot/a$ in eqs. (11 c, d) is the difference of rotation of two tangents to the meridian, $\chi^\cdot d\phi$, divided by the length $a\,d\phi$ of the line element, i. e. the elastic change $\varkappa_\phi$ of the curvature of this element. We may suspect – and we shall soon see that this is true – that $\chi \cot\phi/a$ is the change $\varkappa_\theta$ of the other principal curvature. Eqs. (11 c, d) would then assume the very plausible form

$$M_\phi = K(\varkappa_\phi + \nu\,\varkappa_\theta), \qquad M_\theta = K(\varkappa_\theta + \nu\,\varkappa_\phi). \qquad (47\,c,\,d)$$

These are the same equations which occur in the theory of plane plates, where $\varkappa_\phi$ and $\varkappa_\theta$ stand for the second derivatives of the deflection with respect to a pair of rectangular coordinates. Also the simplified shell equations (9 c, d) agree with this form, if we identify $w^\cdot/r_1$ with χ. But this is an over-simplification which would bar the way to the solution which we seek. What we need is the combination of eqs. (47 c, d) with the definition (12) for χ. To obtain it, we study the elastic change of curvature of the shell.

Since we are dealing with axisymmetric stress systems, the deformed middle surface is also a surface of revolution. Fig. 20 shows two pic-

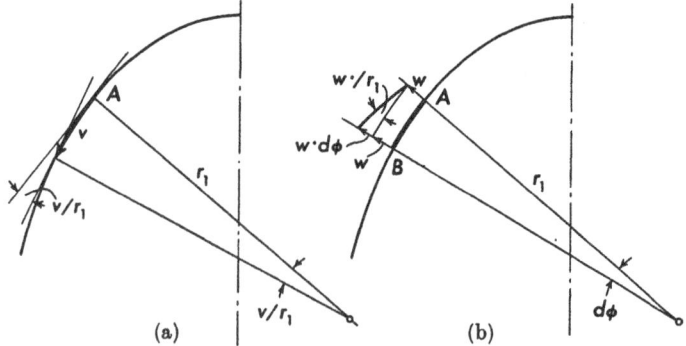

Fig. 20. Meridian of a shell of revolution, showing the rotation of the tangent, (a) due to v, (b) due to $w^\cdot$

tures of a piece of a meridian. In fig. 20a a point A undergoes a displacement v on the meridian. The normal rotates counterclockwise by an angle v/r_1, and so does the tangent. In fig. 20b a line element $A B = r_1 d\phi$ is shown. When both its ends are subjected to the same displacement w there is no rotation of a tangent. But when the point B has an additional displacement $w^{.} d\phi$, the tangent rotates by $w^{.} d\phi/r_1 d\phi = w^{.}/r_1$, but this time in clockwise direction. By linear superposition we find that the total rotation is

$$\chi = \frac{w^{.} - v}{r_1}, \tag{49}$$

which confirms the interpretation given eq. (12).

The tangents at the ends of an arc element $r_1 d\phi$ rotate by χ and $\chi + \chi^{.} d\phi$, respectively, and when we divide the relative rotation $\chi^{.} d\phi$ by the length $r_1 d\phi$ of the element, we have the change of curvature

$$\varkappa_\phi = \frac{\chi^{.}}{r_1}. \tag{50a}$$

It should be noticed that we have divided the angle between the tangents by the *original* length $r_1 d\phi$ of the line element, not by its actual length $(1 + \epsilon_\phi) r_1 d\phi$. This is the reason why, with $r_1 = $ const, the derivative $v^{.}$ makes a contribution to $\varkappa_\phi$, although the radius of curvature does not change with $v^{.}$, whereas a uniform displacement w does not contribute, although it increases the radius from r_1 to $r_1 + w$. As eqs. (11) show, this interpretation of "change of curvature" is a good one for the sphere. The general equations (8) do not confirm it, but we may still consider it a reasonable approximation also in the general case.

To define the second principal curvature we need a line element ds on the middle surface, tangent to a parallel circle but situated in a plane containing the normal. In the undeformed shell its radius of curvature is r_2, its curvature $1/r_2$, and the angle between the tangents at its ends $(1/r_2) ds$.

When the shell is deformed, the radius is again that part of the normal which lies between the middle surface and the axis of revolution. Let its length now be r_2^* and the length of the element $ds^* = (1 + \epsilon_\theta) ds$. The true curvature of the element is $1/r_2^*$ and the angle between the tangents at its ends is ds^*/r_2^*, hence the decrease of this angle, caused by the deformation, is

$$\frac{ds}{r_2} - \frac{ds^*}{r_2^*}.$$

To be consistent, we must divide this decrease by the original length ds and thus find what we should call the elastic decrease of curvature:

$$\varkappa_\theta = \frac{1}{r_2} - \frac{1}{r_2^*} \frac{ds^*}{ds} = \frac{1}{r_2} - \frac{1 + \varepsilon_\theta}{r_2^*}.$$

We shall now work out separately the contributions to $\varkappa_\theta$ made by the displacements v and w (fig. 21).

First we assume that $w \equiv 0$. The displacement v shown in fig. 21 a leads to an increase of radius

$$r_2^* - r_2 = \frac{v}{r_1}(r_1 - r_2)\cot\phi$$

while eq. (48b) yields $\epsilon_\theta = (v/r_2)\cot\phi$. We have therefore

$$\varkappa_\theta = \frac{1}{r_2} - \frac{1 + \dfrac{v}{r_2}\cot\phi}{r_2 + \dfrac{v}{r_1}(r_1 - r_2)\cot\phi}$$

which, by the usual procedure of expanding the denominator, yields

$$\varkappa_\theta = \frac{1}{r_2} - \frac{1}{r_2}\left(1 + \frac{v}{r_2}\cot\phi\right)\left(1 - \frac{v}{r_1 r_2}(r_1 - r_2)\cot\phi\right) = -\frac{v\cot\phi}{r_1 r_2}.$$

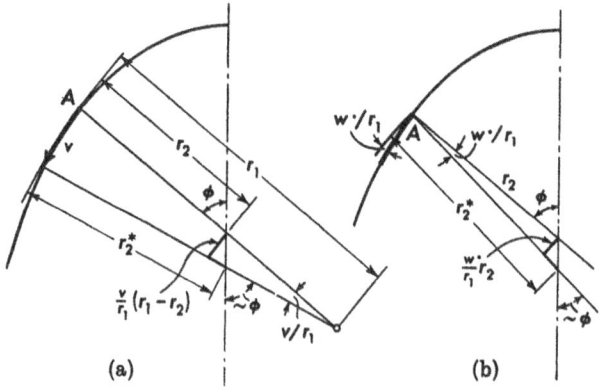

Fig. 21
Meridian of a shell of revolution, showing the change of r_2, (a) due to v, (b) due to $w^{\cdot}$

Now we assume $v = 0$ and consider the influence of w. A uniform displacement w does not make a contribution, since it does not lead to any rotation of tangents. But when the normal rotates by the angle $w^{\cdot}/r_1$, the radius r_2 is increased by

$$r_2^* - r_2 = \frac{w^{\cdot}}{r_1}r_2\cot\phi,$$

while $\epsilon_\theta = 0$. The contribution to $\varkappa_\theta$ is then simply

$$\varkappa_\theta = \frac{1}{r_2} - \frac{1}{r_2^*} = \frac{w^{\cdot}}{r_1 r_2}\cot\phi.$$

When all displacements occur together, we have the total change of curvature

$$\varkappa_\theta = \frac{(-v + w^{\cdot})\cot\phi}{r_1 r_2} = \frac{\chi}{r_2}\cot\phi. \tag{50b}$$

Using eqs. (50 a, b) we may rewrite the elastic law (47 c, d) for the bending moments as

$$M_\phi = K\left(\frac{\dot\chi}{r_1} + \nu\,\frac{\chi}{r_2}\cot\phi\right),$$

$$M_\theta = K\left(\frac{\chi}{r_2}\cot\phi + \nu\,\frac{\dot\chi}{r_1}\right).$$

(51 a, b)

This appears to be a reasonable extension of eqs. (11 c, d) of the sphere. But when we compare eqs. (51) and (8), we see that we have dropped the terms with $(r_2 - r_1)$. As eqs. (48) show, these terms contain ϵ_ϕ and ϵ_θ, respectively. If there is much bending in the shell, these terms are small compared with the $\varkappa$ terms, and if at places the bending moments are small, we do not care to get them with much accuracy. This reasoning justifies the use of eqs. (51) as the elastic law for the bending moments.

6.2.2.2 Differential Equations

6.2.2.2.1 General Equations. The conditions of equilibrium (7), the elastic law (9 a, b), (51 a, b), and the kinematic relation (49) are again a set of 8 equations very similar to those used for the spherical shell on p. 322, and they may be treated in exactly the same way. To obtain all desirable and possible generality, we shall not only allow r_1 and r_2 to be arbitrary functions of ϕ, but we shall also include the possibility that t depends on ϕ.

From eqs. (7 c) and (51) we obtain the equation

$$\frac{r_2}{r_1}\ddot\chi + \left[\frac{r_2}{r_1}\cot\phi + \left(\frac{r_2}{r_1}\right)^{\cdot} + \frac{r_2}{r_1}\frac{\dot K}{K}\right]\dot\chi$$
$$- \left[\frac{r_1}{r_2}\cot^2\phi + \nu - \nu\,\frac{\dot K}{K}\cot\phi\right]\chi = \frac{r_1 r_2 Q_\phi}{K}.$$

(52 a)

The equation replacing (14) reads now

$$r_1\chi = \frac{1}{1-\nu^2}\left[\left(\frac{r_2 N_\theta}{D}\right)^{\cdot} - \nu\left(\frac{r_2 N_\phi}{D}\right)^{\cdot} + (r_2 + \nu\,r_1)\frac{N_\theta}{D}\cot\phi\right.$$
$$\left. - (r_1 + \nu\,r_2)\frac{N_\phi}{D}\cot\phi\right],$$

(53)

and eqs. (15) are replaced by

$$N_\phi = -Q_\phi\cot\phi - \frac{P}{2\pi r_2}\frac{1}{\sin^2\phi},$$

$$N_\theta = -\frac{(r_2 Q_\phi)^{\cdot}}{r_1} + \frac{P}{2\pi r_1}\frac{1}{\sin^2\phi}.$$

(54a, b)

It appears that we should use $r_2 Q_\phi$ rather than Q_ϕ as our second unknown. When we carry the elimination to its end we obtain the following equation replacing (13 b):

$$\frac{r_2}{r_1}(r_2 Q_\phi)^{\cdot\cdot} + \left[\frac{r_2}{r_1}\cot\phi + \left(\frac{r_2}{r_1}\right)^{\cdot} - \frac{r_2}{r_1}\frac{D^{\cdot}}{D}\right](r_2 Q_\phi)^{\cdot} \qquad (52b)$$

$$- \left[\frac{r_1}{r_2}\cot^2\phi - \nu - \nu\frac{D^{\cdot}}{D}\cot\phi\right](r_2 Q_\phi) = -D(1-\nu^2)r_1\chi + \mathsf{P}\,g(\phi).$$

The P term does not drop out in this case. Its coefficient is

$$g(\phi) = \frac{1}{2\pi\sin^2\phi}\left[\frac{r_1^2 - r_2^2}{r_1 r_2}\cot\phi + \left(\frac{r_2}{r_1}\right)^{\cdot} + \frac{\nu r_1 + r_2}{r_1}\frac{D^{\cdot}}{D}\right]. \qquad (55)$$

Eqs. (52 a, b) are now the differential equations of the bending stress problem. For arbitrary shapes of the meridian and an arbitrary variation of the wall thickness it may be necessary to solve them as they stand by numerical integration. There are, however, important groups of shells which allow further simplification, and these we shall now discuss in detail.

6.2.2.2.2 Constant Wall Thickness. If $t = $ const, then $t^{\cdot} = 0$ and consequently also $D^{\cdot}$ and $K^{\cdot}$ vanish. In this case the coefficients of the first derivatives in eqs. (52) become equal, and we may introduce the differential operator

$$L_1(\ldots) = \frac{r_2}{r_1}\frac{(\ldots)^{\cdot\cdot}}{r_1} + \left[\frac{r_2}{r_1}\cot\phi + \left(\frac{r_2}{r_1}\right)^{\cdot}\right]\frac{(\ldots)^{\cdot}}{r_1} - \frac{r_1}{r_2}\cot^2\phi\cdot\frac{(\ldots)}{r_1}. \quad (56)$$

This is a generalization of the operator L of eq. (16). When we put $r_1 = r_2 = a$, we obtain $L_1 = L/a$.

With the operator L_1 eqs. (52) may be written as

$$L_1(\chi) - \frac{\nu}{r_1}\chi = \frac{r_2 Q_\phi}{K},$$
$$L_1(r_2 Q_\phi) + \frac{\nu}{r_1}(r_2 Q_\phi) = -D(1-\nu^2)\chi + \frac{\mathsf{P}\,g(\phi)}{r_1}, \qquad (57\,\text{a, b})$$

and we may now separate χ and $r_2 Q_\phi$ by the same procedure as that used for the spherical shell. The term with P is no major obstacle; however, we shall again drop it to keep our equations simpler, leaving it to the reader to work it in when needed.

Eliminating χ, we obtain for $r_2 Q_\phi$ the equation

$$L_1 L_1(r_2 Q_\phi) + \nu L_1\left(\frac{r_2 Q_\phi}{r_1}\right) - \frac{\nu}{r_1}L_1(r_2 Q_\phi) - \frac{\nu^2}{r_1^2}r_2 Q_\phi = -\frac{D(1-\nu^2)}{K}r_2 Q_\phi. \qquad (58)$$

It differs substantially from eq. (18b) by the presence of the second and third terms, which make it impossible to split it into two second-order equations. In the general case it is necessary to solve the equation as it stands by numerical integration.

We make further progress toward simpler equations when the two embarrassing terms cancel each other. Evidently this is the case when $r_1 = \text{const}$, i. e. for spheres, cones, and toroids. It even happens for any meridian, if we may assume $\nu = 0$, an assumption often considered legitimate for reinforced concrete structures.

In all of these cases eq. (58) assumes the form

$$L_1\, L_1(r_2\, Q_\phi) + \mu^4\, r_2\, Q_\phi = 0 \tag{59}$$

in which

$$\mu^4 = \frac{D(1 - \nu^2)}{K} - \frac{\nu^2}{r_1^2} \tag{60}$$

is a constant. This equation has the same form as (19) and by the same method may be split into two second-order equations

$$L_1(r_2\, Q_\phi) \pm i\, \mu^2 r_2\, Q_\phi = 0 . \tag{61a, b}$$

Again it is enough to solve one of these equations, because the real and imaginary parts of its two solutions are four independent solutions of eq. (59), as explained on p. 324. We shall come back to this equation on p. 371 when treating the conical shell of constant thickness.

6.2.2.2.3 Variable Wall Thickness. We shall now consider shells whose wall thickness t depends on the coordinate ϕ but not on θ. However, we shall assume that the shell is homogeneous, so that eqs. (V–8) for the rigidities apply. We have then

$$\frac{D^{\displaystyle\cdot}}{D} = \frac{t^{\displaystyle\cdot}}{t}, \qquad \frac{K^{\displaystyle\cdot}}{K} = \frac{3 t^{\displaystyle\cdot}}{t} .$$

When we introduce these values in eqs. (52a, b), the coefficients of the first derivatives are no longer identical. Such identity may be brought about by the substitution

$$U = r_2\, Q_\phi , \qquad V = t^2\, \chi ,$$

by which these equations assume the following form:

$$
\left.
\begin{aligned}
\frac{r_2}{r_1}\, V^{\cdot\cdot} &+ \left[\frac{r_2}{r_1} \cot \phi + \left(\frac{r_2}{r_1} \right)^{\cdot} - \frac{r_2}{r_1} \frac{t^{\cdot}}{t} \right] V^{\cdot} - \left[\frac{r_1}{r_2} \cot^2 \phi + \nu + \right. \\
&+ \left(2 \frac{r_2}{r_1} \cot \phi + 2 \left(\frac{r_2}{r_1} \right)^{\cdot} - 3 \nu \cot \phi \right) \frac{t^{\cdot}}{t} + 2 \frac{r_2}{r_1} \frac{t^{\cdot\cdot}}{t} \right] V = \\
&= \frac{r_1 t^2}{K}\, U , \\[2mm]
\frac{r_2}{r_1}\, U^{\cdot\cdot} &+ \left[\frac{r_2}{r_1} \cot \phi + \left(\frac{r_2}{r_1} \right)^{\cdot} - \frac{r_2}{r_1} \frac{t^{\cdot}}{t} \right] U^{\cdot} - \left[\frac{r_1}{r_2} \cot^2 \phi - \nu - \right. \\
&- \nu \frac{t^{\cdot}}{t} \cot \phi \right] U = - \frac{D(1 - \nu^2)\, r_1}{t^2}\, V + \mathsf{P}\, g(\phi) .
\end{aligned}
\right\} \tag{62}
$$

We may now introduce an operator similar to L and L_1. When we make our choice, we try to make the coefficients of U and V on the right-hand sides constant to facilitate the following elimination of one of the variables. We therefore define

$$L_2(\ldots) = \frac{r_2 t}{r_1^2}(\ldots)^{\cdot\cdot} + \left[\frac{r_2 t}{r_1^2}\cot\phi + \left(\frac{r_2}{r_1}\right)^{\cdot}\frac{t}{r_1} - \frac{r_2 t^{\cdot}}{r_1^2}\right](\ldots)^{\cdot}$$
$$- \left[\frac{t}{r_2}\cot^2\phi - \nu\frac{t}{r_1} - \nu\frac{t^{\cdot}}{r_1}\cot\phi\right](\ldots). \tag{63}$$

Most terms of this operator become identical with those of $t L_1$ if we put $t = \text{const}$, but the identity is not perfect. Therefore, eqs. (61) assume now the less symmetric form

$$L_2(V) - 2f(\phi)\cdot V = \frac{12(1-\nu^2)}{E}U,$$
$$L_2(U) \qquad\qquad = -EV + \mathsf{P}\frac{t}{r_1}g(\phi) \tag{64a, b}$$

where $g(\phi)$ is given by eq. (55) and

$$f(\phi) = \nu\frac{t}{r_1} + \left[\frac{r_2}{r_1}\cot\phi + \left(\frac{r_2}{r_1}\right)^{\cdot} - \nu\cot\phi\right]\frac{t^{\cdot}}{r_1} + \frac{r_2 t^{\cdot\cdot}}{r_1^2}. \tag{65}$$

As before, we simplifiy our task by considering only the case $\mathsf{P} = 0$. Elimination of either U or V from eqs. (64) leads then to the two independent equations

$$L_2 L_2(V) - 2 L_2(f V) + 12(1-\nu^2)V = 0,$$
$$L_2 L_2(U) - 2fL_2(U) + 12(1-\nu^2)U = 0. \tag{66a, b}$$

These equations differ from eq. (59) by the presence of the second term with the variable $f(\phi)$ before or under the operator L_2. In general these fourth-order equations cannot be split in pairs of second-order equations. We may find the conditions under which it can be done by simply assuming that the solution U of eq. (66b) satisfies the equation

$$L_2(U) + c U = 0 \tag{67}$$

and by investigating the consequences of this assumption. From eq. (67) we find immediately

$$L_2 L_2(U) = -c L_2(U) = c^2 U$$

and when we introduce this and (67) into eq. (66), we may reduce it to the simple form

$$[c^2 + 2cf + 12(1-\nu^2)]U = 0. \tag{68}$$

This can be fulfilled for all values of ϕ only if the bracket vanishes identically, and this is not possible unless

$$f(\phi) = f = \text{const.}$$

This is the splitting condition. When the meridian shape is given, the condition may be interpreted as a differential equation for the wall thickness $t = t(\phi)$. Since eq. (65) contains t'', it is of the second order. In addition to the two free constants which its solution necessarily contains, it has a third parameter, the arbitrary choice of the constant f.

There are two cases in which eq. (65) has, among others, the trivial solution $t = $ const: where $r_1 = $ const with arbitrary ν and where $\nu = 0$ with any r_1.

For a given value of the constant f, the vanishing of the bracket in eq. (68) yields for c two values

$$c_{1,2} = -f \pm \sqrt{f^2 - 12(1 - \nu^2)}, \qquad (68')$$

and with each of them eq. (67), being of the second order, will yield a pair of independent solutions. As soon as they have been found, V may be obtained from eqs. (64 b) and (67) without an additional integration. Since we assumed $\mathsf{P} = 0$, we have

$$V = -\frac{L_2(U)}{E} = \frac{c}{E} U, \qquad (69)$$

and here the same value for $c(c_1$ or $c_2)$ must be used with which U satisfies eq. (69).

It may be seen from eq. (65) that f is of the order t/r_1, hence a rather small quantity; therefore c_1 and c_2 are conjugate complex. As in the case of a spherical shell, it is enough to solve eqs. (67) and (69) for $c = c_1$ and then to use real and imaginary parts of the solution separately as a fundamental system of solutions of eqs. (66).

In the general case of constant f but variable t according to eq. (65), the operator L_2 is rather involved and there is little hope of solving eq. (67) by analytical means. But even if the complex-valued variable U must be found by numerical integration, it is quite an advantage when this may be done from the second-order equation (67) instead of the fourth-order equation (66 b).

When we have U, eq. (69) will yield V; with U and V we have Q_ϕ and χ. From the latter pair we may find the bending moments from eqs. (51), and the normal forces from eqs. (54), where we must put $\mathsf{P} = 0$.

6.2.2.3 Approximate Theory for Thin Shells

6.2.2.3.1 Differential Equations. In Section 2.1.4 we took advantage of the fact that in a thin spherical shell subjected to an edge load all stresses decrease rapidly with increasing distance from the edge and that, therefore, their higher derivatives are of higher order of magnitude. The laborious computations required by the theory just described suggest tentatively adopting the same simplifying idea in the general case of a meridian of arbitrary shape. If the solutions obtained in this way have the assumed appearance, the procedure is justified; if not, it is not admissible.

If we admit that the lower derivatives of any variable may be neglected compared with the highest and if we exclude the top of the shell where $\cot\phi$ assumes large values, then the operator L_2, eq. (63), is reduced to

$$L_2(\ldots) = \frac{r_2\,t}{r_1^2}\,(\ldots)^{\cdot\cdot},$$

and we may at once neglect the second term in eq. (66b) and write

$$\frac{r_2\,t}{r_1^2}\left(\frac{r_2\,t}{r_1^2}\,U^{\cdot\cdot}\right)^{\cdot\cdot} + 12\,(1 - \nu^2)\,U = 0.$$

In the first term of this equation we differentiate the product and then neglect $U^{\cdot\cdot}$ and $U^{\cdot\cdot\cdot}$ compared with $U^{\cdot\cdot\cdot\cdot}$, and when we when write $r_2\,Q_\phi$ for U, we may again neglect lower derivatives of Q_ϕ. In this way we finally end up with the simple equation

$$Q_\phi^{\cdot\cdot} + 4\varkappa^4\,Q_\phi = 0 \tag{70}$$

in which

$$\varkappa^4 = 3\,(1 - \nu^2)\,\frac{r_1^4}{r_2^2\,t^2} \tag{70'}$$

is a given function of ϕ. We shall now study in detail two cases in which this equation admits simple solutions.

6.2.2.3.2 Constant $\varkappa$, Shell with Deformity. It may, of course, happen that $\varkappa$ is a constant; or $\varkappa$ may vary so little that it may be considered constant in the narrow border zone where stresses of appreciable magnitude occur. In these cases eq. (70) is solved by exponential functions, and we may almost literally repeat everything that has been said for the spherical shell on p. 337.

Having so extended the applicability of the formulas (35), (36), (37), (38) to other than spherical shells, we may use them for an interesting investigation.

Because of the limits of accuracy of all workmanship, the middle surface of an actual shell always deviates a little from its intended shape. Although such deviations should be small compared with the over-all dimensions of the shell, they may easily be of the order of the wall thickness, and the curvature of the middle surface may locally be rather different from that used in computing the membrane forces. The problem is to determine what influence such deviations from the true form have on the stresses.

We investigate this problem in a model case, assuming that the deformity is the same in all meridians, so that the actual shell is also a shell of revolution. Fig. 22 shows the meridian of such an almost hemispherical dome. Between $\phi = 60°$ and $\phi = 75°$ the circle of radius a is replaced by circles of the following radii:

$$60° < \phi < 64°: \quad r_1 = 0.4526a,$$
$$64° < \phi < 71°: \quad r_1 = 1.6230a,$$
$$71° < \phi < 75°: \quad r_1 = 0.4526a.$$

Although these radii are rather different from a, the deviation of the middle surface is not more than $2.5^0/_{00}$ of a, much too small to be visible in fig. 22. However, with $a/t = 200$ this is $0.5t$.

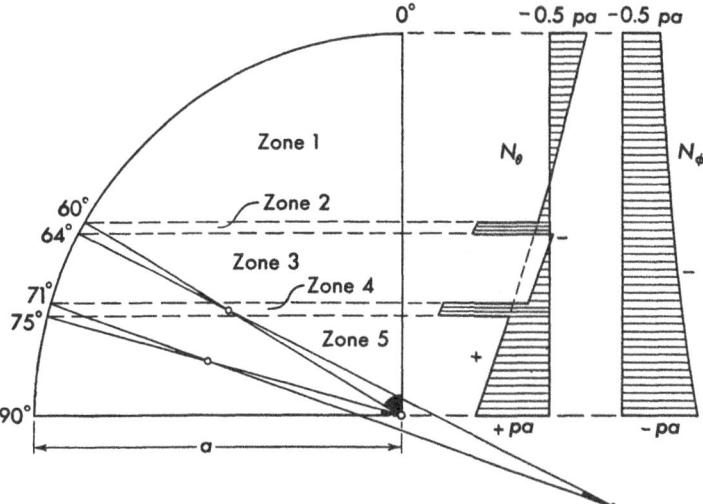

Fig. 22. Meridian of an almost hemispherical dome and membrane forces for uniform dead load

The diagrams in fig. 22 show the membrane forces as calculated from eqs. (II–10) and (II–6c) for a uniform deadload p. The meridional force is not visibly different from that of an exact sphere, but, because of the discontinuity of r_1, the N_θ diagram looks rather wild. Since discontinuities in N_θ are equivalent to discontinuities in ϵ_θ, they cannot exist in the shell. To remove them, we cut the shell into five zones with continuous curvature and apply appropriate forces and moments at their boundaries. The moments are bending moments M_ϕ, and the forces consist of transverse forces Q_ϕ and normal forces $N_\phi = -Q_\phi \cot\phi$, which combine to a horizontal load. Thus each of these loads is self-equilibrating.

These edge loads produce bending stresses. Zones 1 and 5 are parts of a thin spherical shell, and $\varkappa$ is a constant:

$$\varkappa_1 = \varkappa_5 = \sqrt[4]{3} \sqrt{\frac{a}{t}} = \sqrt[4]{3} \cdot \sqrt{200} = 18.61.$$

We may apply here any of the solutions offered on p. 337 and we choose eqs. (37). In zone 1 we drop the B terms and write $\omega_1 = 60° - \phi$; in zone 5 we drop the A terms and write $\omega_2 = \phi - 75°$.

The other three zones are not spherical, and $\varkappa$ must be computed from eq. (70'). At the edges of the zones it has the following values:

$$\text{zone 2:} \quad \phi = 60°: \varkappa_2 = \ \ 8.42; \quad \phi = 64°: \varkappa_2 = \ \ 8.51;$$
$$\text{zone 3:} \quad \phi = 64°: \varkappa_3 = 30.51; \quad \phi = 71°: \varkappa_3 = 30.03;$$
$$\text{zone 4:} \quad \phi = 71°: \varkappa_4 = \ \ 8.37; \quad \phi = 75°: \varkappa_4 = \ \ 8.42.$$

Evidently, $\varkappa$ is almost constant in each zone and we may use average values, viz. $\varkappa_2 = 8.46$, $\varkappa_3 = 30.25$, $\varkappa_4 = 8.39$, and then apply eq. (36) as a solution of eq. (70). We write it in a still different form, introducing as coordinates

$$\omega_2 = \phi - 62°, \quad \omega_3 = \phi - 67.5°, \quad \omega_4 = \phi - 73°$$

in the three zones and putting for $n = 2, 3, 4$:

$$
\begin{aligned}
Q_\phi = {}& C_{1n} \operatorname{Cosh}\varkappa_n \omega_n \cos\varkappa_n \omega_n + C_{2n} \operatorname{Sinh}\varkappa_n \omega_n \sin\varkappa_n \omega_n \\
& + C_{3n} \operatorname{Cosh}\varkappa_n \omega_n \sin\varkappa_n \omega_n + C_{4n} \operatorname{Sinh}\varkappa_n \omega_n \cos\varkappa_n \omega_n.
\end{aligned}
\tag{71a}
$$

This is indeed a solution of eq. (70) if $\varkappa = \text{const}$, but the meaning of the coefficients $C_{1n} \ldots$ is different from that of the factors $C_1 \ldots$ in eq. (36). The following formulas will fit our new solution:

$$
\begin{aligned}
N_\theta = -\frac{r_2}{r_1} Q_\phi^\cdot = -\frac{r_2}{r_1} \varkappa [& (C_{1n} + C_{2n}) \operatorname{Sinh}\varkappa_n \omega_n \cos\varkappa_n \omega_n \\
& - (C_{1n} - C_{2n}) \operatorname{Cosh}\varkappa_n \omega_n \sin\varkappa_n \omega_n \\
& + (C_{3n} + C_{4n}) \operatorname{Cosh}\varkappa_n \omega_n \cos\varkappa_n \omega_n \\
& + (C_{3n} - C_{4n}) \operatorname{Sinh}\varkappa_n \omega_n \sin\varkappa_n \omega_n],
\end{aligned}
$$

$$
D\chi = -\frac{r_2^2}{r_1^2} Q_\phi^{\cdot\cdot}
\tag{71b--d}
$$

$$
\begin{aligned}
= 2\frac{r_2^2}{r_1^2} \varkappa_n^2 [& C_{1n} \operatorname{Sinh}\varkappa_n \omega_n \sin\varkappa_n \omega_n - C_{2n} \operatorname{Cosh}\varkappa_n \omega_n \cos\varkappa_n \omega_n \\
& - C_{3n} \operatorname{Sinh}\varkappa_n \omega_n \cos\varkappa_n \omega_n + C_{4n} \operatorname{Cosh}\varkappa_n \omega_n \sin\varkappa_n \omega_n],
\end{aligned}
$$

$$
\begin{aligned}
M_\phi = \frac{K}{r_1} \chi^\cdot = \frac{(1 - \nu^2) r_1}{2 \varkappa_n} [& (C_{1n} - C_{2n}) \operatorname{Sinh}\varkappa_n \omega_n \cos\varkappa_n \omega_n \\
& + (C_{1n} + C_{2n}) \operatorname{Cosh}\varkappa_n \omega_n \sin\varkappa_n \omega_n \\
& - (C_{3n} - C_{4n}) \operatorname{Cosh}\varkappa_n \omega_n \cos\varkappa_n \omega_n \\
& + (C_{3n} + C_{4n}) \operatorname{Sinh}\varkappa_n \omega_n \sin\varkappa_n \omega_n].
\end{aligned}
$$

We have now 16 unknowns, the constants A_1, A_2 in zone 1, B_1, B_2 in zone 5, and 4 constants C in each of zones 2, 3, 4, and we have 16 conditions from which we can find them, 4 at each of the boundary circles between the zones. They are these: On each of these circles the value of Q_ϕ, M_ϕ, and χ must be the same for the two adjoining zones, and N_θ must have a discontinuity which is equal but opposite in sign to the discontinuity of the membrane force N_θ.

We shall not give here the numerical details of the 16 linear equations which follow from these conditions. Their coefficients form the following pattern in which each dash represents a coefficient which is not zero, each line the left-hand side of an equation, and each column one of the unknowns:

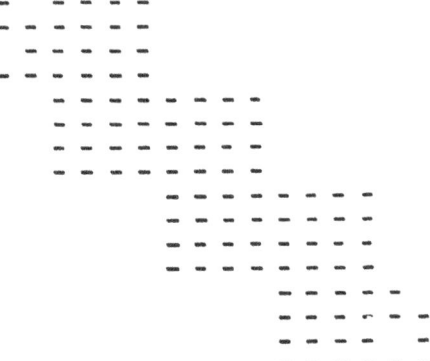

The elimination is not as difficult as it may appear at first glance. With its results and with eqs. (37) and (71) one may compute all stress resultants. The most interesting ones are plotted in fig. 23 over the arc length of the meridian. In the hoop-force diagram, the deformity of the meridian causes considerable waviness, and in the middle of the disturbed part N_θ really reaches the membrane value. But there are no discontinuities left, and the curve makes only a feeble attempt to follow the two high peaks of the membrane force. Fig. 23b shows the bending moment M_ϕ. It is difficult to understand this curve unless some scale of comparison is available. It is therefore more instructive to consider the eccentricity M_ϕ/N_ϕ of the normal force which may be compared with the wall thickness t. In fig. 23c it is represented in the following way: The abscissa is again the arc length of the meridian. The ordinates of the heavy line give the deviation z of the actual meridian from a circle, positive when inward. From this line the eccentricity $e_\phi = M_\phi/(-N_\phi)$ has been plotted so that the thin line gives the position of the resultant force in any cross section. The diagram shows that this force always keeps close to the actual meridian, but that it has a certain tendency to level out the bumpiness of the shell. The greatest eccentricity is slightly more than 5% of t.

The deformity of the shell which we have investigated here, is of a rather special kind, since it has axial symmetry, but it may be assumed that also in shells with locally restricted deviations the essential features of the stress disturbance will be similar.

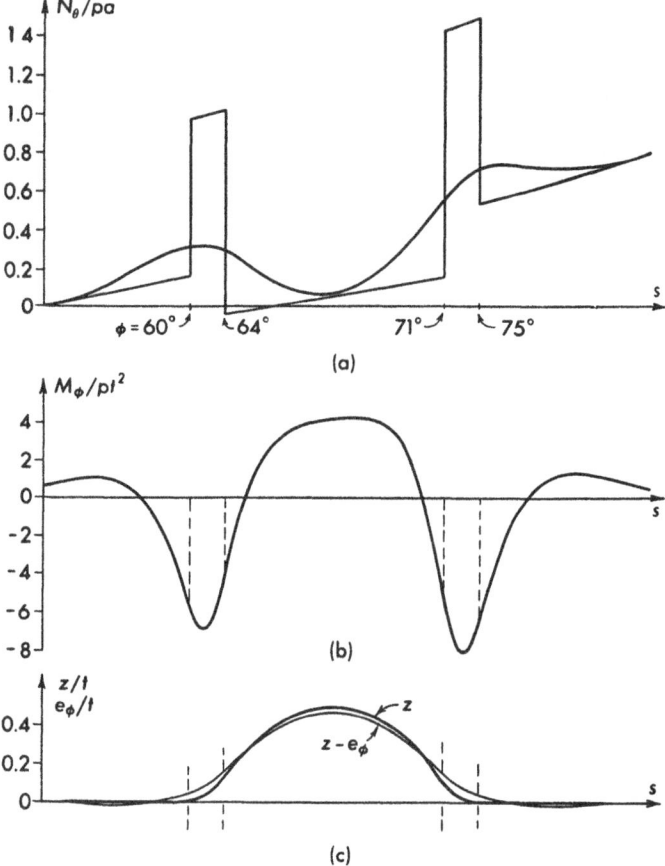

Fig. 23. Stress resultants for the almost hemispherical dome of Fig. 22

6.2.2.3.3 Variable $\varkappa$. When $\varkappa$ is not a constant, eq. (70) has a variable coefficient; but if $\varkappa$ varies slowly we may write

$$(\varkappa^2 Q_\phi)^{\cdot\cdot} = \varkappa^2 Q_\phi^{\cdot\cdot} + 2(\varkappa^2)^{\cdot} Q_\phi^{\cdot} + (\varkappa^2)^{\cdot\cdot} Q_\phi \approx \varkappa^2 Q_\phi^{\cdot\cdot}$$

and therefore

$$Q_\phi^{\cdot\cdot} + 4\varkappa^4 Q_\phi \approx (Q_\phi^{\cdot\cdot} \pm 2i\,\varkappa^2 Q_\phi)^{\cdot\cdot} \mp 2i\,\varkappa^2(Q_\phi^{\cdot\cdot} \pm 2i\,\varkappa^2 Q_\phi).$$

The right-hand side of this equation is zero if

$$Q_\phi^{\cdot\cdot} \pm 2i\,\varkappa^2 Q_\phi = 0, \tag{72}$$

and this pair of second-order equations with variable $\varkappa^2$ may replace eq. (70).

There is one case of particular interest in which these equations may easily be solved. Because of the local bending moments, many shells are reinforced in a border zone by gradually increasing their thickness toward the edge. It is often possible to choose the thickness so that $\varkappa$ may be represented by the formula.

$$\varkappa = \frac{\alpha}{\beta + \phi}, \tag{73}$$

in which α and β are constants, α of the same order of magnitude as $\varkappa$, so that in the following formulas we may neglect 1 compared with α^2, but not necessarily compared with α.

With $\varkappa$ from eq. (73), eq. (72) assumes the following form:

$$\ddot{Q}_\phi \pm \frac{2i\,\alpha^2}{(\beta + \phi)^2}\,Q_\phi = 0,$$

and this equation obviously has solutions of the type

$$Q_\phi = (\beta + \phi)^m.$$

Indeed, when this expression is introduced, the equation

$$m(m - 1) \pm 2i\,\alpha^2 = 0$$

results from which we find

$$m = \frac{1}{2} \pm \sqrt{\frac{1}{4} \pm 2i\,\alpha^2} \approx \frac{1}{2} \pm (1 \pm i)\,\alpha.$$

The four combinations of plus and minus signs yield four independent solutions; when we take them all together we have

$$Q_\phi = (\beta + \phi)^{\alpha + 1/2}[C_1(\beta + \phi)^{i\alpha} + C_2(\beta + \phi)^{-i\alpha}]$$
$$+ (\beta + \phi)^{-(\alpha - 1/2)}[C_3(\beta + \phi)^{i\alpha} + C_4(\beta + \phi)^{-i\alpha}]$$

as a possible form of the general solution. We may replace it by a handier expression. First, we bring the brackets into a real form, writing e. g. for the first one

$$C_1 \exp(i\,\alpha \ln(\beta + \phi)) + C_2 \exp(-i\,\alpha \ln(\beta + \phi))$$
$$= (C_1 + C_2)\cos(\alpha \ln(\beta + \phi)) + i(C_1 - C_2)\sin(\alpha \ln(\beta + \phi)).$$

This is a linear combination of a cosine and a sine, and with other constants it may be written

$$A_1 \cos\left(\alpha \ln\frac{\beta + \phi}{\beta + \phi_1}\right) + A_2 \sin\left(\alpha \ln\frac{\beta + \phi}{\beta + \phi_1}\right)$$

where ϕ_1 is the value of ϕ at the lower edge of the shell. When we remove a constant factor $(\beta + \phi_1)^{-(\alpha + 1/2)}$ from the constants, we arrive at the first line of the following expression:

$$Q_\phi = \left(\frac{\beta+\phi}{\beta+\phi_1}\right)^{\alpha+1/2}\left[A_1 \cos\left(\alpha \ln\frac{\beta+\phi}{\beta+\phi_1}\right) + A_2 \sin\left(\alpha \ln\frac{\beta+\phi}{\beta+\phi_1}\right)\right] \quad (74\,\text{a})$$
$$+ \left(\frac{\beta+\phi}{\beta+\phi_2}\right)^{-(\alpha-1/2)}\left[B_1 \cos\left(\alpha \ln\frac{\beta+\phi}{\beta+\phi_2}\right) + B_2 \sin\left(\alpha \ln\frac{\beta+\phi}{\beta+\phi_2}\right)\right].$$

This is the form which we finally adopt for the solution. The A terms decrease in damped oscillations when ϕ decreases; the B terms do the same when ϕ increases: if the shell is thin enough (i. e. if α is large enough), the two lines of eq. (74 a) represent the two local stress systems at the lower edge and at the upper edge, respectively.

We may now obtain χ from eq. (64 b) which, with our simplifying assumptions including $P = 0$, reads

$$E t \chi = -\frac{r_2^2}{r_1^2}Q_\phi^{\cdot\cdot};$$

then we may find the normal forces from eqs. (54) and the bending moments from eqs. (51). We give here some of the resulting formulas:

$$\chi = \frac{2\alpha^2}{Et}\frac{r_2^2}{r_1^2}\left\{\frac{1}{(\beta+\phi_1)^2}\left(\frac{\beta+\phi}{\beta+\phi_1}\right)^{\alpha-3/2}\left[A_1 \sin\left(\alpha \ln\frac{\beta+\phi}{\beta+\phi_1}\right) - A_2 \cos\left(\alpha \ln\frac{\beta+\phi}{\beta+\phi_1}\right)\right]\right.$$
$$\left.-\frac{1}{(\beta+\phi_2)^2}\left(\frac{\beta+\phi}{\beta+\phi_2}\right)^{-(\alpha+3/2)}\left[B_1 \sin\left(\alpha \ln\frac{\beta+\phi}{\beta+\phi_2}\right) - B_2 \cos\left(\alpha \ln\frac{\beta+\phi}{\beta+\phi_2}\right)\right]\right\},$$

$$N_\theta = -\frac{\alpha\,r_2}{r_1}\left\{\frac{1}{\beta+\phi_1}\left(\frac{\phi+\phi}{\beta+\phi_1}\right)^{\alpha-1/2}\right.$$
$$\left[(A_1 + A_2)\cos\left(\alpha \ln\frac{\beta+\phi}{\beta+\phi_1}\right) - (A_1 - A_2)\sin\left(\alpha \ln\frac{\beta+\phi}{\beta+\phi_1}\right)\right]$$
$$-\frac{1}{\beta+\phi_2}\left(\frac{\beta+\phi}{\beta+\phi_2}\right)^{-(\alpha+1/2)} \quad (74\,\text{b–d})$$
$$\left.\left[(B_1 - B_2)\cos\left(\alpha \ln\frac{\beta+\phi}{\beta+\phi_2}\right) + (B_1 + B_2)\sin\left(\alpha \ln\frac{\beta+\phi}{\beta+\phi_2}\right)\right]\right\},$$

$$M_\phi = \frac{r_1}{2\alpha}\left\{(\beta+\phi_1)\left(\frac{\beta+\phi}{\beta+\phi_1}\right)^{\alpha+3/2}\right.$$
$$\left[(A_1 - A_2)\cos\left(\alpha \ln\frac{\beta+\phi}{\beta+\phi_1}\right) + (A_1 + A_2)\sin\left(\alpha \ln\frac{\beta+\phi}{\beta+\phi_1}\right)\right]$$
$$-(\beta+\phi_2)\left(\frac{\beta+\phi}{\beta+\phi_2}\right)^{-(\alpha-3/2)}$$
$$\left.\left[(B_1 + B_2)\cos\left(\alpha \ln\frac{\beta+\phi}{\beta+\phi_2}\right) - (B_1 - B_2)\sin\left(\alpha \ln\frac{\beta+\phi}{\beta+\phi_2}\right)\right]\right\}.$$

As an example of the application of these formulas we consider a hemispherical concrete dome of $a = 26'$ radius and $t = 1.5''$ wall thickness. This dome carries a dead load $p = 35$ lb/ft^3. At the springing line the shell is assumed to be clamped to an unyielding support (e. g., an extremely strong foot ring or a thick ceiling slab). Because of the bending stresses, the thickness in the border zone increases gradually from 1.5″ at $\phi = 80°$ to 2.25″ at $\phi = 90°$.

If one tries to interpolate $\varkappa$ and consequently t between these limits with the help of eq. (73), he finds $\alpha = 842°$, $\beta = -35.7°$, and the wall thickness obtained for intermediate locations looks reasonable (fig. 24). One may now use eqs. (74) to compute a bending stress system which produces at the edge the rotation $\chi = 0$ and a hoop

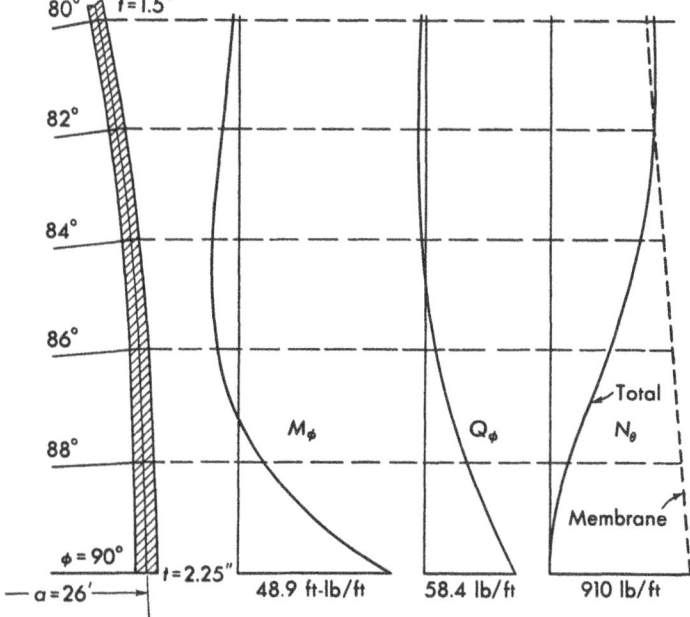

Fig. 24. Reinforced edge of a hemispherical dome; meridional section and stress resultants

force $N_\theta = -910$ lb/ft which just compensates the membrane force $N_\theta = +910$ lb/ft. The ensuing stress resultants are shown in fig. 24. They decrease to insignificant values within the zone of variable wall thickness, so that it is not necessary to follow them farther into the shell of constant thickness.

6.2.3 Conical Shell

6.2.3.1 Constant Wall Thickness

On p. 361 we have seen that the conical shell is among those for which the fourth-order equation (58) may be split into a pair of second-order equations, and we may therefore build the theory on eqs. (61).

However, the cone represents a degenerate case because ϕ is constant along the meridian and cannot serve as a coordinate. We use instead the distance s from the top of the shell, measured along the meridian. It is connected with ϕ by the relation

$$ds = r_1 \, d\phi.$$

When we introduce it into MEISSNER's operator (56) we find the following expression which is also useful for other than conical shells:

$$L(\ldots) = r_2 \frac{d^2(\ldots)}{ds^2} + \cot\phi \cdot \frac{d(\ldots)}{ds} - \frac{1}{r_2} \cot^2\phi \cdot (\ldots).$$

In the particular case of a cone, we have

$$\phi = \alpha, \quad r_1 = \infty, \quad r_2 = s \cot\alpha$$

and therefore

$$L(\ldots) = \left[s \frac{d^2(\ldots)}{ds^2} + \frac{d(\ldots)}{ds} - \frac{(\ldots)}{s} \right] \cot\alpha.$$

When this expression for the operator L is introduced in eqs. (61) and Q_s is written for Q_ϕ, they assume the following form:

$$\frac{d^2(s\,Q_s)}{ds^2} + \frac{1}{s} \frac{d(s\,Q_s)}{ds} + \left(-\frac{1}{s^2} \pm \frac{i\,\mu^2 \tan\alpha}{s} \right) s\,Q_s = 0. \quad \text{(75a, b)}$$

Eqs. (75) are the differential equations of our problem. Their solutions are, of course, conjugate complex, and it will be enough to solve one of the equations, say eq. (75a), and then use the real and imaginary parts of this solution separately as solutions of the fourth-order equation (58) which we do not need to transform to the coordinate s.

Eq. (75a) looks almost like BESSEL's equation, and we can make the likeness complete by a simple transformation of the independent variable. We introduce

$$\eta = 2\mu \sqrt{i \tan\alpha} \sqrt{s}$$

and obtain the equation

$$\frac{d^2(s\,Q_s)}{d\eta^2} + \frac{1}{\eta} \frac{d(s\,Q_s)}{d\eta} + \left(1 - \frac{4}{\eta^2} \right) (s\,Q_s) = 0.$$

The solutions of this equation are BESSEL functions of the second order, e. g. $J_2(\eta)$ and $H_2^{(1)}(\eta)$. With the help of well-known recurrence formulas[1] they may be expressed in terms of the corresponding functions of orders 0 and 1:

$$J_2(\eta) = \frac{2}{\eta} J_1(\eta) - J_0(\eta),$$

$$H_2^{(1)}(\eta) = \frac{2}{\eta} H_1^{(1)}(\eta) - H_0^{(1)}(\eta).$$

In our case the argument η is complex,

$$\eta = y \sqrt{i}$$

---

[1] See WATSON, G. N.: A Treatise on the Theory of BESSEL Functions, Cambridge 1922, pp. 45, 74.

with

$$y = 2\mu\sqrt{\tan\alpha}\,\sqrt{s} = 2\sqrt[4]{3(1-\nu^2)}\sqrt{\frac{2\tan\alpha}{t}}\,\sqrt{s}, \tag{76}$$

and the BESSEL functions have complex values. On p. 289 we saw how the real and imaginary parts of J_0, J_1, $H_0^{(1)}$, $H_1^{(1)}$ may be written in terms of THOMSON's functions of the real variable y. When we use these formulas, we find the following expressions for the functions of order 2:

$$J_2(\eta) = \left(\frac{2}{y}\,\text{bei}\,'y - \text{ber}\,y\right) + i\left(\frac{2}{y}\,\text{ber}\,'y + \text{bei}\,y\right),$$

$$H_2^{(1)}(\eta) = \frac{2}{\pi}\left(\frac{2}{y}\,\text{ker}\,'y + \text{kei}\,y\right) - i\frac{2}{\pi}\left(\frac{2}{y}\,\text{kei}\,'y - \text{ker}\,y\right).$$

These two functions are independent solutions of eq. (75a), and their real and imaginary parts separately will satisfy the fourth-order equation (58) when specialized for a conical shell. We have therefore the following general solution:

$$Q_s = \frac{1}{s}[A_1(\text{ber}\,y - 2y^{-1}\text{bei}\,'y) + A_2(\text{bei}\,y + 2y^{-1}\text{ber}\,'y) \tag{77a}$$
$$+ B_1(\text{ker}\,y - 2y^{-1}\text{kei}\,'y) + B_2(\text{kei}\,y + 2y^{-1}\text{ker}\,'y)].$$

We may now return to eqs. (57b) for χ, (15) for N_ϕ, N_θ, (51a, b) for M_ϕ, M_θ. When we adapt all these formulas to the special situation of a conical shell and drop the terms with P, wherever they appear, we find the following expressions:

$$\chi = \frac{2\sqrt{3(1-\nu^2)}\cot\alpha}{E\,t^2}[A_1(\text{bei}\,y + 2y^{-1}\text{ber}\,'y) - A_2(\text{ber}\,y - 2y^{-1}\text{bei}\,'y)$$
$$+ B_1(\text{kei}\,y + 2y^{-1}\text{ker}\,'y) - B_2(\text{ker}\,y - 2y^{-1}\text{kei}\,'y)]$$

$$N_s = -Q_s\cot\alpha,$$

$$N_\theta = -\frac{\cot\alpha}{2s}[A_1(y\,\text{ber}\,'y - 2\,\text{ber}\,y + 4y^{-1}\text{bei}\,'y)$$
$$+ A_2(y\,\text{bei}\,'y - 2\,\text{bei}\,y - 4y^{-1}\text{ber}\,'y)$$
$$+ B_1(y\,\text{ker}\,'y - 2\,\text{ker}\,y + 4y^{-1}\text{kei}\,'y)$$
$$+ B_2(y\,\text{kei}\,'y - 2\,\text{kei}\,y - 4y^{-1}\text{ker}\,'y)], \tag{77b–f}$$

$$M_s = 2y^{-2}[A_1[y\,\text{bei}\,'y - 2(1-\nu)(\text{bei}\,y + 2y^{-1}\text{ber}\,'y)]$$
$$- A_2[y\,\text{ber}\,'y - 2(1-\nu)(\text{ber}\,y - 2y^{-1}\text{bei}\,'y)]$$
$$+ B_1[y\,\text{kei}\,'y - 2(1-\nu)(\text{kei}\,y + 2y^{-1}\text{ker}\,'y)]$$
$$- B_2[y\,\text{ker}\,'y - 2(1-\nu)(\text{ker}\,y - 2y^{-1}\text{kei}\,'y)]],$$

$$M_\theta = 2y^{-2}[A_1[\nu\,y\,\text{bei}\,'y + 2(1-\nu)(\text{bei}\,y + 2y^{-1}\text{ber}\,'y)]$$
$$- A_2[\nu\,y\,\text{ber}\,'y + 2(1-\nu)(\text{ber}\,y - 2y^{-1}\text{bei}\,'y)]$$
$$+ B_1[\nu\,y\,\text{kei}\,'y + 2(1-\nu)(\text{kei}\,y + 2y^{-1}\text{ker}\,'y)]$$
$$- B_2[\nu\,y\,\text{ker}\,'y + 2(1-\nu)(\text{ker}\,y - 2y^{-1}\text{kei}\,'y)]].$$

In these formulas, the constants A_1, A_2, B_1, B_2 must be determined from the boundary conditions at the two edges of the shell. As one may easily see from a table of the THOMSON functions, they all have an oscillatory character; those appearing in the A terms of our formulas increase beyond all bounds when y increases, while those in the B terms have a singularity at $y = 0$ and decrease in amplitude with increasing distance from the top of the cone. It follows that for a cone which is closed at the top one must put $B_1 = B_2 = 0$, and that in a truncated cone it may easily happen that bending stresses of appreciable magnitude appear only in two border zones. In such cases the constants A_1, A_2 depend only on the conditions at the larger boundary circle, while B_1, B_2 are determined by the boundary conditions at the smaller edge of the shell.

If the variable y is large enough (say 15 or more), each THOMSON function may be replaced by the first term of its asymptotic expansion. If this is done, it is also consistent to neglect all negative powers of y compared with 1, and then simple formulas result which contain only elementary functions. However, they should be used with caution, and if a table of the THOMSON functions is at hand, eqs. (77) should be preferred.

For the vicinity of an outer edge where the B terms may be dropped, the simplified formulas are these:

$$Q_s = \frac{\exp(y/\sqrt{2})}{s\sqrt{2\pi y}}\left[A_1 \cos\left(\frac{y}{\sqrt{2}} - \frac{\pi}{8}\right) + A_2 \sin\left(\frac{y}{\sqrt{2}} - \frac{\pi}{8}\right)\right],$$

$$\chi = \frac{2\sqrt{3(1-\nu^2)}}{E\,t^2}\frac{\exp(y/\sqrt{2})}{\sqrt{2\pi y}}\left[A_1 \sin\left(\frac{y}{\sqrt{2}} - \frac{\pi}{8}\right) - A_2 \cos\left(\frac{y}{\sqrt{2}} - \frac{\pi}{8}\right)\right],$$

$$N_\theta = -\frac{\cot\alpha}{2\sqrt{2\pi}}\frac{\sqrt{y}\exp(y/\sqrt{2})}{s}\left[A_1 \cos\left(\frac{y}{\sqrt{2}} + \frac{\pi}{8}\right) + A_2 \sin\left(\frac{y}{\sqrt{2}} + \frac{\pi}{8}\right)\right],$$

$$M_s = \frac{2\exp(y/\sqrt{2})}{y\sqrt{2\pi y}}\left[A_1 \sin\left(\frac{y}{\sqrt{2}} + \frac{\pi}{8}\right) - A_2 \cos\left(\frac{y}{\sqrt{2}} + \frac{\pi}{8}\right)\right], \qquad (78\,\text{a–e})$$

$$M_\theta = \nu M_s.$$

These formulas are particularly useful for cones which are almost cylindrical.

6.2.3.2 Example: Sludge Digestion Tank

The bottom of a sludge digestion tank (fig. 25) offers a good example to demonstrate the application of eqs. (77) or (78). For the stress analysis we assume that the tank is filled with water ($\gamma = 62.4$ lb/ft^3) and that the surrounding soil has been removed as shown. The water pressure produces a membrane hoop force in the cylinder, which increases from zero at the water level to the maximum $N_\theta = \gamma\, h\, a = 3.745 \times 10^4$ lb/ft at the bottom. The corresponding deformation of the lower edge of the cylindrical shell is described by its radial displacement w and by the rotation dw/dx of the generator. If we assume $\nu = 0$, these two quantities are not affected by the normal force N_x and may be computed from eq. (V–68) which yields $E\, w = 11.23 \times 10^5$ lb/ft, $E\, dw/dx = -0.562 \times 10^5$ lb/ft^2.

The conical bottom carries two loads: the water pressure and the weight of the cylindrical wall and the roof. It may be assumed that the water pressure on the tank bottom is transmitted right across the wall to the ground and hence does not lead to membrane

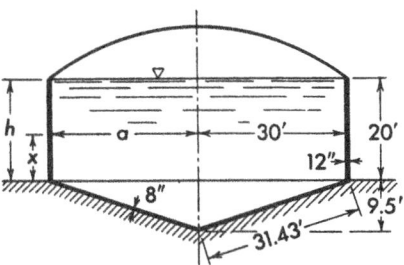

Fig. 25. Sludge-digestion tank

forces. The weight of the cylinder, however, is applied to the cone as a vertical edge load $P = 550$ lb/ft (fig. 26). Since the cone cannot carry a vertical edge load with membrane forces alone, we add a horizontal load $P \cot\alpha = 1737$ lb/ft so that the resultant force $P/\sin\alpha$ has the direction of the meridian. Since actually the horizontal load does not exist, we later compensate it by the difference of the forces H and H' (fig. 27).

The cone is supported by the reaction of the ground, and we assume more or less arbitrarily that the reaction to the edge load consists of a uniform

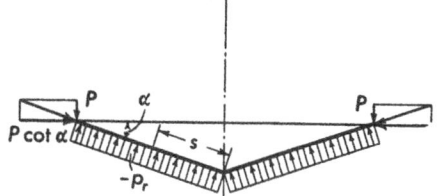

Fig. 26. Bottom of the sludge-digestion tank

load $-p_r$ normal to the shell (fig. 26). This load is $-p_r = 36.67$ lb/ft^2. Then, from eqs. (II–16), (II–17) the stress resultants for the edge of the cone are

$$N_s = -1820 \text{ lb/ft}, \qquad N_\theta = -3635 \text{ lb/ft}.$$

The horizontal component v_r of the displacement due to the membrane forces is directed inward and considered negative. From N_θ one easily finds $E\, v_r = -1.636 \times 10^5$ lb/ft. The rotation of the tangent is negligible ($E\, \chi = -6.10 \times 10^3$ lb/ft^2).

In order to provide continuity of the internal forces and of the deformation, additional forces and moments must be applied to the edges of both shells (fig. 27). For the cylinder there is a radial load H and a moment M. For the cone we have the same moment in the opposite direction and a radial load H' of such magnitude that H, H', and the external force $P \cot \alpha$ shown in fig. 26 add up to zero.

For the cylinder the deformations produced by unit edge loads may be calculated from eqs. (V–76') and (V–77'). They are for $H = 1\,\text{lb/ft}$:

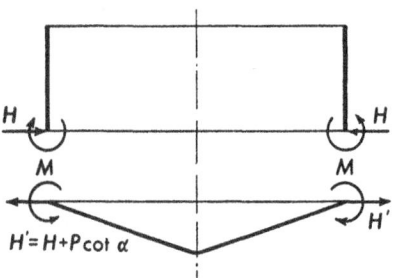

$$Ew = -432\,\text{lb/ft},$$
$$E\,dw/dx = +103.8\,\text{lb/ft}^2,$$

for $M = 1\,\text{ft·lb/ft}$:

$$Ew = +103.8\,\text{lb/ft},$$
$$E\,dw/dx = -49.9\,\text{lb/ft}^2.$$

To compute the corresponding figures for the edge of the cone, we use eqs. (77). First we obtain from eq. (60) $\mu = 2.280\,\text{ft}^{-1/2}$ and from eq. (76) $y = 14.38$. A table of the THOMSON functions yields

Fig. 27. Edge loads at bottom and wall of the sludge-digestion tank

$$\text{ber } 14.38 = -2597, \qquad \text{bei } 14.38 = -931,$$
$$\text{ber}' 14.38 = -1086, \qquad \text{bei}' 14.38 = -2463.$$

Now we may prepare formulas separately for unit loads H' and M.

For $H' = 1\,\text{lb/ft}$, $M = 0$ we have the boundary conditions $M_s = 0$, $Q_s = -\sin\alpha = -0.3020$, and upon introducing all numerical data in eqs. (77e, a) we obtain

$$A_1 = 1.765 \times 10^{-3}\,\text{lb}, \qquad A_2 = 4.98 \times 10^{-3}\,\text{lb}.$$

Now eqs. (77b, d) may be used to find

$$E\chi = 224.5\,\text{lb/ft}^2, \qquad N_\theta = 9.49\,\text{lb/ft},$$

and from N_θ we compute the deflection $E\,v_r = 427\,\text{lb/ft}$.

The load $M = 1\,\text{ft·lb/ft}$ may be handled similarly, and the following results will be obtained:

$$E v_r = 223.3\,\text{lb/ft}, \qquad E\chi = 245.7\,\text{lb/ft}^2.$$

According to MAXWELL's law of reciprocity, the figures 224.5 and 223.3 ought to be equal. The difference lies within the permissible margin of slide-rule errors, and the average 224.0 will be used for both quantities.

We may now formulate the equations which express the continuity of the deformation and from which H and M may be found as redundant quantities of a statically indeterminate structure.

The total deflection of the edge of the cylinder is

$$Ew = 11.23 \times 10^5 - 432H + 103.8M,$$

and the deflection for the edge of the cone is

$$Ev_r = -1.636 \times 10^5 + 433(H + 1737) + 244.1\,M.$$

Both must be equal, and this condition yields the equation

$$865H + 140.3\,M = 5.35 \times 10^5.$$

The second equation needed expresses the equality of the rotations dw/dx and χ of the tangents to the meridians of both shells. It may be compiled from the figures already presented and may be brought into the following form:

$$140.3H + 313.9\,M = -4.79 \times 10^5.$$

We have now two linear equations for H and M which may readily be solved. They yield $H = 934$ lb/ft and $M = -1940$ ft·lb/ft. The negative moment M is due to the heavy load P on the edge of the cone which tends to bend this shell outward. The clamping moment M represents the reaction of the cylinder to this tendency of deformation.

After having determined the redundant quantities one may use eqs. (77) to compute all the stress resultants for the cone and eqs. (V–76), (V–77) for the cylinder. The ensuing meridional bending moments (M_x for the cylinder, M_s for the cone) are shown in fig. 28. The vector at the edge indicates the direction of the resultant force transmitted from one shell to the other. Its magnitude is $(550^2 + 934^2)^{1/2} = 1084$ lb/ft and its eccentricity is 1.79 ft. In the cone the disturbance reaches approximately halfways down the generators, and in the cylinder it has practically died out before it reaches the upper edge. A similar disturbance resulting from the connection of the cylinder with the spherical roof shell is therefore not influenced by it and may be analyzed independently.

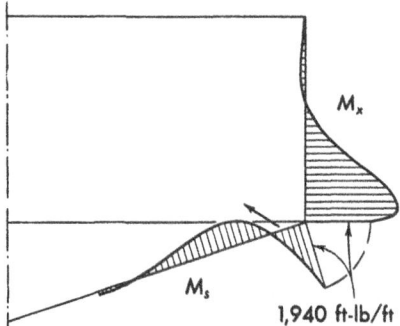

Fig. 28. Bending moments along the generators of the tank of Fig. 25

6.2.3.3 Wall Thickness Proportional to Distance from Apex

There are not many shells whose wall thickness t is proportional to the coordinate s, but since bending moments usually occur in a border zone of limited width, it is sometimes possible to substitute a shell of the type considered here for the real one without too much error. This has the slight advantage that the solution may be expressed in terms of elementary transcendentals. Since sufficient tables of the THOMSON functions are available, this advantage is not as great as it was when this solution was first established, but sometimes it may still be welcome.

The theory of a shell of variable thickness must start from the differential equation (67) and the splitting condition $f = $ const with f from eq. (65). All these equations have been written for an arbitrary meridian and use the angular coordinate ϕ. They must be adapted to the special circumstances of the conical shell by the same limiting process which has been applied on p. 372 to the operator L. When this is done, the splitting condition assumes the simple form

$$\left[(1 - \nu) \frac{dt}{ds} + s \frac{d^2t}{ds^2} \right] \cot \alpha = f = \text{const.}$$

This equation is, of course, satisfied if $t = $ const, but also if t is a linear function of s. If we put in particular

$$t = \delta \cdot s$$

with constant δ, we find that

$$f = (1 - \nu) \, \delta \cot \alpha$$

and from eq. (68') that

$$c_{1,2} = -(1 - \nu) \, \delta \cot \alpha \pm i \sqrt{12(1 - \nu^2) - (1 - \nu)^2 \delta^2 \cot^2 \alpha} \, .$$

Since the theory assumes that the wall thickness is small compared with s, δ is a small number, and unless α is extremely small, the term with δ^2 under the radical may be neglected:

$$c_{1,2} = -(1 - \nu) \, \delta \cot \alpha \pm 2i \sqrt{3(1 - \nu^2)} = -d_1 \pm i d_2 \qquad (79)$$

with

$$d_1 = (1 - \nu) \, \delta \cot \alpha , \qquad d_2 = 2 \sqrt{3(1 - \nu^2)} \, .$$

We may now turn our attention to the differential equation (67). The operator L_2 assumes in the present case the following form:

$$L_2(\ldots) = \left[s^2 \frac{d^2(\ldots)}{ds^2} - (1 - \nu) \, (\ldots) \right] \delta \cot \alpha ,$$

and when this and $c_{1,2}$ are introduced into eq. (67), it reads:

$$s^2 \frac{d^2U}{ds^2} - 2\left(1 - \nu \mp \frac{i}{\delta} \sqrt{3(1 - \nu^2)} \tan \alpha \right) U = 0 . \qquad (80\,\text{a, b})$$

Both equations of this pair have solutions of the form

$$U = C \, s^\lambda$$

with four different values of the exponent λ. These values are all complex and are best written with the help of some auxiliary quantities. When we put

$$\varkappa = \sqrt{\xi + \sqrt{\xi^2 + \eta^2}}, \qquad \mu = \sqrt{-\xi + \sqrt{\xi^2 + \eta^2}}$$

with

$$\xi = \frac{1}{2}(9 - 8\nu), \qquad \eta = \frac{4}{\delta}\sqrt{3(1 - \nu^2)}\tan\alpha,$$

then the general solution of eqs. (80a, b) is

$$U = C_1' s^{(1+\varkappa-i\mu)/2} + C_2' s^{(1-\varkappa+i\mu)/2} + C_3' s^{(1+\varkappa+i\mu)/2} + C_4' s^{(1-\varkappa-i\mu)/2},$$

and here the first two terms satisfy eq. (80a), the other two eq. (80b).

Since U has the dimension of a force, s the dimension of a length, the coefficients $C_1' \ldots C_4'$ have awkward dimensions. It therefore is preferable to choose an arbitrary reference length l and to introduce a dimensionless coordinate

$$y = \left(\frac{s}{l}\right)^{1/2}$$

and to write

$$U = y[C_1 y^{\varkappa-i\mu} + C_2 y^{-\varkappa+i\mu} + C_3 y^{\varkappa+i\mu} + C_4 y^{-\varkappa-i\mu}]. \quad (81\mathrm{a})$$

When we introduce this expression into eq. (69), we must use c_1 with the first two terms, but c_2 with the other two, and thus we get

$$E V = y[c_1 C_1 y^{\varkappa-i\mu} + c_1 C_2 y^{-\varkappa+i\mu} + c_2 C_3 y^{\varkappa+i\mu} + c_2 C_4 y^{-\varkappa-i\mu}]. \quad (81\mathrm{b})$$

Since U and V represent forces and deformations of the shell, they must have real values. It follows that the constants C_1 and C_3, C_2 and C_4 must be pairs of conjugate complex quantities, say

$$C_{1,3} = \frac{1}{2}(A_1 \pm i B_1), \qquad C_{2,4} = \frac{1}{2}(A_2 \mp i B_2).$$

On the other hand, we have

$$y^{\varkappa \pm i\mu} = y^\varkappa \exp(\pm i\mu \ln y) = y^\varkappa[\cos(\mu \ln y) \pm i\sin(\mu \ln y)].$$

When we introduce all this and d_1 and d_2 from eq. (79) into eqs. (81), they read as follows

$$U = Q_s s\cot\alpha = y^{\varkappa+1}[A_1 \cos(\mu \ln y) + B_1 \sin(\mu \ln y)]$$

$$\qquad\qquad + y^{-\varkappa+1}[A_2 \cos(\mu \ln y) + B_2 \sin(\mu \ln y)],$$

$$EV = E\,\delta^2 s^2 \chi = \qquad\qquad\qquad\qquad\qquad\qquad (82\mathrm{a, b})$$

$$\qquad = -y^{\varkappa+1}[(d_1 A_1 + d_2 B_1)\cos(\mu \ln y) + (d_1 B_1 - d_2 A_1)\sin(\mu \ln y)]$$

$$\qquad - y^{-\varkappa+1}[(d_1 A_2 - d_2 B_2)\cos(\mu \ln y) + (d_1 B_2 + d_2 A_2)\sin(\mu \ln y)].$$

Now we may obtain formulas for the normal forces and the bending moments from eqs. (15) and (51) which, of course, must all be adapted to the special case of a conical shell of variable wall thickness. We mention the following results:

$$N_\theta = \frac{y^{\varkappa-1}}{2l}\left[(d_1A_1 + d_2B_1)\left((\varkappa+1)\cos(\mu\ln y) - \mu\sin(\mu\ln y)\right)\right.$$
$$\left. + (d_1B_1 - d_2A_1)\left((\varkappa+1)\sin(\mu\ln y) + \mu\cos(\mu\ln y)\right)\right]$$
$$- \frac{y^{-\varkappa-1}}{2l}\left[(d_1A_2 - d_2B_2)\left((\varkappa-1)\cos(\mu\ln y) + \mu\sin(\mu\ln y)\right)\right.$$
$$\left. + (d_1B_2 + d_2A_2)\left((\varkappa-1)\sin(\mu\ln y) - \mu\cos(\mu\ln y)\right)\right], \qquad (82\,\mathrm{c,\,d})$$

$$M_s = \frac{\delta y^{\varkappa+1}}{24(1-\nu^2)}\left[(d_1A_1 + d_2B_1)\left((\varkappa-3+2\nu)\cos(\mu\ln y) - \mu\sin(\mu\ln y)\right)\right.$$
$$\left. + (d_1B_1 - d_2A_1)\left((\varkappa-3+2\nu)\sin(\mu\ln y) + \mu\cos(\mu\ln y)\right)\right]$$
$$- \frac{\delta y^{-\varkappa+1}}{24(1-\nu^2)}\left[(d_1A_2 - d_2B_2)\left((\varkappa+3-2\nu)\cos(\mu\ln y) + \mu\sin(\mu\ln y)\right)\right.$$
$$\left. + (d_1B_2 + d_2A_2)\left((\varkappa+3-2\nu)\sin(\mu\ln y) - \mu\cos(\mu\ln y)\right)\right].$$

With the help of these formulas problems may be solved similar to the one treated in the preceding Section.

6.3 Solution for the Higher Harmonics

For a shell of revolution the coefficients of all equations are independent of θ. Therefore, when we assume all the stress resultants and all the displacements as FOURIER series in θ, eqs. (1) and (5) are reduced to a set of ordinary differential equations. The solution of this set has been achieved for spherical and for conical shells, and these solutions will be explained in this section.

6.3.1 Spherical Shell

6.3.1.1 Differential Equations

The development of the differential equations starts from the conditions of equilibrium (1) and the elastic law (5). Since we are interested only in edge loads, we put $p_\phi \equiv p_\theta \equiv p_r \equiv 0$ in eqs. (1), and for the spherical shell we still have

$$r_1 = r_2 = a, \qquad r = a\sin\phi.$$

Since eq. (1f) is an identity, we drop it at once. The other moment equations (1d, e) may be used to eliminate Q_ϕ, Q_θ from (1a–c). In this way we obtain the following three conditions of equilibrium

$$\left.\begin{aligned}
&a(N_\theta' + \dot N_{\theta\phi}\sin\phi + 2N_{\theta\phi}\cos\phi) - M_\theta' - \dot M_{\theta\phi}\sin\phi \\
&\hspace{6cm} - 2M_{\theta\phi}\cos\phi = 0, \\
&a(\dot N_\phi\sin\phi + N_\phi\cos\phi - N_\theta\cos\phi + N_{\theta\phi}') - \dot M_\phi\sin\phi - M_\phi\cos\phi \\
&\hspace{6cm} + M_\theta\cos\phi - M_{\theta\phi}' = 0, \\
&a(N_\phi + N_\theta)\sin^2\phi + \ddot M_\phi\sin^2\phi + 2\dot M_\phi\cos\phi\sin\phi - M_\phi\sin^2\phi + M_\theta'' \\
&\quad - \dot M_\theta\cos\phi\sin\phi + M_\theta\sin^2\phi + 2\dot M_{\theta\phi}'\sin\phi + 2M_{\theta\phi}'\cos\phi = 0.
\end{aligned}\right\} \quad (83)$$

The elastic law (5) simplifies considerably for the spherical shell:

$$N_\phi = \frac{D}{a}\left[v\dot{} + w + \frac{\nu u'}{\sin\phi} + \nu v \cot\phi + \nu w\right],$$

$$N_\theta = \frac{D}{a}\left[\frac{u'}{\sin\phi} + v\cot\phi + w + \nu v\dot{} + \nu w\right],$$

$$N_{\phi\theta} = N_{\theta\phi} = \frac{D(1-\nu)}{2a}\left[u\dot{} - u\cot\phi + \frac{v'}{\sin\phi}\right],$$

$$M_\phi = \frac{K}{a^2}\left[-v\dot{} + w\ddot{} - \frac{\nu u'}{\sin\phi} - \nu v\cot\phi + \nu w\dot{}\cot\phi + \frac{\nu w''}{\sin^2\phi}\right],$$

$$M_\theta = \frac{K}{a^2}\left[-\frac{u'}{\sin\phi} - v\cot\phi + w\dot{}\cot\phi + \frac{w''}{\sin^2\phi} - \nu v\dot{} + \nu w\ddot{}\right],$$

$$M_{\phi\theta} = M_{\theta\phi} = \frac{K(1-\nu)}{2a^2}\left[u\cot\phi - u\dot{} - \frac{v'}{\sin\phi} - 2w'\frac{\cos\phi}{\sin^2\phi} + \frac{2w\dot{}'}{\sin\phi}\right].$$

(84 a–f)

The two shearing forces are exactly equal, and so are the two twisting moments.

Eqs. (83) and (84) are together a set of 9 equations for just as many unknowns: 3 forces, 3 moments and 3 displacements. We may use eqs. (84) to eliminate all the stress resultants from (83). This yields a set of three equations for u, v, w which corresponds exactly to eqs. (V–14) of the circular cylinder:

$$(1+k)\left[\frac{1-\nu}{2}\left(u\ddot{}\sin\phi + u\dot{}\cos\phi - u(\cot^2\phi - 1)\sin\phi\right) + \frac{u''}{\sin\phi}\right.$$
$$\left. + \frac{1+\nu}{2}v'\dot{} + \frac{3-\nu}{2}v'\cot\phi + (1+\nu)w'\right]$$
$$- k\left[w''\dot{} + w'\dot{}\cot\phi + 2w' + \frac{w'''}{\sin^2\phi}\right] = 0,$$

$$(1+k)\left[\frac{1+\nu}{2}u'\dot{} - \frac{3-\nu}{2}u'\cot\phi + v\ddot{}\sin\phi + v\dot{}\cos\phi - v\frac{\cos^2\phi + \nu\sin^2\phi}{\sin\phi}\right.$$
$$\left. + \frac{1-\nu}{2}\frac{v''}{\sin\phi} + (1+\nu)w\dot{}\sin\phi\right]$$
$$- k\left[w\ddot{}\sin\phi + w\ddot{}\cos\phi + w\dot{}(1-\cot^2\phi)\sin\phi + \frac{w''\dot{}}{\sin\phi} - 2w''\frac{\cos\phi}{\sin^2\phi}\right] = 0$$

$$1+k)(1+\nu)[u' + v\dot{}\sin\phi + v\cos\phi + 2w\sin\phi]$$

(85 a–c),

$$- k\left[u'\ddot{} - u'\dot{}\cot\phi + u'(3+\cot^2\phi) + \frac{u'''}{\sin^2\phi} + v\ddot{}\sin\phi + 2v\ddot{}\cos\phi\right.$$
$$\left. - v\dot{}\frac{\cos^2\phi}{\sin\phi} + v(3+\cot^2\phi)\cos\phi + \frac{v''\dot{}}{\sin\phi} + v''\frac{\cos\phi}{\sin^2\phi}\right]$$
$$+ k\left[w\ddot{\cdot}\sin\phi + 2w\ddot{}\cos\phi - (1+\nu+\cot^2\phi)w\ddot{}\sin\phi\right.$$
$$+ w\dot{}(2-\nu+\cot^2\phi)\cos\phi - 2(1+\nu)w\sin\phi$$
$$\left. + \frac{2w''\dot{\cdot}}{\sin\phi} - 2w''\dot{}\frac{\cos\phi}{\sin^2\phi} + w''\frac{3-\nu+4\cot^2\phi}{\sin\phi} + \frac{w^{IV}}{\sin^3\phi}\right] = 0.$$

Here the notation

$$k = \frac{K}{Da^2} = \frac{t^2}{12a^2}$$

which was introduced in connection with eqs. (V–14) is used again.

When we now put

$$u = \sum u_n(\phi) \sin n\,\theta, \quad v = \sum v_n(\phi) \cos n\,\theta, \quad w = \sum w_n(\phi) \cos n\,\theta, \quad (86)$$

each of eqs. (85) is split into an infinite number of ordinary differential equations, one for every integer value of n. For an arbitrary n, they are

$$(1 + k)\left[\frac{1-\nu}{2}\left(\ddot{u}_n \sin\phi + \dot{u}_n \cos\phi\right) - u_n\left(\frac{n^2}{\sin\phi} - \frac{1-\nu}{2}\left(1 - \cot^2\phi\right)\sin\phi\right)\right.$$

$$\left. - \frac{1+\nu}{2}\,n\dot{v}_n - \frac{3-\nu}{2}\,n\,v_n \cot\phi - (1+\nu)\,n\,w_n\right]$$

$$+ k\,n\left[\ddot{w}_n + \dot{w}_n \cot\phi + w_n\left(2 - \frac{n^2}{\sin^2\phi}\right)\right] = 0,$$

$$(1 + k)\left[\frac{1+\nu}{2}\,n\,\dot{u}_n - \frac{3-\nu}{2}\,n\,u_n \cot\phi + \ddot{v}_n \sin\phi + \dot{v}_n \cos\phi\right.$$

$$\left. - \frac{v_n}{\sin\phi}\left(\cos^2\phi + \nu\sin^2\phi + \frac{1-\nu}{2}\,n^2\right) + (1+\nu)\,\dot{w}_n \sin\phi\right] \qquad (87\,\text{a–c})$$

$$- k\left[\dddot{w}_n \sin\phi + \ddot{w}_n \cos\phi + \dot{w}_n\left(2 - \frac{1+n^2}{\sin^2\phi}\right)\sin\phi + 2n^2\,w_n\frac{\cos\phi}{\sin^2\phi}\right] = 0,$$

$$(1 + k)(1 + \nu)\left[n\,u_n + \dot{v}_n \sin\phi + v_n \cos\phi + 2w_n \sin\phi\right]$$

$$- k\left[n\,\ddot{u}_n - n\,\dot{u}_n \cot\phi + n\,u_n\left(2 + \frac{1-n^2}{\sin^2\phi}\right) + \dddot{v}_n \sin\phi + 2\ddot{v}_n \cos\phi\right.$$

$$\left. - \frac{\dot{v}_n}{\sin\phi}\left(n^2 + \cos^2\phi\right) + v_n \cos\phi\left(2 + \frac{1-n^2}{\sin^2\phi}\right)\right]$$

$$+ k\left[\ddddot{w}_n \sin\phi + 2\dddot{w}_n \cos\phi - \frac{\ddot{w}_n}{\sin\phi}\left(1 + 2n^2 + \nu\sin^2\phi\right)\right.$$

$$+ \dot{w}_n\left(1 - \nu + \frac{1+2n^2}{\sin^2\phi}\right)\cos\phi - w_n\left(2(1+\nu)\sin\phi + n^2\frac{3-\nu+4\cot^2\phi}{\sin\phi}\right.$$

$$\left.\left. - \frac{n^4}{\sin^3\phi}\right)\right] = 0.$$

These equations look rather involved. However, it is quite feasible to establish an asymptotic solution in general terms when they are brought into a better form. This is a somewhat lengthy procedure, but it may easily be understood if done step by step.

We begin by introducing a special differential operator:

$$H(\ldots) = (\ldots)^{\cdot\cdot} + (\ldots)^{\cdot}\cot\phi + (\ldots)\left(2 - \frac{n^2}{\sin^2\phi}\right), \qquad (88)$$

which is suggested by the second bracket in eq. (87a). From the definition of H we may easily derive the following

$$HH(\ldots) = (\ldots)^{\cdots} + 2(\ldots)^{\cdot\cdot}\cot\phi + (\ldots)^{\cdot\cdot}\left(3 - \frac{1+2n^2}{\sin^2\phi}\right)$$

$$+ (\ldots)^{\cdot}\left(4 + \frac{1+2n^2}{\sin^2\phi}\right)\cot\phi + (\ldots)\left(4 - \frac{2n^2}{\sin^2\phi} - \frac{n^2(4-n^2)}{\sin^4\phi}\right)$$

$$H^{\cdot}(\ldots) \equiv \frac{\partial}{\partial\phi}H(\ldots) = (\ldots)^{\cdots} + (\ldots)^{\cdot\cdot}\cot\phi \qquad\qquad (89\,a, b)$$

$$+ (\ldots)^{\cdot}\left(2 - \frac{1+n^2}{\sin^2\phi}\right) + 2(\ldots)\,n^2\frac{\cot\phi}{\sin^2\phi}.$$

In the last bracket of eq. (87b) we discover $-k\,H^{\cdot}(w_n)\sin\phi$. Encouraged by this success, we may further investigate eq. (87). This leads us to introduce a new variable V_n defined by the relation

$$v_n = V_n^{\cdot},$$

and then it becomes possible to write eqs. (87a, b) in the following form:

$$(1+k)\left[H(u_n\sin\phi) - \frac{1+\nu}{2}(u_n\sin\phi + nV_n)^{\cdot\cdot}\right.$$

$$\left. - \frac{3-\nu}{2}(u_n\sin\phi + nV_n)^{\cdot}\cot\phi - (1+\nu)(u_n\sin\phi + n\,w_n)\right]$$

$$+ k\,n\,H(w_n) = 0, \qquad\qquad (90\,a, b)$$

$$\sin\phi\cdot\frac{\partial}{\partial\phi}\left[(1+k)\left(H(V_n) - (1+\nu)(V_n + w_n) + \frac{1+\nu}{2}n\frac{u_n\sin\phi + nV_n}{\sin^2\phi}\right)\right.$$

$$\left. - kH(w_n)\right] - (1+k)(1-\nu)\,n(u_n\sin\phi + nV_n)\frac{\cot\phi}{\sin\phi} = 0.$$

It is possible to eliminate w_n from these equations by differentiating (90a) with respect to ϕ, then multiplying it by $(\sin\phi)/n$ and adding it to (90b). The resultant relation contains the unknowns u_n and V_n only in the combination $(u_n\sin\phi + n\,V_n)$, and when we introduce

$$\omega_n = \frac{(u_n\sin\phi + nV_n)^{\cdot}}{\sin\phi} = \frac{u_n^{\cdot}\sin\phi + u_n\cos\phi + n\,v_n}{\sin\phi}, \qquad (91)$$

it assumes the simple form

$$H(\omega_n) = 0. \qquad\qquad (92)$$

This equation summarizes eqs. (87a, b). We now introduce the new function ω_n into eq. (87c) and try to make out of it a differential equation for another unknown function. The first step in this program is to eliminate v_n from (87c) by means of (91). This yields the following equation:

$$(1+k)(1+\nu)\left[u_n^{\cdot\cdot}\sin\phi + 3u_n^{\cdot}\cos\phi - (n^2 + 2\sin^2\phi - 1)\frac{u_n}{\sin\phi} - 2n\,w_n\right]$$

$$+k\left[-u_n^{\cdot\cdot\cdot}\sin\phi - 6u_n^{\cdot\cdot}\cos\phi + (2n^2 - 5 + 11\sin^2\phi)\frac{u_n^{\cdot\cdot}}{\sin\phi}\right.$$

$$+ (2n^2 + 1 + 6\sin^2\phi)u_n^{\cdot}\frac{\cot\phi}{\sin\phi} + (n^2 - 1)\left(1 + \cot^2\phi - \frac{n^2}{\sin^2\phi}\right)\frac{u_n}{\sin\phi}\left.\right]$$

$$-n\,k\left[w_n^{\cdot\cdot}+2w_n^{\cdot}\cot\phi - (\nu\sin^2\phi + 1 + 2n^2)\frac{w_n^{\cdot\cdot}}{\sin^2\phi}\right.$$

$$\left.+\left(1-\nu+\frac{1+2n^2}{\sin^2\phi}\right)w_n^{\cdot}\cot\phi + \left(\frac{n^4}{\sin^4\phi} - n^2\frac{3-\nu+4\cot^2\phi}{\sin^2\phi} - 2(1+\nu)\right)w_n\right]$$

$$= (1+k)(1+\nu)\left[\omega_n^{\cdot}\sin\phi + 2\,\omega_n\cos\phi\right]$$

$$-k\left[\omega_n^{\cdot\cdot\cdot}\sin\phi + 5\omega_n^{\cdot\cdot}\cos\phi + \frac{\omega_n^{\cdot}}{\sin\phi}(3 - 6\sin^2\phi - n^2) - 2n^2\,\omega_n\frac{\cot\phi}{\sin\phi}\right].$$

The left-hand side of this relation may be simplified with the help of the operator H. On the right-hand side we make use of eq. (92) to eliminate $\omega_n^{\cdot\cdot}$ and $\omega_n^{\cdot\cdot\cdot}$. In this way the equation assumes the following, much simpler form:

$$(1+k)(1+\nu)\left[H(u_n\sin\phi) - 2(u_n\sin\phi + n\,w_n)\right]$$

$$- k[HH(u_n\sin\phi + n\,w_n) - 2H(u_n\sin\phi + n\,w_n) - (1+\nu)nH(w_n)]$$

$$= [(1+\nu) + (5+\nu)k](\omega_n^{\cdot}\sin\phi + 2\,\omega_n\cos\phi).$$

Besides ω_n, we still have here two unknowns, u_n and w_n, but since they appear almost exclusively in the combination $u_n\sin\phi + n\,w_n$, it seems advisable to use this as a new unknown quantity. In order to rid our equation of the odd terms in u_n and w_n, we prepare eq. (90a) by introducing there the variable ω_n:

$$(1+k)[H(u_n\sin\phi) - (1+\nu)(u_n\sin\phi + n\,w_n)] + k\,n\,H(w_n)$$

$$= (1+k)\left[\frac{1+\nu}{2}\,\omega_n^{\cdot}\sin\phi + 2\,\omega_n\cos\phi\right]. \tag{93}$$

When we multiply this equation by $(1+\nu)$ and then subtract it from the preceding one, we obtain an equation which contains only ω_n and the combination $u_n\sin\phi + n\,w_n$:

$$(1+k)(1-\nu^2)(u_n\sin\phi + n\,w_n)$$

$$+ k[HH(u_n\sin\phi + n\,w_n) - 2H(u_n\sin\phi + n\,w_n)$$

$$= -(1+k)\frac{1-\nu^2}{2}\,\omega_n^{\cdot}\sin\phi - 4k(\omega_n^{\cdot}\sin\phi + 2\,\omega_n\cos\phi).$$

It is possible to go one step further in processing this equation. One may easily verify the relations

$$H(\omega_n^{\cdot} \sin\phi) = -4\,\omega_n \cos\phi,$$
$$HH(\omega_n^{\cdot} \sin\phi) = 8\,\omega_n^{\cdot} \sin\phi + 8\,\omega_n \cos\phi \qquad \text{(94a, b)}$$

by performing the operation indicated on the left-hand side and then using eq. (92). These relations may be used to bring the right-hand side of the preceding equation into the following form:

$$-(1+k)(1-\nu^2)\cdot\frac{1}{2}\,\omega_n^{\cdot}\sin\phi - k\left[\frac{1}{2}HH(\omega_n^{\cdot}\sin\phi) - H(\omega_n^{\cdot}\sin\phi)\right];$$

and now we see that we should introduce

$$U_n = u_n \sin\phi + n\,w_n + \frac{1}{2}\,\omega_n^{\cdot}\sin\phi \qquad (95)$$

as the dependent variable and that U_n must satisfy the fourth-order differential equation

$$HH(U_n) - 2H(U_n) + (1-\nu^2)\frac{1+k}{k}\,U_n = 0. \qquad (96)$$

This and eq. (92) are the differential equations of our problem. They are exact in the sense that we did not lose anything on the way from the fundamental equations (1) and (5).

6.3.1.2 Membrane Forces and Inextensional Bending

There exist two groups of solutions which differ considerably in appearance. Since the bending theory has been established with no other restriction than that the surface load is supposed to be zero, we must expect to find among the solutions some which confirm or correct the homogeneous membrane solution (II–38) and the inextensional deformations (II–65) of the shell; moreover, we may expect solutions which decay exponentially with increasing distance from a loaded edge.

The first group is by far the simpler one, and we shall examine it first.

Eq. (96) certainly has, among others, the solution $U_n \equiv 0$. When we want to find the corresponding displacements, we must go back to eq. (93) and must introduce there the variable U_n. This becomes possible when we apply the relation (94) to the last term on the right. Then all terms but one may be combined to functions of U_n and the equation yields

$$nH(w_n) = (1+k)[H(U_n) - (1+\nu)U_n]. \qquad (97)$$

With $U_n \equiv 0$, this is a homogeneous differential equation for w_n, in detail:

$$w_n^{\cdot\cdot} + w_n^{\cdot} \cot\phi + w_n \left(2 - \frac{n^2}{\sin^2\phi}\right) = 0,$$

and it has the solution

$$w_n = A_1 (n + \cos\phi) \tan^n \frac{\phi}{2} + A_2 (n - \cos\phi) \cot^n \frac{\phi}{2}. \qquad (98\,\text{a})$$

To find u_n and v_n from eqs. (95) and (91), we must first solve eq. (92). For the moment, we consider only the trivial solution $\omega_n \equiv 0$ and obtain

$$u_n = -\frac{n\,w_n}{\sin\phi}, \qquad v_n = -\frac{(u_n \sin\phi)^{\cdot}}{n} = w_n^{\cdot}. \qquad (98\,\text{b, c})$$

When this result is introduced into the elastic law (84), the bending and twisting moments vanish identically and the normal and shearing forces are:

$$N_\phi = -N_\theta = \frac{D(1-\nu)}{a} \frac{n(n^2-1)}{\sin^2\phi} \left(A_1 \tan^n \frac{\phi}{2} + A_2 \cot^n \frac{\phi}{2}\right) \cos n\,\theta,$$

$$N_{\phi\theta} = N_{\theta\phi} = -\frac{D(1-\nu)}{a} \frac{n(n^2-1)}{\sin^2\phi} \left(A_1 \tan^n \frac{\phi}{2} - A_2 \cot^n \frac{\phi}{2}\right) \sin n\,\theta.$$
$$(98\,\text{d, e})$$

Except for the different notation for the constants, these are exactly the formulas (II–38) of the membrane theory. We recognize here that they are not only good approximations but that they happen to be exact solutions of the bending problem.

When we established the last formulas, we used the trivial solution $\omega_n \equiv 0$ of eq. (92). Of course, there are still other solutions, and since eq. (92) has the same form as our last differential equation for w_n, it has the same pair of solutions:

$$\omega_n = B_1 (n + \cos\phi) \tan^n \frac{\phi}{2} + B_2 (n - \cos\phi) \cot^n \frac{\phi}{2}.$$

We might combine this with the trivial solutions $U_n \equiv 0$ and $w_n \equiv 0$, but we arrive at a particularly interesting result when we choose

$$w_n = -\frac{1}{2} B_1 (n + \cos\phi) \tan^n \frac{\phi}{2} + \frac{1}{2} B_2 (n - \cos\phi) \cot^n \frac{\phi}{2}. \quad (99\,\text{a})$$

Then eqs. (95) and (91) yield

$$u_n = \frac{1}{2} \sin\phi \left(B_1 \tan^n \frac{\phi}{2} - B_2 \cot^n \frac{\phi}{2}\right),$$

$$v_n = \frac{1}{2} \sin\phi \left(B_1 \tan^n \frac{\phi}{2} + B_2 \cot^n \frac{\phi}{2}\right). \qquad (99\,\text{b, c})$$

These three formulas are identical with eqs. (II–65a–c) on p. 90, i. e., they represent an inextensional deformation of the shell. It is therefore not surprising that eqs. (5a–d) yield $N_\phi = N_\theta = N_{\phi\theta} = N_{\theta\phi} = 0$. The bending and twisting moments are

$$M_\phi = -M_\theta = -\frac{K(1-\nu)}{2a^2} \frac{n(n^2-1)}{\sin^2\phi}\left(B_1\tan^n\frac{\phi}{2} - B_2\cot^n\frac{\phi}{2}\right)\cos n\theta,$$

$$M_{\phi\theta} = M_{\theta\phi} = \frac{K(1-\nu)}{2a^2} \frac{n(n^2-1)}{\sin^2\phi}\left(B_1\tan^n\frac{\phi}{2} + B_2\cot^n\frac{\phi}{2}\right)\sin n\theta$$

$$(99\,\mathrm{d, e})$$

and from eqs. (1 d, e) we find $Q_\phi = Q_\theta = 0$. Eqs. (99) represent a second pair of exact solutions in the bending theory of the spherical shell.

Of course, it must not be concluded from these results that in any shell of revolution the membrane forces or the inextensional deformations will satisfy the differential equations of the bending theory. But it may be expected that they will come very close to exact solutions if the middle surface is not too different from a sphere.

6.3.1.3 Oscillatory Solutions

Aside from solutions (98) and (99) there exists another set of four, and these are of an oscillatory character. Just as in the case of axial symmetry, these oscillations decay rather rapidly with increasing distance from the edge, the faster, the thinner the shell is. For such edge disturbances the method of asymptotic integration of the differential equations recommends itself. Since this method is based on the fact that $k = t^2/12a^2$ is a very small quantity, we shall neglect k compared with unity wherever an opportunity occurs.

The first step of the procedure is still done in full exactness. In analogy to the splitting of eq. (65b), we split eq. (96) in two equations of the second order of the form

$$H(U_n) + \mu^2 U_n = 0 \tag{100}$$

or in more detail:

$$U_n'' + U_n'\cot\phi + U_n\left(2 + \mu^2 - \frac{n^2}{\sin^2\phi}\right) = 0. \tag{100'}$$

When eq. (100) is introduced into eq. (96), a quadratic equation for μ^2 results:

$$\mu^4 + 2\mu^2 + (1-\nu^2)\frac{1+k}{k} = 0$$

whose roots are

$$\mu^2 = -1 \pm i\sqrt{(1-\nu^2)\frac{1+k}{k} - 1} = -1 \pm 2i\,\varkappa^2.$$

When we now start to drop small quantities we have

$$\varkappa = \sqrt[4]{\frac{1-\nu^2}{4k}}, \qquad \mu^2 = \pm 2i\,\varkappa^2, \qquad \mu = \pm(1\pm i)\,\varkappa, \tag{101}$$

where $\varkappa$ is identical with the parameter used previously if we neglect the small term $\nu^2/4$ in the earlier definition (20).

We now subject both the coordinate ϕ and the dependent variable U_n to transformations with the aim of making out of (100′) an equation whose coefficients are almost constant.

When a new independent variable $\xi = \xi(\phi)$ is introduced, eq. (100′) assumes the form

$$(\xi^{\cdot})^2 \frac{d^2U_n}{d\xi^2} + (\xi^{\cdot\cdot} + \xi^{\cdot}\cot\phi)\frac{dU_n}{d\xi} + \left(2 + \mu^2 - \frac{n^2}{\sin^2\phi}\right)U_n = 0.$$

In this equation the first and the third coefficient become constant if we choose $\xi(\phi)$ so that it satisfies the differential equation

$$(\xi^{\cdot})^2 = 1 - \frac{n^2}{(2 + \mu^2)\sin^2\phi} \tag{102}$$

and then divide the whole equation by $(\xi^{\cdot})^2$. The equation will lose the term with the first derivative if a new dependent variable

$$\eta = U_n\sqrt{\xi^{\cdot}\sin\phi}$$

is introduced. Through this transformation the last term is slightly spoiled but still is almost constant. Indeed, the equation assumes the form

$$\frac{d^2\eta}{d\xi^2} + (2 + \mu^2 + \alpha)\eta = 0 \tag{103}$$

where

$$\alpha = \frac{\sin^2\phi(3\sin^2\phi - 2)}{2\left(\sin^2\phi - \dfrac{n^2}{2 + \mu^2}\right)^2} + \frac{5\sin^4\phi\cos^2\phi}{4\left(\sin^2\phi - \dfrac{n^2}{2 + \mu^2}\right)^3}.$$

Since μ^2 is imaginary, the denominators in this formula can never be zero, and since $|\mu^2| \gg 1$, their absolute values will not become excessively small unless ϕ is very small. We may therefore neglect $2 + \alpha$ compared with μ^2 in eq. (103) and thus we arrive at an extremely simple differential equation.

Before we write and discuss its solution, we must still determine ξ as a function of ϕ, i. e., we must solve the differential equation (102). Evidently, this may be done by quadrature, and since we decided earlier to neglect 2 compared with μ^2, we should do the same here and write

$$\xi = -\int\sqrt{1 - \frac{n^2}{\mu^2\sin^2\phi}}\,d\phi.$$

The minus sign before the integral is arbitrary (because of the root) but useful in order to obtain later positive ξ_1, ξ_2 for $\phi < \pi/2$. The evaluation of the integral is not an easy task. We give the result here and leave it to the reader to check it by differentiation:

$$\xi = \arctan\frac{\cos\phi}{\sqrt{\sin^2\phi - \dfrac{n^2}{\mu^2}}} - \frac{n}{2\mu}\arctan\frac{\dfrac{2n}{\mu}\cos\phi\sqrt{\sin^2\phi - \dfrac{n^2}{\mu^2}}}{\left(1 - \dfrac{n^2}{\mu^2}\right) - \left(1 + \dfrac{n^2}{\mu^2}\right)\cos^2\phi}. \tag{104}$$

This formula gives ξ as a function of the coordinate ϕ and of the parameter n^2/μ^2 which, because of μ^2, is imaginary and double-valued [see eq. (101)]. Therefore, ξ is complex, say

$$\xi = \xi_r + i\,\xi_i \quad \text{for} \quad \mu^2 = +2i\,\varkappa^2.$$

When we choose the other sign for μ^2, the arguments of the inverse tangents will be changed into their conjugate complex values and so will ξ:

$$\xi = \xi_r - i\,\xi_i \quad \text{for} \quad \mu^2 = -2i\,\varkappa^2.$$

After this preparation we may return to the differential equation (103). When we neglect the small quantity $2 + \alpha$ compared with μ^2 and write for the latter its value from eq. (101), the equation will read

$$\frac{d^2\eta}{d\xi^2} \pm 2i\,\varkappa^2\,\eta = 0,$$

and it has the solution

$$\eta = E_1 \exp\big((1-i)\,\varkappa(\xi_r + i\,\xi_i)\big) + E_2 \exp\big(-(1-i)\,\varkappa(\xi_r + i\,\xi_i)\big)$$
$$+ E_3 \exp\big((1+i)\,\varkappa(\xi_r - i\,\xi_i)\big) + E_4 \exp\big(-(1+i)\,\varkappa(\xi_r - i\,\xi_i)\big).$$

This may as well be written in the following form:

$$\eta = E_1 e^{\varkappa(\xi_1 - i\,\xi_2)} + E_2 e^{-\varkappa(\xi_1 - i\,\xi_2)} + E_3 e^{\varkappa(\xi_1 + i\,\xi_2)} + E_4 e^{-\varkappa(\xi_1 + i\,\xi_2)} \quad (105)$$

with

$$\xi_1 = \xi_r + \xi_i, \quad \xi_2 = \xi_r - \xi_i.$$

The numerical evaluation of ξ_1 and ξ_2 as functions of ϕ and of n^2/μ^2 is a cumbersome task. It has been done by A. HAVERS, and his tables are reproduced here.

The variable η which we now have found, is connected with U_n by the relation (see p. 388):

$$U_n = \frac{\eta}{\sqrt{\xi \cdot \sin\phi}} = \frac{\eta}{\sqrt[4]{\sin^2\phi - n^2/\mu^2}}.$$

When we introduce here the solution (105) we must keep in mind that its first and second terms belong to $\mu^2 = +2i\,\varkappa^2$, the third and fourth to $\mu^2 = -2i\,\varkappa^2$. We have therefore

$$U_n = \frac{E_1 e^{\varkappa(\xi_1 - i\,\xi_2)} + E_2 e^{-\varkappa(\xi_1 - i\,\xi_2)}}{\sqrt[4]{\sin^2\phi + i\,n^2/2\varkappa^2}} + \frac{E_3 e^{\varkappa(\xi_1 + i\,\xi_2)} + E_4 e^{-\varkappa(\xi_1 + i\,\xi_2)}}{\sqrt[4]{\sin^2\phi - i\,n^2/2\varkappa^2}}.$$

Now the time has come to leave the complex variables behind us and to write the solution in a real form. We put

$$\sin^2\phi + \frac{i\,n^2}{2\varkappa^2} = \Lambda^4 e^{4i\psi} \quad (106)$$

with

$$\Lambda^8 = \frac{n^4}{4\varkappa^4} + \sin^4\phi = \frac{k\,n^4}{1-\nu^2} + \sin^4\phi, \quad \tan 4\psi = \frac{n^2}{2\varkappa^2\sin^2\phi} \quad (106')$$

Table 8a. Values of ξ_1

$\dfrac{n^2}{2\varkappa^2}$	$\phi =$ 80°	70°	60°	50°	40°	30°	25°	20°	15°	10°	5°
0.00	0.1745	0.3491	0.5236	0.6981	0.8727	1.0472	1.1345	1.223	1.309	1.396	1.484
0.01	0.1754	0.3509	0.5265	0.7024	0.8787	1.0559	1.1453	1.236	1.328	1.425	1.544
0.02	0.1763	0.3527	0.5294	0.7066	0.8847	1.0647	1.1563	1.250	1.347	1.455	1.604
0.03	0.1772	0.3546	0.5323	0.7109	0.8908	1.0736	1.1674	1.264	1.366	1.485	1.661
0.04	0.1781	0.3564	0.5353	0.7152	0.8969	1.0826	1.1786	1.278	1.385	1.516	1.716
0.05	0.1790	0.3583	0.5382	0.7195	0.9031	1.0916	1.1899	1.293	1.405	1.547	1.770
0.06	0.1799	0.3602	0.5412	0.7238	0.9092	1.1007	1.2012	1.308	1.425	1.578	1.821
0.07	0.1808	0.3620	0.5442	0.7281	0.9154	1.1099	1.2126	1.323	1.445	1.609	1.871
0.08	0.1817	0.3639	0.5472	0.7325	0.9217	1.1191	1.2241	1.338	1.465	1.639	1.918
0.09	0.1826	0.3658	0.5502	0.7369	0.9280	1.1283	1.2356	1.353	1.485	1.668	1.964
0.10	0.1836	0.3677	0.5532	0.7413	0.9343	1.1375	1.2471	1.368	1.505	1.697	2.009
0.15	0.1883	0.3773	0.5685	0.7636	0.9663	1.1847	1.3059	1.443	1.605	1.837	2.219
0.20	0.1930	0.3871	0.5841	0.7865	0.9990	1.2329	1.3657	1.519	1.702	1.969	2.409
0.25	0.1978	0.3970	0.5999	0.8098	1.0321	1.2816	1.4259	1.594	1.797	2.086	2.583
0.30	0.2027	0.4071	0.6159	0.8333	1.0656	1.3303	1.4858	1.667	1.888	2.202	2.746
0.35	0.2076	0.4173	0.6321	0.8569	1.0992	1.3787	1.5448	1.739	1.977	2.318	2.903
0.40	0.2125	0.4276	0.6484	0.8806	1.1329	1.4267	1.6026	1.810	2.063	2.431	3.054
0.45	0.2175	0.4379	0.6648	0.9043	1.1666	1.4742	1.6592	1.874	2.148	2.541	3.201
0.50	0.2225	0.4483	0.6813	0.9281	1.2002	1.5212	1.7148	1.946	2.230	2.649	3.342
0.55	0.2275	0.4586	0.6978	0.9519	1.2336	1.5677	1.7695	2.0123	2.3101	2.750	3.479
0.60	0.2325	0.4689	0.7143	0.9757	1.2667	1.6136	1.8234	2.0772	2.3881	2.848	3.611
0.65	0.2375	0.4791	0.7307	0.9994	1.2995	1.6589	1.8766	2.1408	2.4642	2.944	3.739
0.70	0.2424	0.4893	0.7470	1.0230	1.3320	1.7036	1.9291	2.2032	2.5386	3.037	3.862
0.75	0.2473	0.4995	0.7632	1.0464	1.3642	1.7477	1.9808	2.2643	2.6114	3.127	3.982
0.80	0.2522	0.5097	0.7793	1.0696	1.3961	1.7912	2.0318	2.3242	2.6826	3.215	4.097
0.85	0.2572	0.5199	0.7953	1.0925	1.4276	1.8341	2.0821	2.3830	2.7523	3.300	4.209
0.90	0.2622	0.5301	0.8112	1.1152	1.4588	1.8764	2.1317	2.4407	2.8206	3.384	4.318
0.95	0.2672	0.5403	0.8270	1.1376	1.4897	1.9181	2.1805	2.4974	2.8876	3.466	4.423
1.00	0.2722	0.5505	0.8426	1.1598	1.5203	1.9592	2.2286	2.5532	2.9532	3.5461	4.5268
1.05	0.2771	0.5605	0.8581	1.1818	1.5505	1.9997	2.2748	2.6081	3.0175	3.6249	4.6287
1.10	0.2819	0.5704	0.8735	1.2035	1.5804	2.0396	2.3204	2.6621	3.0806	3.7023	4.7291
1.15	0.2866	0.5801	0.8887	1.2250	1.6099	2.0789	2.3658	2.7152	3.1427	3.7784	4.8281
1.20	0.2913	0.5897	0.9038	1.2463	1.6390	2.1177	2.4103	2.7674	3.2039	3.8532	4.9257
1.25	0.2959	0.5992	0.9187	1.2674	1.6677	2.1559	2.4543	2.8198	3.2642	3.9268	5.0219
1.30	0.3005	0.6085	0.9335	1.2883	1.6959	2.1935	2.4978	2.8694	3.3237	3.9993	5.1167
1.35	0.3051	0.6179	0.9481	1.3090	1.7237	2.2306	2.5409	2.9192	3.3824	4.0707	5.2102
1.40	0.3096	0.6272	0.9626	1.3295	1.7511	2.2671	2.5835	2.9683	3.4403	4.1410	5.3023
1.45	0.3141	0.6364	0.9769	1.3498	1.7781	2.3030	2.6256	3.0167	3.4975	4.2103	5.3931
1.50	0.3185	0.6455	0.9911	1.3698	1.8047	2.3384	2.6672	3.0644	3.5539	4.2786	5.4825
1.55	0.3229	0.6545	1.0051	1.3896	1.8310	2.3733	2.7082	3.1115	3.6096	4.3460	5.5706
1.60	0.3272	0.6634	1.0190	1.4092	1.8570	2.4078	2.7486	3.1580	3.6645	4.4125	5.6574
1.65	0.3317	0.6723	1.0328	1.4286	1.8828	2.4419	2.7884	3.2039	3.7186	4.4781	5.7428
1.70	0.3360	0.6811	1.0464	1.4478	1.9084	2.4757	2.8276	3.2493	3.7719	4.5428	5.8268
1.75	0.3403	0.6898	1.0599	1.4668	1.9338	2.5092	2.8663	3.2941	3.8244	4.6066	5.9094
1.80	0.3445	0.6984	1.0733	1.4856	1.9590	2.5424	2.9044	3.3384	3.8761	4.6695	5.9907
1.85	0.3487	0.7069	1.0865	1.5042	1.9840	2.5753	2.9420	3.3821	3.9271	4.7315	6.0706
1.90	0.3527	0.7153	1.0996	1.5225	2.0088	2.6079	2.9791	3.4253	3.9773	4.7926	6.1492
1.95	0.3568	0.7237	1.1126	1.5406	2.0334	2.6403	3.0156	3.4680	4.0267	4.8528	6.2264
2.00	0.3609	0.7320	1.1255	1.5585	2.0578	2.6724	3.0516	3.5102	4.0753	4.9121	6.3022

Table 8b. Values of ξ_2

$\frac{n^2}{2\varkappa^2}$	$\phi =$ 80°	70°	60°	50°	40°	30°	25°	20°	15°	10°	5°
0.00	0.1745	0.3491	0.5236	0.6981	0.8727	1.0472	1.1345	1.222	1.309	1.396	1.484
0.01	0.1736	0.3472	0.5207	0.6939	0.8667	1.039	1.124	1.208	1.292	1.369	1.430
0.02	0.1727	0.3454	0.5178	0.6898	0.8608	1.030	1.113	1.195	1.273	1.343	1.390
0.03	0.1719	0.3436	0.5150	0.6856	0.8550	1.022	1.103	1.182	1.256	1.320	1.357
0.04	0.1710	0.3419	0.5122	0.6815	0.8492	1.013	1.093	1.169	1.240	1.298	1.329
0.05	0.1701	0.3401	0.5094	0.6775	0.8434	1.005	1.083	1.157	1.224	1.277	1.306
0.06	0.1693	0.3383	0.5066	0.6734	0.8377	0.997	1.073	1.145	1.209	1.258	1.285
0.07	0.1684	0.3366	0.5038	0.6694	0.8321	0.989	1.063	1.133	1.196	1.240	1.266
0.08	0.1676	0.3348	0.5010	0.6654	0.8265	0.981	1.053	1.121	1.181	1.223	1.249
0.09	0.1667	0.3331	0.4983	0.6615	0.8209	0.973	1.044	1.110	1.168	1.208	1.233
0.10	0.1659	0.3314	0.4956	0.6575	0.8154	0.965	1.035	1.098	1.156	1.193	1.218
0.15	0.1618	0.3229	0.4822	0.6384	0.7887	0.928	0.991	1.046	1.096	1.126	1.148
0.20	0.1579	0.3147	0.4694	0.6199	0.7633	0.894	0.951	1.001	1.041	1.071	1.089
0.25	0.1541	0.3068	0.4571	0.6024	0.7393	0.862	0.914	0.959	0.994	1.025	1.039
0.30	0.1504	0.2992	0.4453	0.5857	0.7166	0.832	0.881	0.922	0.953	0.984	0.996
0.35	0.1469	0.2919	0.4339	0.5697	0.6952	0.804	0.850	0.888	0.918	0.946	0.957
0.40	0.1435	0.2849	0.4230	0.5544	0.6750	0.779	0.822	0.858	0.886	0.912	0.922
0.45	0.1402	0.2782	0.4125	0.5398	0.6560	0.755	0.797	0.831	0.858	0.880	0.891
0.50	0.1370	0.2718	0.4025	0.5260	0.6381	0.733	0.773	0.806	0.832	0.851	0.862
0.55	0.1340	0.2657	0.3929	0.5129	0.6213	0.7133	0.7507	0.783	0.808	0.824	0.835
0.60	0.1311	0.2598	0.3837	0.5005	0.6055	0.6945	0.7301	0.762	0.785	0.798	0.810
0.65	0.1283	0.2542	0.3750	0.4887	0.5906	0.6768	0.7108	0.742	0.764	0.775	0.788
0.70	0.1256	0.2488	0.3667	0.4775	0.5766	0.6601	0.6927	0.723	0.744	0.753	0.766
0.75	0.1230	0.2436	0.3588	0.4669	0.5634	0.6443	0.6757	0.705	0.725	0.734	0.747
0.80	0.1206	0.2387	0.3514	0.4568	0.5509	0.6293	0.6597	0.688	0.707	0.716	0.729
0.85	0.1183	0.2340	0.3444	0.4473	0.5390	0.6150	0.6446	0.672	0.691	0.699	0.712
0.90	0.1161	0.2296	0.3378	0.4383	0.5277	0.6014	0.6304	0.656	0.675	0.684	0.697
0.95	0.1140	0.2254	0.3316	0.4298	0.5169	0.5885	0.6171	0.642	0.660	0.671	0.682
1.00	0.1120	0.2214	0.3257	0.4218	0.5065	0.5762	0.6046	0.628	0.647	0.658	0.668
1.05	0.1100	0.2175	0.3201	0.4141	0.4966	0.5646	0.5928	0.615	0.634	0.646	0.655
1.10	0.1081	0.2138	0.3147	0.4066	0.4872	0.5536	0.5816	0.603	0.621	0.633	0.642
1.15	0.1063	0.2102	0.3095	0.3994	0.4783	0.5432	0.5709	0.591	0.601	0.621	0.630
1.20	0.1045	0.2067	0.3044	0.3925	0.4698	0.5334	0.5607	0.581	0.599	0.610	0.619
1.25	0.1028	0.2034	0.2995	0.3859	0.4617	0.5242	0.5510	0.570	0.588	0.599	0.608
1.30	0.1012	0.2002	0.2947	0.3796	0.4541	0.5157	0.5417	0.561	0.578	0.589	0.597
1.35	0.0996	0.1971	0.2901	0.3736	0.4469	0.5076	0.5328	0.552	0.569	0.580	0.588
1.40	0.0981	0.1942	0.2856	0.3679	0.4401	0.4998	0.5243	0.543	0.560	0.571	0.578
1.45	0.0967	0.1914	0.2812	0.3625	0.4337	0.4923	0.5161	0.535	0.551	0.562	0.569
1.50	0.0954	0.1888	0.2772	0.3575	0.4277	0.4851	0.5083	0.528	0.543	0.554	0.560
1.55	0.0942	0.1863	0.2732	0.3527	0.4220	0.4782	0.5008	0.520	0.535	0.546	0.552
1.60	0.0930	0.1839	0.2694	0.3481	0.4165	0.4716	0.4936	0.513	0.527	0.538	0.544
1.65	0.0918	0.1815	0.2657	0.3436	0.4112	0.4652	0.4868	0.506	0.520	0.531	0.536
1.70	0.0907	0.1792	0.2622	0.3392	0.4060	0.4590	0.4803	0.499	0.513	0.524	0.529
1.75	0.0896	0.1770	0.2588	0.3349	0.4009	0.4530	0.4739	0.493	0.507	0.517	0.521
1.80	0.0885	0.1748	0.2556	0.3306	0.3958	0.4472	0.4678	0.486	0.500	0.510	0.515
1.85	0.0874	0.1727	0.2525	0.3264	0.3907	0.4416	0.4618	0.480	0.494	0.504	0.508
1.90	0.0864	0.1706	0.2495	0.3223	0.3857	0.4361	0.4561	0.474	0.488	0.497	0.502
1.95	0.0854	0.1685	0.2466	0.3183	0.3807	0.4307	0.4507	0.468	0.481	0.491	0.496
2.00	0.0844	0.1665	0.2439	0.3144	0.3756	0.4254	0.4456	0.462	0.475	0.485	0.490

and have then

$$U_n = \frac{1}{\Lambda}[E_1 e^{\varkappa\xi_1 - i(\varkappa\xi_2 + \psi)} + E_2 e^{-\varkappa\xi_1 + i(\varkappa\xi_2 - \psi)}$$

$$+ E_3 e^{\varkappa\xi_1 + i(\varkappa\xi_2 + \psi)} + E_4 e^{-\varkappa\xi_1 - i(\varkappa\xi_2 - \psi)}].$$

This may easily be written in real terms:

$$U_n = C_1 U_{1n} + C_2 U_{2n} + D_1 U_{3n} + D_2 U_{4n} \qquad (107)$$

with

$$U_{1n}(\phi) = \frac{1}{\Lambda} e^{\varkappa\xi_1} \cos(\varkappa\xi_2 + \psi),$$

$$U_{2n}(\phi) = \frac{1}{\Lambda} e^{\varkappa\xi_1} \sin(\varkappa\xi_2 + \psi),$$

$$\qquad\qquad (107')$$

$$U_{3n}(\phi) = \frac{1}{\Lambda} e^{-\varkappa\xi_1} \cos(\varkappa\xi_2 - \psi),$$

$$U_{4n}(\phi) = \frac{1}{\Lambda} e^{-\varkappa\xi_1} \sin(\varkappa\xi_2 - \psi).$$

In these formulas only $\varkappa$ is a constant, and Λ, ξ_1, ξ_2, and ψ depend on ϕ according to eqs. (106′) and to Table 8. We must therefore know the derivatives of all these variables with respect to ϕ before we can differentiate U_n.

From the definition (106) we find by differentiating

$$\Lambda^{\cdot} + i \Lambda \psi^{\cdot} = \frac{1}{4\Lambda^3} e^{-4i\psi} \sin 2\phi,$$

and when we here separate real and imaginary parts we obtain

$$\Lambda^{\cdot} = \frac{\sin 2\phi \cos 4\psi}{4\Lambda^3}, \qquad \psi^{\cdot} = -\frac{\sin 2\phi \sin 4\psi}{4\Lambda^4}.$$

When we combine eqs. (106) and (102) (neglecting 2 compared with μ^2), we may find the relation

$$-\frac{\Lambda^2 e^{2i\psi}}{\sin\phi} = \xi^{\cdot} = \frac{\xi_1^{\cdot} - i\xi_2^{\cdot}}{1 - i},$$

and in separating again the real from the imaginary, we obtain formulas for the derivatives:

$$\xi_1^{\cdot} = -\frac{\Lambda^2 \sqrt{2}}{\sin\phi} \sin\left(2\psi + \frac{\pi}{4}\right).$$

$$\xi_2^{\cdot} = -\frac{\Lambda^2 \sqrt{2}}{\sin\phi} \cos\left(2\psi + \frac{\pi}{4}\right).$$

After these preparations, we may differentiate $U_{1n} \ldots U_{4n}$. The result is this:

$$
\left.
\begin{aligned}
U_{1n}^{\cdot} &= -\frac{e^{\varkappa\xi_1}}{\varLambda}\left[\frac{\sin 2\phi}{4\varLambda^4}\cos(\varkappa\xi_2 + 5\psi) + \frac{\varkappa\varLambda^2\sqrt{2}}{\sin\phi}\cos\left(\varkappa\xi_2 - \psi + \frac{\pi}{4}\right)\right], \\
U_{2n}^{\cdot} &= -\frac{e^{\varkappa\xi_1}}{\varLambda}\left[\frac{\sin 2\phi}{4\varLambda^4}\sin(\varkappa\xi_2 + 5\psi) + \frac{\varkappa\varLambda^2\sqrt{2}}{\sin\phi}\sin\left(\varkappa\xi_2 - \psi + \frac{\pi}{4}\right)\right], \\
U_{3n}^{\cdot} &= -\frac{e^{-\varkappa\xi_1}}{\varLambda}\left[\frac{\sin 2\phi}{4\varLambda^4}\cos(\varkappa\xi_2 - 5\psi) - \frac{\varkappa\varLambda^2\sqrt{2}}{\sin\phi}\sin\left(\varkappa\xi_2 + \psi + \frac{\pi}{4}\right)\right], \\
U_{4n}^{\cdot} &= -\frac{e^{-\varkappa\xi_1}}{\varLambda}\left[\frac{\sin 2\phi}{4\varLambda^4}\sin(\varkappa\xi_2 - 5\psi) + \frac{\varkappa\varLambda^2\sqrt{2}}{\sin\phi}\cos\left(\varkappa\xi_2 + \psi + \frac{\pi}{4}\right)\right].
\end{aligned}
\right\} \quad (107'')
$$

Instead of repeating the procedure and thus getting the second derivatives, we now aim at $H(U_n)$. We remember that each one of the functions $U_{1n} \ldots U_{4n}$ is the sum of two terms which satisfy eq. (100) for different values of μ^2, for instance:

$$
\left.\begin{aligned}U_{1n}\\i\,U_{2n}\end{aligned}\right\} = \frac{e^{\varkappa\xi_1}}{2\varLambda}\left(\pm e^{-i(\varkappa\xi_2 + \psi)} + e^{i(\varkappa\xi_2 + \psi)}\right),
$$

and when we apply eq. (100) to each term with the correct μ^2, we obtain

$$
\left.\begin{aligned}H(U_{1n})\\i\,H(U_{2n})\end{aligned}\right\} = 2\,i\,\varkappa^2\,\frac{e^{\varkappa\xi_1}}{2\varLambda}\left(\mp e^{-i(\varkappa\xi_2 + \psi)} + e^{i(\varkappa\xi_1 + \psi)}\right) = \begin{cases} -2\varkappa^2\,U_{2n} \\ 2\varkappa^2\,i\,U_{1n} \end{cases}.
$$

. In this way we arrive at the following relations:

$$
\begin{aligned}
H(U_{1n}) &= -2\,\varkappa^2\,U_{2n}, & H(U_{3n}) &= 2\varkappa^2\,U_{4n}, \\
H(U_{2n}) &= 2\,\varkappa^2\,U_{1n}, & H(U_{4n}) &= -2\varkappa^2\,U_{3n}.
\end{aligned} \quad (108)
$$

With the help of these formulas and of the definition (88) of the operator H it is possible to express the second and all higher derivatives of $U_{1n} \ldots$ by $U_{1n} \ldots$ and their first derivatives. When this has been done we retrace our steps and find formulas for the displacements and for the stress resultants.

We begin with eq. (97). Since we have already studied its homogeneous solutions, we may now content ourselves with finding one particular solution. With the help of eq. (96) it may easily be verified that

$$
w_n = \frac{1}{n}\left[\frac{k}{1 - \nu}H(U_n) + U_n\right]
$$

satisfies eq. (97). When we introduce here eq. (107), we have

$$
\begin{aligned}
w_n = \frac{1}{n}\,&(C_1\,U_{1n} + C_2\,U_{2n} + D_1\,U_{3n} + D_2\,U_{4n}) \\
&-\frac{1 + \nu}{2n\,\varkappa^2}\,(C_1\,U_{2n} - C_2\,U_{1n} - D_1\,U_{4n} + D_2\,U_{3n}).
\end{aligned}
$$

Compared with the first line, the second line of this formula is of the order $\varkappa^{-2}$. When we wrote eqs. (101) we neglected 1 compared with $\varkappa^2$, and we are now obliged to neglect $\varkappa^{-2}$ compared with 1. Since it turns out to be very necessary to keep track carefully of what must be dropped, we shall replace this second line by the symbol $+O(\varkappa^{-2})$ indicating that a contribution of this order has been dropped:

$$w_n = \frac{1}{n}(C_1\,U_{1n} + C_2\,U_{2n} + D_1\,U_{3n} + D_2\,U_{4n}) + O(\varkappa^{-2}). \quad (109\text{a})$$

We now attempt to find u_n and v_n. For u_n eq. (95) seems to be available, and since we have studied before the influence of ω_n, we may now put $\omega_n \equiv 0$. When we introduce U_n from eq. (107) and w_n from (109a), we have

$$u_n \sin\phi = U_n - n w_n$$
$$= (C_1\,U_{1n} + \ldots) - (C_1\,U_{1n} + \ldots) + O(\varkappa^{-2}) = O(\varkappa^{-2}),$$

i. e. we get nothing but a statement of the order of magnitude. We have no better luck when we turn to eq. (91) for v_n. Again putting $\omega_n \equiv 0$ we find

$$n v_n = -\dot{u}_n \sin\phi - u_n \cos\phi = -(u_n \sin\phi)^{\cdot}.$$

As may be seen from the second term in each bracket of eq. (107''), differentiating U_n raises the order of magnitude by one step. Therefore,

$$v_n = O(\varkappa^{-1}).$$

In other cases, when we had found the displacements, we turned to the elastic law to find expressions for the stress resultants. Here, because of the failure of finding expressions for u_n and v_n, this way is only partially open.

First, we introduce the FOURIER series (86) into eqs. (84) and see that the forces and moments are also FOURIER series:

$$N_\phi = \sum N_{\phi n} \cos n\theta, \qquad N_\theta = \sum N_{\theta n} \cos n\theta, \qquad N_{\phi\theta} = \sum N_{\phi\theta n} \sin n\theta,$$
$$M_\phi = \sum M_{\phi n} \cos n\theta, \qquad M_\theta = \sum M_{\theta n} \cos n\theta, \qquad M_{\phi\theta} = \sum M_{\phi\theta n} \sin n\theta.$$

When we now inspect the elastic law for the moments, eqs. (84d–f), we see that we have enough information to derive formulas for $M_{\phi n}$, $M_{\theta n}$, $M_{\phi\theta n}$. Introducing all we know in eq. (84d), we obtain, for example,

$$M_{\phi n} = \frac{K}{a^2}\left[O(\varkappa^0) + \ddot{w}_n + O(\varkappa^{-2}) + O(\varkappa^{-1}) + \nu\,\dot{w}_n \cot\phi - \frac{\nu\,n^2\,w_n}{\sin^2\phi}\right].$$

Since $\ddot{w}_n$ is of the order $\varkappa^2$, we may evaluate terms of this and of the next lower order, before our lack of knowledge about $\dot{v}_n$ interferes.

When going through the details, we find

$$M_{\phi n} = -\frac{K}{n\,a^2}\,[2\varkappa^2(C_1\,U_{2n} - \ldots)$$
$$+ (1 - \nu)\,(C_1\,U_{1n}^{\cdot} + \ldots)\cot\phi + O(\varkappa^0)]. \quad (110\text{a})$$

Since the functions $U_{1n}, \ldots U_{4n}$ or their derivatives appear in all formulas only in two different linear combinations we adopt here the abbreviations

$$(C_1\,U_{1n} + \ldots) = C_1\,U_{1n} + C_2\,U_{2n} + D_1\,U_{3n} + D_2\,U_{4n},$$
$$(C_1\,U_{2n} - \ldots) = C_1\,U_{2n} - C_2\,U_{1n} - D_1\,U_{4n} + D_2\,U_{3n}.$$

By similar procedures eq. (84e, f) yield

$$M_{\theta n} = -\frac{K}{n\,a^2}\,[2\,\nu\,\varkappa^2(C_1\,U_{2n} - \ldots)$$
$$- (1 - \nu)\,(C_1\,U_{1n}^{\cdot} + \ldots)\cot\phi + O(\varkappa^0)],$$
$$M_{\phi\theta n} = -\frac{K(1-\nu)}{a^2\sin\phi}\,[(C_1\,U_{1n}^{\cdot} + \ldots)$$
$$- (C_1\,U_{1n} + \ldots)\cot\phi + O(\varkappa^{-1})]. \quad (110\text{b, c})$$

When we try to use eqs. (84a–c) in the same way, we get only order-of-magnitude statements:

$$N_{\phi n} = \frac{D}{a}\,O(\varkappa^0), \qquad N_{\theta n} = \frac{D}{a}\,O(\varkappa^0), \qquad N_{\phi\theta n} = \frac{D}{a}\,O(\varkappa^{-1}),$$

and even these are uncertain in that the order of magnitude may be lower than indicated. To obtain more, we must make use of the equations of equilibrium (83). In these equations the M terms are now known and are brought to the right-hand side. Upon careful inspection of these right-hand sides we discover that $N_{\phi n}$ cannot be of the order $\varkappa^0$, but must be of order $\varkappa^{-1}$. It is then possible to find from eq. (83c) the term of order $\varkappa^0$ of $N_{\theta n}$ and then from eq. (83b) the term with $\varkappa^{-1}$ of $N_{\phi n}$. Returning to eq. (83c) we then get $N_{\theta n}$ complete and by some more steps between the three equations we get all we want:

$$N_{\phi n} = \frac{D(1-\nu^2)}{2\,a\,\varkappa^2\sin\phi}\left[\frac{1}{n}(C_1\,U_{2n}^{\cdot} - \ldots)\cos\phi - \frac{n}{\sin\phi}(C_1\,U_{2n} - \ldots) - O(\varkappa^{-1})\right],$$
$$N_{\theta n} = \frac{D(1-\nu^2)}{2\,n\,a}\left[2(C_1\,U_{1n} + \ldots) - \frac{\cot\phi}{\varkappa^2}(C_1\,U_{2n}^{\cdot} - \ldots) + O(\varkappa^{-2})\right],$$
$$N_{\phi\theta n} = \frac{D(1-\nu^2)}{2\,a\,\varkappa^2\sin\phi}\left[(C_1\,U_{2n}^{\cdot} - \ldots) - (C_1\,U_{2n} - \ldots)\cot\phi + O(\varkappa^{-1})\right].$$
$$(110\text{d–f})$$

On the way to these formulas several integrals of the type $\int(C_1\,U_{1n} + \ldots)\cos\phi\,d\phi$ must be evaluated. Within the necessary range of accuracy this can be done by assuming the solution in the form

$$a(\phi)\, U_{1n} + b(\phi)\, U_{2n} + c(\phi)\, U_{1n}' + d(\phi)\, U_{2n}',$$

expanding the coefficients $a(\phi) \ldots$ in series of negative powers of $\varkappa$, and collecting like powers of $\varkappa$. Where necessary, eqs. (108) must be used to eliminate second derivatives of $U_{1n} \ldots U_{4n}$.

It is now a straightforward procedure to derive formulas for the transverse shears from eqs. (1d, e):

$$Q_{\phi n} = -\frac{2K}{n\,a^3}\,[\varkappa^2(C_1\,U_{2n}' - \ldots) + O(\varkappa)],$$

$$Q_{\theta n} = \frac{2K}{a^3 \sin\phi}\,[\varkappa^2(C_1\,U_{2n} - \ldots) + O(\varkappa^0)]. \tag{110g, h}$$

As may easily be seen, Q_ϕ is a cosine series and Q_θ a sine series in θ.

For an edge $\phi = \text{const}$ of the shell we may define an effective shear T_ϕ and an effective transverse force S_ϕ similar to those introduced for the cylinder by eqs. (V–32). However, it so happens that the contributions of $M_{\phi\theta}$ are of·negligible order and that we simply have $T_\phi = N_{\phi\theta}$ and $S_\phi = Q_\phi$.

Last of all, we may now attempt to establish formulas for the tangential displacements u_n and v_n, using the elastic law (84a–c). The first two of these equations may be solved for $(v_n' + w_n)$ and $(n\,u_n/\sin\phi + v_n \cot\phi + w_n)$. Since w_n is known, we find at once

$$v_n = -\frac{1+\nu}{2n\,\varkappa^2}\,[(C_1\,U_{2n}' - \ldots) + O(\varkappa^{-1})]. \tag{109b}$$

When we try also to calculate u_n, we again arrive only at a statement of its order of magnitude, but eq. (84c) yields

$$u_n = \frac{1+\nu}{2\,\varkappa^2 \sin\phi}\,[(C_1\,U_{2n} - \ldots) - \frac{2}{\varkappa^2}(C_1\,U_{1n} + \ldots)\cot\phi + O(\varkappa^{-2})]. \tag{109c}$$

This concludes the solution of the problem. The complete stress system is represented by three sets of formulas: eqs. (98), (99), and the equations just derived. All together they contain 8 free constants of integration, which allow satisfying 4 boundary conditions at each of two edges of a shell. These conditions must be formulated in the same way as for the cylindrical shell (p. 232).

6.3.2 Conical Shell

6.3.2.1 Differential Equations

The conditions of equilibrium (1) and the elastic law (5) use the slope angle ϕ as a meridional coordinate. Since on a cone $\phi = \alpha$ is a constant, we must switch over to the arc length coordinate s

(fig. II–16). Consequently, we write for the dot-derivatives in eqs. (1) and (5)

$$(\ldots)^{\cdot} \equiv \frac{\partial (\ldots)}{\partial \phi} = r_1 \frac{\partial (\ldots)}{\partial s}$$

and change the subscript ϕ into s wherever it occurs. Then we put

$$\phi = \alpha, \quad r_1 = \infty, \quad r_2 = s \cot \alpha, \quad r = s \cos \alpha.$$

Since our equations will be a good·deal more complicated than those of the axisymmetric case, we at last introduce a new dot symbol

$$(\ldots)^{\cdot} = \frac{\partial (\ldots)}{\partial s}.$$

Different from the dot symbols used elsewhere in this book, this one represents differentiation with respect to a length and thus changes the dimension of the quantity to which it is attached. We shall see that this fact yields the key to the solution of our equations.

The details of the limiting process just described are tiresome but not difficult. It is sufficient to give the result here.

The elastic law assumes the following form:

$$N_s = D\left[v^{\cdot} + \frac{v}{s}\left(\frac{u'}{\cos \alpha} + v + w \tan \alpha\right)\right] - K\frac{w^{\cdot\cdot}}{s}\tan \alpha,$$

$$N_\theta = D\left[\frac{1}{s}\left(\frac{u'}{\cos \alpha} + v + w \tan \alpha\right) + v\, v^{\cdot}\right]$$
$$\cdot + K\left[\frac{v}{s^3}\tan \alpha + \frac{w}{s^3}\tan^2 \alpha + \frac{w''}{s^3 \cos^2 \alpha} + \frac{w^{\cdot}}{s^2}\right]\tan \alpha,$$

$$N_{s\theta} = D\frac{1-v}{2}\left[u^{\cdot} - \frac{u}{s} + \frac{v'}{s \cos \alpha}\right]$$
$$+ K\frac{1-v}{2}\left[\frac{u^{\cdot}}{s^2} - \frac{u}{s^3} - \frac{w^{\cdot\prime}}{s^2 \sin \alpha} + \frac{w'}{s^3 \sin \alpha}\right]\tan^2 \alpha,$$

$$N_{\theta s} = D\frac{1-v}{2}\left[u^{\cdot} - \frac{u}{s} + \frac{v'}{s \cos \alpha}\right] \qquad\qquad (111\,\text{a–h})$$
$$+ K\frac{1-v}{2}\left[\frac{v'}{s^3 \cos \alpha} + \frac{w^{\cdot\prime}}{s^2 \sin \alpha} - \frac{w'}{s^3 \sin \alpha}\right]\tan^2 \alpha,$$

$$M_s = K\left[w^{\cdot\cdot} - \frac{v^{\cdot}}{s}\tan \alpha + v\left(\frac{w''}{s^2 \cos^2 \alpha} + \frac{w^{\cdot}}{s} - \frac{u'}{s^2}\frac{\tan \alpha}{\cos \alpha}\right)\right],$$

$$M_\theta = K\left[\frac{w''}{s^2 \cos^2 \alpha} + \frac{w^{\cdot}}{s} + \frac{w}{s^2}\tan^2 \alpha + \frac{v}{s^2}\tan \alpha + v\, w^{\cdot\cdot}\right],$$

$$M_{s\theta} = K(1-v)\left[\frac{w^{\cdot\prime}}{s} - \frac{w'}{s^2} - \frac{u^{\cdot}}{s}\sin \alpha + \frac{u}{s^2}\sin \alpha\right]\frac{1}{\cos \alpha},$$

$$M_{\theta s} = K(1-v)\left[\frac{w^{\cdot\prime}}{s} - \frac{w'}{s^2} - \frac{u^{\cdot}}{2s}\sin \alpha + \frac{u}{2s^2}\sin \alpha + \frac{v'}{2s^2}\tan \alpha\right]\frac{1}{\cos \alpha}.$$

Like eqs. (5) from which they have been derived, these equations are valid for constant and for variable wall thickness.

When the conditions of equilibrium (1) are subjected to the same treatment, the following equations are obtained:

$$(s N_s)\dot{} + N'_{\theta s} \sec\alpha - N_\theta = -p_s\, s,$$

$$(s N_{s\theta})\dot{} + N'_\theta \sec\alpha + N_{\theta s} - Q_\theta \tan\alpha = -p_\theta\, s,$$

$$N_\theta \tan\alpha + Q'_\theta \sec\alpha + (s Q_s)\dot{} = p_r\, s, \qquad \text{(112a–f)}$$

$$(s M_s)\dot{} + M'_{\theta s} \sec\alpha - M_\theta = s Q_s,$$

$$(s M_{s\theta})\dot{} + M'_\theta \sec\alpha + M_{\theta s} = s Q_\theta,$$

$$s(N_{\theta s} - N_{s\theta}) = M_{\theta s} \tan\alpha.$$

Since the last of these equations does not contain any derivatives, we may there introduce the elastic law (111) without specifying any assumption regarding the wall thickness. When we do so, we find again that the equation is an identity and may be dropped as worthless for the analysis.

Two more equations, (112d, e), may be used to eliminate the transverse shearing forces from the remaining three. This yields the following set, in which we have exchanged the first and the second equation to obtain the order which we later shall find most desirable:

$$s(s N_{s\theta})\dot{} + s N'_\theta \sec\alpha + s N_{\theta s} - (s M_{s\theta})\dot{}\tan\alpha$$
$$- M_{\theta s}\tan\alpha - M'_\theta \tan\alpha \sec\alpha = -p_\theta\, s^2,$$

$$(s N_s)\dot{} + N'_{\theta s}\sec\alpha - N_\theta = -p_s\, s, \qquad \text{(113a–c)}$$

$$s N_\theta \tan\alpha + s(s M_s)\dot{\dot{}} + (s M'_{s\theta})\dot{}\sec\alpha + (s M'_{\theta s})\dot{}\sec\alpha$$
$$+ M''_\theta \sec^2\alpha - s M'_\theta = p_r\, s^2.$$

These equations correspond to eqs. (V–2a–c) of the cylinder theory which may be derived from them as a limiting case.

From here on, we restrict our attention to the edge load problem and put $p_\theta = p_s = p_r = 0$. Equations (113) then become homogeneous in the stress resultants, and these homogeneous equations will be meant when, on the following pages, reference is made to eqs. (113).

In the elastic law (111) as in all similar equations in this book, the wall thickness t is hidden in the rigidities D and K. When the shell has a variable thickness, then D and K vary from place to place. We shall here consider a shell whose thickness is independent of θ but proportional to s:

$$t = \delta \cdot s.$$

In this case we have

$$D\dot{} = \frac{E\,\delta}{1 - \nu^2}, \qquad K\dot{} = \frac{E\,\delta^3}{4(1 - \nu^2)}\, s^2.$$

When we introduce the stress resultants from eqs. (111) into the homogeneous equations (113), every term will have a factor D, $D\dot{}$, K, or $K\dot{}$. After dividing everything by D we find that the terms are either free of δ or have a factor $\qquad \dfrac{\delta^2}{12} = k$.

With this abbreviation the resulting equations assume the following form:

$$\frac{1-\nu}{2}\,s^2\,u^{\cdot\cdot} + u''\sec^2\alpha + (1-\nu)\,s\,u^{\cdot} - (1-\nu)\,u + \frac{1+\nu}{2}\,s\,v'^{\cdot}\sec\alpha$$

$$+ (2-\nu)\,v'\sec\alpha + w'\tan\alpha\sec\alpha + k\Big[\frac{3}{2}(1-\nu)\,s^2\,u^{\cdot\cdot}\tan\alpha$$

$$+ 3(1-\nu)\,s\,u^{\cdot}\tan\alpha - 3(1-\nu)\,u\tan\alpha - \frac{3-\nu}{2}\,s^2\,w'^{\cdot\cdot}\sec\alpha$$

$$- 3(1-\nu)\,s\,w'^{\cdot}\sec\alpha + 3(1-\nu)\,w'\sec\alpha\Big]\tan\alpha = 0,$$

$$\frac{1+\nu}{2}\,s\,u'^{\cdot}\sec\alpha - \frac{3}{2}(1-\nu)\,u'\sec\alpha + s^2\,v^{\cdot\cdot} + \frac{1-\nu}{2}\,v''\sec^2\alpha$$

$$+ 2s\,v^{\cdot} - (1-\nu)\,v + \nu\,s\,w^{\cdot}\tan\alpha - (1-\nu)\,w\tan\alpha$$

$$+ k\Big[\frac{1-\nu}{2}\,v''\tan\alpha\sec^2\alpha - v\tan\alpha - s^3\,w^{\cdot\cdot} + \frac{1-\nu}{2}\,s\,w'^{\cdot\cdot}\sec^2\alpha$$

$$- 3s^2\,w^{\cdot\cdot} - \frac{3-\nu}{2}\,w''\sec^2\alpha - s\,w^{\cdot} - w\tan^2\alpha\Big]\tan\alpha = 0, \qquad \text{(114a-c)}$$

$$[u'\sec\alpha + \nu\,s\,v^{\cdot} + v + w\tan\alpha]\tan\alpha + k\Big[-\frac{3-\nu}{2}\,s^2\,u'^{\cdot\cdot}\sec\alpha$$

$$- (3+\nu)\,s\,u'^{\cdot}\sec\alpha + (3-5\nu)\,u'\sec\alpha - s^3\,v^{\cdot\cdot} + \frac{1-\nu}{2}\,s\,v'^{\cdot\cdot}\sec^2\alpha$$

$$- 6s^2\,v^{\cdot\cdot} + (2-\nu)\,v''\sec^2\alpha - 7s\,v^{\cdot} - v(1-\tan^2\alpha)\Big]\tan\alpha$$

$$+ k[s^4\,w^{\cdot\cdot} + 2s^2\,w''^{\cdot\cdot}\sec^2\alpha + w^{\mathrm{IV}}\sec^4\alpha + 8s^3\,w^{\cdot\cdot} + 4s\,w'''\sec^2\alpha$$

$$+ (11+3\nu)\,s^2\,w^{\cdot\cdot} + 2w''\tan^2\alpha\sec^2\alpha - (5-6\nu)\,w''\sec^2\alpha$$

$$- 2(1-3\nu)\,s\,w^{\cdot} - w(1-\tan^2\alpha)\tan^2\alpha] = 0.$$

The equations (114) are three simultaneous partial differential equations for the displacements u, v, w and they represent the fundamental differential equations of our problem. Our next task is to find their general solution.

6.3.2.2 Solution

Since the coefficients do not depend on θ we may certainly expect a product solution of the type

$$u = u_n(s)\sin n\,\theta, \qquad v = v_n(s)\cos n\,\theta, \qquad w = w_n(s)\cos n\,\theta, \quad \text{(115)}$$

and it is easily shown that this choice of sines and cosines is acceptable. When we introduce this solution into eqs. (114), we obtain three ordinary differential equations for u_n, v_n, w_n. Without writing these equations, we may recognize that they have a peculiar structure. Since the dot represents the differentiation with respect to s, it produces a change of dimension equivalent to division by a length. For dimensional homogeneity of the equations this change must be compensated by a factor which has the dimension of a length. Now, on an unbounded cone, there is only one such factor available, the coordinate s. Indeed, we see that in every term of eqs. (114) the power of s is equal to the number of dots attached to u, v, or w, and this property of the equations is preserved when the expressions (115) are introduced. Such equations are solved by assuming u_n, v_n, $w_n = \text{const} \cdot s^\lambda$. Since the exponent λ may not be an integer (possibly even a complex number), the constants have awkward dimensions. We avoid this by introducing now a dimensionless coordinate. After experience gained with the axisymmetric problem we choose

$$y = \sqrt{\frac{s}{l}},$$

where l is an arbitrary reference length, and put

$$
\begin{aligned}
u &= A f_n(s) \sin n\theta, \\
v &= B f_n(s) \cos n\theta, \\
w &= C f_n(s) \cos n\theta
\end{aligned}
\tag{116}
$$

with

$$f_n(s) = y^{\lambda-1}. \tag{116'}$$

When this is introduced in eqs. (114) they become three ordinary linear equations for the constants A, B, C:

$$
\begin{aligned}
d_{11} A + d_{12} B + d_{13} C &= 0, \\
d_{21} A + d_{22} B + d_{23} C &= 0, \\
d_{31} A + d_{32} B + d_{33} C &= 0,
\end{aligned}
\tag{117}
$$

with the following expressions for their coefficients:

$$
\begin{aligned}
d_{11} &= \frac{1-\nu}{8}(1 + 3k\tan^2\alpha)(9 - \lambda^2) + n^2 \sec^2\alpha, \\
d_{12} &= \frac{1}{4}[(7 - 5\nu) + (1 + \nu)\lambda] n \sec\alpha, \\
d_{13} &= \left[1 + \frac{k}{8}(3(9 - 11\nu) + 8\nu\lambda - (3 - \nu)\lambda^2)\right] n \tan\alpha \sec\alpha, \\
d_{22} &= \frac{1}{4}(1 - \lambda^2) + (1 - \nu)\left(1 + \frac{1}{2} n^2 \sec^2\alpha\right) \\
&\quad + k\tan^2\alpha\left(1 + \frac{1-\nu}{2} n^2 \sec^2\alpha\right), \\
d_{23} &= \frac{1}{2}\tan\alpha[(2 - \nu) - \nu\lambda] \\
&\quad - \frac{1}{k}\tan\alpha[1 - 8\tan^2\alpha + 2(7 - 3\nu)n^2\sec^2\alpha \\
&\quad - (3 + 2(1 - \nu)n^2\sec^2\alpha)\lambda + 3\lambda^2 - \lambda^3], \\
d_{33} &= \tan^2\alpha + \frac{1}{16}k[(13 - 12\nu) - 16(1 - \tan^2\alpha)\tan^2\alpha \\
&\quad + 8((11 - 12\nu) - 4\tan^2\alpha)n^2\sec^2\alpha + 16n^4\sec^4\alpha \\
&\quad - 2((7 - 6\nu) + 4n^2\sec^2\alpha)\lambda^2 + \lambda^4].
\end{aligned}
\tag{118}
$$

The expressions for d_{21}, d_{31}, d_{32} are obtained by replacing λ with $-\lambda$ in d_{12}, d_{13}, d_{23}, respectively.

The linear equations (117) are homogeneous and cannot have a nontrivial solution unless their determinant vanishes. Since the coefficients contain λ, the vanishing of the determinant yields an algebraic equation for this quantity. It turns out to be of the form

$$\lambda^8 - g_6\,\lambda^6 + g_4\,\lambda^4 - g_2\,\lambda^2 + g_0 = 0, \tag{119}$$

and its coefficients are these:

$$
\begin{aligned}
g_6 &= 4(7 - 4\nu) - 8\nu\tan^2\alpha + 16n^2\sec^2\alpha, \\
g_4 &= 16k^{-1}(1 - \nu^2)\tan^2\alpha + 2[(127 - 136\nu + 24\nu^2) \\
&\quad - 4(8 + 3\nu)\tan^2\alpha + 8(4 - 3\nu^2)\tan^4\alpha] \\
&\quad + 16[(17 - 12\nu) - 6\tan^2\alpha]n^2\sec^2\alpha + 96n^4\sec^4\alpha, \\
g_2 &= -160k^{-1}(1 - \nu^2)\tan^2\alpha + 4[(203 - 316\nu + 120\nu^2) \\
&\quad - 2(80 - 61\nu)\tan^2\alpha + 40(4 - 3\nu^2)\tan^4\alpha] \\
&\quad + 16[(71 - 72\nu) - 4(13 - 10\nu)\tan^2\alpha \\
&\quad + 8(2 - \nu)\tan^4\alpha]n^2\sec^2\alpha \\
&\quad + 64[(13 - 12\nu) - 2(4 - \nu)\tan^2\alpha]n^4\sec^4\alpha + 256n^6\sec^6\alpha, \\
g_0 &= 144k^{-1}(1 - \nu^2)\tan^2\alpha \\
&\quad + 9[(13 - 12\nu)(5 - 4\nu) - 8(8 - 7\nu)\tan^2\alpha + 16(4 - 3\nu^2)\tan^4\alpha] \\
&\quad + 16[(215 - 412\nu + 192\nu^2) + 2(89 - 172\nu + 96\nu^2)\tan^2\alpha \\
&\quad + 40(2 - \nu)\tan^4\alpha]n^2\sec^2\alpha \\
&\quad - 32[(81 - 184\nu + 96\nu^2) + 4(16 - 13\nu)\tan^2\alpha - 8\tan^4\alpha]n^4\sec^4\alpha \\
&\quad + 256[(3 - 4\nu) - 2\tan^2\alpha]n^6\sec^6\alpha + 256n^8\sec^8\alpha.
\end{aligned}
\tag{120}
$$

Eq. (119) has 8 roots λ_j $(j = 1, 2, \ldots, 8)$, and to each one of them there belongs one solution (116) with constants A_j, B_j, C_j. These three constants depend on each other through the linear equations (117). Since the determinant of these equations vanishes, any two of them may be used to express A_j and B_j as multiples of C_j:

$$A_j = \alpha_j\,C_j, \qquad B_j = \beta_j\,C_j.$$

The ratios α_j, β_j depend on the coefficients $d_{11}\ldots d_{33}$ and hence on λ_j and are different for each of the eight solutions.

The roots λ_j may be real or complex. When they are complex they always occur in groups of four: $\lambda = \pm\varkappa \pm i\,\mu$, while real roots come in pairs of opposite sign. We must discuss both cases, and we begin with the simpler one.

Table 9. *Coefficients for Conical Shells*

f	c	ϱ	a_1
u	1	-1	$\bar{\alpha}_1$
v	1	-1	$\bar{\beta}_1$
w	1	-1	1
$w^{\cdot}$	$1/2l$	-3	$\varkappa_1 - 1$
N_s	$\dfrac{E\,\delta}{1-\nu^2}$	-1	$\nu\, n\, \bar{\alpha}_1 \sec\alpha + \dfrac{1}{2}(\varkappa_1 - 1 + 2\nu)\,\bar{\beta}_1 - \dfrac{1}{2}\mu_1\bar{\beta}_2 + \nu\tan\alpha$ $-\dfrac{1}{4}k[(\varkappa_1 - 1)(\varkappa_1 - 3) - \mu_1^2]\tan\alpha$
N_θ	$\dfrac{E\,\delta}{1-\nu^2}$	-1	$n\,\bar{\alpha}_1\sec\alpha + [1 + \dfrac{1}{2}\nu(\varkappa_1 - 1) + k\tan^2\alpha]\bar{\beta}_1 - \dfrac{1}{2}\nu\mu_1\bar{\beta}_2$ $+\tan\alpha + k[\tan^2\alpha - n^2\sec^2\alpha + \dfrac{1}{2}(\varkappa_1 - 1)]\tan\alpha$
$N_{s\theta}$	$\dfrac{E\,\delta}{4(1+\nu)}$	-1	$[(\varkappa_1 - 3)\,\bar{\alpha}_1 - \mu_1\bar{\alpha}_2](1 + k\tan^2\alpha) - 2n\bar{\beta}_1\sec\alpha$ $+ k\,n(\varkappa_1 - 3)\tan\alpha\sec\alpha$
$N_{\theta s}$	$\dfrac{E\,\delta}{4(1+\nu)}$	-1	$[(\varkappa_1 - 3)\,\bar{\alpha}_1 - \mu_1\bar{\alpha}_2] - 2n\bar{\beta}_1\sec\alpha(1 + k\tan^2\alpha)$ $- k\,n(\varkappa_1 - 3)\tan\alpha\sec\alpha$
M_s	$\dfrac{E\,\delta^3 l}{12(1-\nu^2)}$	$+1$	$\dfrac{1}{4}[(\varkappa_1 - 1)(\varkappa_1 - 3) - \mu_1^2] + \dfrac{1}{2}\nu(\varkappa_1 - 1 - 2n^2\sec^2\alpha)$ $-\nu\,n\,\bar{\alpha}_1\tan\alpha\sec\alpha - \dfrac{1}{2}[(\varkappa_1 - 1)\bar{\beta}_1 - \mu_1\bar{\beta}_2]\tan\alpha$
M_θ	$\dfrac{E\,\delta^3 l}{12(1-\nu^2)}$	$+1$	$-n^2\sec^2\alpha + \dfrac{1}{2}(\varkappa_1 - 1)$ $+ \tan^2\alpha + \dfrac{1}{4}\nu[(\varkappa_1 - 1)(\varkappa_1 - 3) - \mu_1^2] + \bar{\beta}_1\tan\alpha$
$M_{s\theta}$	$\dfrac{E\,\delta^3 l}{24(1+\nu)}$	$+1$	$-n(\varkappa_1 - 3)\sec\alpha - [(\varkappa_1 - 3)\,\bar{\alpha}_1 - \mu_1\bar{\alpha}_2]\tan\alpha$
$M_{\theta s}$	$\dfrac{E\,\delta^3 l}{24(1+\nu)}$	$+1$	$-n(\varkappa_1 - 3)\sec\alpha - \dfrac{1}{2}[(\varkappa_1 - 3)\,\bar{\alpha}_1 - \mu_1\bar{\alpha}_2]\tan\alpha$ $- n\,\bar{\beta}_1\tan\alpha\sec\alpha$

Table 9. (Continued)

a_2	θ factor
$\bar{\alpha}_2$	sin
$\bar{\beta}_2$	cos
0	cos
μ_1	cos
$\nu n \bar{\alpha}_2 \sec \alpha + \dfrac{1}{2} \mu_1 \bar{\beta}_1 + \dfrac{1}{2}(\varkappa_1 - 1 + 2\nu) \bar{\beta}_2 - \dfrac{1}{2} k(\varkappa_1 - 2) \mu_1 \tan \alpha$	cos
$n \bar{\alpha}_2 \sec \alpha + \dfrac{1}{2} \nu \mu_1 \bar{\beta}_1 + [1 + \dfrac{1}{2} \nu(\varkappa_1 - 1) + k \tan^2 \alpha] \bar{\beta}_2$ $+ \dfrac{1}{2} k \mu_1 \tan \alpha$	cos
$[\mu_1 \bar{\alpha}_1 + (\varkappa_1 - 3) \bar{\alpha}_2](1 + k \tan^2 \alpha) - 2n \bar{\beta}_2 \sec \alpha + k n \mu_1 \tan \alpha \sec \alpha$	sin
$[\mu_1 \bar{\alpha}_1 + (\varkappa_1 - 3) \bar{\alpha}_2] - 2n \bar{\beta}_2 \sec \alpha(1 + k \tan^2 \alpha) - k n \mu_1 \tan \alpha \sec \alpha$	sin
$\dfrac{1}{2}(\varkappa_1 - 2) \mu_1 + \dfrac{1}{2} \nu \mu_1 - \nu n \bar{\alpha}_2 \tan \alpha \sec \alpha$ $\qquad - \dfrac{1}{2}[\mu_1 \bar{\beta}_1 + (\varkappa_1 - 1) \bar{\beta}_2] \tan \alpha$	cos
$\dfrac{1}{2} \mu_1 + \dfrac{1}{2} \nu(\varkappa_1 - 2) \mu_1 + \bar{\beta}_2 \tan \alpha$	cos
$- n \mu_1 \sec \alpha - [\mu_1 \bar{\alpha}_1 + (\varkappa_1 - 3) \bar{\alpha}_2] \tan \alpha$	sin
$- n \mu_1 \sec \alpha - \dfrac{1}{2}[\mu_1 \bar{\alpha}_1 + (\varkappa_1 - 3) \bar{\alpha}_2] \tan \alpha - n \bar{\beta}_2 \tan \alpha \sec \alpha$	sin

Table 9. (Continued)

f	c	ϱ	a_1
Q_s	$\dfrac{E\,\delta^3}{12(1-\nu^2)}$	-1	$\dfrac{1}{8}(\varkappa_1-1)(\varkappa_1^2-13+12\nu)-\dfrac{1}{8}(3\varkappa_1-1)\mu_1^2-\tan^2\alpha$ $-\dfrac{1}{2}(\varkappa_1-5+6\nu)\,n^2\sec^2\alpha$ $-\dfrac{1}{4}[(1+\nu)\,\varkappa_1-3\,(1-3\nu)]\,n\,\bar\alpha_1\tan\alpha\sec\alpha$ $+\dfrac{1}{4}(1+\nu)\,\mu_1\,n\,\bar\alpha_2\tan\alpha\sec\alpha$ $-\dfrac{1}{4}[(\varkappa_1+1)^2-\mu_1^2+2(1-\nu)\,n^2\sec^2\alpha]\,\bar\beta_1\tan\alpha$ $+\dfrac{1}{2}(\varkappa_1+1)\,\mu_1\bar\beta_2\tan\alpha$
Q_θ	$\dfrac{E\,\delta^3}{12(1-\nu^2)}$	-1	$-\dfrac{1}{4}[\varkappa_1^2+2(2-3\nu)\,\varkappa_1-17+18\nu-\mu_1^2]\,n\sec\alpha$ $+n^3\sec^3\alpha-n\tan^2\alpha\sec\alpha$ $-\dfrac{1}{4}(1-\nu)[(\varkappa_1-3)\,(\varkappa_1+4)-\mu_1^2]\bar\alpha_1\tan\alpha$ $+\dfrac{1}{4}(1-\nu)(2\varkappa_1+1)\,\mu_1\bar\alpha_2\tan\alpha-\dfrac{1}{2}(3-\nu)\,n\bar\beta_1\tan\alpha\sec\alpha$
S_s	$\dfrac{E\,\delta^3}{12(1-\nu^2)}$	-1	$\dfrac{1}{8}(\varkappa_1-1)(\varkappa_1^2-13+12\nu)-\dfrac{1}{8}(3\varkappa_1-1)\mu_1^2-\tan^2\alpha$ $-\dfrac{1}{2}[(2-\nu)\,\varkappa_1-8+9\nu]\,n^2\sec^2\alpha$ $-\dfrac{1}{4}[(3-\nu)\,\varkappa_1-3(3-5\nu)]\,n\,\bar\alpha_1\tan\alpha\sec\alpha$ $+\dfrac{1}{4}(3-\nu)\,n\,\mu_1\bar\alpha_2\tan\alpha\sec\alpha$ $-\dfrac{1}{4}[(\varkappa_1+1)^2-\mu_1^2+2(1-\nu)\,n^2\sec^2\alpha]\,\bar\beta_1\tan\alpha$ $+\dfrac{1}{2}(\varkappa_1+1)\,\mu_1\bar\beta_2\tan\alpha$
T_s	$\dfrac{E\,\delta}{4(1+\nu)}$	-1	$[(\varkappa_1-3)\,\bar\alpha_1-\mu_1\bar\alpha_2](1+3k\tan^2\alpha)-2n\,\bar\beta_1\sec\alpha$ $+3kn(\varkappa_1-3)\tan\alpha\sec\alpha$

To obtain formulas for real roots, put $\varkappa_1=\lambda_1$, $\mu_1=0$, $\bar\alpha_1=\alpha_1$, $\bar\beta_1=\beta_1$, $\bar\alpha_2=\bar\beta_2=0$. Then $a_2=0$, and a_1 is the coefficient to be used in eq. (121).

Table 9. (Continued)

a_2	θ factor
$\frac{1}{8}\mu_1(3\varkappa_1^2 - 2\varkappa_1 - 13 + 12\nu) - \frac{1}{8}\mu_1^3 - \frac{1}{2}\mu_1 n^2 \sec^2\alpha$ $-\frac{1}{4}(1+\nu)\mu_1 n\,\bar\alpha_1 \tan\alpha \sec\alpha$ $-\frac{1}{4}[(1+\nu)\varkappa_1 - 3(1-3\nu)]\,n\,\bar\alpha_2 \tan\alpha \sec\alpha$ $-\frac{1}{2}(\varkappa_1 + 1)\mu_1\bar\beta_1 \tan\alpha$ $-\frac{1}{4}[(\varkappa_1 + 1)^2 - \mu_1^2 + 2(1-\nu)n^2\sec^2\alpha]\,\bar\beta_2 \tan\alpha$	cos
$-\frac{1}{2}(\varkappa_1 - 2 + 3\nu)n\,\mu_1 \sec\alpha - \frac{1}{4}(1-\nu)(2\varkappa_1 + 1)\mu_1\bar\alpha_1 \tan\alpha$ $-\frac{1}{4}(1-\nu)[(\varkappa_1 - 3)(\varkappa_1 + 4) - \mu_1^2]\,\bar\alpha_2 \tan\alpha$ $-\frac{1}{2}(3-\nu)n\,\bar\beta_2 \tan\alpha \sec\alpha$	sin
$\frac{1}{8}\mu_1(3\varkappa_1^2 - 2\varkappa_1 - 13 + 12\nu) - \frac{1}{8}\mu_1^3 - \frac{1}{2}(2-\nu)n^2\mu_1 \sec^2\alpha$ $-\frac{1}{4}(3-\nu)\mu_1 n\,\bar\alpha_1 \tan\alpha \sec\alpha$ $-\frac{1}{4}[(3-\nu)\varkappa_1 - 3(3-5\nu)]\,n\,\bar\alpha_2 \tan\alpha \sec\alpha - \frac{1}{2}(\varkappa_1 + 1)\mu_1\bar\beta_1 \tan\alpha$ $-\frac{1}{4}[(\varkappa_1 + 1)^2 - \mu_1^2 + 2(1-\nu)n^2\sec^2\alpha]\,\bar\beta_2 \tan\alpha$	cos
$[\mu_1\bar\alpha_1 + (\varkappa_1 - 3)\bar\alpha_2](1 + 3k\tan^2\alpha) - 2n\,\bar\beta_2 \sec\alpha$ $+ 3k\,n\,\mu_1 \tan\alpha \sec\alpha$	sin

Let λ_1 be a real root. The corresponding solution of eqs. (114) is

$$u = C_1 \alpha_1 \, y^{\lambda_1 - 1} \sin n \, \theta, \qquad v = C_1 \beta_1 \, y^{\lambda_1 - 1} \cos n \, \theta,$$
$$w = C_1 \, y^{\lambda_1 - 1} \cos n \, \theta.$$

These expressions must be introduced into the elastic law (111) to obtain the stress resultants N and M, and then the transverse forces Q_s and Q_θ may be obtained from eqs. (112d, e). The resulting formulas may be obtained from Table 9 by applying the substitutions indicated at the end.

As a rule, we have to expect at least one if not two groups of complex roots λ. Let

$$\lambda_1 = +\varkappa_1 + i \, \mu_1, \qquad \lambda_3 = -\varkappa_1 + i \, \mu_1,$$
$$\lambda_2 = +\varkappa_1 - i \, \mu_1, \qquad \lambda_4 = -\varkappa_1 - i \, \mu_1$$

be one such group. Since λ_1 and λ_2 are conjugate complex, the corresponding coefficients $d_{11} \ldots d_{33}$ are also conjugate complex, and so are the values α_1, β_1 and α_2, β_2 calculated from them. We may therefore write

$$\alpha_1 = \bar{\alpha}_1 + i \, \bar{\alpha}_2, \qquad \alpha_3 = \bar{\alpha}_3 + i \, \bar{\alpha}_4,$$
$$\alpha_2 = \bar{\alpha}_1 - i \, \bar{\alpha}_2, \qquad \alpha_4 = \bar{\alpha}_3 - i \, \bar{\alpha}_4$$

and similar expressions for β_j. The corresponding part of the solution (116) is

$$u = [\alpha_1 \, C_1 \, y^{\lambda_1 - 1} + \alpha_2 \, C_2 \, y^{\lambda_2 - 1} + \alpha_3 \, C_3 \, y^{\lambda_3 - 1} + \alpha_4 \, C_4 \, y^{\lambda_4 - 1}] \sin n \, \theta,$$
$$v = [\beta_1 \, C_1 \, y^{\lambda_1 - 1} + \beta_2 \, C_2 \, y^{\lambda_2 - 1} + \beta_3 \, C_3 \, y^{\lambda_3 - 1} + \beta_4 \, C_4 \, x^{\lambda_4 - 1}] \cos n \, \theta,$$
$$w = [\quad C_1 \, y^{\lambda_1 - 1} + \quad C_2 \, y^{\lambda_2 - 1} + \quad C_3 \, y^{\lambda_3 - 1} + \quad C_4 \, y^{\lambda_4 - 1}] \cos n \, \theta.$$

The displacements are real quantities. The brackets are real when C_1, C_2 and C_3, C_4 are conjugate complex pairs:

$$C_1 = \frac{1}{2} (\bar{C}_1 - i \bar{C}_2), \qquad C_3 = \frac{1}{2} (\bar{C}_3 - i \bar{C}_4),$$
$$C_2 = \frac{1}{2} (\bar{C}_1 + i \bar{C}_2), \qquad C_4 = \frac{1}{2} (\bar{C}_3 + i \bar{C}_4).$$

Following the same procedure as on p. 369, we may now bring the solution into real form:

$$u = y^{-1} \{ y^{\varkappa_1} [(\bar{\alpha}_1 \, \bar{C}_1 + \bar{\alpha}_2 \, \bar{C}_2) \cos (\mu_1 \ln y) + (\bar{\alpha}_1 \, \bar{C}_2 - \bar{\alpha}_2 \, \bar{C}_1) \sin (\mu_1 \ln y)$$
$$+ \, y^{-\varkappa_1} [(\bar{\alpha}_3 \, \bar{C}_3 + \bar{\alpha}_4 \, \bar{C}_4) \cos (\mu_1 \ln y)$$
$$+ (\bar{\alpha}_3 \, \bar{C}_4 - \bar{\alpha}_4 \, \bar{C}_3) \sin (\mu_1 \ln y)] \} \sin n \, \theta,$$
$$w = y^{-1} \{ y^{\varkappa_1} [\bar{C}_1 \cos (\mu_1 \ln y) + \bar{C}_2 \sin (\mu_1 \ln y)]$$
$$+ \, y^{-\varkappa_1} [\bar{C}_3 \cos (\mu_1 \ln y) + \bar{C}_4 \sin (\mu_1 \ln y)] \} \cos n \, \theta.$$

When these solutions are introduced in eqs. (111) and (112d, e), it is found that all stress resultants may be brought into the form:

$$f = c\, y^{\varrho}\{y^{\varkappa_1}[(a_1\overline{C}_1 + a_2\overline{C}_2)\cos(\mu_1\ln y) + (a_1\overline{C}_2 - a_2\overline{C}_1)\sin(\mu_1\ln y)] \quad (121)$$

$$+ y^{-\varkappa_1}[(a_3\overline{C}_3 + a_4\overline{C}_4)\cos(\mu_1\ln y) + (a_3\overline{C}_4 - a_4\overline{C}_3)\sin(\mu_1\ln y)]\}{\textstyle{\cos\atop\sin}}\, n\theta.$$

The values of c, ϱ, and the expressions for a_1 and a_2 are given in Table 9. To obtain expressions for a_3 and a_4, one has to replace $\bar{\alpha}_1, \bar{\alpha}_2$ by $\bar{\alpha}_3, \bar{\alpha}_4$; $\bar{\beta}_1, \bar{\beta}_2$ by $\bar{\beta}_3, \bar{\beta}_4$; and $\varkappa_1$ by $-\varkappa_1$, while μ_1 remains unaltered. When there is another set of four complex roots, they are dealt with in the same way. When there are some complex and some real roots, each part of the solution must be handled in its own way, but when they are all superposed there are always 8 real constants of integration available, whether they are real constants C_j or the real and imaginary parts $\overline{C}_j$ of complex constants.

Chapter 7.

BUCKLING OF SHELLS

7.1 Introduction

In many examples in the preceding chapters we have seen that shells can be very thin-walled and that they very often are subjected to compressive stresses in extensive areas. The question arises whether the elastic equilibrium of such shells is stable. To answer this question, one of the standard methods of the theory of elastic stability must be applied: the method of adjacent equilibrium or the energy method. We shall explain here the basic ideas of both methods in the terminology of shells and then consider an EULER column to demonstrate their use simply.

7.1.1 Adjacent Equilibrium

We consider a shell carrying a certain load, which we shall call the *basic load*. It produces the *basic stresses* and the *basic displacements*.

We disturb the elastic equilibrium by imposing a small additional deformation, say, some lateral deflection. Every such deformation is connected with strains and hence with stresses, and we may expect that certain external forces will be needed to produce it. When these forces are removed, the whole disturbance vanishes. If this situation prevails, the elastic equilibrium is stable.

When the basic load is increased, it may happen that less force is needed to produce the same disturbance and that, at last, a certain disturbance becomes possible without any disturbing forces. In such a case the elastic equilibrium is neutral with respect to this particular disturbance. It may be shown quite generally[1] that the elastic equilibrium is always stable when the basic load is small enough. The smallest value which the load must assume to reach a neutral equilibrium is called the critical load or buckling load. The disturbance, i. e. a system of additional stresses and displacements, may then occur spontaneously, and this phenomenon is the buckling of the shell.

When the basic load is increased beyond its critical limit, the elastic equilibrium becomes unstable, and any incidental disturbance causes the shell to leave entirely its initial position of equilibrium. Whether or not this leads to a collapse is a question still to be discussed (see the papers on post-buckling behavior mentioned in the bibliography).

Actually to find the buckling load, we proceed in this way: We formulate the differential equations for the disturbed equilibrium without a disturbing load and ask whether these equations, together with appropriate boundary conditions, admit a solution. These equations contain, of course, all the terms which occur in the equations for the undisturbed equilibrium. They also contain terms with the additional stresses (or stress resultants). Since the disturbance is supposed to be very small (infinitesimal, if we wish), these new terms are very small, and since they are essential for our problem, we must take all terms of the same order of magnitude. There is a second group of such terms resulting from the fact that the basic load is now acting on a slightly deformed element. As we shall see later in more detail, these terms consist of products of a basic force or stress resultant with an additional displacement or its derivative.

Both groups of small terms are proportional to the disturbance: the first to the stress resultants and the others to the displacements which are added to the basic state. Since the conditions of equilibrium are satisfied without all these terms (i. e. for the undisturbed case), the small terms by themselves must add up to zero in every equation. And since HOOKE's law expresses the stress resultants in terms of the displacements, we arrive at last at a set of *homogeneous* linear differential equations for these displacements u, v, w.

Now let us look at the boundary conditions to which these buckling displacements are subjected. Whatever conditions we impose on the basic state, the same conditions will be imposed on the buckled state. When we subtract the one from the other, we see that the buckling

---

[1] v. MISES, R.: Über die Stabilitätsprobleme der Elastizitätstheorie. Z. angew. Math. Mech. vol. 3 (1923), p. 406.

displacements have to satisfy *nomogeneous* boundary conditions. The mathematical problem is, therefore, to solve a set of homogeneous differential equations with homogeneous boundary conditions. In general, such a problem has only the trivial solution $u \equiv v \equiv w \equiv 0$. But the coefficients of the equations depend on the magnitude of the basic load, and it is our task to find values of this load for which a non-trivial solution is possible. This is the typical formulation of an eigenvalue problem, and, mathematically speaking, all buckling theory is eigenvalue theory[1].

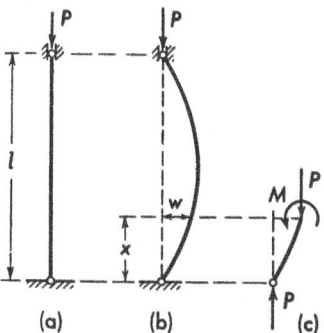

We shall now study the technique of the described procedure on the simplest example, the EULER column, fig. 1a. The force P is the basic load, and the basic stress system consists only of the axial force $N = -P$ in all sections of the bar.

We disturb the equilibrium by imposing a lateral deflection w (fig. 1b).

Fig. 1. EULER column, (a) straight equilibrium, (b) deflected equilibrium, (c) equilibrium of external and internal forces

In the case of a shell, the next step will always be to cut out a differential element. In the present case, it is possible to formulate the equilibrium for the finite portion of length x. There are two conditions of equilibrium. The one for the vertical forces is $P = P$ and does not contain any disturbance term. It is without interest. The condition for the moments, on the other hand, contains only disturbance terms:

$$M - P w = 0. \tag{1a}$$

The two terms of this equation are representative of the two types which always occur. The first one, M, is a stress resultant which comes into being through the disturbance. The second term is the product of the basic force P and the displacement w of the disturbance, and it is caused by the fact that the equilibrium is formulated for the disturbed shape of the elastic body under consideration.

The elastic law, i. e. the equation which corresponds to eqs. (V–8) for the cylindrical shell, is here

$$M = - E I \frac{d^2 w}{d x^2}. \tag{1b}$$

---

[1] A comprehensive study of the mathematical problem is found in COLLATZ, L.: Eigenwertprobleme und ihre numerische Behandlung. Leipzig 1945.

We use it to express M in eq. (1a) in terms of the displacement and obtain

$$E I \frac{d^2w}{dx^2} + P w = 0.$$ (2)

This is a homogeneous differential equation for w. The boundary conditions are $w = 0$ for both ends of the column. They are also homogeneous.

The problem may now be solved in the well-known way. It is found that, in general, no deflection w is possible. Only if

$$P = n^2 \frac{E I \pi^2}{l^2}$$

with an integer value of n is there a solution

$$w = C \sin \frac{n \pi x}{l}$$

with an undetermined amplitude C. Each of these loads P is a buckling load, representing a possible form of neutral elastic equilibrium. The corresponding displacement w is the buckling mode connected with this critical load. The lowest buckling load (for $n = 1$) is the EULER load P_{cr}.

7.1.2 Energy Method

When applying the energy method, we start from the same situation: a shell carrying a basic load and having a basic stress system and a basic deformation. Again we consider a small deviation from this state of stress and deformation. This time, instead of asking for the conditions under which a zero load can produce the deviation, we ask for the energy which must be put into the shell to produce it. This energy consists of two parts: the work which must be done against the external forces (gain in potential energy of the basic load) and the increase in strain energy. We call it the variation of the potential energy.

We choose a certain distribution of the additional displacements u, v, w and then assume that they all may still be varied by multiplying them by a common factor λ. We may then write the variation of the potential energy as a power series in λ. For $\lambda = 0$ there is no disturbance, hence no variation of energy, and therefore the power series has no constant term. The principle of virtual displacements states that the linear term must also be zero for every conceivable displacement; otherwise the basic loads and stresses would not be in equilibrium. It follows that the first term is the one with λ^2.

When the basic load is small enough, the quadratic term of the variation of potential energy is always positive, no matter how the eviations are chosen. It is then impossible for any deviation to occur unless additional forces are applied which can do work and supply the required energy. If, for some critical value of the basic load, there exists a set of displacements which makes the quadratic term vanish, then such displacements may occur spontaneously, i. e. the elastic equilibrium is neutral. If the basic load is increased beyond the critical value, the same set of displacements must be expected to yield a negative quadratic term in the variation of energy. When the disturbance of the specific type is produced accidentally, energy will be set free; the larger the displacement, the more energy is released, and this energy cannot be transformed into anything but kinetic energy of motion. The shell will move with increasing velocity away from its initial position of equilibrium, which is unstable.

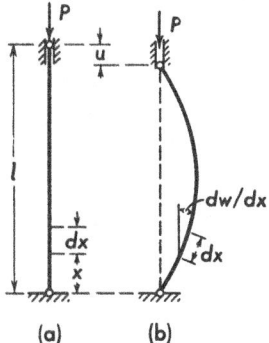

These statements are limited to small positive or negative values of the parameter λ which measures the magnitude of the deviation. For larger deviations the higher powers of λ come into play and influence the behavior of the shell.

The critical load obtained by the energy method is identical with the value obtained by the equilibrium method.

Fig. 2. EULER column, (a) straight equilibrium, (b) deflected equilibrium

To demonstrate the technique, we again consider the EULER column (fig. 2). When the column is deflected (fig. 2b) there is some strain energy stored in it. In the theory of bending of beams it is shown that this strain energy is

$$V = \frac{E\,I}{2} \int\limits_0^l \left(\frac{d^2 w}{dx^2}\right)^2 dx.$$

This expression is quadratic in w and therefore of the kind we need.

Besides the strain energy there is the work T which must be done against the external forces. In our case it is negative, since the displacement u and the force P have the same direction:

$$T = -P\,u.$$

The displacement u depends on the deflection w, and since we need energy terms which are quadratic in w, we must have the relation between u and w up to the term with w^2.

The elements dx of the bar are originally vertical and together they form the length

$$l = \int\limits_0^l dx.$$

In the deflected state (fig. 2b) the element dx is measured on the curved axis. It makes the angle dw/dx with the vertical, and its vertical projection is

$$dx \cos \frac{dw}{dx} = dx \left[1 - \frac{1}{2} \left(\frac{dw}{dx} \right)^2 + \cdots \right].$$

The sum of these projections is the chord length $l - u$. Up to the second-order term,

$$l - u = \int_0^l \left[1 - \frac{1}{2} \left(\frac{dw}{dx} \right)^2 \right] dx = \int_0^l dx - \frac{1}{2} \int_0^l \left(\frac{dw}{dx} \right)^2 dx,$$

whence

$$u = \frac{1}{2} \int_0^l \left(\frac{dw}{dx} \right)^2 dx.$$

The total energy needed to produce the deflection w is

$$V + T = \frac{E I}{2} \int_0^l \left(\frac{d^2 w}{dx^2} \right)^2 dx - \frac{P}{2} \int_0^l \left(\frac{dw}{dx} \right)^2 dx.$$

It is zero if

$$P = \frac{E I \int_0^l \left(\frac{d^2 w}{dx^2} \right)^2 dx}{\int_0^l \left(\frac{dw}{dx} \right)^2 dx}. \tag{3}$$

Since the numerator and the denominator are quadratic in derivatives of w, the quotient does not depend on the absolute magnitude of the deflection but only on its distribution. We ask for that special distribution $w = w(x)$ which yields the smallest value of P. This is P_{cr}.

There are different ways of finding this smallest value. We may derive the differential equation (2) from eq. (3) and then solve and discuss it as in the method of adjacent equilibrium. We may also assume a very general expression for the unknown function $w(x)$, say a FOURIER series,

$$w(x) = \sum_{n=1}^{\infty} C_n \sin \frac{n \pi x}{l},$$

which satisfies the boundary conditions, and then determine the coefficients so that P from eq. (3) becomes a minimum. When this is done, one finds that $C_n = 0$ for $n > 1$ and that C_1 is arbitrary, while P becomes equal to the EULER load.

It is a special merit of the energy method that it allows finding approximate values of P_{cr} by introducing into eq. (3) some plausible function $w(x)$. Since this is not necessarily the one which makes P a minimum, the approximate value of P is always higher than the correct one. Therefore, the result is not on the safe side, but it usually comes close to the correct one if the function $w(x)$ is skillfully chosen. However, the method should be used with some caution, in particular with shells, since the buckling displacements cannot easily be guessed in advance.

7.2 Cylindrical Shell

7.2.1 Differential Equations for Compression and Shear

7.2.1.1 Basic Concepts

We now consider a shell shaped as a circular cylinder of length l (fig. 3), and subject it simultaneously to three simple loads:

1) a uniform normal pressure on its wall, $p_r = -p$,

2) an axial compression applied at the edges, the force per unit of circumference being P,

3) a shear load applied at the same edges so as to produce a torque in the cylinder. The shearing force (shear flow) is T.

These loads are the basic loads, and the stress resultants which they produce are the basic stress system. We distinguish them by a subscript I from the additional forces appearing when the shell buckles.

The normal pressure p produces the hoop force

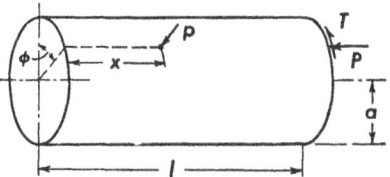

Fig. 3. Cylindrical shell; coordinates and basic loads

$$N_{\phi I} = -p\,a$$

[see eq. (II–8a)]. The axial load produces the longitudinal force

$$N_{xI} = -P,$$

and the shear load T produces shearing forces

$$N_{x\phi I} = N_{\phi xI} = -T.$$

This is a membrane stress system, and it is uniform all over the shell. The corresponding deformation is also uniform and may either be neglected or – in a more rigorous procedure – be eliminated from our considerations by tracing the coordinate lines on the cylinder after the basic deformation has taken place.

We impose now an additional deformation, described by the displacements u, v, w. The additional stress resultants which accompany

this disturbance of the original state are denoted by N_ϕ, N_x, ... M_ϕ,
... Q_x. Except for Q_ϕ and Q_x, they are connected with the displacements
by the elastic law of the cylinder, eqs. (V–8). The normal and shearing
forces are additional to the basic forces of the same kind. The total
forces are

$$\begin{aligned}
\overline{N}_\phi &= N_{\phi I} + N_\phi = -p\,a + N_\phi, \\
\overline{N}_x &= N_{xI} + N_x = -P + N_x, \\
\overline{N}_{x\phi} &= N_{x\phi I} + N_{x\phi} = -T + N_{x\phi}, \\
\overline{N}_{\phi x} &= N_{\phi x I} + N_{\phi x} = -T + N_{\phi x}.
\end{aligned} \tag{4}$$

These quantities are forces per unit length of certain line elements.
Since we are interested in products of the basic forces N_{xI}, ... with
such quantities as ϵ_x or ϵ_ϕ, we must raise the question as to which
length the stress resultants should be referred, to the original length
of the line element or to its
length after deformation.

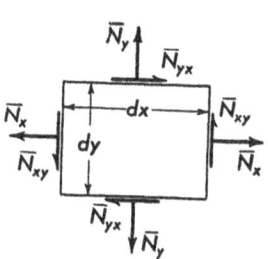
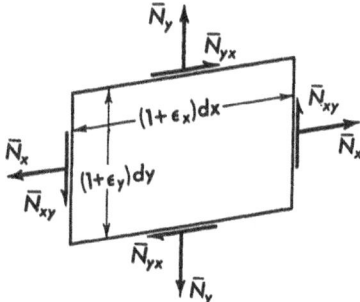

Fig. 4. Plane plate element   Fig. 5. Deformed plane plate element

To simplify the issue, we shall discuss it for an element $dx \cdot dy$
of a plane plate, acted upon by the normal forces $\overline{N}_x$ and $\overline{N}_y$ and by
the shearing forces $\overline{N}_{xy}$ and $\overline{N}_{yx}$ (fig. 4).

Let us consider one of the normal forces as an example. Before
buckling, the force acting on a line element dy is $N_{xI}\,dy$. If we decide
that after buckling it shall be $(N_{xI} + N_x)\,dy$, then the increase is
$N_x\,dy$, and this is the definition of the quantity N_x which we shall
later connect by a set of equations like (V–9) with the buckling dis-
placements u, v, w and the strains ϵ_x, ϵ_y, γ_{xy}. On the other hand, if
we decide to write the force after buckling as $(N_{xI} + N_x)(1 + \epsilon_y)\,dy$,
then the increase is

$$N_{xI}\,\epsilon_y\,dy + N_x(1 + \epsilon_y)\,dy \approx (N_{xI}\,\epsilon_y + N_x)\,dy,$$

and $N_x\,dy$ is only a part of the total increase. Now only this part would
be related to u, v, w, and this amounts to a different elastic law. It

looks here as if the decision had to be found from an experiment, measuring the strain connected with a small increase in the normal stress.

This, however, would not settle the issue. What we want is not the most exact description of the facts but a *linear* theory which comes as close to the facts as possible. In this respect, one assumption would be as good as the other. The first one linearizes the elastic behavior between zero load and N_{xI}, while the second assumption is a linear approximation of the elastic law in the immediate vicinity of the basic stress.

The decision must therefore be sought in another field. A first part of it comes from the equilibrium of moments with respect to a normal to the element. Fig. 5 shows the deformed element. If we were to resolve the force on its right-hand side in orthogonal components, the normal force would enter the moment equation, as discussed on p. 15. We prefer to use skew components as shown in the figure. The moment equation then reads either

$$\overline{N}_{xy}\,dy \cdot (1 + \epsilon_x)\,dx = \overline{N}_{yx}\,dx \cdot (1 + \epsilon_y)\,dy$$

or

$$\overline{N}_{xy}(1 + \epsilon_y)\,dy \cdot (1 + \epsilon_x)\,dx = \overline{N}_{yx}(1 + \epsilon_x)\,dx \cdot (1 + \epsilon_y)\,dy.$$

In these equations we put

$$\overline{N}_{xy} = N_{xyI} + N_{xy}, \qquad \overline{N}_{yx} = N_{yxI} + N_{yx}$$

and drop all products of two buckling quantities ($\epsilon_x\,\epsilon_y$, $N_{xy}\,\epsilon_x$ etc.). We obtain either

$$N_{xyI}(1 + \epsilon_x) + N_{xy} = N_{yxI}(1 + \epsilon_y) + N_{yx}$$

or

$$N_{xyI}(1 + \epsilon_x + \epsilon_y) + N_{xy} = N_{yxI}(1 + \epsilon_x + \epsilon_y) + N_{yx}.$$

Now the moment equilibrium of the element before buckling requires that $N_{xyI} = N_{yxI}$, and an elastic law similar to eqs. (V–9 c, d) yields $N_{xy} = N_{yx}$. When we introduce this into the preceding equations, we see that the last one is identically satisfied, while the first version yields

$$N_{xyI}(\epsilon_x - \epsilon_y) = 0,$$

a relation which does not make sense. This indicates that we must discard the idea of using the stress resultants $\overline{N}$ with the undeformed elements.

This decision, however, does not settle the question completely. To see this, we consider the same element $dx \cdot dy$, subjected only to normal forces N_{xI} and N_{yI} (fig. 6a). We assume now that each stress resultant $\overline{N}$ must be multiplied by the deformed length of the line element in which it is transmitted. We shall go with the forces $\overline{N}_x$ and $\overline{N}_y$ through a cycle of increasing and decreasing loads and follow up the strain energy which during this cycle is deposited in the element. Since the lateral contraction is immaterial for the issue at hand, we shall simplify the discussion by putting $\nu = 0$.

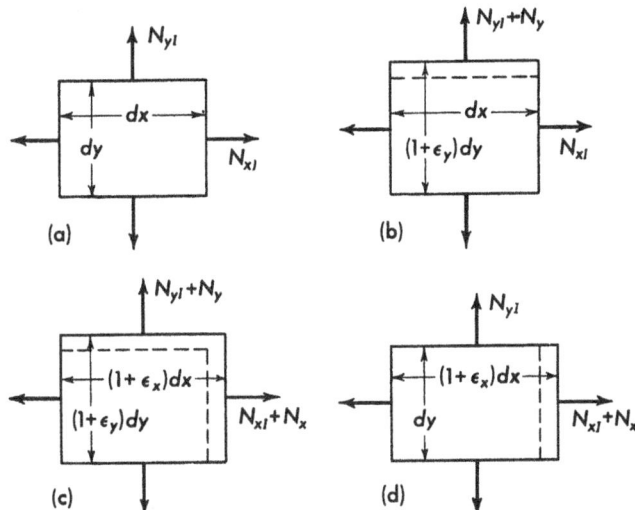

Fig. 6. Successive stages of deformation of a plane plate element

The first step is to increase N_{yI} to $N_{yI} + N_y$ (fig. 6b). The work done on the element is

$$\left(N_{yI} + \frac{1}{2} N_y\right) dx \cdot \epsilon_y \, dy.$$

Since the length of the sides dy is increased to $(1 + \epsilon_y)\, dy$, the forces acting on them must be increased from $N_{xI}\, dy$ to $N_{xI}(1 + \epsilon_y)\, dy$, but they do no work. Next we increase N_{xI} to $N_{xI} + N_x$. In this step the additional work

$$\left(N_{xI} + \frac{1}{2} N_x\right)(1 + \epsilon_y)\, dy \cdot \epsilon_x \, dx$$

is done, and the forces on the horizontal sides are increased to $(N_{yI} + N_y)\,(1 + \epsilon_x)\, dx$. The third step is to decrease these forces to $N_{yI}(1 + \epsilon_x)\, dx$. The work done in this step is negative:

$$-\left(N_{yI} + \frac{1}{2} N_y\right)(1 + \epsilon_x)\, dx \cdot \epsilon_y \, dy.$$

In this step the forces on the vertical sides are decreased to $(N_{xI}+N_x)\cdot dy$. With the concluding step we return to the original state of stress (fig. 6a), decreasing the latter force to $N_{xI}\,dy$. The work done in this step is

$$-\left(N_{xI} + \frac{1}{2}N_x\right)dy \cdot \epsilon_x\,dx.$$

When we add up all the work, we find that most terms cancel, but we are left with the energy

$$(N_{xI} - N_{yI})\,\epsilon_x^{\cdot}\,\epsilon_y\,dx\,dy$$

which has been deposited in the element. It ought to be zero, but evidently it is not.

We see here that we must also reject the idea of multiplying each stress resultant $\overline{N}$ or N_I with the deformed length of the element in which it is transmitted, since this would allow closed cycles of loading and unloading in which energy is created or destroyed.

The dilemma can be solved by using the deformed element for the shearing forces and the undeformed element for the normal forces. This, however, would not be a consistent system of notations.

An acceptable representation can be found in the following way: After the basic load has been placed on the plate or shell, we engrave on the element two unit vectors, pointing in the x and y directions (fig. 7a). When the element is deformed during

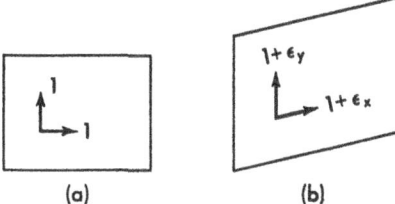

(a)   (b)

Fig. 7. Undeformed and deformed element showing the deformation of the reference vectors

buckling, these vectors participate in the deformation. In the deformed state they have lengths $1 + \epsilon_x$ and $1 + \epsilon_y$, respectively, and make an angle of $\pi/2 - \gamma_{xy}$. These vectors are chosen as a reference frame for the forces acting on the four sides of the element. This leads automatically to the use of the skew components shown in fig. 5, but the magnitudes of the stress resultants $\overline{N}$ are now such that they must still be multiplied by the magnitude of the reference vector to give the force per unit length of the *undeformed* element dx or dy. Hence, we have on the sides dy the forces $\overline{N}_x(1 + \epsilon_x)\,dy$ and $\overline{N}_{xy}(1 + \epsilon_y)\,dy$ and on the sides dx the forces $\overline{N}_y(1 + \epsilon_y)\,dx$ and $\overline{N}_{yx}(1 + \epsilon_x)\,dx$. For the shearing forces this amounts exactly to the use of the deformed elements $(1 + \epsilon_x)\,dx$ and $(1 + \epsilon_y)\,dy$ but not for the normal forces. One may easily check that, with the new definition of the stress resultants, the work done during the four steps of the loading cycle is, respectively:

$$\left(N_{yI} + \frac{1}{2} N_{yI} \epsilon_y + \frac{1}{2} N_y\right) \epsilon_y \, dx \, dy,$$

$$\left(N_{xI} + \frac{1}{2} N_{xI} \epsilon_x + \frac{1}{2} N_x\right) \epsilon_x \, dx \, dy,$$

$$-\left(N_{yI} + \frac{1}{2} N_{yI} \epsilon_y + \frac{1}{2} N_y\right) \epsilon_y \, dx \, dy,$$

$$-\left(N_{xI} + \frac{1}{2} N_{xI} \epsilon_x + \frac{1}{2} N_x\right) \epsilon_x \, dx \, dy,$$

and this cycle ends, as it should, with zero energy left.

7.2.1.2 Differential Equations

We are now prepared to write the six conditions for the equilibrium of a shell element as shown in figs. (V–1a, b).

In the x direction we have on one side of the element the force $\overline{N}_x(1 + \epsilon_x) \, a \, d\phi$, where ϵ_x is the longitudinal strain in the middle surface and identical with $\overline{\epsilon}_x$ in eqs. (V–10). On the opposite side the force is larger by a differential, and this differential

$$\frac{\partial}{\partial x} \left[\overline{N}_x(1 + \epsilon_x) \, a \, d\phi\right] dx = \left[\overline{N}'_x(1 + \epsilon_x) + \overline{N}_x \epsilon'_x\right] a \, d\phi \, dx$$

makes a contribution to the condition of equilibrium. Since $N'_{xI} = 0$, we have

$$N'_x(1 + \epsilon_x) = \overline{N}'_x(1 + \epsilon_x),$$

and for this we may simply write N'_x, since the term $N'_x \epsilon_x$ is quadratic in the disturbance quantities. For the same reason we write $\overline{N}_x \epsilon'_x = -P \epsilon'_x$ and drop the term $N_x \epsilon'_x$. If we now introduce u'/a for ϵ_x from eqs. (V–10) we finally have

$$\left[N'_x - P \frac{u''}{a}\right] d\phi \, dx.$$

It may still be observed that this force does not point exactly in the x direction but makes a small angle with it (fig. 5). To project it into the x direction, the force must be multiplied by the cosine of this angle, but the difference of this cosine from unity is small of the second order and therefore negligible.

The shearing forces $\overline{N}_{\phi x}$ on the sides dx of the element may be handled in exactly the same way. They yield the contribution

$$\frac{\partial}{\partial \phi} \left[\overline{N}_{\phi x}(1 + \epsilon_x) \, dx\right] d\phi = \left[N'_{\phi x} - T \frac{u''}{a}\right] dx \, d\phi.$$

These two contributions are modifications of the first two terms in eq. (V–1a). In the stability equation we have two more terms, containing $\bar{N}_\phi$ and $\bar{N}_{x\phi}$. The force $\bar{N}_\phi(1 + \epsilon_\phi)\,dx$ at one side of the element makes the small angle u'/a with the ϕ direction and therefore has a component $\bar{N}_\phi(1 + \epsilon_\phi)\,dx \cdot u'/a$ in the x direction. At the opposite side this component is larger by the increment

$$\frac{\partial}{\partial\phi}\left[\bar{N}_\phi(1 + \epsilon_\phi)\,dx\,\frac{u^\cdot}{a}\right]d\phi,$$

and this is a contribution to our equation. Since there is already the small factor $u^\cdot$, we may neglect ϵ_ϕ compared with 1 and also replace $\bar{N}_\phi$ by $N_{\phi I}$. All that is then left is $-p\,u^{\cdot\cdot}\,dx\,d\phi$. In a similar way $\bar{N}_{x\phi}$ yields the contribution

$$\frac{\partial}{\partial x}\left[\bar{N}_{x\phi}(1 + \epsilon_\phi)\,a\,d\phi\,\frac{u^\cdot}{a}\right]dx = -T\,\frac{u^{\cdot\cdot}}{a}\,d\phi\,dx.$$

Finally, there is a contribution of the external force $p\cdot(1 + \epsilon_x)\,dx$ $\cdot\,(1 + \epsilon_\phi)\,a\,d\phi$. Because of the deflection w the shell element is slightly tilted, and the external force has a component

$$p(1 + \epsilon_x)\,dx(1 + \epsilon_\phi)\,a\,d\phi\cdot\frac{w'}{a}$$

in the x direction. If we again drop everything that is of higher order in the disturbance quantities, this simplifies to $p\,w'\,dx\,d\phi$.

We may now write the first condition of equilibrium. We drop the common factor $dx\,d\phi$ and rearrange the terms:

$$N_x' + N_{\phi x}^\cdot - p(u^{\cdot\cdot} - w') - P\,\frac{u''}{a} - 2\,T\,\frac{u^{\cdot\cdot}}{a} = 0. \tag{5a}$$

The equilibrium of the forces in the circumferential direction may be treated in the same way. There is the force $\bar{N}_\phi(1 + \epsilon_\phi)\,dx$ which is treated like $\bar{N}_x$, and the force $\bar{N}_{x\phi}(1 + \epsilon_\phi)\,a\,d\phi$, which is treated like $\bar{N}_{\phi x}$. They yield

$$\left[N_\phi^\cdot - p(v^{\cdot\cdot} + w^\cdot) + N_{x\phi}' - T\,\frac{v'^\cdot + w'}{a}\right]dx\,d\phi.$$

The forces $\bar{N}_x(1 + \epsilon_x)\,a\,d\phi$ on either side of the element make angles v'/a and $v'/a + (v''/a^2)\,dx$, respectively, with the x direction and yield the contribution

$$-P\,\frac{v''}{a}\,dx\,d\phi.$$

In a similar way the forces $\bar{N}_{\phi x}(1 + \epsilon_x)\,dx$ make angles v'/a and $v'/a + (v'^\cdot/a)\,d\phi$, respectively, and yield the contribution $-T\,(v'^\cdot/a)\,dx\,d\phi$. But this is not all. When the edges dx are tilted in radial planes by an angle w/a, radial components $\bar{N}_{\phi x}(1 + \epsilon_x)\,dx \cdot w'/a$ are generated at opposite sides of the element, inward at one side and outward at the other. They make an angle $d\phi$ with each other and, therefore, have a resultant in the ϕ direction which belongs in our equation. The total contribution of the shear $\bar{N}_{\phi x}$ is then

$$- T \frac{v^{\cdot\prime} + w^\prime}{a}\, dx\, d\phi.$$

The transverse forces $Q_\phi\, dx$ on either side of the element make the same angle $d\phi$ with each other and, therefore, also have a resultant in the ϕ direction, which is $-Q_\phi\, dx\, d\phi$. This term has already occurred in eq. (V–1b) of the bending theory.

Finally, one might expect a contribution of the load due to the tilting of the element by the angle $w^{\cdot}/a$, but this is compensated by the tilting of the forces $N_{\phi I}$ which participate in this rotation and remain in equilibrium with the load. The second condition of equilibrium is, therefore, this:

$$N_\phi^{\cdot} + N_{x\phi}^\prime - Q_\phi - p(v^{\cdot\cdot} + w^{\cdot}) - P \frac{v^{\cdot\prime}}{a} - 2\,T \frac{v^{\cdot\prime} + w^\prime}{a} = 0. \qquad (5\,\mathrm{b})$$

The third equation contains all forces normal to the axis of the cylinder. The transverse forces make the same contributions as they did to eq. (V–1c):

$$(Q_\phi^{\cdot} + Q_x^\prime)\, dx\, d\phi.$$

With the hoop force we have to be more careful. Since $\bar{N}_\phi$ contains the large part $N_{\phi I}$, we must apply the exact value of the reference vector, $(1 + \epsilon_\phi)$, and must take into account that the angle between the two forces $\bar{N}_\phi (1 + \epsilon_\phi)\, dx$ is not simply $d\phi$, but $[1 + (v^{\cdot} - w^{\cdot\cdot})/a]\, d\phi$. When we drop quadratic terms, we find the contribution of the hoop force to be

$$\bar{N}_\phi (1 + \epsilon_\phi)\, dx \left(1 + \frac{v^{\cdot} - w^{\cdot\cdot}}{a}\right) d\phi = \left[N_\phi - p\,a \left(1 + \frac{2v^{\cdot} + w - w^{\cdot\cdot}}{a}\right)\right] dx\, d\phi.$$

The longitudinal forces $\bar{N}_x (1 + \epsilon_x)\, a\, d\phi$ make a contribution because of the curvature $w^{\prime\prime}/a^2$ of the generators:

$$- \bar{N}_x (1 + \epsilon_x)\, a\, d\phi \cdot \frac{w^{\prime\prime}}{a}\, dx = P \frac{w^{\prime\prime}}{a}\, d\phi\, dx.$$

The shearing forces make a contribution similar to the shear term in eq. (IV–3c), but careful analysis is needed to get it correctly. The force $\bar{N}_{x\phi}(1 + \epsilon_\phi)\, a\, d\phi$ is rotated through the angle $(v - w^\prime)/a$ and thus gets a component in the direction of the original radius at the point x, ϕ. At the opposite side of the shell element the corresponding component is larger by the differential

$$\frac{\partial}{\partial x}\left[\bar{N}_{x\phi}(1 + \epsilon_\phi)\, a\, d\phi \cdot \frac{v - w^\prime}{a}\right] dx = -T \frac{v^\prime - w^{\prime\cdot}}{a}\, d\phi\, dx.$$

The forces at the other edges contribute in two ways. First, the force $\overline{N}_{\phi x}(1 + \epsilon_x)\, dx$ is rotated by w'/a in a plane normal to the shell. This yields a radial component whose increment

$$-\frac{\partial}{\partial \phi}\left[\overline{N}_{\phi x}(1 + \epsilon_x)\, dx \cdot \frac{w'}{a}\right] d\phi = T\,\frac{w''}{a}\, dx\, d\phi$$

belongs in our equation. Second, the tangential components $\overline{N}_{\phi x}(1 + \epsilon_x)\, dx \cdot (v'/a)$ whose difference we included in eq. (5b), make an angle $d\phi$ with each other and combine to a radial force

$$\overline{N}_{\phi x}(1 + \epsilon_x)\, dx\, \frac{v'}{a}\, d\phi = -T\,\frac{v'}{a}\, dx\, d\phi.$$

The last term in the equation comes from the load $p \cdot (1 + \epsilon_x)\, dx \cdot (1 + \epsilon_\phi)\, a\, d\phi$. It is to be introduced in full size since the tilting of the element and hence of the pressure requires only multiplication by a cosine of a small angle, i. e. by a factor which may be replaced by unity. The contribution is, therefore,

$$p\left(1 + \frac{u' + v^{\cdot} + w}{a}\right) a\, d\phi\, dx.$$

We are now ready to collect the terms and to write the equation:

$$Q_{\phi}^{\cdot} + Q_x' + N_\phi + p(u' - v^{\cdot} + w^{\cdot\cdot}) + P\,\frac{w''}{a} - 2T\,\frac{v' - w^{\cdot\cdot}}{a} = 0. \quad (5c)$$

The conditions for the moment equilibrium are much easier to obtain. For the axes in the directions x and ϕ they are exactly the same as in Chapter V,

$$M_\phi^{\cdot} + M_{x\phi}' - a\,Q_\phi = 0,$$

$$M_x' + M_{\phi x}^{\cdot} - a\,Q_x = 0,$$

$$(5d, e)$$

because none of the basic forces has a moment of sufficient magnitude with respect to these axes. The moment equation about a normal to the shell (strictly speaking, about the normal to the undeformed middle surface) contains the shearing forces. Their contributions are now

$$\overline{N}_{x\phi}(1 + \epsilon_\phi)\, a\, d\phi \cdot (1 + \epsilon_x)\, dx - \overline{N}_{\phi x}(1 + \epsilon_x)\, dx \cdot (1 + \epsilon_\phi)\, a\, d\phi$$

$$= -T(1 + \epsilon_\phi + \epsilon_x)\, a\, d\phi\, dx + N_{x\phi}\, a\, d\phi\, dx$$

$$+ T(1 + \epsilon_x + \epsilon_\phi)\, dx\, a\, d\phi - N_{\phi x}\, dx\, a\, d\phi.$$

The terms with T cancel, and when we add the contribution of the twisting moment $M_{\phi x}$ we arrive exactly at eq. (V–1f). We already know that this equation becomes an identity when the stress resultants are expressed in terms of the displacements, and we may, there-fore, drop this equation.

In Chapter 5 we used eqs. (V–1d, e) to eliminate the transverse forces from eqs. (V–1b, c), and we now do the same in eqs. (5). This yields a set of three equations:

$$a\,N_x' + a\,N_{\phi x}^{\cdot} - p\,a(u^{\cdot\cdot} - w^{\cdot}) - P\,u'' - 2T\,u'^{\cdot} = 0,$$

$$a\,N_{\phi}^{\cdot} + a\,N_{x\phi}' - M_{\phi}^{\cdot} - M_{x\phi}' - p\,a(v^{\cdot\cdot} + w^{\cdot}) - P\,v'' - 2T(v'^{\cdot} + w') = 0,$$

$$M_{\phi}^{\cdot\cdot} + M_{x\phi}^{\prime\cdot} + M_{\phi x}^{\prime\cdot} + M_{x}'' + a\,N_{\phi} + p\,a(u' - v^{\cdot} + w^{\cdot\cdot})$$
$$+ P\,w'' - 2T(v' - w'^{\cdot}) = 0.$$

Now we may use the elastic law (V–9) to express all the forces N and moments M by u, v, w and their derivatives. When we do so, the rigid-ities D and K defined by eqs. (V–8) enter the equations. It is advisable to divide everything by D and to introduce the dimensionless param-eters

$$k = \frac{K}{D\,a^2} = \frac{t^2}{12\,a^2}, \qquad q_1 = \frac{p\,a}{D}, \qquad q_2 = \frac{P}{D}, \qquad q_3 = \frac{T}{D}. \qquad (6)$$

The differential equations of the buckling problem then appear in the following form:

$$u'' + \frac{1-\nu}{2}\,u^{\cdot\cdot} + \frac{1+\nu}{2}\,v'^{\cdot} + \nu\,w' + k\left(\frac{1-\nu}{2}\,u^{\cdot\cdot} - w''' + \frac{1-\nu}{2}\,w'^{\cdot\cdot}\right)$$
$$- q_1(u^{\cdot\cdot} - w') - q_2\,u'' - 2q_3\,u'^{\cdot} = 0,$$

$$\frac{1+\nu}{2}\,u'^{\cdot} + v^{\cdot\cdot} + \frac{1-\nu}{2}\,v'' + w^{\cdot} + k\left(\frac{3}{2}(1-\nu)\,v'' - \frac{3-\nu}{2}\,w'^{\cdot\cdot}\right)$$
$$- q_1(v^{\cdot\cdot} + w^{\cdot}) - q_2\,v'' - 2q_3(v'^{\cdot} + w') = 0,$$

$$\nu\,u' + v^{\cdot} + w + k\left(\frac{1-\nu}{2}\,u'^{\cdot\cdot} - u''' - \frac{3-\nu}{2}\,v'^{\cdot\cdot} + w^{\mathrm{IV}} + 2w''^{\cdot\cdot}\right.$$
$$\left. + w^{\cdot\cdot\cdot\cdot} + 2w'' + w\right) + q_1(u' - v^{\cdot} + w^{\cdot\cdot}) + q_2\,w'' - 2q_3(v' - w'^{\cdot}) = 0.$$
$$(7\,a\text{–}c)$$

They describe the buckling of a cylindrical shell under the most general homogeneous membrane stress action.

It may be observed that the parameters defined by eqs. (6) are small quantities. For k this is obvious, since we are concerned with thin shells where $t \ll a$. The three load parameters q are approximately the elastic strains caused by the corresponding basic loads, and since all our theory is based on the assumption that such strains are small compared with unity, we shall neglect q_1, q_2, q_3 compared with 1 wherever the opportunity occurs.

7.2.2 Solution for Shells without Shear Load

7.2.2.1 Two-way Compression

When there is no shear load on the shell ($T = 0$, hence $q_3 = 0$), eqs. (7) admit a solution of the form

$$
\left.
\begin{aligned}
u &= A \cos m\phi \cos \frac{\lambda x}{a}, \\
v &= B \sin m\phi \sin \frac{\lambda x}{a}, \\
w &= C \cos m\phi \sin \frac{\lambda x}{a},
\end{aligned}
\right\} \tag{8}
$$

where

$$
\lambda = \frac{n \pi a}{l} \tag{9}
$$

and n is an integer. The solution (8) describes a buckling mode with n half waves along the length of the cylinder and with $2m$ half waves around its circumference. Although this is far from being the most general solution, it is one which fulfills reasonable boundary conditions. We assume the edges of the cylinder to be at $x = 0$ and $x = l$, and we see at a glance that there $v = w = 0$. From the elastic law (V–9 b, f) we see also that $N_x = 0$, $M_x = 0$. Thus the solution (8) represents the buckling of a shell whose edges are supported in tangential and radial directions but are neither restricted in the axial direction nor clamped.

When we introduce the solution (8) into the differential equations (7) (with $q_3 = 0$), the trigonometric functions drop out entirely and we are left with the following equations:

$$
\left.
\begin{aligned}
&A\left[\lambda^2 + \frac{1-\nu}{2} m^2(1+k) - q_1 m^2 - q_2 \lambda^2\right] + B\left[-\frac{1+\nu}{2}\lambda m\right] \\
&\qquad + C\left[-\nu\lambda - k\left(\lambda^3 - \frac{1-\nu}{2}\lambda m^2\right) - q_1\lambda\right] = 0, \\
&A\left[-\frac{1+\nu}{2}\lambda m\right] + B\left[m^2 + \frac{1-\nu}{2}\lambda^2(1+3k) - q_1 m^2 - q_2\lambda^2\right] \\
&\qquad + C\left[m + \frac{3-\nu}{2}k\lambda^2 m - q_1 m\right] = 0, \\
&A\left[-\nu\lambda - k\left(\lambda^3 - \frac{1-\nu}{2}\lambda m^2\right) - q_1\lambda\right] + B\left[m + \frac{3-\nu}{2}k\lambda^2 m - q_1 m\right] \\
&\qquad + C[1 + k(\lambda^4 + 2\lambda^2 m^2 + m^4 - 2m^2 + 1) - q_1 m^2 - q_2\lambda^2] = 0.
\end{aligned}
\right\} \tag{10}
$$

These are three linear equations with the buckling amplitudes A, B, C as unknowns and with the brackets as coefficients. Since these equations are homogeneous, they admit, in general; only the solution $A = B = C$

$= 0$, indicating that the shell is not in neutral equilibrium. Only if the determinant of the nine coefficients equals zero is a nonvanishing solution A, B, C possible. Thus the vanishing of this determinant is the buckling condition of the shell. Whenever the buckling condition is fulfilled, any two of the three equations (10) determine the ratios A/C and B/C and thus the buckling mode according to eqs. (8). As in all cases of neutral equilibrium, the magnitude of the possible deformation remains arbitrary.

The buckling condition contains four unknowns: the dimensionless loads q_1 and q_2 and the modal parameters m and λ. Of m we know that it must be an integer $(0, 1, 2, \ldots)$; of λ, that it must be an integer multiple $(n = 1, 2, \ldots)$ of $\pi a/l$. We may write the buckling condition separately for every pair m, λ fulfilling these requirements and consider it as a relation between q_1 and q_2 which describes those combi-

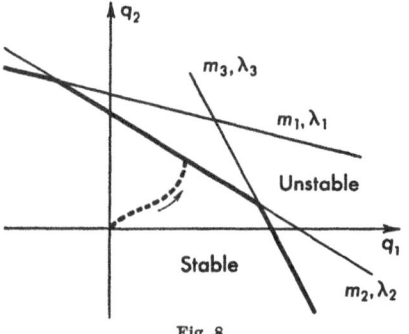

nations of the two loads for which the shell is in neutral equilibrium. When we plot these equations as curves in a q_1, q_2 plane, we obtain a diagram like fig. 8, which can be interpreted as follows. The origin $q_1 = q_2 = 0$ represents the unloaded shell. When a load is gradually applied, the corresponding diagram point moves along some path, as shown by the dotted line. As long as it does not meet any of the curves, the shell is in stable equilibrium; but as soon as one of the curves is reached equilibrium becomes neutral, with the buckling mode defined by the parameters m, λ of this curve. The stable domain in the q_1, q_2 plane is, therefore, bounded by the envelope of all the curves which is shown in fig. 8 by a heavy line.

The coefficients of eqs. (10) are linear functions of k, q_1, q_2. The expanded determinant is, therefore, a polynomial of the third degree in these parameters. Since they are very small quantities, it is sufficient to keep only the linear terms and to write the buckling condition as

$$c_1 + c_2\,k = c_3\,q_1 + c_4\,q_2. \tag{11}$$

This equation describes a straight line in the q_1, q_2 plane, and the limit of the stable domain, as shown in fig. 8, is a polygon consisting of sections of straight lines for different pairs m, λ.

The coefficients $c_1 \ldots c_4$ of eq. (11) may easily be found by really expanding the determinant. Since c_1 turns out to be proportional to λ^4, we may drop the term with λ^4 in all other coefficients, and we obtain

$$
\left.\begin{aligned}
c_1 &= (1 - \nu^2)\, \lambda^4, \\
c_2 &= (\lambda^2 + m^2)^4 - 2\,(\nu\, \lambda^6 + 3\,\lambda^4\, m^2 + (4 - \nu)\, \lambda^2\, m^4 + m^6) \\
&\quad + 2\,(2 - \nu)\, \lambda^2\, m^2 + m^4, \\
c_3 &= m^2\,(\lambda^2 + m^2)^2 - m^2\,(3\,\lambda^2 + m^2), \\
c_4 &= \lambda^2\,(\lambda^2 + m^2)^2 + \lambda^2\, m^2.
\end{aligned}\right\} \quad (12)
$$

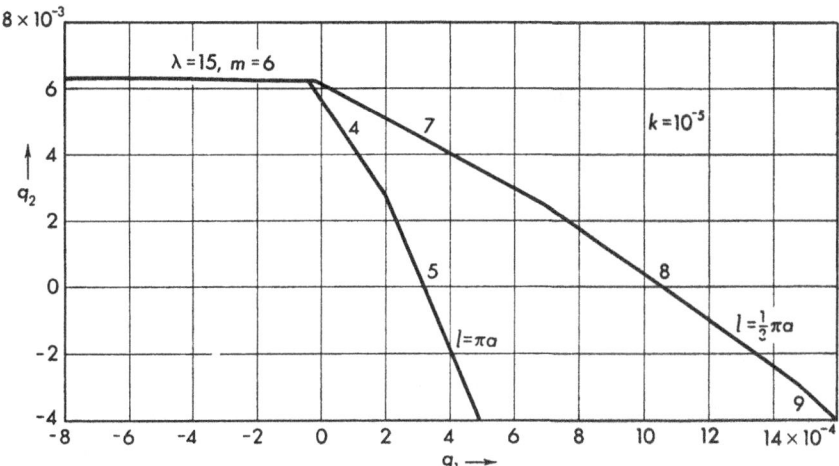

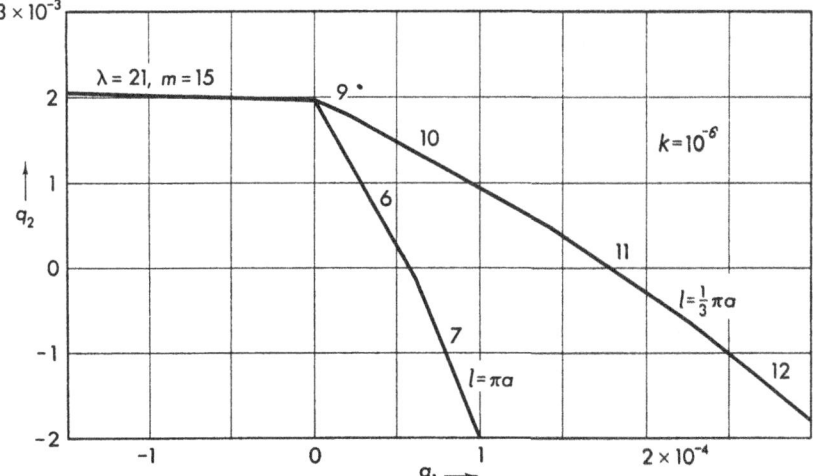

Fig. 9. Buckling diagrams for a cylindrical shell subjected to two-way thrust

From these formulas the stability curve may easily be constructed when l and k are given. Some examples are shown in fig. 9, and the following conclusions may be drawn from them:

Although the load and the basic stress system have axial symmetry, the buckling mode does not ($m \neq 0$) but develops nodal generators. Their number increases as q_1 increases and is higher for thinner shells.

In the right-hand part of the diagrams the curves for shells of different lengths are so arranged that the shorter shell has higher critical loads. For this reason nodal circles cannot occur. Somewhere close to the q_2 axis the curves for different λ intersect, and from there on toward the left long shells can buckle at the smaller load of shorter shells by adopting a mode with nodal circles. As a consequence, an internal pressure ($q_1 < 0$) does not perceptibly increase the axial load q_2, while an axial tension ($q_2 < 0$) increases considerably the resistance offered to an external pressure.

7.2.2.2 Axial Compression Only

7.2.2.2.1 Exact Solution. In fig. 9 we used the dimensionless loads q_1 and q_2 as coordinates and l/a or λ as curve parameter, and we had to draw a special diagram for every k. If only one of the loads is present, it is possible to plot it as a function of l/a or λ and to use k as a parameter, thus compressing the results of the theory in one single diagram. We shall do this separately for the two cases of axial and circumferential loading. It might just as well be done for cases of two-way compression when q_1 and q_2 depend on each other, as, for example, in the case of a closed vessel subjected to an external pressure, where $q_2 = q_1/2$.

If $q_1 = 0$, we have from eqs. (11) and (12)

$$q_2 = \{(1 - \nu^2)\,\lambda^4 + k[(\lambda^2 + m^2)^4 - 2(\nu\,\lambda^6 + 3\lambda^4\,m^2 + (4 - \nu)\,\lambda^2\,m^4 + m^6)$$

$$+ 2(2 - \nu)\,\lambda^2\,m^2 + m^4]\}\,[\lambda^2(\lambda^2 + m^2)^2 + \lambda^2\,m^2]^{-1}. \qquad (13)$$

This has been plotted in fig. 10 against $l/na = \pi/\lambda$ as the abscissa. In order to cover a wide range of values in both variables, logarithmic scales have been used.

In this diagram there is one curve for every integer m. For $m = 0$ it has a minimum in the left half of the diagram and then rises indefinitely. The curve $m = 1$ has a similar minimum but then reaches a maximum and from there drops to $q_2 = 0$ for $l/na \to \infty$. The next curves, $m = 2, 3, \ldots$ each have a second minimum. When m is increased, the two minima move closer together and finally coincide, as may clearly be seen for $m = 10$. For still higher m this minimum moves upward, and the curves are no longer interesting.

Now let us consider a shell whose k is the one for which fig. 10 or a similar diagram has been made and which has a certain ratio l/a. To judge its stability, we have to watch the points with the common ordinate q_2 and with the abscissas l/na with $n = 1, 2, 3$, etc. As long as all these points lie well below the area covered by the many curves, the shell is certainly stable. When the load and hence q_2 is increased, all the points move upward. As soon as the first of them reaches one of the curves, the shell is in neutral-equilibrium and ready to buckle.

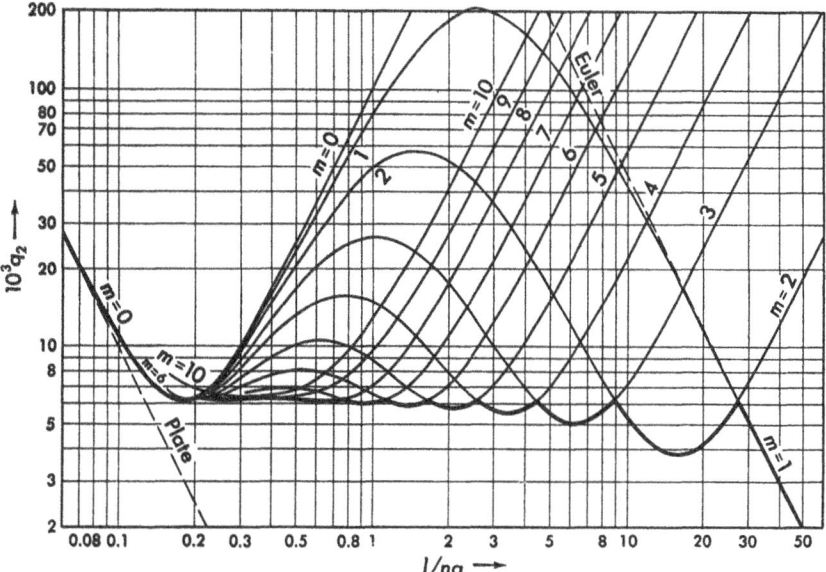

Fig. 10. Construction of the buckling diagram for axial compression, $k = 10^{-5}$

A further increase of load is not possible. The buckling mode is determined by the n connected with the abscissa of the critical point and by the m found at the curve passing through it.

It is obvious that, in this way, only that part of each curve which has no other curve between it and the base line of the diagram can be reached. These parts combine to the festoon-shaped curve shown by a heavier line in fig. 10. All other parts of the curves may at once be discarded as not contributing to the solution of the buckling problem.

This festoon curve is plotted over l/na. For practical use, it would be more convenient to have l/a as the abscissa. If we want this, we must plot several replicas of this festoon curve, one for $n = 1$, another one for $n = 2$, etc. In the logarithmic plot of fig. 10 all these curves can be derived from the first one by just shifting it by different amounts toward the right so that the point with the abscissa 1 gets the abscissas 2,

3, etc. From these overlapping festoon curves we derive a new festoon curve by again discarding everything which cannot be reached from the base line of the diagram without first crossing another branch of the curve. In this way the final festoon curve for $k = 10^{-5}$ is obtained. It is shown in fig. 11, and there similar curves for other values of k

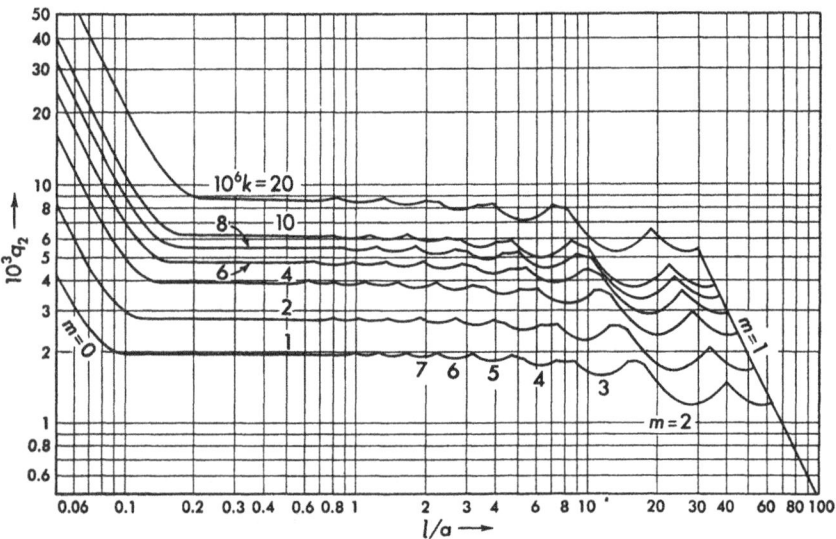

Fig. 11. Final form of the buckling diagram for axial compression

have been added. This set of curves is the final buckling diagram for cylindrical shells with simply supported ends, subjected to an axial thrust.

7.2.2.2.2 Limiting Cases. The various values of l/a may be obtained by keeping the radius a constant and varying the length of the cylinder or by fixing l and varying a. If the latter is done, the left-hand part of the diagrams belongs to cylinders of very small curvature (fig. 12). The distance b of the nodal generators is $b = \pi\,a/m$ and goes to infinity with a. The buckling load of a plane plate strip of length (or better width) l/n is

$$P = \frac{K\,\pi^2}{(l/n)^2},$$

and when we introduce this in the definition (6) of our load parameter q_2, we find

$$q_2 = \frac{K\,\pi^2\,n^2}{D\,l^2} = k\,\pi^2\left(\frac{n\,a}{l}\right)^2. \tag{14}$$

In a logarithmic diagram this is represented by a straight line. It has been entered in fig. 10, and it is readily seen how all the curves approach it asymptotically. It should, however, be noticed that for $l/na = 0.1$ the distance from the asymptote is still rather large, indicating that even a small curvature can substantially raise the buckling resistance of a plate.

On the right-hand side of fig. 10 we see that for very long cylinders buckling takes place with $m = 1$. If $B = -C$ [(see eqs. (8)], this would mean that the circular cross sections of the cylinder remain circular and undergo only rigid-body displacements. The deformation would then be the same as that of a slender bar of tubular shape, buckling as an EULER column. Instead of investigating whether really $B = -C$, we simply transcribe the EULER for-mula in our shell notations and compare it with the line $m = 1$ of our diagram.

Since P is the load per unit of circumference, the total axial force in the shell is $P \cdot 2\pi a$, and EULER's formula for a tube of length l/n yields

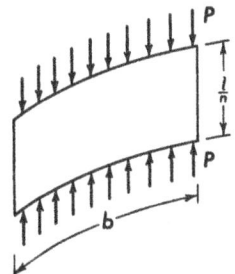

Fig. 12. Rectangular cy-lindrical panel, bounded by pairs of nodal circles and nodal generators

$$2\pi a P = \frac{E \cdot \pi a^3 t \cdot \pi^2}{(l/n)^2}.$$

When we introduce this into the definition (6) of q_2, we find

$$q_2 = \frac{E t \pi^2 n^2 a^2}{2 D l^2} = (1 - \nu^2) \frac{\pi^2}{2} \left(\frac{n a}{l}\right)^2. \qquad (15)$$

This is again a straight line, parallel to the one we had before, and it is also shown in fig. 10. The curve $m = 1$ comes remarkably close to it.

7.2.2.2.3 **Approximate Formulas.** When we compare the different festoon curves in fig. 11, we see that the arcs $m = 2$ all look very much alike, and so do the arcs $m = 3, 4$, etc., until they become indistinguish-able. This suggests that it should be possible to make all the festoon curves coincide by choosing the right quantities for plotting. Such a change of coordinates would shift the curves toward each other in the direction of the EULER line, and it would certainly not make the left-hand parts of the curves coincide.

On the other hand, these left-hand parts also show a remarkable similarity, suggesting that there might still be another way of plotting which would bring them together.

We find the solution when we assume that λ^2 is either very large or very small compared with unity.

On the right-hand side of fig. 11, l/na is large, hence $\lambda^2 \ll 1$. We may, therefore, in eq. (13) neglect λ^2 compared with 1 or m^2, but in the first brackets we must do so separately for the terms with and without the small factor k. We thus obtain the approximate relation

$$q_2 = \frac{(1 - \nu^2)\,\lambda^4 + k\,(m^2 - 1)^2\,m^4}{\lambda^2\,(m^2 + 1)\,m^2}$$

from which we see that

$$\varrho_2 = \frac{q_2}{\sqrt{(1 - \nu^2)\,k}} = \frac{1}{(m^2 + 1)\,m^2}\,\frac{\lambda^2\,\sqrt{1 - \nu^2}}{\sqrt{k}} + \frac{(m^2 - 1)^2\,m^2}{m^2 + 1}\,\frac{\sqrt{k}}{\lambda^2\,\sqrt{1 - \nu^2}} \tag{16}$$

may be plotted as a unique curve over $k^{1/2}\,\lambda^{-2}\,(1 - \nu^2)^{-1/2}$ as the abscissa. Because of the parameter m this will be a festoon curve like those in fig. 11.

From the definitions of k and λ we find

$$k^{1/2}\,\lambda^{-2}(1 - \nu^2)^{-1/2} = \frac{1}{\pi^2\,\sqrt{12\,(1 - \nu^2)}}\,\frac{(l/n)^2}{a^2}\,\frac{t}{a}\,,$$

and we may as well use any multiple or any power thereof as the abscissa of the plot. In fig. 13b the simplest such variable has been

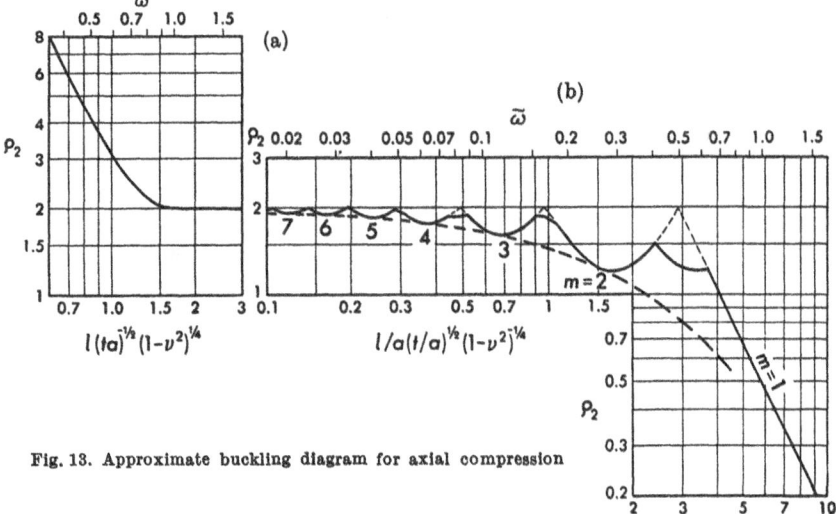

Fig. 13. Approximate buckling diagram for axial compression

chose, namely, $(l/a)\,(t/a)^{1/2}\,(1 - \nu^2)^{-1/4}$ but we might as well have chosen the form parameter $\tilde{\omega}$ defined later by eq. (38). A scale in this variable has been added at the upper edge of the diagram.

The numbers at the major arcs of the curve give values of m. The smaller arcs in between which have not been numbered, belong to buckling modes with $n > 1$. They occur in the diagram because we dropped the factor n from our abscissa.

When we deal with a complete cylinder, m in eq. (8) must be an integer. One may, however, apply the solution (8) and all formulas

drawn from it to cylindrical segments (fig. 14) covering less than 360°
circumference, if suitable boundary conditions can be established at
the straight edges of the segment. If its width is b, one must choose m
so that nodal generators appear at distances b,
that is, m must be equal to or an integer multiple of $\pi\,a/b$.

With such a choice of values m we arrive
again at a festoon curve, and when we draw
the envelope of all such festoon curves, for
all possible choices of a/b, we obtain the broken line in fig. 13b. As may readily be seen,
it keeps on the safe side of the true buckling
curve.

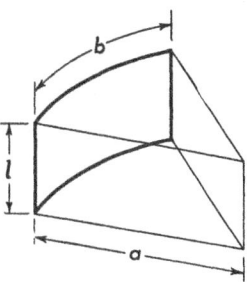

Fig. 14
Rectangular cylindrical panel

We now turn our attention to the left-hand
side of fig. 11. There l/na is small and hence
$\lambda^2 \gg 1$. We may then simplify eq. (13) by neglecting 1 compared
with λ^2. This yields the approximate formula

$$q_2 = \frac{(1 - \nu^2)\,\lambda^4 + k(\lambda^2 + m^2)^4}{\lambda^2(\lambda^2 + m^2)^2},$$

which may be written as

$$\frac{q_2}{\sqrt{(1 - \nu^2)\,k}} = \frac{\lambda^2}{(\lambda^2 + m^2)^2}\sqrt{\frac{1 - \nu^2}{k}} + \frac{(\lambda^2 + m^2)^2}{\lambda^2}\sqrt{\frac{k}{1 - \nu^2}}. \qquad (17)$$

On the left-hand side of this equation we have the same quantity ϱ_2
as in eq. (16), and on the right-hand side one term is the reciprocal
of the other.

We must now distinguish two cases: In the region $0.2 < l/na < 0.5$
the mode parameter m is rather large, and its exact value does not
have much influence on q_2. We may then permit m to vary continuously
and may find the minimum of q_2 by differentiating eq. (17) with respect
to m [or better with respect to $(\lambda^2 + m^2)^2$, which is a function of m
when λ is given]. The result is that we must put

$$(\lambda^2 + m^2)^2 = \lambda^2 \sqrt{\frac{1 - \nu^2}{k}}, \qquad (18)$$

and upon introducing this into eq. (17) we find the simple result

$$\varrho_2 = \frac{q_2}{\sqrt{(1 - \nu^2)\,k}} = 2. \qquad (19\,\mathrm{a})$$

This corresponds to the horizontal lines which all the festoon curves
in figs. 10 and 11 approach toward the left before they start their
final rise.

Eq. (19a) is acceptable as long as eq. (18) yields a reasonable value of m^2. Now it may easily be checked that when λ is increased, m^2 finally becomes negative. From there on we must dismiss eq. (18) as inapplicable and must choose that non-negative value of m with which eq. (17) yields the lowest q_2. Since with increasing λ the second term in this equation becomes dominating, we evidently have to put $m = 0$ and thus obtain

$$\varrho_2 = \frac{q_2}{\sqrt{(1-\nu^2)\,k}} = \frac{1}{\lambda^2}\sqrt{\frac{1-\nu^2}{k}} + \lambda^2\sqrt{\frac{k}{1-\nu^2}}. \tag{19b}$$

This may be plotted over such abscissas as

$$\frac{1}{\lambda}\sqrt[4]{\frac{1-\nu^2}{k}} = \frac{l}{n\,\pi\,a}\sqrt[4]{\frac{12(1-\nu^2)\,a^2}{t^2}}$$

or simply $(1-\nu^2)^{1/4}\,l/\sqrt{at}$, and this latter variable has been used in fig. 13a. The form parameter ω defined by eq. (42) is a constant multiple of this quantity, and an ω scale has been added at the upper edge of the diagram. The curve corresponds to the sharp rise towards the left end of all the curves in fig. 11. For $\omega = 1$ the right-hand side of eq. (19b) assumes a minimum value, which is $\varrho_2 = 2$. This is the point of transition between eqs. (19a) and (19b).

The two terms in eq. (19b) may be traced back through the analysis. The second term represents the influence of the bending stiffness K and hence of the bending and twisting moments, while the first term represents the extensional stiffness D and hence the normal and shearing forces connected with the buckling deformation of the shell. The predominance of the second term in extremely short cylinders simply indicates that we are approaching the situation prevailing in a plane plate, which derives its buckling strength entirely from its bending stiffness.

7.2.2.3 External Pressure Only

We consider now the other case of one-parametric loading and assume that $q_2 = 0$. Eqs. (11) and (12) then yield

$$q_1 = \{(1-\nu^2)\,\lambda^4 + k[(\lambda^2+m^2)^4 - 2(\nu\,\lambda^6 + 3\,\lambda^4\,m^2$$
$$+ (4-\nu)\,\lambda^2\,m^4 + m^6) + 2(2-\nu)\,\lambda^2\,m^2 + m^4]\} \tag{20}$$
$$[m^2(\lambda^2+m^2)^2 - m^2(3\,\lambda^2+m^2)]^{-1}.$$

This formula may be treated in the same way as eq. (13). With l/na as abscissa and q_1 as ordinate every integer m yields one curve, and from these a festoon curve is derived by the same reasoning as in the preceding case. The result is shown in fig. 15. Since the curves rise monotonically toward the left, there is no doubt as to the choice of $n = 1$, and the abscissa may be written simply as l/a.

The festoon curves of fig. 15 end with the arc $m = 2$ and on the right approach a horizontal asymptote giving the buckling load for a cylinder of infinite length. When we put in eq. (20) $m = 2$ and $\lambda \rightarrow 0$,

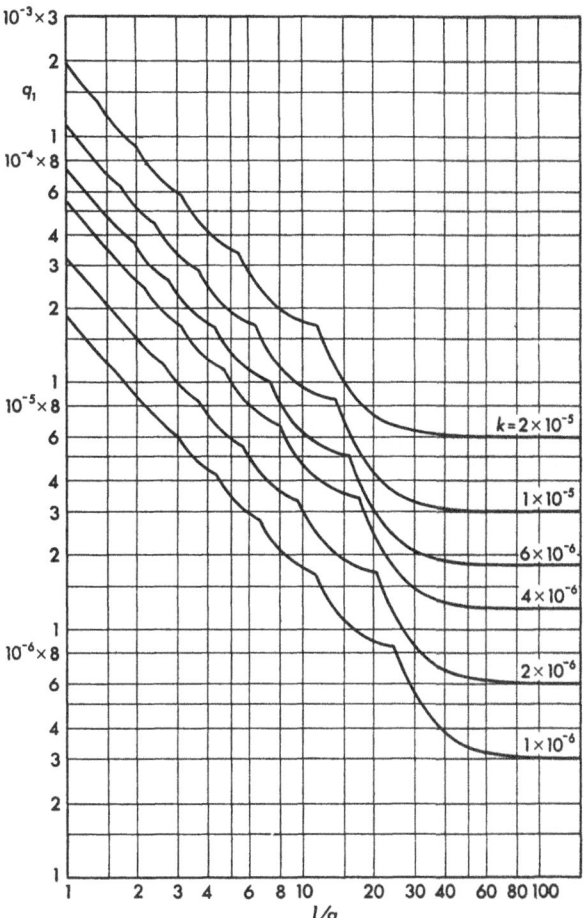

Fig. 15. Buckling diagram for external pressure

we find the asymptotic value to be $q_1 = 3k$. This result may easily be interpreted in the following way:

The solution (8) from which all preceding buckling formulas have been derived, assumes that at the ends of the cylinder $w = 0$, i. e. that there are bulkheads which prevent deflection. If the cylinder is very long, its central part is little influenced by the stiffening effect of these bulkheads. A circular strip of width dx which we may isolate here, behaves much like a circular ring of cross section $t \cdot dx$. The moment of inertia of this cross section is $I = t^3\, dx/12$, and the load per unit of circumference is $p\, dx$. When such a ring buckles in its plane, it gets ovalized according to eqs. (8) with $m = 2$ and $B = -C$, and the buckling load[1] is

$$p\, dx = \frac{3\,E\,I}{a^3} = \frac{3\,E\,t^3\, dx}{12\,a^3}.$$

When we replace E here by $E/(1 - \nu^2)$, we may bring this into the form $q_1 = 3\,k$ in perfect agreement with the asymptotic value for the long shell.

It may easily be seen that $m = 0$ cannot yield a finite q_1 since the denominator in eq. (20) contains a factor m^2; but the case $m = 1$ yields a surprise. With $m = 1$ eq. (20) simplifies considerably because many terms cancel:

$$q_1 = \frac{(1 - \nu^2)\,\lambda^2 + k\,\lambda^4(\lambda^2 + 2(2 - \nu))}{\lambda^2 - 1}. \tag{21}$$

For $\lambda > 1$ this yields values of q_1 which are larger than those obtained with $m = 1$ and therefore there is no arc $m = 1$ in the festoon curves of fig. 15. But for $\lambda < 1$, i. e. for $l > \pi\,a$, eq. (21) yields a negative q_1, corresponding to an *internal* pressure in the shell. If λ is small enough we may neglect λ^2 compared with 1 and have

$$q_1 = -(1 - \nu^2)\,\lambda^2 - k\,\lambda^6 \approx -(1 - \nu^2)\,\lambda^2,$$

i. e.

$$p = -\pi^2\,E\,\frac{a\,t}{l^2}. \tag{22}$$

This buckling of a shell whose basic stress system consists of nothing but a tensile hoop force $N_{\phi I}$ appears at first sight rather improbable. It can, however, be easily explained.

Let us assume that the shell is part of a pressure vessel with very rigid bulkheads. Since we assumed $N_{x I} \equiv 0$, we must connect the centers of these bulkheads by a rod with a turnbuckle and adjust it so that it just transmits the force acting on either bulkhead, keeping the shell entirely free of axial load (fig. 16). To simplify the following formulas, we shall denote here by p the *internal* pressure, in contradiction to the sign convention underlying eq. (22).

We now make some simplifying assumptions for the buckling displacements. Since $m = 1$, the overall deformation of the cylinder is

---

[1] TIMOSHENKO, S.: Theory of Elastic Stability, New York 1936, p. 218.

that of a tubular bar bent into a sinusoidal shape. We assume that the deformation pattern is exactly that of elementary beam theory, i. e., that the circular sections are not deformed but only shifted laterally and rotated to such a degree that they remain normal to the deformed center line of the tube.

The shape of the cross section is preserved when we put $B = -C$ in eqs. (8). The conservation of normals yields, as in elementary beam theory, the relation

$$-u = \frac{\partial w}{\partial x} a,$$

valid for $\phi = 0$. Introducing this into eqs. (8), we find that

$$A = -\frac{\pi a}{l} C.$$

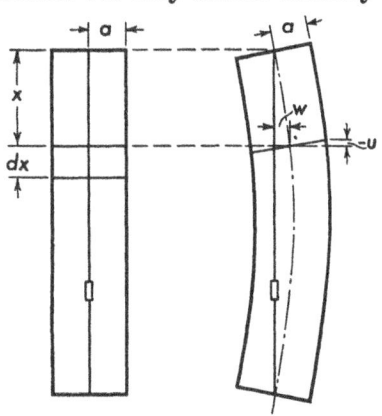

Fig. 16. Buckling of a cylindrical shell under internal pressure

We now cut a ring-shaped element of length dx from the shell. During buckling its right half is stretched while its left half gets shorter. Therefore, the right half picks up more of the pressure p than the left half does, and a resultant force acts on the ring, pushing it toward the right. We may easily figure the magnitude of this force.

We consider an arbitrary element $dx \cdot a\, d\phi$ on this ring (fig. 17). Its strain is

$$\epsilon_x = \frac{\partial u}{\partial x} = -A \frac{\lambda}{a} \cos\phi \sin\frac{\lambda x}{a}$$

$$= \frac{\pi^2 a}{l^2} C \cos\phi \sin\frac{\lambda x}{a},$$

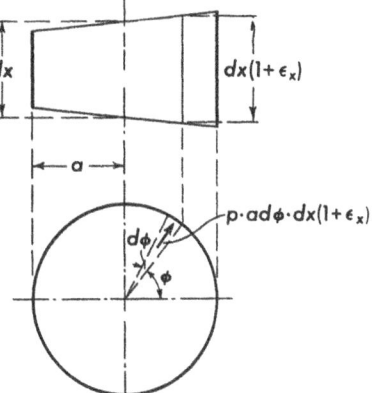

Fig. 17
Deformed element of the tube of Fig. 16b

and the force acting on it is indicated in the figure. We are interested in its component in the direction of the horizontal diameter, which is $p\, a\, d\phi\, dx (1 + \epsilon_x) \cos\phi$. When we integrate this component over the entire circumference of the ring, we obtain the resultant of the pressure p on the ring:

$$p\, a\, dx \int_0^{2\pi} (1 + \epsilon_x) \cos\phi\, d\phi = \frac{p\, \pi^3 a^2\, dx}{l^2} C \sin\frac{\lambda x}{a}.$$

Except for the factor dx, this is the lateral load per unit length of the deflected tube. We can now compute axial stresses σ_x in two different ways: Either we compute the bending moment M which the lateral load produces in a bar of span l, and from it the bending stress in a circular cross section of radius a and thickness t; or we start from the strain ϵ_x and apply Hooke's law. When both results agree, the deflected tube is in equilibrium, and the shell is ready to buckle.

Simple beam analysis yields

$$\mathsf{M} = p\,\pi\,a^2\,C\sin\frac{\lambda\,x}{a},$$

and then

$$\sigma_x = \frac{p}{t}\,C\cos\phi\sin\frac{\lambda\,x}{a}.$$

On the other hand, Hooke's law yields

$$\sigma_x = E\,\epsilon_x = E\,\frac{\pi^2 a}{l^2}\,C\cos\phi\sin\frac{\lambda\,x}{a}.$$

Both results agree if

$$E\,\frac{\pi^2 a}{l^2} = \frac{p}{t},$$

and this is identical with the buckling condition (22), except for the missing minus sign, caused by the changed sign convention for p.

This analysis' explains in one way the strange buckling phenomenon of a shell with internal pressure. Another explanation may be found when the tube plus the air in it is considered as a built-up Euler column subjected to an axial compressive force $p \cdot \pi\,a^2$, exerted by the tie rod. The air carries the axial force and is the destabilizing element, while the tube has the bending rigidity $E\,I$ and is the stabilizing element.

7.2.3 Solution for Shells with Shear Load

7.2.3.1 Torsion of a Long Tube

The solution (8) of the differential equations (7) is possible only because the dash and dot derivatives in (7) are distributed according to a simple pattern. The q_3 terms do not fit into this pattern, and as soon as $q_3 \neq 0$, the expressions (8) no longer satisfy the differential equations. However, there exists a solution which is applicable even in this case:

$$\left.\begin{aligned}
u &= A\sin\left(\frac{\lambda\,x}{a} + m\,\phi\right),\\[4pt]
v &= B\sin\left(\frac{\lambda\,x}{a} + m\,\phi\right),\\[4pt]
w &= C\cos\left(\frac{\lambda\,x}{a} + m\,\phi\right).
\end{aligned}\right\} \tag{23}$$

It may be used for any combination of loads p, P, and T. The zeros of u, v, w and of their derivatives are found on lines $m\phi + \lambda\, x/a = \text{const}$, winding around the cylinder (fig. 18). It is, therefore, not possible to satisfy reasonable boundary conditions on lines $x = \text{const}$, and the solution (23) cannot be used to deal with cylinders of finite length. We shall use it here to study the buckling of an infinitely long cylinder subjected to a shear load T only (fig. 18).

When we introduce the expressions (23) for u, v, w into eqs. (7), we again obtain a set of linear equations for A, B, C:

$$
\begin{aligned}
&A\left[\lambda^2 + \frac{1-\nu}{2}\, m^2(1+k) - 2q_3\lambda m\right] + B\left[\frac{1+\nu}{2}\,\lambda m\right] \\
&\qquad + C\left[\nu\lambda + k\left(\lambda^3 - \frac{1-\nu}{2}\,\lambda m^2\right)\right] = 0, \\
&A\left[\frac{1+\nu}{2}\,\lambda m\right] + B\left[m^2 + \frac{1-\nu}{2}\,\lambda^2(1+3k) - 2q_3\lambda m\right] \\
&\qquad + C\left[m + \frac{3-\nu}{2}\, k\lambda^2 m - 2q_3\lambda\right] = 0, \\
&A\left[\nu\lambda + k\left(\lambda^3 - \frac{1-\nu}{2}\,\lambda m^2\right)\right] + B\left[m + \frac{3-\nu}{2}\, k\lambda^2 m - 2q_3\lambda\right] \\
&\qquad + C\left[1 + k(\lambda^4 + 2\lambda^2 m^2 + m^4 - 2m^2 + 1) - 2q_3\lambda m\right] = 0.
\end{aligned}
\tag{24}
$$

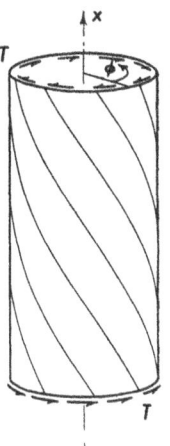

These equations are similar to eqs. (10), but besides the absence of the q_1 and q_2 terms and the presence of the q_3 terms, they differ also in some minus signs. In the same way as we did there, we conclude here that the coefficient determinant must vanish, and thus we find the condition of neutral equilibrium:

$$
c_1 + c_2\, k = c_5\, q_3
\tag{25}
$$

which corresponds to eq. (11). The quantities c_1 and c_2 are the same as given in eqs. (12), and

$$
c_5 = 2\lambda m(\lambda^2 + m^2)(\lambda^2 + m^2 - 1).
\tag{26}
$$

We see at once that neither λ nor m can be zero, because in both cases $c_5 = 0$ and hence $q_3 = \infty$. It is also without interest to consider negative values of λ or m. When both are negative, nothing is changed in eq. (25). When either λ or m alone is negative, the buckling mode (23) is altered insofar as the nodal lines (fig. 18) become right handed screws. One would

Fig. 18. Nodal lines for w on a cylinder buckling according to eqs. (23)

expect that then the buckling load T must be applied in the opposite sense, and this is exactly what happens. In eq. (25) the left-hand side remains the same while c_5 changes sign and hence q_3.

The discussion of the buckling formulas (25) is now restricted to positive values λ and to positive integers m. One might think of solving for q_3, differentiating the expression with respect to λ and m, and putting the first partial derivatives equal to zero. This would yield two algebraic equations for λ and m, and their solution (or one of them) would lead to the smallest possible q_3. This procedure, however, is rather tiresome and may be avoided. By some trial computations one may find out that any $m > 2$ yields a higher buckling load than does $m = 2$ and that λ must be chosen rather small, $\lambda \ll 1$, to obtain a low q_3.

With this in mind, we now investigate separately the two cases $m = 1$ and $m = 2$.

For $m = 2$ eq. (25) reads

$$q_3 = \frac{(1 - \nu^2)\,\lambda^4 + k[\lambda^8 + 2(8 - \nu)\,\lambda^6 + 72\,\lambda^4 + 24\,(6 + \nu)\,\lambda^2 + 144]}{4\,\lambda(\lambda^2 + 4)\,(\lambda^2 + 3)},$$

and when we neglect everywhere λ^2 compared with 1, we have

$$q_3 = \frac{1 - \nu^2}{48}\,\lambda^3 + \frac{3\,k}{\lambda}.$$

Now it is easy to find from

$$\frac{\partial q_3}{\partial \lambda} = \frac{1 - \nu^2}{16}\,\lambda^2 - \frac{3\,k}{\lambda^2} = 0$$

that

$$\lambda^4 = \frac{48\,k}{1 - \nu^2}$$

yields the lowest possible value of q_3, which is

$$q_3 = 2\,k\,\sqrt[4]{\frac{1 - \nu^2}{3\,k}} = \frac{\sqrt[4]{1 - \nu^2}}{3\,\sqrt{2}}\left(\frac{t}{a}\right)^{3/2}. \tag{27a}$$

Through the last of eqs. (6) we may now return to the real shear load T and find its critical value

$$T_{\mathrm{cr}} = \frac{1}{3\,\sqrt{2}\,(1 - \nu^2)^{3/4}}\,\frac{E\,t^{5/2}}{a^{3/2}}, \tag{27b}$$

and we may compute the total torque $M = T \cdot 2\pi\,a \cdot a$ applied to the tube. Its critical value is

$$M_{\mathrm{cr}} = \frac{\pi\,\sqrt{2}}{3\,(1 - \nu^2)^{3/4}}\,E\,\sqrt{t^5\,a}. \tag{27c}$$

All these results have been derived for an infinitely long cylinder. Since they do not contain any wavelength, there is the temptation of applying them to cylinders of finite length. However, such a cylinder usually has some kind of stiffening at the end, say a bulkhead requiring $w = 0$. Any such condition is in contradiction to eqs. (23), and the additional constraint imposed by the bulkhead will increase the buckling load beyond the one given by the preceding formulas. One may expect that the difference is not much if the cylinder is rather long.

With $m = 1$ eq. (25) reads

$$q_3 = \frac{(1 - \nu^2)\,\lambda + k\,\lambda^3[\lambda^2 + 2(2 - \nu)]}{2\,(\lambda^2 + 1)},$$

and when we now neglect λ^2 compared with unity, we may drop the k term entirely and obtain

$$q_3 = \frac{1 - \nu^2}{2}\,\lambda. \tag{28}$$

If we can choose λ arbitrarily, we may choose it as small as we like and thus make q_3 approach zero. This shows that there is no finite buckling load for the infinite shell unless we prevent the buckling mode with $m = 1$. In this mode the axis of the tube is deformed to a steep helical curve, while the circular cross sections remain circular and normal to the deformed axis. Since every such cross section rotates about one of its diameters, this mode may be excluded by preventing such a rotation of the terminal cross sections of a long cylinder. On the other hand, one may think up experimental arrangements in which the length $2\pi\,a/\lambda$ of one turn of the helical axis is related in a definite way to the length l of the cylinder. If eq. (28) with $\lambda = 2\pi\,a/l$ yields a lower value q_3 than eq. (27a), the latter one should not be trusted, but how far these formulas really can be applied cannot be decided without investigating the buckling problem of the cylinder of finite length.

7.2.3.2 Shear and Axial Compression in a Cylinder of Finite Length

7.2.3.2.1 General Theory. When we want to solve the buckling problem for a shell of finite length l, we must resort to a FOURIER series solution. We consider the following expressions for the displacements:

$$u = \cos m\phi \sum_{n=1,3,\ldots}^{\infty} A_{1n} \cos\frac{n\pi x}{l} + \sin m\phi \sum_{n=0,2,\ldots}^{\infty} A_{2n} \cos\frac{n\pi x}{l},$$

$$v = \sin m\phi \sum_{n=1,3,\ldots}^{\infty} B_{1n} \sin\frac{n\pi x}{l} - \cos m\phi \sum_{n=2,4,\ldots}^{\infty} B_{2n} \sin\frac{n\pi x}{l},$$

$$w = \cos m\phi \sum_{n=1,3,\ldots}^{\infty} C_{1n} \sin\frac{n\pi x}{l} + \sin m\phi \sum_{n=2,4,\ldots}^{\infty} C_{2n} \sin\frac{n\pi x}{l}.$$

$$\tag{29a--c}$$

The restriction of n to odd or to even values appears here to be rather arbitrary. It will soon be seen that it is reasonable and advisable.

Before we introduce the solution (29) into the differential equations (7), we must discuss the boundary conditions which it fulfills at the edges $x = 0$ and $x = l$. Evidently we have there $v \equiv w \equiv 0$, and also all the derivatives appearing in eq. (V-8f) are zero, so that $M_x \equiv 0$. This means that the edges are supported in radial and tangential directions and that the shell is free to rotate about a tangent to the edge. These are exactly the boundary conditions one would want to prescribe for a hinged (simple) support, and they are equally applicable if the shell under consideration is part of an infinitely long tube stiffened by rigid diaphragms spaced at regular intervals of length l.

The fourth boundary condition needs closer inspection. From eq. (29a) we see that u does not vanish, i. e. that the edge is not supported in axial direction. The question is whether the displacement u is entirely unconstrained. When we check N_x from eq. (V-8b) we see that it vanishes. But there is still the large force $N_{x\phi I} = -T$ at the edges, and if a line element rotates by an angle u'/a, then this force develops an axial component $N_{x\phi I} u'/a$. This external force depends on the angle u'/a, not on the displacement u, and this situation is typical of a nonconservative system in which a problem of static stability does not make sense[1]. It therefore would not be legitimate to postulate $N_x = 0$ as a fourth boundary condition on an edge $x = \text{const}$, but the condition should read

$$N_x - N_{x\phi I} \frac{u^{\cdot}}{a} = 0.$$

Our solution (29) evidently does not satisfy this condition. But when we consider the cylinder of length l as part of a much longer cylinder stiffened by diaphragms, then the force $N_{x\phi I}$ at the edge $x = 0$ comes from the adjoining part of the shell, and there is no external force at all. Our boundary condition is then simply that there is no support in the direction of the displacement u, and this condition is satisfied by eqs. (29). It is for this boundary condition that we now shall solve the buckling problem.

When we introduce u, v, w from eqs. (29) into the differential equations (7), each of them will consist of two terms, one with $\cos m\,\phi$, the other with $\sin m\,\phi$. Each of them must vanish, and this yields

---

[1] ZIEGLER, H.: Stabilitätsprobleme bei geraden Wellen. Z. ang. Math. Phys., vol. 2 (1951), 265–289.

six equations:

$$\sum_{n=1,3,\ldots}^{\infty} \{a_{11,n} A_{1n} + a_{12,n} B_{1n} + a_{13,n} C_{1n}\} \cos\frac{n\pi x}{l}$$

$$-\frac{2\pi a}{l} q_3 \sum_{n=2,4,\ldots}^{\infty} n\, m\, A_{2n} \sin\frac{n\pi x}{l} = 0,$$

$$\sum_{n=0,2,\ldots}^{\infty} \{a_{11,n} A_{2n} + a_{12,n} B_{2n} + a_{13,n} C_{2n}\} \cos\frac{n\pi x}{l}$$

$$+\frac{2\pi a}{l} q_3 \sum_{n=1,3,\ldots}^{\infty} \cdot\; n\, m\, A_{1n} \sin\frac{n\pi x}{l} = 0,$$

$$\sum_{n=1,3,\ldots}^{\infty} \{a_{12,n} A_{1n} + a_{22,n} B_{1n} + a_{23,n} C_{1n}\} \sin\frac{n\pi x}{l}$$

$$+\frac{2\pi a}{l} q_3 \sum_{n=2,4,\ldots}^{\infty} (n\, m\, B_{2n} + n\, C_{2n}) \cos\frac{n\pi x}{l} = 0,$$

$$\sum_{n=2,4,\ldots}^{\infty} \{a_{12,n} A_{2n} + a_{22,n} B_{2n} + a_{23,n} C_{2n}\} \sin\frac{n\pi x}{l}$$

$$-\frac{2\pi a}{l} q_3 \sum_{n=1,3,\ldots}^{\infty} (n\, m\, B_{1n} + n\, C_{1n}) \cos\frac{n\pi x}{l} = 0,$$

$$\sum_{n=1,3,\ldots}^{\infty} \{a_{13,n} A_{1n} + a_{23,n} B_{1n} + a_{33,n} C_{1n}\} \sin\frac{n\pi x}{l}$$

$$+\frac{2\pi a}{l} q_3 \sum_{n=2,4,\ldots}^{\infty} (n\, B_{2n} + n\, m\, C_{2n}) \cos\frac{n\pi x}{l} = 0,$$

$$\sum_{n=2,4,\ldots}^{\infty} \{a_{13,n} A_{2n} + a_{23,n} B_{2n} + a_{33,n} C_{2n}\} \sin\frac{n\pi x}{l}$$

$$-\frac{2\pi a}{l} q_3 \sum_{n=1,3,\ldots}^{\infty} (n\, B_{1n} + n\, m\, C_{1n}) \cos\frac{n\pi x}{l} = 0.$$

$$(30)$$

The coefficients $a_{11,n} \ldots a_{33,n}$ in the braces are identical with the bracketed coefficients of A, B, C in eqs. (10).

The sines and cosines of $n\pi x/l$ which appear in these equations are shown in fig. 19. If we were to extend them over the double range, they would (together with $\cos 0 \equiv 1$) be a complete set of orthogonal functions. We are, however, interested only in the domain l as shown, and there the cosines may be expanded in terms of the sines, and vice versa:

$$\cos\frac{n\pi x}{l} = \frac{4}{\pi} \sum_j \frac{j}{j^2 - n^2} \sin\frac{j\pi x}{l},$$

$$\sin\frac{n\pi x}{l} = \frac{4n}{\pi} \sum_j \frac{1}{n^2 - j^2} \cos\frac{j\pi x}{l}, \qquad n = \text{even},$$

$$\sin\frac{n\pi x}{l} = \frac{2}{n\pi} + \frac{4n}{\pi} \sum_j \frac{1}{n^2 - j^2} \cos\frac{j\pi x}{l}, \qquad n = \text{odd}.$$

$$(31\,\text{a--c})$$

In these formulas the subscript j must run either through all odd integers or through all even integers, so that for the expansion of a symmetric function only symmetric functions are used, i. e.

if $n = 2, 4, 6, \ldots$, then $j = 1, 3, 5, \ldots$,
if $n = 1, 3, 5, \ldots$, then $j = 2, 4, 6, \ldots$.

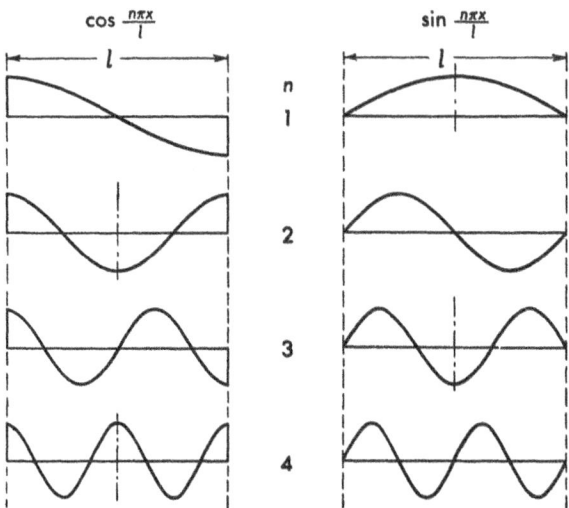

Fig. 19. Cosines and sines appearing in eqs. (30)

We use eqs. (31) to transform the second sum in each of the preceding six equations. This may be shown for the first one:

$$\sum_{n=2,4,\ldots}^{\infty} n\, m\, A_{2n} \sin\frac{n\pi x}{l} = \frac{4m}{\pi} \sum_{n=2,4,\ldots}^{\infty} n^2 A_{2n} \sum_{j=1,3,\ldots}^{\infty} \frac{1}{n^2-j^2} \cos\frac{j\pi x}{l}$$

$$= \frac{4m}{\pi} \sum_{j=1,3,\ldots}^{\infty} \sum_{n=2,4,\ldots}^{\infty} \frac{n^2}{n^2-j^2} A_{2n} \cos\frac{j\pi x}{l}$$

$$= \frac{4m}{\pi} \sum_{n=1,3,\ldots}^{\infty} \sum_{j=2,4,\ldots}^{\infty} \frac{j^2}{j^2-n^2} A_{2j} \cos\frac{n\pi x}{l}.$$

In this way we arrive at a summation over the same functions and over the same choice of n which occurs in the first term of each equation, and we may combine them under one summation sign, e. g.

$$\sum_{n=1,3,\ldots}^{\infty} \left\{ a_{11,n} A_{1n} + a_{12,n} B_{1n} + a_{13,n} C_{1n} \right.$$

$$\left. - \frac{8aq_3}{l} m \sum_{j=2,4,\ldots}^{\infty} \frac{j^2}{j^2-n^2} A_{2j} \right\} \cos\frac{n\pi x}{l} = 0,$$

and similarly for the other five equations. Now we may conclude that such a FOURIER series cannot vanish identically unless each of its coefficients, i. e. the expression in the braces, vanishes. This yields for every n six equations, the first two of which are these:

$$a_{11,n} A_{1n} + a_{12,n} B_{1n} + a_{13,n} C_{1n} - \frac{8 a q_3}{l} m \sum_{j=2,4,\ldots}^{\infty} \frac{j^2}{j^2 - n^2} A_{2j} = 0,$$
$$n = 1, 3, \ldots$$

$$a_{11,n} A_{2n} + a_{12,n} B_{2n} + a_{13,n} C_{2n} + \frac{8 a q_3}{l} m \sum_{j=1,3,\ldots}^{\infty} \frac{j^2}{j^2 - n^2} A_{1j} = 0,$$
$$n = 2, 4, \ldots$$

For $n = 0$ this last equation is irregular:

$$a_{11,0} A_{20} + \frac{4 a q_3}{l} m \sum A_{1j} = 0.$$

Since this is the only equation in the whole set which contains A_{20}, it may be set aside for calculating this unknown, and it is not necessary to carry it in the following discussion of the other equations.

It may be seen from eqs. (29) that the coefficients A_{1n}, B_{1n}, C_{1n} exist only for odd values of n and that the coefficients A_{2n}, B_{2n}, C_{2n} exist only for even n. We may therefore omit the subscripts 1 and 2 and combine the last two equations in a single one:

$$a_{11,n} A_n + a_{12,n} B_n + a_{13,n} C_n + (-1)^n \frac{8 a q_3}{l} m \sum_j \frac{j^2}{j^2 - n^2} A_j = 0. \quad (32\,\mathrm{a})$$

This equation may now be written for every integer $n = 1, 2, 3, \ldots$, and in every such equation the summation index j must run through either even or odd integers only, so that $n + j$ is always an odd number.

Exactly in the same way the remaining four equations can be contracted to two for every n:

$$a_{12,n} A_n + a_{22,n} B_n + a_{23,n} C_n$$
$$+ (-1)^n \frac{8 a q_3}{l} \sum_j \frac{nj}{j^2 - n^2} (m B_j + C_j) = 0,$$

$$a_{13,n} A_n + a_{23,n} B_n + a_{33,n} C_n \quad\quad (32\,\mathrm{b, c})$$
$$+ (-1)^n \frac{8 a q_3}{l} \sum_j \frac{nj}{j^2 - n^2} (B_j + m C_j) = 0.$$

Since there are three equations (32) for every integer n, we have here an infinite set of linear equations for an infinite number of unknowns A_n, B_n, C_n. Before we discuss the solution of this set, we reduce it to one third its size by eliminating all unknowns A_n and B_n. This is done in the following way:

We pick one pair of equations (32a, b) and solve for A_n, B_n:

$$
\left.
\begin{aligned}
A_n &= \frac{-1}{a_{11,n}\,a_{22,n} - a_{12,n}^2} \left\{ (a_{13,n}\,a_{22,n} - a_{23,n}\,a_{12,n})\,C_n \right. \\
&\quad + (-1)^n \frac{8a\,q_3}{l} \left[a_{22,n}\,m \sum_j \frac{j^2}{j^2 - n^2}\,A_j \right. \\
&\quad \left. \left. - a_{12,n}\,m \sum_j \frac{nj}{j^2 - n^2}\,B_j - a_{12,n}\,m \sum_j \frac{nj}{j^2 - n^2}\,C_j \right] \right\}, \\[6pt]
B_n &= \frac{-1}{a_{11,n}\,a_{22,n} - a_{12,n}^2} \left\{ (a_{11,n}\,a_{23,n} - a_{12,n}\,a_{13,n})\,C_n \right. \\
&\quad + (-1)^n \frac{8a\,q_3}{l} \left[-a_{12,n}\,m \sum_j \frac{j^2}{j^2 - n^2}\,A_j \right. \\
&\quad \left. \left. + a_{11,n}\,m \sum_j \frac{nj}{j^2 - n^2}\,B_j + a_{11,n}\,m \sum_j \frac{nj}{j^2 - n^2}\,C_j \right] \right\}.
\end{aligned}
\right\} \quad (33)
$$

This does not look much like a success since there are also unknowns A_j, B_j on the right-hand side, though not A_n and B_n, because of $j \neq n$. However, we can use eqs. (33) for an iteration which aims at obtaining A_n and B_n in terms of all the C_j in the form of power series in q_3. We have already seen on p. 422 that it does not make sense to keep squares and higher powers of q_1 and q_2, and the same applies to q_3. Therefore, those power series may be stopped after the linear term.

We start the iteration by dropping the q_3 terms from eqs. (33), obtaining

$$
A_n = -\frac{a_{13,n}\,a_{22,n} - a_{23,n}\,a_{12,n}}{a_{11,n}\,a_{22,n} - a_{12,n}^2}\,C_n
$$

and a similar expression for B_n. These are now introduced on the right-hand side of eqs. (33):

$$
\begin{aligned}
A_n &= \frac{-1}{a_{11,n}\,a_{22,n} - a_{12,n}^2} \left\{ (a_{13,n}\,a_{22,n} - a_{23,n}\,a_{12,n})\,C_n \right. \\
&\quad + (-1)^n \frac{8a\,q_3}{l} \left[a_{22,n}\,m \sum_j \frac{j^2}{j^2 - n^2} \frac{a_{12,j}\,a_{23,j} - a_{22,j}\,a_{13,j}}{a_{11,j}\,a_{22,j} - a_{12,j}^2}\,C_j \right. \\
&\quad \left. \left. + a_{12,n}\,m \sum_j \frac{nj}{j^2 - n^2} \frac{a_{11,j}\,a_{23,j} - a_{12,j}\,a_{13,j}}{a_{11,j}\,a_{22,j} - a_{12,j}^2}\,C_j - a_{12,n}\,m \sum_j \frac{nj}{j^2 - n^2}\,C_j \right] \right\}, \\[6pt]
B_n &= \frac{-1}{a_{11,n}\,a_{22,n} - a_{12,n}^2} \left\{ (a_{11,n}\,a_{23,n} - a_{12,n}\,a_{13,n})\,C_n \right. \\
&\quad + (-1)^n \frac{8a\,q_3}{l} \left[-a_{12,n}\,m \sum_j \frac{j^2}{j^2 - n^2} \frac{a_{12,j}\,a_{23,j} - a_{22,j}\,a_{13,j}}{a_{11,j}\,a_{22,j} - a_{12,j}^2}\,C_j \right. \\
&\quad \left. \left. - a_{11,n}\,m \sum_j \frac{nj}{j^2 - n^2} \frac{a_{11,j}\,a_{23,j} - a_{12,j}\,a_{13,j}}{a_{11,j}\,a_{22,j} - a_{12,j}^2}\,C_j + a_{11,n}\,m \sum_j \frac{nj}{j^2 - n^2}\,C_j \right] \right\}.
\end{aligned}
$$

These relations are final, because upon their introduction into eqs. (33) only a term with $q_3{}^2$ will be added, which we decided to neglect. We may now introduce these expressions into eq. (32c), again neglecting higher than the first powers of q_3. This yields

$$
\begin{vmatrix} a_{11,n} & a_{12,n} & a_{13,n} \\ a_{12,n} & a_{22,n} & a_{23,n} \\ a_{13,n} & a_{23,n} & a_{33,n} \end{vmatrix} C_n
$$

$$
+ (-1)^n \frac{8 a q_3}{l} \Bigg\{ \begin{vmatrix} a_{12,n} & a_{13,n} \\ a_{22,n} & a_{23,n} \end{vmatrix} m \sum_j \frac{j^2}{j^2 - n^2} \frac{a_{12,j} a_{23,j} - a_{22,j} a_{13,j}}{a_{11,j} a_{22,j} - a_{12,j}^2} C_j
$$

$$
+ \begin{vmatrix} a_{11,n} & a_{13,n} \\ a_{12,n} & a_{23,n} \end{vmatrix} m \sum_j \frac{nj}{j^2 - n^2} \frac{a_{11,j} a_{23,j} - a_{12,j} a_{13,j}}{a_{11,j} a_{22,j} - a_{12,j}^2} C_j
$$

$$
+ \begin{vmatrix} a_{11,n} & a_{12,n} \\ a_{12,n} & a_{22,n} \end{vmatrix} m \sum_j \frac{nj}{j^2 - n^2} C_j - \begin{vmatrix} a_{11,n} & a_{13,n} \\ a_{12,n} & a_{23,n} \end{vmatrix} \sum_j \frac{nj}{j^2 - n^2} C_j
$$

$$
- \begin{vmatrix} a_{11,n} & a_{12,n} \\ a_{12,n} & a_{22,n} \end{vmatrix} \sum_j \frac{nj}{j^2 - n^2} \frac{a_{11,j} a_{23,j} - a_{12,j} a_{13,j}}{a_{11,j} a_{22,j} - a_{12,j}^2} C_j \Bigg\} = 0. \tag{34}
$$

Here, as in all preceding equations, the subscript j in the sums is subject to the restriction that $n + j$ must be an odd integer.

The coefficient of C_n in eq. (34) is exactly the coefficient determinant of eqs. (10) and may be written as

$$
\frac{1 - \nu}{2} (c_1 + c_2 k - c_3 q_1 - c_4 q_2) \tag{35}
$$

with the notation used in eqs. (11) and (12). The factor $(1 - \nu)/2$ appearing here was dropped in eq. (12) as a common factor to all terms of eq. (11), but here it must, of course, be carried.

Eqs. (34), one for every integer $n = 1, 2, \ldots$, are an infinite set of linear equations for the unknowns C_n. Under certain conditions, it may be shown that a non-trivial solution exists, if the infinite determinant of coefficients vanishes[1]. If this is the case, $C_n \neq 0$ for every n, but some of the C_n are larger than the other ones and dominate in eqs. (29).

The vanishing of the determinant of eqs. (34) yields a transcendental equation for the critical value of the load parameter q_3. It has an infinite number of solutions, among which only the smallest is of interest. It still depends on the choice of the wave number m which appears visibly in eq. (34) and is also hidden in the quantities $a_{11,n}, \ldots, a_{33,n}$. As in the case of compressive forces, it is necessary to try different integers m and to find the one which makes q_3 a minimum.

---

[1] A survey of the mathematical problem with many references is found in Enzyklopädie Math. Wiss., Vol. II, 3, p. 1417, Leipzig 1927.

In the general case thus far considered it is very difficult to evaluate the buckling condition numerically. We may expect to simplify the problem if we assume that the cylinder is either very long or very short. In the next two sections we shall study these special cases.

7.2.3.2.2 Long Cylinder. We assume now that $l \gg a$. If the cylinder is very long and if the shear load predominates ($q_3 \gg q_1, q_2$), we must expect that the constants C_n for low values of n predominate. We may then safely neglect $n\pi a/l$, that is, λ, compared with unity wherever the opportunity occurs.

When we do so in eqs. (12) and introduce the simplified expressions into (35), we have

$$\frac{1-\nu}{2}\left[(1-\nu^2)\left(\frac{n\pi a}{l}\right)^4\right.$$
$$\left. + k\,m^4(m^2-1)^2 - q_1\,m^4(m^2-1) - q_2\,m^2(m^2+2)\left(\frac{n\pi a}{l}\right)^2\right]$$

as the coefficient of C_n.

It may be left to the reader to work out the details of the q_3 term in eq. (34). He will find that the first term inside the braces is negligible because of a factor $(\pi a/l)^2$, that the second and fourth terms cancel, and that the third and fifth terms combine to

$$\{\,\} = \left(\frac{1-\nu}{2} - \frac{3-\nu}{2}\,q_1 - q_2\right)m^3(m^2-1)\sum_j \frac{nj}{j^2-n^2}\,C_j.$$

Since this is to be multiplied by q_3, the terms with q_1 and q_2 must be dropped, because they would yield quadratic terms in the small load parameters. Eq. (34) may then be written as

$$\left[(1-\nu^2)\left(\frac{n\pi a}{l}\right)^4 + k\,m^4(m^2-1)^2 - q_1\,m^4(m^2-1) - q_2\,m^2(m^2+2)\left(\frac{n\pi a}{l}\right)^2\right]C_n$$
$$+ (-1)^n\,\frac{8a\,q_3}{l}\,n\,m^3(m^2-1)\sum_j \frac{j}{j^2-n^2}\,C_j = 0. \qquad (36)$$

If we use $n\,C_n$, $j\,C_j$ as unknowns and divide each equation by $(1-\nu^2)\,n^3\,(\pi a/l)^4$, the determinant of the coefficients may be written as

$$\begin{vmatrix} 1+\epsilon_{11} & \epsilon_{12} & \epsilon_{13} & \cdots \\ \epsilon_{21} & 1+\epsilon_{22} & \epsilon_{23} & \cdots \\ \epsilon_{31} & \epsilon_{32} & 1+\epsilon_{33} & \cdots \\ \vdots & \vdots & \vdots & \end{vmatrix}$$

and $\displaystyle\sum_{n,j} \epsilon_{nj}$ is absolutely convergent. This is a sufficient condition for the convergence of the infinite determinant, and we may set it equal to zero and use this equation to find the dimensionless buckling load q_3.

For actual computations one may, of course, use the coefficients as they stand in eq. (36).

If we were to start our numerical work immediately from eq. (36), we would obtain q_3 depending on the form parameters k and a/l, on POISSON's ratio ν, and on the load parameters q_1 and q_2. These are too many to allow presentation of the result in a graph.

To make the solution more manageable, we first drop one load parameter, assuming $q_1 = 0$. We are then dealing with a shell which is subjected only to an axial thrust P and a shear T [see eqs. (6)]. For these two we define new dimensionless parameters:

$$\varrho_2 = \sqrt{12(1 - \nu^2)}\, \frac{P\,a}{E\,t^2}, \qquad \varrho_3 = \frac{8[12(1 - \nu^2)]^{5/8}}{\pi^{3/2}} \frac{T}{E\,t} \sqrt[4]{\frac{a^3\,l^2}{t^5}}, \qquad (37)$$

and we introduce

$$\tilde{\omega}^4 = \frac{1}{12(1 - \nu^2)\,\pi^4} \frac{l^4\,t^2}{a^6} \qquad (38)$$

as a form parameter. When we multiply eq. (36) by a suitable factor, it finally assumes the following form:

$$\frac{1}{m\sqrt{\tilde{\omega}}} \left[\frac{n^2}{\tilde{\omega}^2\,m^2(m^2 - 1)} + \frac{\tilde{\omega}^2\,m^2(m^2 - 1)}{n^2} - \varrho_2 \frac{m^2 + 2}{m^2 - 1} \right] n\,C_n$$
$$+ (-1)^n\,\varrho_3 \sum_j \frac{j\,C_j}{j^2 - n^2} = 0. \qquad (39)$$

Such an equation may be written for every integer n.

If we denote the coefficient of $n\,C_n$ by T_n, the first four of these equations are:

$$\varrho_3^{-1} T_1 \cdot C_1 - \quad \frac{1}{3} \cdot 2 C_2 \qquad\qquad -\frac{1}{15} \cdot 4 C_4 \qquad\qquad \ldots = 0,$$

$$-\frac{1}{3} \cdot C_1 + \varrho_3^{-1} T_2 \cdot 2 C_2 + \quad \frac{1}{5} \cdot 3 C_3 \qquad\qquad +\frac{1}{21} \cdot 5 C_5 \ldots = 0,$$

$$\frac{1}{5} \cdot 2 C_2 + \varrho_3^{-1} T_3 \cdot 3 C_3 - \quad \frac{1}{7} \cdot 4 C_4 \qquad\qquad \ldots = 0,$$

$$-\frac{1}{15} \cdot C_1 \qquad\qquad -\frac{1}{7} \cdot 3 C_3 + \varrho_3^{-1} T_4 \cdot 4 C_4 + \frac{1}{9} \cdot 5 C_5 \ldots = 0.$$

This is an infinite system of homogeneous linear equations for the unknowns $n\,C_n$, and since its coefficients meet certain requirements, we conclude that the determinant of these coefficients must vanish. This determinant may be approximated by segments of increasing size as indicated by the dotted lines in fig. 20. In this figure the double lines indicate the diagonal coefficients $\varrho_3^{-1}\,T_n$, the single lines the other nonvanishing coefficients, and the dots the zero coefficients. If we start with a third order determinant we obtain a third degree equation for ϱ_3. Proceeding to determinants of higher order, we obtain algebraic equations of higher degree, from which we find additional solutions ϱ_3 and improved values for those already obtained. This process may

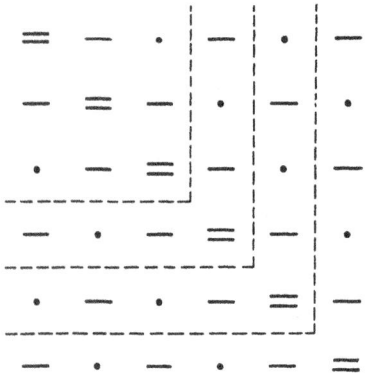

Fig. 20.   Schematic representation of the
buckling determinant

be continued until the smallest ϱ_3 is stabilized at a certain value, which then determines the buckling shear of the shell.

There is an easier way to evaluating the lowest eigenvalue ϱ_3 of this problem, using matrix iteration. The details of this method may be found in the literature on matrices[1]. It is first applied to a segment of moderate size, say to the 4×4 matrix indicated in fig. 20, and with the modal column so obtained a larger segment may easily be treated. Practical computations have

shown that the 4×4 segment gives good results that are only slightly improved when a fifth row and column are added to the matrix.

The diagonal coefficients T_n depend not only on the shape of the shell ($\tilde{\omega}$) and on the axial tension or compression (ϱ_2) but also on the choice of the wave number m. It is necessary to repeat the entire

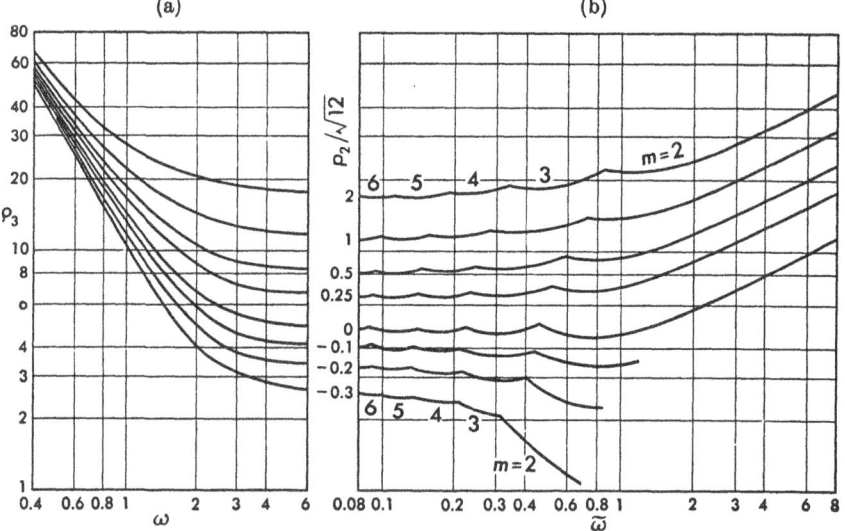

Fig. 21. Buckling diagram for combined shear and axial compression

---

[1] FRAZER, R. A., W. J. DUNCAN and A. R. COLLAR: Elementary Matrices, Cambridge, 1957, p. 140. – ZURMÜHL, R.: Matrizen, 2nd ed. Berlin 1958, p. 273.

computation for several integers m and finally to choose the one for which the lowest ϱ_3 has been obtained.

Results of computations are shown in fig. 21b. This diagram has been taken from KROMM's paper mentioned in the bibliography, and the curve $\varrho_2 = 0$ has been checked by matrix iteration. Because of the discontinuous variation of the parameter m the curves in this diagram are festoon curves like those obtained previously (figs. 10, 11, 13, 15) for the shell under compressive loads.

7.2.3.2.3 Short Cylinder. We consider a cylinder as being short if $l/a \ll 1$. The limit $l/a = 0$ is reached when the cylinder of length l degenerates into a plate strip of width l. When such a strip is sub-

jected to a shear load applied to its edges, it buckles as indicated in fig. 22[1]. In the x direction the wavelength is l, and in the y direction (corresponding to the circumferential direction of the cylinder) the wavelength is roughly the same. It may be assumed that in a short cylindrical shell a similar situation

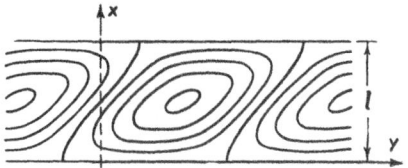

Fig. 22. Buckling mode of a flat plate strip subjected to a shear load

prevails and that in eqs. (29) the terms with $n = 1, 2, 3$ predominate while m must be chosen so that l and the circumferential wavelength $\pi a/m$ are of the same order of magnitude, hence

$$m \approx \frac{\pi a}{l} \gg 1.$$

For low values of n in which we are primarily interested, we have then

$$\lambda = \frac{n \pi a}{l} \approx m.$$

We may now discuss eq. (34). The coefficient of C_n is again given by the expression (35). When we substitute in (35) the values for $c_1 \ldots c_4$ from eqs. (12), we may now neglect all terms but those carrying the highest power in λ and m together. This yields

$$c_1 = (1 - \nu^2)\,\lambda^4, \qquad c_2 = (\lambda^2 + m^2)^4,$$
$$c_3 = m^2(\lambda^2 + m^2)^2, \qquad c_4 = \lambda^2(\lambda^2 + m^2)^2,$$

and the coefficient of C_n reduces to

$$\frac{1 - \nu}{2}[(1 - \nu^2)\lambda^4 + k(\lambda^2 + m^2)^4 - (q_1 m^2 + q_2 \lambda^2)\,(\lambda^2 + m^2)^2]. \qquad (40)$$

Again it is left to the reader to work out the determinants in the q_3 term of eq. (34). He will find that of the five terms in the braces only

---

[1] SOUTHWELL, R. V., and S. W. SKAN: On the stability under shearing forces of a flat elastic strip. Proc. Roy. Soc. London, A, vol. 105 (1924) p. 587.

the third one contains λ^5 and m^5, while the other four are only of the order λ^3, m^3 and, therefore, may be neglected. Consequently,

$$\left\{ \ \right\} = \frac{1-\nu}{2}\,(\lambda^2 + m^2)^2\,m \sum_j \frac{nj}{j^2 - n^2}\,C_j,$$

and eq. (34) assumes the simple form

$$[(1 - \nu^2)\,\lambda^4 + k(\lambda^2 + m^2)^4 - (q_1 m^2 + q_2 \lambda^2)\,(\lambda^2 + m^2)^2]\,C_n$$
$$+ (-1)^n \frac{8a\,q_3}{l}\,(\lambda^2 + m^2)^2\,m \sum_j \frac{nj}{j^2 - n^2}\,C_j = 0. \qquad (41)$$

Since $m \gg 1$, it is of little importance for the buckling load that m must be an integer. We therefore put

$$m = \mu\,\frac{\pi a}{l} \quad \text{and} \quad \lambda = n\,\frac{\pi a}{l}$$

and admit for the parameter μ whatever value makes q_3 a minimum. We again use the load parameters ϱ_2 and ϱ_3 defined by eqs. (37), but we need a different form parameter

$$\omega^4 = \frac{12(1 - \nu^2)}{\pi^4}\,\frac{l^4}{a^2\,t^2}. \qquad (42)$$

With all these notations eq. (41) may be brought into the following form:

$$\frac{\sqrt{\omega}}{\mu}\left[\frac{\omega^2\,n^2}{(n^2 + \mu^2)^2} + \frac{(n^2 + \mu^2)^2}{\omega^2\,n^2} - \varrho_2\right] n\,C_n + (-1)^n \varrho_3 \sum_j \frac{j}{j^2 - n^2}\,C_j = 0. \quad (43)$$

Again such an equation may be written for every integer n, and these equations form an infinite set for an infinite number of unknowns $j\,C_j$. The occurrence of powers of n and j is the same as in eq. (39) and so is the convergence of the infinite determinant. If we now use the symbol T_n for the coefficient of $n\,C_n$ in eq. (43), the equations are exactly represented by the set shown after eq. (39) and may be treated in the same way as those for the long cylinder.

There is, however, one difference. In eq. (39), the coefficient T_n depended on the parameter m which must be an integer. The necessity for choosing in each case that integer which makes ϱ_3 a minimum led to the festoon curves of fig. 21 b. Here T_n depends on the parameter μ which is allowed to vary continuously. It is out of the question to differentiate ϱ_3 formally with respect to μ and to find the zero of the derivative. Instead, one has to try several choices of μ and to find by trial and error the one which yields the least ϱ_3. Since the minimum happens to be shallow, it is neither necessary nor possible to determine this μ with high accuracy. Since a theory like this one is scarcely ever used to solve numerically an isolated case, but rather to compute diagrams, the search for the correct μ is still simplified by the fact that during the computation the computer soon gains much experience as to where to search the next time.

The result of such a computation is shown in fig. 21 a, which also has been taken from KROMM's paper. Since μ varies continuously, there are no festoon curves in this diagram. Since both parts of fig. 21 use the same quantity as ordinates, they match in the same way as do the parts of fig. 13.

7.2.4 Nonuniform Axial Compression

Thus far we have restricted our attention to the stability of uniform stress systems. There is, however, no major difficulty involved if the basic stresses depend on one of the coordinates, either x or ϕ. As an example we consider here the case of a nonuniform axial compression

$$N_{xI} = -P = -P_0 - P_1 \cos\phi. \tag{44}$$

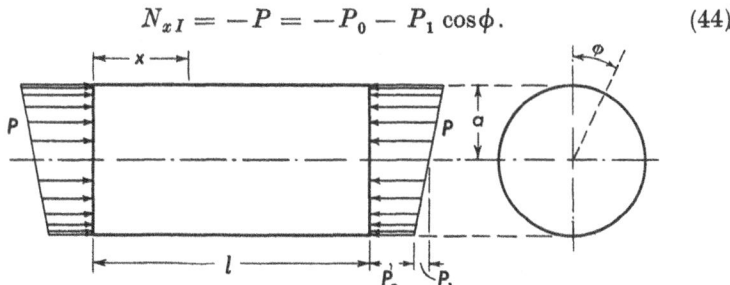

Fig. 23. Cylinder under nonuniform axial compression

When loads corresponding to this equation are applied to the edges of the shell (fig. 23), the force N_{xI} has the same distribution at any cross section $x = $ const of the shell, while $N_{\phi I} \equiv N_{x\phi I} \equiv 0$.

In analogy to eqs. (6) we introduce dimensionless load parameters

$$q_0 = P_0/D, \qquad q_1 = P_1/D, \tag{45}$$

and we obtain the differential equations of the problem by replacing q_2 in eqs. (7) by $q_0 + q_1 \cos\phi$ while dropping the original terms with q_1 and q_3. This simple procedure is, of course, possible only because we never needed to differentiate any load with respect to ϕ when deriving eqs. (7). The differential equations thus obtained are these:

$$
\left.
\begin{aligned}
&u'' + \frac{1-\nu}{2} u^{\cdot\cdot} + \frac{1+\nu}{2} v^{\prime\cdot} + \nu w' \\
&\qquad + k\left(\frac{1-\nu}{2} u^{\cdot\cdot} - w''' + \frac{1-\nu}{2} w^{\prime\cdot\cdot}\right) - q_0 u'' = q_1 u'' \cos\phi, \\
&\frac{1+\nu}{2} u^{\prime\cdot} + v'' + \frac{1-\nu}{2} v'' + w^{\cdot} \\
&\qquad + k\left(\frac{3}{2}(1-\nu) v'' - \frac{3-\nu}{2} w^{\prime\prime\cdot}\right) - q_0 v'' = q_1 v'' \cos\phi, \\
&\nu u' + v^{\cdot} + w + k\left(\frac{1-\nu}{2} u^{\prime\cdot\cdot} - u''' - \frac{3-\nu}{2} v^{\prime\prime\cdot} + w^{IV} + 2w^{\prime\prime\cdot\cdot} \right. \\
&\qquad\left. + w^{\cdot\cdot} + 2w^{\cdot\cdot} + w\right) + q_0 w'' = - q_1 w'' \cos\phi.
\end{aligned}
\right\} \tag{46}
$$

They are still homogeneous in u, v, w, but they contain terms with variable coefficients. Therefore eqs. (8) are on longer a solution, but we may put

$$u = \cos\frac{\lambda x}{a} \sum_{m=0}^{\infty} A_m \cos m\,\phi,$$

$$v = \sin\frac{\lambda x}{a} \sum_{m=1}^{\infty} B_m \sin m\,\phi, \qquad (47\,\text{a–c})$$

$$w = \sin\frac{\lambda x}{a} \sum_{m=0}^{\infty} C_m \cos m\,\phi.$$

Upon introducing this into eqs. (46) we obtain the following set:

$$
\left.
\begin{aligned}
&\sum_{m=0}^{\infty}\left\{A_m\left[\lambda^2 + \frac{1-\nu}{2}m^2(1+k) - q_0\lambda^2\right] + B_m\left[-\frac{1+\nu}{2}\lambda m\right]\right.\\
&\quad\left. + C_m\left[-\nu\lambda - k\left(\lambda^3 - \frac{1-\nu}{2}\lambda m^2\right)\right]\right\}\cos m\,\phi\\
&= q_1\lambda^2\sum_{m=0}^{\infty}A_m\cos m\,\phi\cos\phi\\
&= \frac{1}{2}q_1\lambda^2\left[\sum_{m=0}^{\infty}(A_{m-1}+A_{m+1})\cos m\,\phi + A_0\cos\phi\right],\\[4pt]
&\sum_{m=1}^{\infty}\left\{A_m\left[-\frac{1+\nu}{2}\lambda m\right] + B_m\left[m^2 + \frac{1-\nu}{2}\lambda^2(1+3k) - q_0\lambda^2\right]\right.\\
&\quad\left. + C_m\left[m + \frac{3-\nu}{2}k\lambda^2 m\right]\right\}\sin m\,\phi\\
&= q_1\lambda^2\sum_{m=1}^{\infty}B_m\sin m\,\phi\cos\phi\\
&= \frac{1}{2}q_1\lambda^2\sum_{m=1}^{\infty}(B_{m-1}+B_{m+1})\sin m\,\phi,\\[4pt]
&\sum_{m=0}^{\infty}\left\{A_m\left[-\nu\lambda - k\left(\lambda^3 - \frac{1-\nu}{2}\lambda m^2\right)\right] + B_m\left[m + \frac{3-\nu}{2}k\lambda^2 m\right]\right.\\
&\quad\left. + C_m[1 + k(\lambda^4 + 2\lambda^2 m^2 + m^4 - 2m^2 + 1) - q_0\lambda^2]\right\}\cos m\,\phi\\
&= q_1\lambda^2\sum_{m=0}^{\infty}C_m\cos m\,\phi\cos\phi\\
&= \frac{1}{2}q_1\lambda^2\left[\sum_{m=0}^{\infty}(C_{m-1}+C_{m+1})\cos m\,\phi + C_0\cos\phi\right].
\end{aligned}
\right\} \quad (48)
$$

In the second version of the right-hand sides of these equations it is understood that $A_{-1} = B_0 = C_{-1} = 0$. When this second version is used, each of the equations states that two FOURIER series in ϕ must be identically equal. This is possible only if their coefficients are the same, and thus each of the three differential equations yields an infinite number of linear, algebraic equations for the coefficients A_m, B_m, C_m. There are three equations corresponding to each integer m, and the general triplet is as follows:

$$
A_{m-1}\left[-\frac{1}{2} q_1 \lambda^2\right] + A_m\left[\lambda^2 + \frac{1-\nu}{2} m^2(1+k) - q_0 \lambda^2\right]
$$
$$
+ B_m\left[-\frac{1+\nu}{2} \lambda m\right] + C_m\left[-\nu\lambda - k\left(\lambda^3 - \frac{1-\nu}{2}\lambda m^2\right)\right]
$$
$$
+ A_{m+1}\left[-\frac{1}{2} q_1 \lambda^2\right] = 0,
$$
$$
B_{m-1}\left[-\frac{1}{2} q_1 \lambda^2\right] + A_m\left[-\frac{1+\nu}{2}\lambda m\right]
$$
$$
+ B_m\left[m^2 + \frac{1-\nu}{2}\lambda^2(1+3k) - q_0\lambda^2\right] \qquad (49)
$$
$$
+ C_m\left[m + \frac{3-\nu}{2} k\lambda^2 m\right] + B_{m+1}\left[-\frac{1}{2} q_1 \lambda^2\right] = 0,
$$
$$
C_{m-1}\left[-\frac{1}{2} q_1 \lambda^2\right] + A_m\left[-\nu\lambda - k\left(\lambda^3 - \frac{1-\nu}{2}\lambda m^2\right)\right]
$$
$$
+ B_m\left[m + \frac{3-\nu}{2} k\lambda^2 m\right]
$$
$$
+ C_m[1 + k(\lambda^4 + 2\lambda^2 m^2 + m^4 - 2m^2 + 1) - q_0\lambda^2]
$$
$$
+ C_{m+1}\left[-\frac{1}{2} q_1 \lambda^2\right] = 0.
$$

The first equations ($m = 0$, $m = 1$) have some irregularities, but since we shall see that they are usually of little importance, we leave it to the reader to work them out if the need should arise.

An infinite set of equations does not make much sense unless its coefficient determinant satisfies certain convergence requirements. We may try to satisfy the condition used on p. 446. In order to come at least close to it, we must use $m A_m$, $m B_m$, $m^2 C_m$ as unknowns and must divide the equations by m, m, m^2, respectively. However, this does not lead to success, since some terms of the order m^{-1} will be left outside the diagonal where they should be of order m^{-2} to assure convergence.

To handle this problem, a technique similar to the one used in Section 7.2.3.2.1 may be employed. We first introduce an abbreviated notation writing eqs. (49) in the following form:

$$
a_{11,m} A_m + a_{12,m} B_m + a_{13,m} C_m = \frac{1}{2} q_1 \lambda^2 (A_{m-1} + A_{m+1}),
$$
$$
a_{21,m} A_m + a_{22,m} B_m + a_{23,m} C_m = \frac{1}{2} q_1 \lambda^2 (B_{m-1} + B_{m+1}), \qquad (49')
$$
$$
a_{31,m} A_m + a_{32,m} B_m + a_{33,m} C_m = \frac{1}{2} q_1 \lambda^2 (C_{m-1} + C_{m+1}).
$$

We then use the first two of these equations to express A_m and B_m in terms of C_m and of $A_{m-1} \ldots B_{m+1}$. This involves certain determinants of the coefficients $a_{11,m}$ etc., and for brevity we shall put the subscript m only at the end of these determinants and not at all the individual coefficients:

$$A_m = \frac{1}{(a_{11}a_{22} - a_{12}^2)_m}\left[(a_{12}a_{23} - a_{13}a_{22})_m\, C_m\right.$$
$$\left. + \frac{1}{2}\, q_1\, \lambda^2\big(a_{22}(A_{m-1} + A_{m+1}) - a_{12}(B_{m-1} + B_{m+1})\big)\right],$$

$$B_m = \frac{1}{(a_{11}a_{22} - a_{12}^2)_m}\left[(a_{21}a_{13} - a_{11}a_{23})_m\, C_m\right.$$
$$\left. + \frac{1}{2}\, q_1\, \lambda^2\big(a_{11}(B_{m-1} + B_{m+1}) - a_{21}(A_{m-1} + A_{m+1})\big)\right].$$

These equations are used first to eliminate A_m and B_m from the third equation (49'). This introduces there terms with A_{m-1}, A_{m+1}, B_{m-1}, B_{m+1}. The same two equations may now be used again to eliminate these terms. This introduces new terms with A_{m-2}, A_m, ..., but all these terms carry a factor q_1^2 and may be neglected. We thus arrive at the following equation:

$$\begin{vmatrix} a_{11} & a_{12} & a_{13} \\ a_{21} & a_{22} & a_{23} \\ a_{31} & a_{32} & a_{33} \end{vmatrix}_m \cdot C_m - \frac{1}{2}\, q_1\, \lambda^2[\cdots C_{m-1} + \cdots C_{m+1}] = 0. \quad (50)$$

The coefficients of C_{m-1} and C_{m+1}, which have been indicated by a row of dots, are rather lengthy expressions of determinants of $a_{11}, \ldots, a_{33}$. If they are worked out in detail, it is found that they increase as m^4, while the coefficient of C_m has the form $m^8 + m^6 + \ldots$. Therefore, if we divide each equation by m^4 and use $m^4\, C_m$ as an unknown, the determinant will fulfill the convergence requirement as stated on p. 446. It follows that the infinite determinant of eqs. (50) must vanish and that this condition is fulfilled with increasing degrees of approximation if finite systems of increasing size of the determinant are set equal to zero.

Since the coefficients of eqs. (50) contain the load parameters q_0 and q_1, the vanishing of the determinant is a relation between these two quantities. The situation is similar to the one encountered on p. 424 for the two-way compression: We may assume values of q_0 and calculate from the determinantal equation the corresponding values of q_1 or vice versa.

The numerical work may be done in different ways. One may work out the expressions which in eq. (50) were indicated by dots and so obtain the elements of the determinant in general form; or one may write a sufficient segment of eqs. (49), introducing numerical values for everything but q_1, and then perform numerically the elimination of A_m and B_m as described; or one may just use eqs. (49) as they stand. In each case, one has to solve the eigenvalue problem of a large determinant, and this again can be done in several ways. One may expand the determinant and solve the ensuing algebraic equation, or one may fix a value of q_0 and find by trial and error the value of q_1 which makes the determinant vanish, or one may use the method of matrix iteration.

After a pair q_0, q_1 has been found, .eqs. (50) may be solved for the constants C_m, which, as in all buckling problems, are determined except for a common factor. Calculations of this kind have shown that in the series (47c) – and consequently also in (47a, b) – there is somewhere, but usually not at the beginning, a sequence of terms with large coefficients C_m, while all the other C_m are small; the farther away from the big ones, the smaller they are. The biggest terms in the series describe the essential features of the buckling deformation, and a reasonable approximation of the buckling load is obtained if the infinite determinant is approximated by a segment containing just those rows and columns which are associated with the biggest values C_m.

An example may illustrate this. For a shell with $l/a = \pi$, $k = 10^{-6}$, $\nu = 1/6$ it was found that $q_0 = 8.0 \times 10^{-4}$ and $q_1 = 1.67 \times 10^{-3}$ represent a buckling load. Solving eqs. (50) and introducing the values C_m thus obtained into eq. (47c) yielded the following deflection:

$$w = \sin\frac{x}{a}(\cdots + 0.0320\cos3\phi + 0.320\cos4\phi + 1.000\cos5\phi$$

$$+ 1.100\cos6\phi + 0.530\cos7\phi + 0.1283\cos8\phi + \cdots).$$

This suggests that it would be sufficient to use only a four by four determinant made up of the coefficients of C_4, C_5, C_6, C_7 in the four equations (50) for which $m = 4, 5, 6, 7$.

The buckling displacement as computed from the formula just given is shown in fig. 24. Although the deflection is not exactly zero in the tensile zone of the shell, it is there exceedingly small, and the largest deflection occurs where N_{xI} has its largest negative value.

For the same shell a number of critical pairs q_0, q_1 have been computed, and the result of this computation is shown in fig. 25. When the cylinder is considered as a tubular bar subjected to an eccentric axial force, then q_0 represents the direct stress and q_1 the bending stress. The ratio of both has been chosen as she abscissa of fig. 25, while the ordinate $q_0 + q_1$ represents the greatest compressive stress occurring in the cross section of the tube. The diagram shows that the latter does not very much depend on the former and that it increases

as the compressive zone of the cylinder decreases in width. The details
of this result, however, should not be indiscriminately generalized.

The plane $\phi = 0$, $\phi = \pi$ is a plane of symmetry for the basic
stress pattern (44). From fig. 24 it may be seen that the buckling

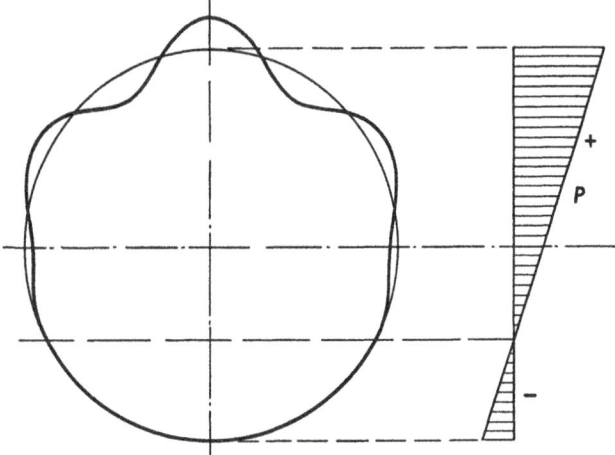

Fig. 24. Buckling mode of a cylinder under nonuniform axial compression; cross section and
load distribution

deformation has the same symmetry. This is a necessary consequence
of the form (47) which we adopted for the solution of eqs. (46). There
exists, however, another solution of these equations, which is antimetric

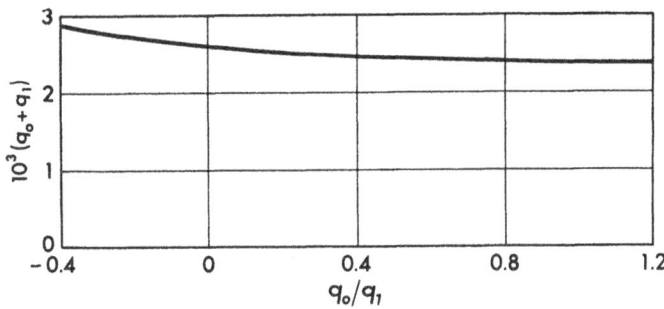

Fig. 25. Dimensionless buckling loads for a cylinder with $l/a = \pi$, $k = 10^{-6}$, $\nu = 1/6$

with respect to the same plane. We obtain it from eqs. (47) by inter-
changing there $\sin m\,\phi$ and $\cos m\,\phi$. In the numerical example underlying fig. 24 the antimetric mode has the same buckling load, but in
other cases it may well happen that it leads to a higher or to a lower
buckling load than does the symmetric mode. It is therefore necessary
to keep an eye on both.

7.2.5 The Beam-Column Problem

A bar is said to be a beam when it carries a lateral load and thus is subject to bending, and it is called a column when it carries an axial compressive load. The column has a stability problem, the beam has none. When loads of both kinds act at the same time, a new problem arises. The bending load produces a lateral deflection, and this deflection provides a lever arm for the axial load which now produces additional bending. It is well known that in this case stresses and deformations increase linearly with the lateral load but that they increase faster than linear when the axial load is raised and that they tend toward infinity when the axial load approaches the buckling load of the column. This stress problem is known as the beam-column problem, and we shall discuss here under the same name analogous problems for cylindrical shells.

7.2.5.1 The Axisymmetric Problem

We consider a cylindrical shell put between the plates of a testing machine (fig. 26a). When the load P (per unit of circumference) is applied, a negative stress $\sigma_x = -P/t$ is produced and the length of the cylinder decreases. Due to POISSON's ratio, the diameter increases,

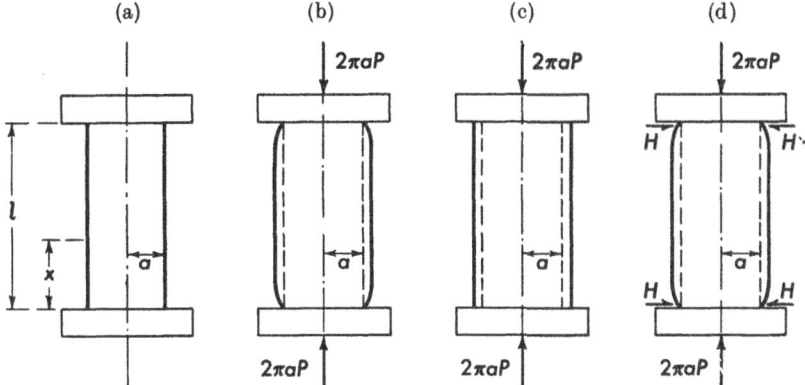

(a)   (b)   (c)   (d)

Fig. 26. Cylindrical shell subjected to a compression test

but this increase is prevented at the edges because of the friction between the cylinder and the plates of the press (fig. 26b). Evidently, bending stresses will appear, and we must find out whether they or a possible instability will determine the strength of the shell.

The deformation shown in fig. 26b may be produced in two steps. First we assume that the ends of the cylinder are so well lubricated that the edges can expand without constraint, according to a hoop strain $\epsilon_\phi = \nu\, P/Et$ (fig. 26c). Then we apply a radial load H (fig. 26d) of sufficient magnitude to bring the edges back to zero deflection.

The first part of this deformation is trivial. In the second part we have the normal force $N_{x\,I} = -P$ as a large basic force and the additional small load H which produces the small displacements u, v, w. To this deformation we may apply eqs. (7) which we established under similar circumstances. The essential difference is that the solution we seek now is not an incidental deformation which becomes possible when P assumes a certain critical value but a deformation which is produced by the load H.

In our special case eqs. (7) simplify greatly. Since there is only the axial load P, we have $q_1 = q_3 = 0$. And since the deformation must be expected to have axial symmetry, we must put $v = 0$ and must drop all the dot derivatives. Eq. (7b) vanishes then altogether, and the other two simplify to

$$u'' + \nu\, w' - k\, w''' - q_2\, u'' = 0,$$
$$\nu\, u' + w + k(-u''' + w^{IV} + w) + q_2\, w'' = 0. \qquad \text{(51a, b)}$$

There are two boundary conditions at each end of the shell, say at $x = 0$ and at $x = l$. First, the radial displacement w must cancel the displacement $a\,\epsilon_\phi = \nu\, Pa/Et$ of fig. 26c, i. e. we must have

$$w = -\frac{\nu\, P a}{E t} = -\frac{\nu}{1 - \nu^2}\, q_2\, a. \qquad \text{(52a)}$$

Second, we want to have hinged edges, $M_x = 0$, hence from eq. (V–9f):

$$w'' - u' = 0. \qquad \text{(52b)}$$

Then we have the condition that the load H has no component in the axial direction. In fig. 26c the axial force per unit of undeformed circumference is $-P$; after the deformation it is (see p. 417):

$$-P\left(1 + \frac{u'}{a}\right) + N_x,$$

where N_x is connected with u, w by eq. (V–9b). We have, therefore, the condition that at each edge there is $N_x = Pu'/a$, hence

$$u' + \nu\, w - k\, w'' = q_2\, u'. \qquad \text{(52c)}$$

From eq. (51a) it is evident that it is enough to enforce the condition (52c) at $x = 0$ and that it then will be fulfilled everywhere, including the other edge $x = l$. Lastly, we may exclude or fix a rigid-body displacement of the whole shell by prescribing u for one value of the coordinate x.

The differential equations (51) have constant coefficients and may be solved by exponential functions:

$$u = A\,e^{\lambda\,x/a}, \qquad w = C\,e^{\lambda\,x/a}.$$

When these are introduced, we get two linear equations for A and C:

$$
\begin{aligned}
A \cdot \lambda^2(1 - q_2) + C \cdot \lambda(\nu - k\,\lambda^2) &= 0, \\
A \cdot \lambda(\nu - k\,\lambda^2) + C \cdot (1 + k\,\lambda^4 + k + q_2\,\lambda^2) &= 0.
\end{aligned}
\tag{53}
$$

Since these equations are homogeneous, their determinant must vanish, and this yields an equation for λ. When the small quantities k and q_2 are neglected compared with 1, it reads

$$k\,\lambda^6 + (2\,\nu\,k + q_2)\,\lambda^4 + (1 - \nu^2)\,\lambda^2 = 0.$$

It has the double root

$$\lambda_5 = \lambda_6 = 0$$

and four nontrivial roots

$$\lambda_{1,2,3,4} = \pm\sqrt{-\frac{q_2}{2k} - \nu \pm \frac{1}{2k}\sqrt{q_2^2 - 4(1 - \nu^2)\,k}}\,. \tag{54}$$

While these last four roots lead to true exponential solutions, the fifth and sixth solutions degenerate into linear functions of x, and we have

$$
\begin{aligned}
u &= \sum_{n=1}^{4} A_n\,e^{\lambda_n\,x/a} + A_5 + A_6\,\frac{x}{a}, \\
w &= \sum_{n=1}^{4} C_n\,e^{\lambda_n\,x/a} + C_5 + C_6\,\frac{x}{a}.
\end{aligned}
\tag{55}
$$

For $n = 1, \ldots, 4$ the constants A_n and C_n are connected by eqs. (53), and since the determinant of these equations is zero, we may use either one to formulate the relation. We choose the first one and have

$$A_n = -C_n\,\frac{\nu - k\,\lambda_n^2}{\lambda_n}. \tag{56a}$$

The degenerated solutions, $n = 5, 6$, must be introduced in the differential equations (51) to make sure that they really are solutions and to determine how their constants are interconnected. We find that

$$C_6 = 0, \qquad A_6 = -C_5\,\frac{1 + k}{\nu}. \tag{56b}$$

This indicates that the term $C_6\,x/a$ is no solution at all and that $u = A_6\,x/a$ and $w = C_5$ together are the fifth independent solution of eqs. (51). The sixth solution is $u = A_5$, $w = 0$.

This last solution evidently represents a rigid-body movement of the cylinder, and we may simply discard it. The remaining free constants $C_1, \ldots, C_5$ can then be determined from the boundary conditions (52a–c). We begin with eq. (52c). When we introduce there the solution (55) and make use of eqs. (56), we find that the exponential solutions cancel out, and we are left with

$$C_5 = 0.$$

Thus we are entirely rid of the linear terms in eqs. (55). The boundary conditions (52a, b), written for $x = 0$ and for $x = l$, yield four equations for the remaining four unknown coefficients. They may easily be brought into the following form:

$$C_1 + \quad C_2 + \quad C_3 + \quad C_4 = -\frac{\nu}{1 - \nu^2} q_2 a,$$

$$\lambda_1^2 C_1 + \quad \lambda_2^2 C_2 + \quad \lambda_3^2 C_3 + \quad \lambda_4^2 C_4 = \frac{\nu^2}{1 - \nu^2} q_2 a,$$

$$e^{\lambda_1 l/a} C_1 + \quad e^{\lambda_2 l/a} C_2 + \quad e^{\lambda_3 l/a} C_3 + \quad e^{\lambda_4 l/a} C_4 = -\frac{\nu}{1 - \nu^2} q_2 a,$$

$$\lambda_1^2 e^{\lambda_1 l/a} C_1 + \lambda_2^2 e^{\lambda_2 l/a} C_2 + \lambda_3^2 e^{\lambda_3 l/a} C_3 + \lambda_4^2 e^{\lambda_4 l/a} C_4 = \frac{\nu^2}{1 - \nu^2} q_2 a.$$

$$(57\,\text{a–d})$$

We shall not go into the details of solving these equations, but we shall discuss the solution as it is obtained when the results are introduced into eqs. (55).

We see from eq. (54) that for small values of q_2 all four values λ_n^2 are complex but that they are real and negative if q_2 grows beyond the limit

$$q_2^2 = 4(1 - \nu^2)\, k. \tag{58}$$

If q_2 is smaller than this limit we may write

$$\lambda_1 = -\lambda_3 = -\alpha + i\beta, \quad \lambda_2 = -\lambda_4 = -\alpha - i\beta$$

with real, positive quantities α, β. When q_2 is small and the cylinder long, then $e^{-\alpha l/a}$ is a very small quantity. In this case it turns out that $C_1, C_2 \gg C_3, C_4$, so that C_3, C_4 may be neglected in eqs. (57a, b) and C_1, C_2 in eqs. (57c, d). The solution then becomes extremely simple. For small values of x/a only the terms with C_1 and C_2 make appreciable contributions and yield

$$w = -\frac{\nu}{1 - \nu^2} q_2 a\, e^{-\alpha x/a} \left(\cos\frac{\beta x}{a} + \frac{\nu + \alpha^2 - \beta^2}{2\alpha\beta} \sin\frac{\beta x}{a} \right), \tag{59}$$

and near $x = l$ only C_3 and C_4 are essential:

$$w = -\frac{\nu}{1 - \nu^2} q_2 a\, e^{-\alpha(l - x)/a} \left(\cos\frac{\beta(l - x)}{a} + \frac{\nu + \alpha^2 - \beta^2}{2\alpha\beta} \sin\frac{\beta(l - x)}{a} \right).$$

These formulas represent two identical end disturbances, produced by the constraint imposed upon the circumferential expansion of the shell, and these disturbances affect only two narrow border zones.

When q_2 is increased, these disturbances penetrate deeper into the shell. This is shown by the first three diagrams of fig. 27. In the fourth diagram the two disturbance zones have reached each other, and in the last diagram they overlap completely. At this loading stage α has decreased so far that it is no longer possible to neglect anything in eqs. (51). The solution is then best written in terms of hyperbolic and trigonometric functions, but we shall leave it to the reader to work out by himself the somewhat bulky formula. Very soon the limit (58)

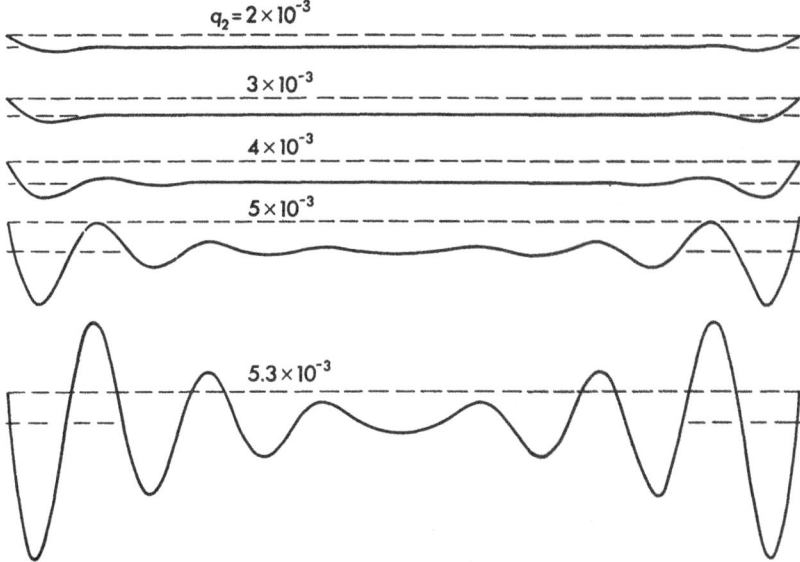

Fig. 27. Deformed generator of as cylinder loaded a shown in Fig. 26b. The axis of the cylinder lies above the generators shown. The broken line indicates the undeformed position of the generator; the dashes at the end give the deformed position for a cylinder without edge disturbance

is reached, in fig. 27 at $q_2 = 5.39 \times 10^{-3}$. Beyond this all four values λ_n are purely imaginary, say

$$\lambda_1 = -\lambda_3 = i\,\mu_1\,, \qquad \lambda_2 = -\lambda_4 = i\,\mu_2\,,$$

and the solution assumes the following form:

$$w = -\frac{\nu}{1-\nu^2}\,\frac{q_2\,a}{\mu_2^2-\mu_1^2}\left[\frac{\mu_2^2-\nu}{\cos\dfrac{\mu_1 l}{2a}}\cos\frac{\mu_1(l-2x)}{2a} - \frac{\mu_1^2-\nu}{\cos\dfrac{\mu_2 l}{2a}}\cos\frac{\mu_2(l-2x)}{2a}\right].$$

$$\tag{60}$$

When this is plotted, diagrams similar to the last one of fig. 27 are obtained. As q_2 grows, one of the cosine denominators very rapidly

approaches zero. Consequently, one of the terms in the brackets out-
grows the other one, and the deflection approaches a pure sinusoidal
shape, but at the same time its amplitude increases beyond bounds.

This is in strict analogy to the well-known phenomenon that in
an ordinary beam-column the deflection tends toward infinity as the
axial load approaches the EULER load. Also the mechanical interpretation is the same in both cases. For the shell, the infinite deflection
occurs when

$$\mu_{1,2}^2 = \frac{q_2}{2k} + \nu \mp \frac{1}{2k}\sqrt{q_2^2 - 4(1-\nu^2)k} = \frac{n\pi a}{l}$$

where n is an *odd* integer. When this equation is solved for q_2, one
arrives at eq. (13), written for $m = 0$ and with $\lambda = n\pi a/l$ such that n
is an odd integer (except that in eq. (13) two small terms have been
neglected). We conclude that the bending stresses in our shell grow
beyond bounds when the load P approaches a critical value connected
with a buckling mode which is axisymmetric ($m = 0$) and also symmetric to the plane $x = l/2$ ($n =$ odd).

On its way to infinity, the bending stress will sooner or later pass
the yield limit. As soon as this happens, our theory ceases to be valid,
and the first and largest bulge of the cylinder will be squeezed flat.
Except for this, the elastic theory is still applicable, and if the test is
continued, the next folds will grow until they also start to yield and
are squeezed flat[1].

The equations (51) which govern the bending collapse just described
are a special case of the general buckling equations (7). We derived
the former from the latter essentially by putting $m = 0$. This seems
reasonable but is by no means necessary. Our solutions (59) and (60)
are just as well solutions of the general equations (7) with the special
boundary conditions (52). When the load P (or the dimensionless
parameter q_2) reaches the buckling load given by fig. 11 or eq. (13),
then eqs. (7) permit a certain deflection of arbitrary amplitude, and
since this deflection satisfies homogeneous boundary conditions it
may be superposed on the solution (59) or (60), and the sum will still
be a solution of the differential equations (7) with the boundary conditions (52) and hence of the beam-column problem. The existence of
such a solution in which an amplitude coefficient can be varied at
discretion, indicates a neutral equilibrium, which always stands at
the threshold to instability. The developing bending collapse may, therefore, be interrupted by a true and sudden buckling if the shell reaches
a buckling load, either for a mode with $m > 0$ or for one with $m = 0$

---

[1] For pictures of such specimens, see the paper by GECKELER, mentioned
in the bibliography.

and an *even* value of n. The former may be expected to happen, if the shell lies toward the right-hand side in figs. 10 and 11, but the antimetric-axisymmetric modes have their critical loads always so close to the symmetric ones that they are without importance in this respect.

7.2.5.2 Imperfections of Shape

Shells are remarkably stiff, and elastic buckling occurs only in shells that are rather thin. This is particularly true for axial compression. The diagram, fig. 11, has been drawn for values k from 2×10^{-5} to 10^{-6}, corresponding to thickness ratios t/a between anproximately 0.016 and 0.003. Such thin shells, whether test specimens or parts of real structures, can hardly be expected to be very perfectly cylindrical. Deviations from the exact shape may easily be of the order of the wall thickness or even more. On p. 367 when studying the bending stresses of an almost spherical shell, we have seen that such deviations are not a matter of great concern. This, however, is quite different when it comes to buckling. When a structure, whether column or shell, approaches a neutral equilibrium, little causes have grave consequences, and it is necessary to go into this problem to some extent. As an example, we again choose the shell under uniform compression, i. e. the case $q_2 \neq 0$, $q_1 = q_3 = 0$ in the notation of eqs. (6).

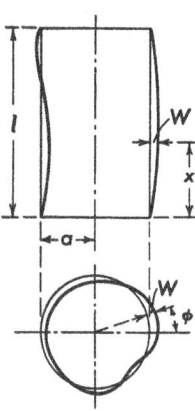

Fig. 28
Section through a shell deviating slightly from a cylindrical shell

The deviation of the real middle surface from a true cylinder of radius a may be described by the radial distance W of each point of the middle surface from the reference cylinder (fig. 28). This deviation W may be an arbitrary function and, most probably, will not show any symmetry. However, it is reasonable to assume that $W \equiv 0$ along the edges $x = 0$ and $x = l$, since there the shell is supported by more rigid members which will be closer to perfect circular shape. A function fulfilling these conditions may be expanded into a double FOURIER series, containing sine and cosine terms in ϕ. However, the sine terms would contribute nothing essential to the conclusions at which we shall arrive and, therefore, may be omitted. Thus we are led to assume the deviation in the form

$$W = \sum_{m=0}^{\infty} \sum_{n=1}^{\infty} W_{mn} \cos m\phi \sin \frac{n\pi x}{l} \qquad (61)$$

To get started, we shall be content with a much simpler case and assume that the deviation is represented by only one term of this series, say

$$W = W_{mn} \cos m\,\phi \sin \frac{\lambda\,x}{a}, \tag{62}$$

where $\lambda = n\,\pi\,a/l$, as before, and where m and n are two arbitrary but definitely chosen integers.

When the compressive edge load P is applied to the shell, each point of the middle surface undergoes elastic displacements u, v, w, and its normal distance from the reference cylinder is then $W + w$. We assume, of course, that W is of the order of an elastic deformation, and then the element of the shell looks like the deformed elements which we used to establish the differential equations (7) of the buckling problem. When we again go through the same procedure we find that the terms of those equations belong in two groups. In those terms which contain the factor q_2 (or q_1, q_3 if we admit such loads), the quantities u, v, w describe the difference in shape between the deformed element and an element of a true cylinder. In these terms w must now be replaced by $W + w$. On the other hand, all terms which do not have a factor q_2 can be traced back to terms of the elastic law (V–8) and represent the stress resultants acting on the shell element. Before the application of the load the shell has been free of stress, and the stress resultants depend only on the elastic displacements u, v, w. Consequently, in all these terms w is just w and nothing else. Thus we arrive at the following set of differential equations:

$$\left. \begin{aligned}
&u'' + \frac{1-\nu}{2}\,u^{\cdot\cdot} + \frac{1+\nu}{2}\,v'^{\cdot} + \nu\,w' \\
&\qquad + k\Big(\frac{1-\nu}{2}\,u^{\cdot\cdot} - w''' + \frac{1-\nu}{2}\,w'^{\cdot\cdot}\Big) - q_2\,u'' = 0, \\[2mm]
&\frac{1+\nu}{2}\,u'^{\cdot} + v^{\cdot\cdot} + \frac{1-\nu}{2}\,v'' + w^{\cdot} \\
&\qquad + k\Big(\frac{3}{2}(1-\nu)\,v'' - \frac{3-\nu}{2}\,w''^{\cdot}\Big) - q_2\,v'' = 0, \\[2mm]
&\nu\,u' + v^{\cdot} + w + k\Big(\frac{1-\nu}{2}\,u'^{\cdot\cdot} - u''' - \frac{3-\nu}{2}\,v''^{\cdot} + w^{IV} + 2w''^{\cdot\cdot} \\
&\qquad + w^{\cdot\cdot\cdot\cdot} + 2w^{\cdot\cdot} + w\Big) + q_2\,w'' = q_2\,W_{mn}\,\lambda^2\cos m\,\phi\sin\frac{\lambda\,x}{a}.
\end{aligned} \right\} \tag{63}$$

We see here that we again have a problem of the beam-column type. The left-hand sides of the differential equations are the same as those of the associated buckling problem, but the equations are no longer homogeneous. While in the preceding problem the nonvanishing right-hand sides appeared in some of the boundary conditions, these conditions are here homogeneous and the same as in the buckling problem.

We can easily check that

$$u = A_{mn} \cos m\phi \cos \frac{\lambda x}{a},$$

$$v = B_{mn} \sin m\phi \sin \frac{\lambda x}{a}, \qquad\qquad (64)$$

$$w = C_{mn} \cos m\phi \sin \frac{\lambda x}{a}$$

is a particular solution of eqs. (63), and when we introduce it there, we obtain three linear equations for A_{mn}, B_{mn}, C_{mn}. Their left-hand sides are identical with those of eqs. (10) (except for $q_1 = 0$ and for the subscripts which we have now attached to the constants A, B, C), but their right-hand sides are 0, 0, $q_2 \lambda^2 W_{mn}$, respectively. Let us give these equations the number (10′).

We shall not attempt here to solve these equations (which can easily be done numerically if W_{mn} is given), but we shall review the situation in the whole and draw some general conclusions.

As long as the coefficient determinant of eqs. (10′), and that is of eqs. (10), does not vanish, there is always a unique solution A_{mn}, B_{mn}, C_{mn}, and these constants are proportional to the amplitude W_{mn} of the deviation. But we know that the determinant is equal to zero if q_2 has that critical value which belongs to the buckling mode m, n. In this case the alternative theorem for linear equations states that A_{mn}, B_{mn}, $C_{mn} = \infty$. Mechanically speaking, this means that the deflections and hence the stresses in the shell grow beyond bounds when q_2 approaches this particular critical value.

We may now come back to the general form (62) of the deviations. For every term of this FOURIER series we have a solution (64) and a set of eqs (10′) to find from them the amplitudes A_{mn}, B_{mn}, C_{mn} of the displacements. Since these amplitudes depend linearly on the coefficients W_{mn}, we may superpose all the displacements to form three FOURIER series for u, v, w, for example,

$$W = \sum_{m=0}^{\infty} \sum_{n=1}^{\infty} C_{mn} \cos m\phi \sin \frac{n \pi x}{l}, \qquad\qquad (65)$$

and thus obtain the resultant deformation of the shell. As long as the load on the shell is below any critical value all the coefficients in the series (65) are finite. When q_2 approaches the lowest critical value, i. e., the buckling load of the shell, then those coefficients A_{mn}, B_{mn}, C_{mn} which belong to this particular buckling mode tend toward infinity, and the corresponding terms in the series outgrow all the others. The deflection then becomes more and more sinusoidal, and the nodal lines $w = 0$ approach the rectangular pattern known from the buckling deformation.

A glance at fig. 10 shows that in many shells there exist several critical values immediately above the lowest one, belonging to different modes m, n. If this is the case, several terms in the series (65) grow almost in the same way and, although one of them reaches infinity first, they may all be large but still of comparable magnitude when the shell collapses. This detail is different from the otherwise similar behavior of an imperfect column for which all the critical loads are far apart from each other. Therefore, if one wants to apply to a shell the method of SOUTHWELL and LUNDQUIST[1] for the determination of the critical load from a non-destructive test, he has to explore the situation carefully and to measure a displacement which is large for the lowest buckling mode but small for all competitive modes. If such care is taken, the method is well applicable, as A. KROMM in cooperation with the author has proved in a series of tests which, unfortunately, never have been published.

7.2.6 Nonlinear Theory of Shell Buckling

The theory of the instability of cylindrical shells presented in Sections 7.2.1 to 7.2.4 has been built up following the general pattern of the theory of elastic stability. There is little doubt that it gives the correct answer to the question: "When does the elastic equilibrium become neutral?" However, the load under which this neutral equilibrium is reached — the classical critical load — is not necessarily the collapse load of the shell. Ever since the results of the theory have been compared with those of buckling tests, the large discrepancy has been puzzling to investigators. The "beam-column theory" of Section 7.2.5 explains to some extent why the actual collapse load should be smaller than the classical buckling load, but it does not say how much smaller it is, and it does not explain the details of the actual buckling process.

When a column is carefully made and loaded, it behaves exactly as the beam-column theory predicts. First there is scarcely any deflection, and when the critical load is approached, the deflection grows at an increasing rate until the column is overstressed in bending and collapses. Quite to the contrary, a cylindrical shell subjected to axial compression jumps suddenly into a deflected state, and it does so at a load which may be a quarter of the classical buckling load. Observations of this kind suggest that a phenomenon of large deformation is at play and that a nonlinear theory could give a deeper insight into what really is happening.

---

[1] SOUTHWELL, R. V.: On the analysis of experimental observations in problems of elastic stability. Proc. Roy. Soc. A, vol. 135 (1932), 601. — LUNDQUIST, E. E.: Generalized analysis of experimental observations in problems of elastic stability. N.A.C.A., Techn. Note 658 (1938).

We shall give here an outline of such a theory for a cylinder under axial compression (fig. 29). Since it is mathematically rather involved, we must simplify the basic equations as much as possible and be content just to trace the essential features of the buckling process.

There are always three fundamental groups of equations from which the solution of an elastic problem must be started: (1) the conditions of equilibrium, connecting the stresses or stress resultants with the loads, (2) the elastic law (usually HOOKE's law), connecting the stresses or stress resultants with the strains (the usual strains ϵ, γ, or a change of curvature or of twist), (3) the kinematic relations, connecting the strains and the displacements.

In the bending theory of shells we used HOOKE's law for the individual stresses [eqs. (V-6), (II-59)], and we eliminated the strains between it and a set of kinematic relations before we integrated the stresses to form the stress resultants. Ultimately, therefore, we had to deal with only two groups of equations, the conditions of equilibrium, eqs. (V-1), (5) and an "elastic law" connecting the stress resultants immediately with the displacements and their derivatives [eqs. (V-9), (V-12), (VI-5)].

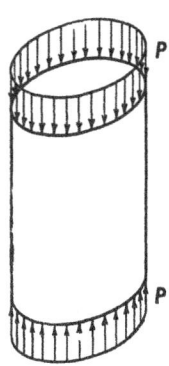

Fig. 29
Cylindrical shell under axial compression

All these equations are linear if the basic forces in eqs. (5) are considered as given coefficients. The linearity of the kinematic relations is based on the assumption that the displacements are so small that their squares can be neglected. When we want to consider large deflections, we must drop this assumption. The first step in this direction is to admit squares of the displacements, and since w is larger than u and v, it may be enough to admit only the squares of w and its derivatives.

The orthodox way would now be to find the quadratic terms which must be introduced into eqs. (V-5), to find from them the improved form of eqs. (V-9), and then to discuss possible simplifications. We prefer an easier way starting from eqs. (V-12). The right-hand sides of eqs. (V-12a-c) may easily be interpreted as combinations of strains of the middle surface, writing

$$N_\phi = D(\epsilon_\phi \dot{} + \nu\,\epsilon_x), \qquad N_x = D(\epsilon_x + \nu\,\epsilon_\phi), \qquad N_{x\phi} = D\frac{1-\nu}{2}\,\gamma_{x\phi} \quad (66)$$

and
$$\epsilon_\phi = \frac{v\dot{} + w}{a}, \qquad \epsilon_x = \frac{u'}{a}, \qquad \gamma_{x\phi} = \frac{u\dot{} + v'}{a}. \quad (67)$$

Eqs. (66) represent HOOKE's law and will be used as they are, and eqs. (67) are the linearized kinematic relations, which we now must supplement by large-deflection terms.

Since we know the linear terms in u and v and are interested only in quadratic terms in w, it is enough to study only those deformations for which $u = v = 0$. Products of u or v with w escape in this way, but they are not in the sphere of our interest.

Fig. 30a shows a line element dx before and after such a deformation. The deformed length is

$$ds_x = \sqrt{dx^2 + \left(\frac{w'}{a}\, dx\right)^2} \approx dx\left[1 + \frac{1}{2}\left(\frac{w'}{a}\right)^2\right],$$

and hence the strain

$$\epsilon_x = \frac{ds_x - dx}{dx} = \frac{1}{2}\frac{w'^2}{a^2}. \tag{68a}$$

Fig. 30b shows the deformation of a circumferential element $a\, d\phi$. When the angle $d\phi$ is small enough (i. e. when we neglect terms of

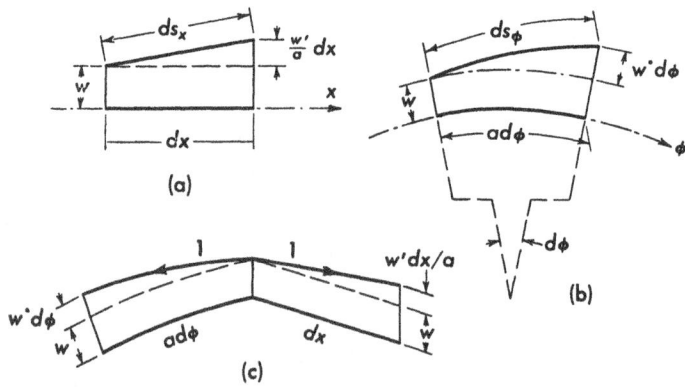

Fig. 30. Line elements of the middle surface before and after deformation, (a) generator element, (b) circumferential element, (c) both elements

higher order in the *differential*), we may apply PYTHAGORAS' theorem to the curvilinear triangle and have

$$ds_\phi = \sqrt{((a + w)d\phi)^2 + (w^{\cdot} d\phi)^2} = a\, d\phi\, \sqrt{1 + 2\frac{w}{a} + \left(\frac{w}{a}\right)^2 + \left(\frac{w^{\cdot}}{a}\right)^2}$$

$$\approx a\, d\phi\left[1 + \frac{w}{a} + \frac{1}{2}\frac{w^2}{a^2} + \frac{1}{2}\frac{w^{\cdot 2}}{a^2} - \frac{1}{8}\left(2\frac{w}{a} + \cdots\right)^2\right]$$

$$= a\, d\phi\left[1 + \frac{w}{a} + \frac{1}{2}\frac{w^{\cdot 2}}{a^2}\right].$$

It follows that the hoop strain is

$$\epsilon_\phi = \frac{ds_\phi - a\, d\phi}{a\, d\phi} = \frac{w}{a} + \frac{1}{2}\frac{w^{\cdot 2}}{a^2}. \tag{68b}$$

To find the corresponding expression for $\gamma_{x\phi}$ we consider fig. 30c, which shows both line elements dx and $a\, d\phi$ before and after deformation. Again we may disregard the curvature except for the fact that the triangle to the left has the base length $(a + w)\, d\phi$. The angle between the two deformed elements is $(\pi/2) - \gamma_{x\phi}$. The cosine of this angle is equal to the scalar product of two unit vectors located on the deformed elements as shown. Hence

$$\cos\left(\frac{\pi}{2} - \gamma_{x\phi}\right) = \sin\gamma_{x\phi} = 1 \times 0 + 0 \times 1 + \frac{w'}{a} \times \frac{w^{\cdot}}{a+w}$$

and in sufficient approximation

$$\gamma_{x\phi} = \frac{w'\, w^{\cdot}}{a^2}. \tag{68 c}$$

Since we assumed $u = v = 0$, eqs. (68) are not the complete expressions for the strains, but we must still add the linear terms taken from eqs. (67). The final form of the kinematic relations is therefore this:

$$\epsilon_x = \frac{u'}{a} + \frac{w'^2}{2a^2}, \qquad \epsilon_\phi = \frac{v^{\cdot} + w}{a} + \frac{w^{\cdot 2}}{2a^2},$$

$$\gamma_{x\phi} = \frac{u^{\cdot} + v'}{a} + \frac{w'\, w^{\cdot}}{a^2}. \tag{69 a–c}$$

These expressions may now be introduced on the right-hand sides of eqs. (66) to obtain the revised form of eqs. (V–12a–c):

$$N_\phi = \frac{D}{a}\left[v^{\cdot} + w + \nu\, u'\right] + \frac{D}{2a^2}\left[w^{\cdot 2} + \nu\, w'^2\right],$$

$$N_x = \frac{D}{a}\left[u' + \nu(v^{\cdot} + w)\right] + \frac{D}{2a^2}\left[w'^2 + \nu\, w^{\cdot 2}\right], \tag{70 a–c}$$

$$N_{x\phi} = \frac{D(1 - \nu)}{2a}\left[u^{\cdot} + v' + \frac{w'\, w^{\cdot}}{a}\right].$$

Eqs. (V–12d–f) may be used as they stand, and they represent, together with eqs. (70), the elastic law for large deformations in the form which already incorporates the kinematic relations.

We still have to write the conditions of equilibrium. We may start from eqs. (5) and there put $p = T = 0$, since we want to study a cylinder subjected to axial load only. However, on p. 220 we have seen that it is reasonable to neglect Q_ϕ in eq. (V–1b), and consequently we neglect the Q_ϕ term in eq. (5b) of the buckling theory. There are two more terms which we may drop from eqs. (5). When we introduce N_x from eq. (70b) into eq. (5a), a term $D\,u''/a$ results. Since $D \approx Et \gg P = \sigma_x t$, the term $P\,u''/a$ in this equation is evidently of very little importance and may be dropped, and the same can be shown for the term $P\,v''/a$ in eq. (5b). The conditions of equilibrium then assume the following form:

$$N_x' + N_{\phi x}^{\cdot} = 0, \qquad N_\phi^{\cdot} + N_{x\phi}' = 0,$$

$$Q_\phi^{\cdot} + Q_x' + N_\phi + P\,\frac{w''}{a} = 0. \tag{71 a–c}$$

The elastic law (70), (V–12d–f) and the conditions of equilibrium (71), (5d, e) are 11 equations for the following 11 unknowns: the stress resultants N_ϕ, N_x, $N_{x\phi}$, M_ϕ, M_x, $M_{x\phi}$, Q_ϕ, Q_x, and the displacements u, v, w. The first step on the way to a solution is to reduce the number of unknowns and equations. This may be accomplished in the following way:

Eqs. (71a, b) are identically satisfied when we introduce AIRY's stress function $\varPhi$ in analogy to the plane stress problem, writing

$$N_x = \varPhi^{\cdot\cdot}, \quad N_\phi = \varPhi'', \quad N_{\phi x} = N_{x\phi} = -\varPhi^{\cdot'}. \tag{72}$$

When we use the stress function on the left-hand sides of eqs. (70) and eliminate u and v, we obtain a differential equation relating $\varPhi$ and w:

$$\varPhi^{\mathrm{IV}} + 2\varPhi''^{\cdot\cdot} + \varPhi^{\cdot\cdot} = \frac{D(1 - \nu^2)}{a}\left[w'' - \frac{1}{a}(w''\,w^{\cdot\cdot} - w'^{\cdot 2})\right]. \tag{73a}$$

To obtain a second equation in the same variables, we use eqs. (5d, e) to express Q_ϕ, Q_x in eq. (71c) in terms of the moments. These in turn may be expressed by derivatives of w, according to eqs. (V–12d–f), while N_ϕ is replaced by $\varPhi''$. Thus we obtain:

$$w^{\mathrm{IV}} + 2w''^{\cdot\cdot} + w^{\cdot\cdot} + \frac{P\,a^2}{K}\,w'' = -\frac{a^3}{K}\,\varPhi''. \tag{73b}$$

Since this equation has been derived from eq. (71c), the assumption that $P \gg N_x$, N_ϕ, $N_{\phi x}$ has been worked into it. This assumption is certainly correct for the linearized buckling theory, which considers $N_x, \ldots$ as infinitesimal quantities. It may, however, be challenged in the present case. If a refinement is desired, one must go back to the origin of eq. (5c) and introduce there terms with $N_x\,w''$ etc.

Eqs. (73a, b) – or whatever may be used instead – are two partial differential equations for two unknown functions of x and ϕ. Since the first of these equations contains some nonlinear terms, it is difficult to solve them. Solving eqs. (73) may be avoided by using the principle of virtual work. The differential equations describe the mechanical concept, and it is easy to write expressions for the strain energy of the direct stresses as an integral of certain derivatives of $\varPhi$, for the bending energy as a similar integral of w, and for the work done by the external forces P as another integral involving w. If a reasonable assumption is now made for w as a function of the coordinates, containing a number of free constants, eq. (73a) becomes a linear differential equation for $\varPhi$, which, with a suitable set of boundary conditions, yields a unique solution. This and the assumed w are introduced into the energy integrals, which then depend entirely on the free constants of w. The principle of virtual work requires that the derivative of the potential energy with respect to each of these constants be zero. This leads to a set of equations for these constants and thus to an approximate solution of the problem. Since only w is varied, the principle replaces only the condition of normal equilibrium, eq. (71c), while eqs. (71a, b) are taken care of by solving eq. (73a).

The tedious calculations have been made by T. v. KARMAN and H.-S. TSIEN and have been described in their paper of 1941, mentioned in the bibliography. The essential result may be seen in fig. 31 which has been extracted from their paper. The ordinate is a dimensionless representation of the load P. The quantity ϵ in the abscissa is the over-all compressive strain of the cylinder, i. e. the amount a unit length of the cylinder is shorter than before loading.

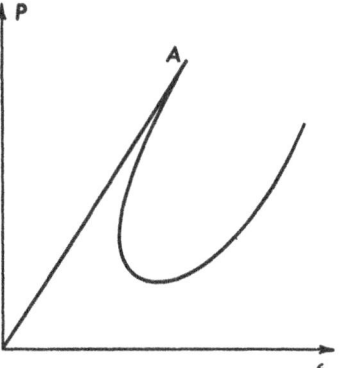

Fig. 31. Load-deformation curve for a cylinder loaded as in Fig. 29

When the load P is gradually applied, the relation of P to ϵ is first represented by the straight line corresponding to the unbuckled compression. At the point A the classical buckling load has been reached. Here an adjacent equilibrium exists with infinitesimal lateral deflections but with the same ϵ. However, for loads smaller than this one, there exists an elastic equilibrium, involving finite lateral deflections. It may even happen

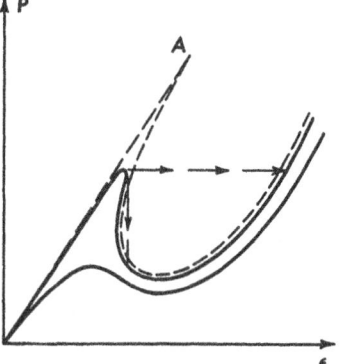

Fig. 32. Load-deformation curve for an imperfect cylinder loaded as in Fig. 29

that in this deflected state the potential energy of the shell is less than in the undeflected form. The stability of the undeflected shell is then as precarious as that of a pencil standing on its flat end; a very small disturbance suffices to bring it over the threshold and to make it drop into a position of lower energy.

One may easily imagine that the diagram for a shell with slight imperfections looks like one of the curves in fig. 32. We shall discuss these curves for two typical cases of loading, weight loading and loading in a rigid testing machine.

When the cylinder is put in an upright position and then loaded by packing weights on its upper edge, the load P is given as any instant, and the deformation ϵ is not restricted. The loading will proceed without incident, until the maximum point on the curve has been reached. When the load is further increased there is no equilibrium available except on the right branch of the curve, and the diagram point must jump in the direction of the horizontal arrows. This indicates a sudden buckling of the shell. It is seen from the figure that this occurs far below the classical buckling load and that it leads at once to a very large deflection. It may be so large that the shell locally undergoes plastic deformation. If this should not happen, elastic unloading would be possible. The diagram point would then descend the right branch until reaches the minimum point. Upon further reduction of the load it has to jump again, back to the left branch, and the buckling deflection disappears as suddenly as it came.

The phenomenon is slightly different when the cylinder is loaded in a rigid testing machine. Then ϵ is given by the position of the end plates of the machine, and the load P is determined by one of the curves of fig. 32. If the upper curve is applicable (very small imperfections), the load reaches a maximum and then drops slightly until the right-most point of the curve has been reached. When ϵ is still increased, the diagram point jumps vertically downward, indicating the sudden appearance of large deformations connected with a sharp drop in load. If the imperfections of the shell are larger, the curve looks like the lower one in fig. 32. In this case there is no rightmost point, no jumping takes place, and the buckling phenomenon develops gradually.

The diagrams of figs. 31 and 32 are drawn for one certain number m of circumferential waves forming on the buckling shell. Of course, similar curves may be drawn for other integer values of m. Each of them begins at the corresponding classical load of higher order; however, they do not lie inside each other like onion peels but intersect each other. This brings additional complications into the buckling process. The shell may (or perhaps may not) jump from one buckling configuration into another one, when this is possible with the available energy and when there is some disturbance to help it through transitional stages.

7.3 Spherical Shell

We consider a complete sphere of radius a, subjected to a uniform external pressure p. This load produces the membrane forces

$$N_{\phi I} = N_{\theta I} = -\frac{1}{2} p a,$$

which form the basic stress system. For the buckling displacements u, v, w and the corresponding stress resultants N_ϕ, N_θ, etc. we use the notation established in Chapters II and VI (see figs. VI-1, II-52).

The basic stress system has a remarkably high degree of symmetry. It is axially symmetric not only with respect to the polar axis of our coordinate system, but with respect to every diameter of the sphere. As we have seen with other shells of revolution, the buckling deformations may have a lower degree of symmetry. However, in this particular case it is advantageous to consider first only axially symmetric deformations and to postpone the discussion of less symmetric modes until later.

The condition of equilibrium of the deformed shell element must contain all those terms of eqs. (VI–1) which are not ruled out by the assumed axial symmetry of the buckling stresses. This restriction eliminates completely eqs. (VI–1 b, e, f), and in the other three we have to drop all the prime derivatives and to put $p_\phi = 0$, $p_r = -p$.

Additionally, the equations must contain terms which express the action of the load and the basic forces on a *deformed* shell element. It is our first task to find these terms.

In an axisymmetric deformation the shell element rotates by an angle $\chi = (w^\cdot - v)/a$, and the force $N_{\phi1} \, r \, d\theta$ at the upper edge develops a radial component

$$N_{\phi I} \, r \, d\theta \cdot \frac{w^\cdot - v}{a}$$

pointing inward (fig. 33). The corresponding force at the opposite edge has a similar component pointing outward, but it is larger by the differential

$$\frac{\partial}{\partial \phi} \left(N_{\phi I} \, r \, d\theta \, \frac{w^\cdot - v}{a} \right) d\phi = N_{\phi I} \left[(w^\cdot - v) \sin\phi \right]^\cdot d\phi \, d\theta \,.$$

This differential yields a new term for eq. (VI–1 c), and since we there counted forces positive when pointing inward and since we dropped the factor $d\phi \, d\theta$, the new term is

$$-N_{\phi I} \left[(w^\cdot - v) \sin\phi \right]^\cdot = \frac{1}{2} \, p \, a \left[(w^{\cdot\cdot} - v^\cdot) \sin\phi + (w^\cdot - v) \cos\phi \right].$$

As we have already seen in the membrane theory (p. 21), the forces $N_{\phi I}$ on opposite edges of the shell element have a resultant $N_{\phi I} \, r \, d\theta \cdot d\phi$ normal to the shell. When the element is rotated, this resultant participates in this rotation and develops a tangential component

$$N_{\phi I} \, r \, d\theta \, d\phi \cdot \frac{w^\cdot - v}{a} = N_{\phi I} (w^\cdot - v) \sin\phi \, d\phi \, d\theta \,.$$

It points down the meridian, and after stripping it of the factors $d\phi \, d\theta$ we must introduce it without change of sign in eq. (VI–1 a).

The hoop forces $N_{\theta I}$ do not make a similar contribution. Since the deformation is axisymmetric, the meridional edges of the element stay in meridional planes, and the hoop forces never change direction.

The load applied to the shell element is originally $p \cdot a \, d\phi \cdot r \, d\theta$. We assume that it is produced by a gas pressure. Such a load is always normal to the surface and proportional to the actual size

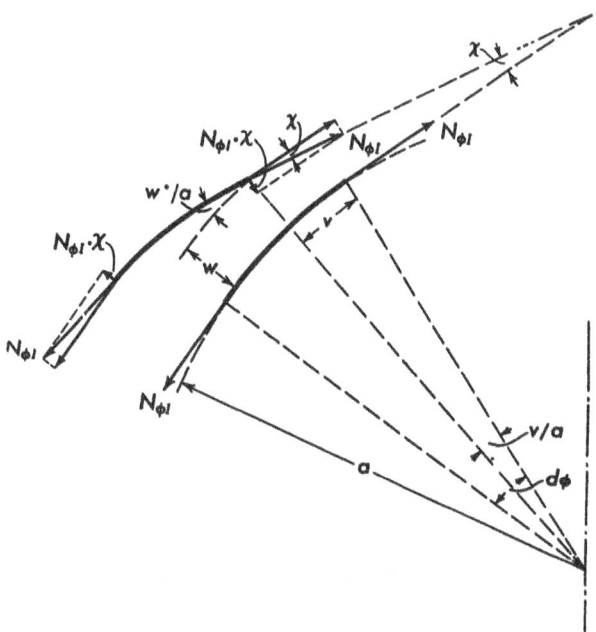

Fig. 33. Meridian of a spherical shell before and after buckling

of the element. During the buckling deformation the area $dA = a \, d\phi \cdot r \, d\theta$ of the element increases by $(\epsilon_\phi + \epsilon_\theta) \, dA$, where

$$\epsilon_\phi = \frac{v^{\cdot} + w}{a}, \qquad \epsilon_\theta = \frac{v \cot\phi + w}{a}$$

are the strains of the middle surface. The radial load on the element increases correspondingly by

$$p(\epsilon_\phi + \epsilon_\theta) \, dA = p \, a(v^{\cdot} + v \cot\phi + 2w) \sin\phi \, d\phi \, d\theta,$$

and this yields a positive contribution to eq. (VI–1c). On the other hand, the rotation χ of the element makes the load rotate and thus generates a tangential component (tangential to the original meridian!):

$$p \cdot a \, d\phi \cdot r \, d\theta \cdot \chi = p \, a(w^{\cdot} - v) \sin\phi \, d\phi \, d\theta.$$

This component yields a positive contribution to eq. (VI–1a).

Taking all these simplifications and additions together, we arrive at the following conditions of equilibrium:

$$
\left.\begin{aligned}
&(N_\phi \sin\phi)^{\boldsymbol{\cdot}} - N_\theta \cos\phi - Q_\phi \sin\phi + \frac{p}{2}(w^{\boldsymbol{\cdot}} - v)\sin\phi = 0, \\
&N_\theta \sin\phi + N_\phi \sin\phi + (Q_\phi \sin\phi)^{\boldsymbol{\cdot}} + \frac{p}{2}(v^{\boldsymbol{\cdot}}\sin\phi + v\cos\phi \\
&\qquad\qquad + w^{\boldsymbol{\cdot\cdot}}\sin\phi + w^{\boldsymbol{\cdot}}\cos\phi + 4w\sin\phi) = 0, \\
&(M_\phi \sin\phi)^{\boldsymbol{\cdot}} - M_\theta \cos\phi = a Q_\phi \sin\phi.
\end{aligned}\right\} \quad (74)
$$

We use the third of these equations to eliminate Q_ϕ from the other two, and then we use the elastic law (VI–11) to express the forces and moments in terms of v and w. When we now introduce the dimensionless parameters

$$
k = \frac{K}{D a^2}, \qquad q = \frac{p a}{2 D}, \qquad (75)
$$

we obtain at last the following pair of equations:

$$
\begin{aligned}
&(1 + k)[v^{\boldsymbol{\cdot\cdot}} + v^{\boldsymbol{\cdot}}\cot\phi - v(\nu + \cot^2\phi)] + (1 + \nu)w^{\boldsymbol{\cdot}} \\
&\quad - k[w^{\boldsymbol{\cdot\cdot}} + w^{\boldsymbol{\cdot\cdot}}\cot\phi - w^{\boldsymbol{\cdot}}(\nu + \cot^2\phi)] - q(v - w^{\boldsymbol{\cdot}}) = 0,
\end{aligned}
$$

$$
\begin{aligned}
&(1 + \nu)[v^{\boldsymbol{\cdot}} + v\cot\phi + 2w] \\
&\quad + k[-v^{\boldsymbol{\cdot\cdot\cdot}} - 2v^{\boldsymbol{\cdot\cdot}}\cot\phi + v^{\boldsymbol{\cdot}}(1 + \nu + \cot^2\phi) \qquad\qquad (76\text{a, b}) \\
&\qquad - v(2 - \nu + \cot^2\phi)\cot\phi + w^{\boldsymbol{\cdot\cdot\cdot}} + 2w^{\boldsymbol{\cdot\cdot}}\cot\phi \\
&\qquad - w^{\boldsymbol{\cdot\cdot}}(1 + \nu + \cot^2\phi) + w^{\boldsymbol{\cdot}}(2 - \nu + \cot^2\phi)\cot\phi] \\
&\quad + q[v^{\boldsymbol{\cdot}} + v\cot\phi + w^{\boldsymbol{\cdot\cdot}} + w^{\boldsymbol{\cdot}}\cot\phi + 4w] = 0.
\end{aligned}
$$

Since $k \ll 1$, we may drop the factor $(1 + k)$ from eq. (76a), and in the following calculations we shall repeatedly make similar simplifications without announcing them in each case.

When dealing with the bending theory of spherical shells, we used the operators L and H defined by eqs. (VI–16) and (VI–88). Since we want to extend our theory to cover also buckling modes which are not axisymmetric, we prefer here the operator H. In the axisymmetric case ($n = 0$) it simplifies to

$$
H(\ ) = (\ )^{\boldsymbol{\cdot\cdot}} + (\ )^{\boldsymbol{\cdot}}\cot\phi + 2(\ ).
$$

Before we introduce it into eqs. (76), we realize that the unknown variables occur in many places in the combination $v - w^{\boldsymbol{\cdot}}$, due largely to the angular displacement χ. We make good use of this situation by introducing an auxiliary variable V such that

$$
v = -V^{\boldsymbol{\cdot}}.
$$

With the help of eq. (VI–89b) it is then easily seen that eq. (76a) may be brought into the following form:

$$\frac{\partial}{\partial \phi} [H(V) + k H(w) - (1 + \nu)(V + w) - q(V + w)] = 0.$$

We conclude that the bracket does not depend on ϕ, and since v is not affected when we add a constant to V, we may simply put the bracket equal to zero:

$$H(V) + k H(w) - (1 + \nu)(V + w) - q(V + w) = 0. \quad (77a)$$

When we introduce the operator H in eq. (76b), we make use of eq. (VI–89a) and find the relation

$$k H H(V + w) - (1 + \nu) H(V) - (3 + \nu) k H(w)$$
$$+ 2(1 + \nu)(V + w) + q[-H(V) + H(w) + 2(V + w)] = 0. \quad (77b)$$

Now, any regular function of $\cos\phi$ in the interval $-1 \leq \cos\phi \leq 1$ may be expanded in a series of LEGENDRE functions $P_n(\cos\phi)$ and, therefore, the solution V, w of eqs. (77) may be assumed in the form

$$V = \sum_{n=0}^{\infty} b_n P_n(\cos\phi), \qquad w = \sum_{n=0}^{\infty} c_n P_n(\cos\phi). \quad (78)$$

Each one of the spherical harmonics satisfies the differential equation

$$P_n^{\cdot\cdot} + P_n^{\cdot} \cot\phi + n(n + 1) P_n = 0,$$

which, with our operator H, may be written as

$$H(P_n) = -\lambda_n P_n, \qquad \lambda_n = n(n + 1) - 2.$$

When we introduce the series (78) into eqs. (77), we make use of this relation and obtain:

$$\sum_{n=0}^{\infty} \{b_n [\lambda_n + 1 + \nu + q] + c_n [k \lambda_n + 1 + \nu + q]\} P_n(\cos\phi) = 0,$$

$$\sum_{n=0}^{\infty} \{b_n [k \lambda_n^2 + (1 + \nu)(\lambda_n + 2) + q(\lambda_n + 2)]$$
$$+ c_n [k \lambda_n^2 + (3 + \nu) k \lambda_n + 2(1 + \nu) - q(\lambda_n - 2)]\} P_n(\cos\phi) = 0.$$

The LEGENDRE functions form a complete set of functions. Therefore, the two series cannot vanish identically in ϕ unless each coefficient vanishes. Thus we obtain for every n a pair of homogeneous linear equations for b_n and c_n:

$$b_n [\lambda_n + 1 + \nu + q] + c_n [k \lambda_n + 1 + \nu + q] = 0,$$
$$b_n [k \lambda_n^2 + (1 + \nu)(\lambda_n + 2) + q(\lambda_n + 2)] \qquad (79)$$
$$+ c_n [k \lambda_n^2 + (3 + \nu) k \lambda_n + 2(1 + \nu) - q(\lambda_n - 2)] = 0.$$

In general, these equations have the solution $b_n = c_n = 0$. Only when, for a certain n, the determinant of the four bracketed coefficients vanishes, is a finite solution b_n, c_n possible. Eqs. (78) then each have one nonvanishing term, describing a buckling mode of the shell. The corresponding critical value of q is found from the determinantal equations. If k and q are again neglected compared with unity, this equation has the following form:

$$(1 - \nu^2)\,\lambda_n + k\,[\lambda_n^3 + 2\lambda_n^2 + (1 + \nu)^2\,\lambda_n] - q\,\lambda_n^2 = 0. \qquad (80)$$

It is satisfied for any load q, if $\lambda_n = 0$. This is a degenerated case. It leads to $n = 1$, and eqs. (79) yield $b_1 = -c_1$. Since $P_1(\cos\phi) \equiv \cos\phi$, this "buckling" mode is the rigid-body displacement

$$v = -c_1 \sin\phi, \qquad w = c_1 \cos\phi.$$

This case must, of course, be excluded when we wish to investigate the elastic instability of the shell.

If we now assume that $\lambda_n \neq 0$, we may solve eq. (80) for q and find

$$q = \frac{1 - \nu^2}{\lambda_n} + k\left[\lambda_n + 2 + \frac{(1 + \nu)^2}{\lambda_n}\right]. \qquad (81)$$

For every integer n there is a certain $\lambda_n = n(n + 1) - 2$ and a corresponding buckling mode. The shell will buckle with that mode which yields the lowest value q. To find it, we differentiate eq. (81) with respect to λ_n and set the derivative equal to zero. This yields

$$\lambda_n^2 = \frac{1 - \nu^2}{k} + (1 + \nu)^2. \qquad (82)$$

Strictly speaking, this is not the true λ_n of the buckling mode, unless it corresponds accidentally to an integer n. But since λ_n turns out to be rather large, the requirement of an integer n is not of much importance, and we may simply introduce λ_n from eq. (82) into eq. (81) to find a good approximation for the lowest critical value of q, which then is

$$q_{\min} = 2k + 2\sqrt{k}\,\sqrt{1 - \nu^2 + k(1 + \nu)^2} \approx 2\sqrt{k(1 - \nu^2)}.$$

From the definition of q, eq. (75), we find the critical pressure

$$p_{\mathrm{cr}} = \frac{2E\,t}{(1 - \nu^2)\,a}\,q_{\min} = \frac{2E\,t^2}{\sqrt{3(1 - \nu^2)}\,a^2}. \qquad (83)$$

The critical pressure is proportional to the square of the wall thickness. As may easily be seen from eq. (83), a metal shell must be pretty thin to buckle within the elastic range.

Having solved the axisymmetric problem, we can easily deal with the general case. Because of the high symmetry of the sphere and of the basic stress system the choice of the axis $\phi = 0$ of the coordinate system is entirely arbitrary. We may pick one of the axisymmetric modes and then choose a coordinate system whose axis does not coincide with the axis of symmetry of the mode. With respect to the new coordinate system all three displacements components u, v, w are different from zero and are periodic functions of θ which may be written as

$$w(\phi, \theta) = w_0(\phi) + \sum w_m(\phi) \cos m\theta + \sum \overline{w}_m(\phi) \sin m\theta$$

and similarly for u and v. They must be a solution of the differential equations for the general case of unsymmetric buckling. The coefficients of these equations may and really do depend on ϕ but certainly not on θ. Therefore, when we introduce the FOURIER series for u, v, w, we find that every single term of these series is a solution for the same eigenvalue q. We see here that the buckling modes of the type $w = w_m(\phi) \cos m\,\theta$ do not yield new eigenvalues but again those covered by eq. (81). Hence, this equation gives the complete set of eigenvalues of the problem.

The situation is, of course, less simple if the shell consists only of a spherical cap, or if the load is less symmetrical, e. g. the weight of a spherical dome. In such cases a complete treatment of the general equations would be necessary.

APPENDIX

Forces and Deformations in Circular Rings

Shells of revolution are frequently connected with circular rings, to which they transmit forces and moments. The theory of stresses and deformations of such rings is part of the theory of structures. While a few formulas are found almost everywhere, it is not easy to find the complete set in books of this kind. Therefore they have been compiled here.

In all formulas, we assume that the ring is thin, i. e., that the dimensions of its cross section are small compared with the radius. The axis of the ring is supposed to pass through the centroids of all cross sections. One principal axis of these sections lies in the plane of the ring.

Besides those explained in the figures, the following notations have been used:

$A =$ area of cross section,
$I_1 =$ moment of inertia for the centroidal axis in the plane of the ring,
$I_2 =$ moment of inertia for the centroidal axis normal to the plane of the ring,
$J_T =$ torsional rigidity factor of the section.

Where moments and angular displacements have been represented by arrows, the corkscrew rule applies for the interpretation.

1. Radial Load. (Fig. 1)

The load per unit length of the axis is assumed to be

$$p = p_n \cos n\,\theta.$$

Stress resultants:

$$N = -\frac{p_n\,a}{n^2 - 1}\cos n\,\theta, \qquad M_2 = -\frac{p_n\,a^2}{n^2 - 1}\cos n\,\theta.$$

Displacements:

$$v = -\frac{p_n\,a^2}{n(n^2 - 1)}\left(\frac{a^2}{(n^2 - 1)\,E\,I_2} + \frac{1}{E\,A}\right)\sin n\,\theta,$$

$$w = \frac{p_n\,a^4}{(n^2 - 1)^2\,E\,I_2}\cos n\,\theta.$$

These formulas are not valid for $n = 0$ and $n = 1$. For $n = 0$ (axisymmetric load) they must be replaced by the well-known formulas

$$N = p\,a, \qquad M_2 = 0,$$

$$v = 0, \qquad w = \frac{p\,a^2}{E\,A}.$$

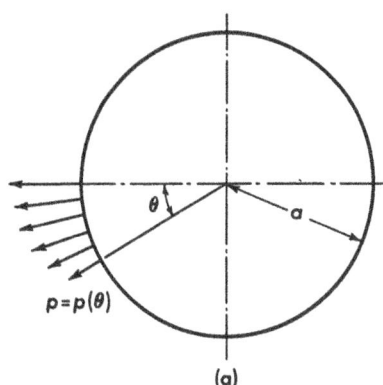

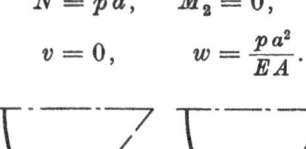

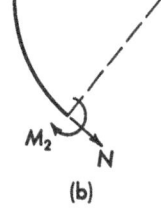

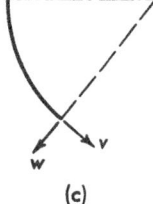

(a) (b) (c)

Fig. 1. Radial load

For $n = 1$ the problem does not exist, since a load of this type is not self-equilibrating.

2. Tangential Load. (Fig. 2)

The load per unit length of the axis is assumed to be

$$p = p_n \sin n\,\theta.$$

Stress resultants:

$$N = \frac{n\,p_n\,a}{n^2 - 1}\cos n\,\theta,$$

$$M_2 = \frac{p_n\,a^2}{n(n^2 - 1)}\cos n\,\theta.$$

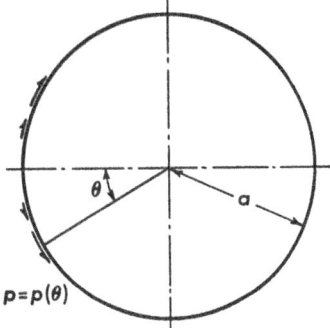

Fig. 2. Tangential load

Displacements:

$$v = \frac{p_n a^2}{n^2 - 1} \left[\frac{a^2}{n^2 (n^2 - 1) E I_2} + \frac{1}{E A} \right] \sin n\, \theta,$$

$$w = -\frac{p_n a^4}{n(n^2 - 1)^2 E I_2} \cos n\, \theta.$$

For $n = 0$ and $n = 1$ this problem does not exist, because the external forces would not be in equilibrium.

3. Load Normal to the Plane of the Ring. (Fig. 3)

The load per unit length of the axis is assumed to be

$$p = p_n \cos n\, \theta.$$

Stress resultants:

$$M_1 = \frac{p_n a^2}{n^2 - 1} \cos n\, \theta, \qquad M_T = -\frac{p_n a^2}{n(n^2 - 1)} \sin n\, \theta.$$

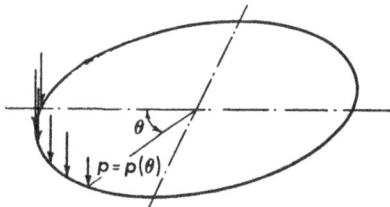

Displacements:
deflection

$$u = \frac{p_n a^4}{(n^2 - 1)^2}$$

$$\times \left[\frac{1}{E I_1} + \frac{1}{n^2 G J_T} \right] \cos n\, \theta,$$

rotation of the cross section in its plane

$$\psi = \frac{p_n a^3}{(n^2 - 1)^2}$$

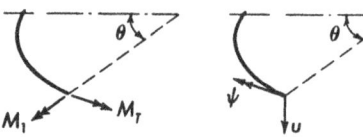

Fig. 3. Load normal to the plane of the ring

$$\times \left[\frac{1}{E I_1} + \frac{1}{G J_T} \right] \cos n\, \theta.$$

For $n = 0$ and $n = 1$ this problem does not exist, since the loads would not be self-equilibrating.

4. External Moments, Turning about the Ring Axis. (Fig. 4)

The couple applied per unit length of the axis of the ring is assumed to be

$$m = m_n \cos n\, \theta.$$

Stress resultants:

$$M_1 = -\frac{m_n a}{n^2 - 1} \cos n\, \theta,$$

$$M_T = \frac{n\, m_n a}{n^2 - 1} \sin n\, \theta.$$

Displacements:

$$u = -\frac{m_n a^3}{(n^2 - 1)^2}\left[\frac{1}{E\,I_1} + \frac{1}{G\,J_T}\right]\cos n\,\theta,$$

$$\psi = \frac{m_n a^2}{(n^2 - 1)^2}\left[\frac{1}{E\,I_1} + \frac{n^2}{G\,J_T}\right]\cos n\,\theta.$$

The case $n = 1$ does not exist, since this special load would not be in equilibrium. In the case $n = 0$ the formulas for the moments yield correctly

$$M_1 = m\,a, \qquad M_T = 0,$$

i. e. we are dealing with pure bending. Also the formula for ψ is correct and yields

$$\psi = \frac{m\,a^2}{E\,I_1},$$

but the displacement u, a rigid-body movement normal to the plane of the ring, may assume any desired value.

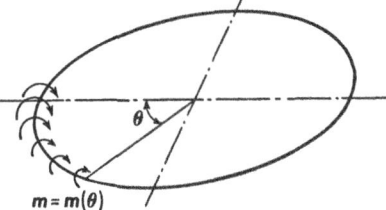

Fig. 4. Moment load

$m = m(\theta)$

In practical problems the loads are usually not applied at the ring axis but at some other circular fiber. It is then necessary to replace them by an equivalent set of loads of the kind used in the preceding formulas. This substitution must be done with some care.

BIBLIOGRAPHY

Since the author's first book on the subject appeared, the literature on stresses in shells has grown immensely and is still growing at a rapid pace. An almost complete listing of modern papers may be obtained by consulting the Zentralblatt für Mechanik, vols. 1–12 (1933–1945) and the Applied Mechanics Reviews, which cover the time since about 1947. The following bibliography contains a small choice of books and papers which either have been used as source material for this book or are considered particularly useful for those desiring more detailed information. The bibliography also contains a few papers on special problems of shell theory which are not covered in this book. Although it was unavoidable to trace the history of some pioneering thoughts, it has not been intended to make this bibliography a historical review of shell theory. This field of knowledge is much too young to be fit for historical considerations.

Chapter 1. General Properties of Stress Systems in Shells

The theory of shells has been treated in much detail in the following books: W. FLÜGGE: Statik und Dynamik der Schalen, 2nd ed., Berlin 1957; S. TIMOSHENKO, S. WOINOWSKY-KRIEGER: Theory of Plates and Shells, 2nd ed., New York 1959, pp. 429–568; K. GIRKMANN: Flächentragwerke, 5th ed., Wien 1959, pp. 352–582; R. L'HERMITE: Résistance des Matériaux, vol. 1: Théorie de l'Elasticité et des Structures Elastiques, Paris 1954, pp. 616–831; V. Z. VLASOV: General Theory of Shells (in Russian), Moscow 1949, German translation: Allgemeine Schalentheorie und ihre Anwendung in der Technik, by A. KROMM, Berlin 1958; V. V. NOVOZHILOV: The Theory of Thin Shells, English translation by P. G. LOWE and J. R. M. RADOK, Groningen, 1959; C. B. BIEZENO, R. GRAMMEL: Technische Dynamik, 2nd ed., vol. 1, Berlin 1953, pp. 502–561; O. BELLUZZI: Scienza delle Costruzioni, vol. 3, Bologna 1951, pp. 241–546. General treatments, based on three-dimensional theory, are given in A. E. H. LOVE: A Treatise on the Mathematical Theory of Elasticity, 4th ed., Cambridge, England, 1927, pp. 499–613; A. E. GREEN, W. ZERNA: Theoretical Elasticity, Oxford 1954, pp. 375–437.

Chapter 2. Direct Stresses in Shells of Revolution

Sections 2.1, 2.2. Additional examples may be found in many books on tanks and shell roofs. A nonlinear problem of large deformations has been treated by E. BROMBERG, J. J. STOKER: Non-linear theory of curved elastic sheets, Qu. Appl. Math. 3 (1945), 246–265. The authors show that in the realm of large deflections a membrane shell can satisfy a boundary condition prescribing the radial deflection. The nonlinear effect is restricted to a boundary zone, while elsewhere the linear theory applies. The subject has been studied in more detail by P. M. RIPLOG: Large deformations of symmetrically loaded shell membranes of revolution, Diss. Stanford 1956.

Section 2.3. The shape of an axisymmetric water drop has been studied by C. RUNGE, H. KÖNIG: Vorlesungen über numerisches Rechnen, Berlin 1924, p. 320. The use of Bessel functions for starting the solution was suggested by E. E. ZAJAC when he studied at Stanford. Figs. 20 and 21 have been prepared from numerical material offered by C. CODEGONE: Serbatoi a involucro uniformamente teso, Ann. Lavori Pubbl. **79** (1941), 179–183. Test results were reported by C. A. BOUMAN: Strength tests on a spheroid tank (in Dutch), De Ingenieur **53**, P (1938), 39–46. Tanks subjected to uniform gas pressure have been studied by F. TÖLKE: Über Rotationsschalen gleicher Festigkeit für konstanten Innen- oder Außendruck, Z. angew. Math. Mech. **19** (1939), 338–343. Except for the sphere, these shells need additional bulkheads, which must be acted upon by an axial force different from the resultant of the gas pressure. For the dome problem see G. MEGAREUS: Die Kuppel gleicher Festigkeit, Bauing. **20** (1939), 232–234. For the simultaneous action of dead-load and water pressure see V. DASEK: Zur Berechnung von Behälterböden gleicher Festigkeit, Beton u. Eisen **36** (1937), 54–55. A general survey was given by K. FEDERHOFER: Über Schalen gleicher Festigkeit, Bauing. **20** (1939), 366–370. The problem has been generalized by H. ZIEGLER: Kuppeln gleicher Festigkeit, Ing.-Arch. **26** (1958), 378–382. In this paper the condition $\sigma_\phi = \sigma_\theta$ has been replaced by TRESCA's yield condition.

Section 2.4. The membrane theory of shells of revolution under nonsymmetric loads begins with the paper by H. REISSNER: Spannungen in Kugelschalen (Kuppeln), MÜLLER-BRESLAU-Festschr., Leipzig 1912, pp. 181–193. The subject has been taken up by F. DISCHINGER in his article Schalen und Rippenkuppeln in F. EMPERGER: Handbuch für Eisenbeton, 4th ed., vol. 6, Berlin 1928, and by the author in his earlier book. In another paper by F. DISCHINGER: Die Rotationsschalen mit unsymmetrischer Form und Belastung, Bauing. **16** (1935), 374–381, 393–398, the first harmonic is used to find the dead-load stresses in shells with horizontal axis, and a solution is given for a hemispherical dome on an even number of supports. This paper also contains a discussion of a shell dome erected over a polygonal area and of the halfdome problem. For the first of these problems see also A. A. JAKOBSEN: Kugelschalen über vier- und vieleckigem Grundriß, Int. Assoc. Bridge Struct. Engg., Proc. **5** (1938), 1–17. A critical discussion of the general theory with many interesting details is found in two papers by C. TRUESDELL: The membrane theory of shells of revolution, Trans. Am. Math. Soc. **58** (1945), 96–166; On the reliability of the membrane theory of shells of revolution, Bull. Am. Math. Soc. **54** (1948), 994–1008. The content of section 2.4.2.4 is based on a paper by F. MARTIN: Die Membran-Kugelschale unter Einzellasten, Ing.-Arch. **17** (1949), 167–186. MARTIN's approach has been generalized by V. Z. VLASOV: Membrane theory of thin shells of revolution (in Russian), Appl. Math. Mech. (Prikl. Mat. Mekh.) **11** (1947), 397–408.

The auxiliary variable used in section 2.4.4.1 is P. NEMÉNYI's stress function, see his paper: Beiträge zur Berechnung der Schalen unter unsymmetrischer und unstetiger Belastung, Bygningsstat. Med. **8** (1936), 53–72. Further information may be found in TRUESDELL's papers just mentioned; see also W. ZERNA: Zur Membrantheorie der allgemeinen Rotationsschalen, Ing.-Arch. **17** (1949), 223–232. The connection between NEMÉNYI's and PUCHER's stress functions has been shown by E. REISSNER: Note on the membrane theory of shells of revolution, J. Math. Phys. **26** (1948), 290–293.

The discussion of the consequences of negative curvature was opened by W. FLÜGGE: Zur Membrantheorie der Drehschalen negativer Krümmung, Z. angew. Math. Mech. **25/27** (1947), 65–70. A geometric discussion of the subject was given by R. SAUER: Geometrische Bemerkungen zur Membrantheorie der negativ gekrümmten Schalen, Z. angew. Math. Mech. **28** (1948), 198–204. The details for several cases of loading have been worked out by R. RABICH: Die Membrantheorie der einschalig hyperbolischen Rotationsschalen, Bauplanung u. Bautechn. **7** (1953), 310–318, 320. A combination of several cones approximating a hyperboloid has been studied by R. E. PAULSEN: Shells of negative curvature, Diss. Stanford 1953. In this case the propagation of edge loads along characteristics is missing, but the loads at both edges depend on each other in a similar way as in a hyperboloid.

Section 2.5. The theory of the inextensional deformation of curved surfaces is found in every text on differential geometry. For shells of revolution the harmonic components of this deformation have been used as approximate vibration modes; see e. g. Lord RAYLEIGH: Theory of Sound, vol. 1, 2nd ed., London 1894, p. 402; M. J. O. STRUTT: Eigenschwingungen einer Kegelschale, Ann. Phys. **V, 17** (1933), 729–735. In a paper by W. R. DEAN: The distortion of a curved tube due to internal pressure, Phil. Mag. **VII, 28** (1939), 452–464, the extensional deformation of a toroid is studied, and it is found that in the vicinity of the top circles no such deformation exists.

Chapter 3. Direct Stresses in Cylindrical Shells

Section 3.1. Membrane forces in circular cylinders have been studied in the following papers: D. THOMA: Die Beanspruchung freitragender gefüllter Rohre durch das Gewicht der Flüssigkeit, Z. ges. Turbinenwes. **17** (1920), 49–52; D. THOMA: Spannungen in dünnen zylindrischen Gefässwänden, Föppl-Festschr., Berlin 1924, pp. 42–51; K. MIESEL: Über die Festigkeit von Kreiszylinderschalen mit nicht-achsensymmetrischer Belastung, Ing.-Arch. **1** (1930), 22–71. The theory of the barrel vault is found in two papers by U. FINSTERWALDER: Die Schalendächer des Elektrizitätswerkes in Frankfurt am Main, Beton u. Eisen **27** (1928), 205–208; Die querversteiften zylindrischen Schalengewölbe mit kreissegmentförmigem Querschnitt, Ing.-Arch. **4** (1933), 43–65; and in DISCHINGER's handbook article (see chapter 2, section 2.4). Cylinders of various cross sections have been studied in the same handbook article and by E. WIEDEMANN: Ein Beitrag zur Frage der Formgebung räumlich tragender Tonnenschalen, Ing.-Arch. **8** (1937), 301–310.

Section 3.2. Deformations connected with the membrane stress system have been studied in the papers by MIESEL (Ing.-Arch. **1**) and FINSTERWALDER (Ing.-Arch. **4**), just mentioned, and in F. DISCHINGER: Das durchlaufende ausgesteifte zylindrische Rohr oder Zeiss-Dywidag-Dach, Int. Assoc. Bridge Struct. Engg., Publ. **4** (1936), 227–248.

Section 3.3. A statically indeterminate problem has been treated in two papers by K. GIRKMANN: Zur Berechnung zylindrischer Flüssigkeitsbehälter auf Winddruck, S.-B. Akad. Wiss., Wien **141** (1932), 651–672; Berechnung zylindrischer Flüssigkeitsbehälter auf Winddruck unter Zugrundelegung beobachteter Lastverteilungen, Stahlbau **6** (1933), 45–48.

Section 3.4. The concept of the polygonal dome is due to F. DISCHINGER, see his handbook article and his paper: Theorie der Vieleckskuppeln und der Zusammenhang mit den einbeschriebenen Rotationsschalen, Beton u. Eisen **28** (1929), 100–107, 119–122, 150–156, 169–175; see also F. DISCHINGER, H. RÜSCH: Die Grossmarkthalle in Leipzig, Beton u. Eisen **28** (1929), 325–329, 341–346, 422–429, 437–442. Our presentation of the unsymmetric case follows that in the author's earlier book.

Section 3.5. The membrane theory of folded structures begins with the papers by H. CRAEMER: Allgemeine Theorie der Faltwerke, Beton u. Eisen 29 (1930), 276–281, and G. EHLERS: Die Spannungsermittlung in Flächentragwerken, Beton u. Eisen 29 (1930), 281–286, 291–296. For further information see E. GRUBER: Die Berechnung pyramidenartiger Scheibenwerke und ihre Anwendung auf Kaminkühler, Int. Assoc. Bridge Struct. Engg., Publ. 2 (1934), 206–222; J. GOLDENBLATT, E. RATZ: Berechnung von Faltwerken, welche aus Scheiben mit verschiedenen statischen Systemen bestehen, Beton u. Eisen 33 (1934), 369–371; G. WINTER, M. PEI: Hipped plate construction, J. Am. Concr. Inst. 18 (1947), 505–531; H. CRAEMER: Prismatic structures with transverse stiffeners, Concrete Constr. Engg. 45 (1950), 81–86. A comprehensive presentation of the subject with many numerical examples is found in the book by J. BORN: Faltwerke, Stuttgart 1954.

Chapter 4. Direct Stresses in Shells of Arbitrary Shape

Sections 4.1, 4.2. The usefulness of AIRY's stress function for the solution of membrane shell problems was discovered by A. PUCHER: Über den Spannungszustand in gekrümmten Flächen, Beton u. Eisen 33 (1934), 298–304. For further information see: A. PUCHER: Die Berechnung der Dehnungsspannungen von Rotationsschalen mit Hilfe von Spannungsfunktionen, Int. Assoc. Bridge Struct. Engg., Publ. 5 (1938), 275–299; A. PUCHER: Über die Spannungsfunktion beliebig gekrümmter dünner Schalen, Proc. 5th Int. Congr. Appl. Mech., Cambridge, Mass., 1938., pp. 134–139; E. REISSNER: On some aspects of the theory of thin elastic shells, Boston Soc. Civ. Eng. 42 (1955), 100–133. Finite-difference equations are used by A. PUCHER: Die Berechnung von doppelt gekrümmten Schalen mittels Differenzengleichungen, Bauing. 18 (1937), 118–123; W. FLÜGGE: Das Relaxationsverfahren in der Schalenstatik, in: Beiträge zur Angewandten Mechanik (FEDERHOFER-GIRKMANN Anniv. Vol.) Wien 1950, pp. 17–35. A method avoiding the stress function has been given by the author in his earlier book, 1st ed., pp. 97–100.

Section 4.3. The use of shells of negative curvature as roof structures originated in France. Many details of their membrane stress problem are found in the following papers: B. LAFFAILLE: Mémoire sur l'étude générale des surfaces gauches minces, Int. Assoc. Bridge Struct. Engg., Publ. 3 (1935), 295–332; F. AIMOND: Etude statique des voiles minces en paraboloide hyperbolique travaillant sans flexion, Int. Assoc. Bridge Struct. Engg., Publ. 4 (1936), 1–112; K. TESTER: Beitrag zur Berechnung der hyperbolischen Paraboloidschale, Ing.-Arch. 16 (1947), 39–44. Shells with zero GAUSSian curvature other than cylinders and right circular cones are treated in the following papers: E. TORROJA: Un nuovo tipo di muro di sostegno e le sue possibilità di calcolo, Ric. Ing. 9 (1941), 29–43 (conoid); V. MORKOVIN: Membrane stresses in shells of constant slope, Qu. Appl. Math. 2 (1944), 102–112; L. FÖPPL: Die schiefe Kreiskegelschale bei zwei Beanspruchungsarten, Z. angew. Math. Mech. 24 (1944), 195–203 (skew cone).

The deformations of shells of arbitrary shape have been studied by F. T. GEYLING: A general theory of deformations of membrane shells, Diss. Stanford 1953, see also two papers of the same title by W. FLÜGGE, F. T. GEYLING, Proc. 9th Intern. Congr. Appl. Mech., Bruxelles 1956, vol. 6, pp. 250–262, and Intern. Assoc. Bridge Struct. Engg., Publ. 17 (1957), 23–46. The basic equations have also been developed by E. BÖLCSKEI: Déformation des voiles minces, Acta Techn. Acad. Sci. Hungaricae 5 (1952), 489–506. Further papers on the subject: M. G. SALVADORI: Analysis and testing of translational shells, J. Am. Concrete Inst. 27 (1956), 1099–1114; M. W. JOHNSON, E. REISSNER: On inextensional deformations of shallow elastic shells, J. Math. Phys. 34 (1956), 335–346.

Section 4.4. The relations between the membrane forces of affine shells have been studied first by F. DISCHINGER in his handbook article (see chapter 2, section 2.4) and later in his paper: Der Spannungszustand in affinen Schalen und Raumfachwerken unter Anwendung des statischen Massenausgleichs, Bauing. **17** (1936), 228–231, 260–267, 289–295.

Chapter 5. Bending of Circular Cylindrical Shells

Section 5.1. The equations (5) and (13) were derived in this form by W. FLÜGGE: Die Stabilität der Kreiszylinderschale, Ing.-Arch. **3** (1932), 463–506. The radical simplifications underlying eqs. (12) and (15) have been introduced by L. H. DONNELL: Stability of thin-walled tubes under torsion, NACA, Techn. Rep. 479 (1933). For a critical comparison see J. KEMPNER: Remarks on DONNELL's equations, J. Appl. Mech. **22** (1955), 117–118; N. J. HOFF: The accuracy of DONNELL's equations, J. Appl. Mech. **22** (1955), 329–334; J. MOE: On the theory of cylindrical shells, explicit solution of the characteristic equation and discussion of the accuracy of various shell theories, Int. Assoc. Bridge Struct. Engg., Publ. **13** (1953), 283–296. Other forms of the cylinder equations have been suggested in the following papers: R. BYRNE: Theory of small deformations of thin elastic shells, Univ. of Calif. Publ. in Math., n. s. **2** (1944), 103–152; L. S. D. MORLEY: An improvement on DONNELL's approximation for thin-walled circular cylinders, Qu. J. Mech. Appl. Math. **12** (1959), 89–99. A thorough study of the basic relations underlying shell theory has been made by HILDEBRAND, REISSNER and THOMAS in their paper cited under chapter 6, section 1; see also M. W. JOHNSON, E. REISSNER: On the foundations of the theory of thin elastic shells, J. Math. Phys. **37** (1959), 371–392. The equations for a cylinder of arbitrary cross section were given by H. PARKUS: Die Grundgleichungen der allgemeinen Zylinderschale, Österr. Ing.-Arch. **6** (1951), 30–35. Singular solutions of the shell equations are found in W. FLÜGGE, D. A. CONRAD: Thermal singularities for cylindrical shells, Proc. 3rd US Nat. Congr. Appl. Mech., Providence, R. I., 1958, pp. 321–328.

Section 5.2. The solution in the form (19) is due to H. REISSNER: Formänderungen und Spannungen einer dünnwandigen, an den Rändern frei aufliegenden Zylinderschale, Z. angew. Math. Mech. **13** (1933), 133–138; see also C. B. BIEZENO, J. J. KOCH: Some explicit formulae, of use in the calculation of arbitrarily loaded, thin-walled cylinders, Akad. Wetensch. Amsterdam, Proc. **44** (1941), 505–512; N. J. HOFF: Boundary value problems of the thin-walled circular cylinder, J. Appl. Mech. **21** (1954), 343–350.

Section 5.3. A solution for $m = 0$ and $m = 1$ was given by E. SCHWERIN: Über die Spannungen in freitragenden, gefüllten Rohren, Z. angew. Math. Mech. **2** (1922), 340–353; and the complete solution was given by K. MIESEL: Über die Festigkeit von Kreiszylinderschalen mit nicht-achsensymmetrischer Belastung, Ing.-Arch. **1** (1930), 22–71. This latter paper also gives an approximate solution omitting the weakly damped part. The opposite approximation which drops the strongly damped part of the solution, has been treated by E. GRUBER: Die Berechnung zylindrischer, biegungssteifer Schalen unter beliebigem Lastangriff. Int. Assoc. Bridge Struct. Engg., Publ. **2** (1934), 196–204; H. WAGNER, H. SIMON: Über die Krafteinleitung in dünnwandige Zylinderschalen, Luftf.-Forschg. **13** (1936), 293–308. The boundary conditions of a plane plate have been formulated by G. KIRCHHOFF: Über das Gleichgewicht und die Bewegungen einer elastischen Scheibe, J. reine angew. Math. **40** (1850), 51–88. For the corresponding shell problem see A. B. BASSETT: On the extension and flexure of cylindrical and spherical thin elastic shells, Phil. Trans. Roy. Soc. London, A, **181** (1890), 433–480.

Section 5.4. The application of eqs. (13) to the barrel-vault problem is due to F. DISCHINGER: Die strenge Berechnung der Kreiszylinderschale in ihrer Anwendung auf die Zeiss-Dywidag-Schalen, Beton u. Eisen **34** (1935), 257–264, 283–294. In this paper orthotropic shells are also considered. The tediousness of the numerical work has led to many attempts toward simplification. A. A. JAKOBSEN: Über das Randstörungsproblem an Kreiszylinderschalen, Bauing. **20** (1939), 394–405, applied a procedure of succesive approximations, which replaces eqs. (35) by one single equation and thus avoids the constants α_j, β_j. E.-R. BERGER: Die Auflösung der charakteristischen Gleichung für Zylinderschalen durch Iteration, Beton u. Stahlbet. **48** (1953), 62–64, devised an iteration method for solving the characteristic equation.

The barrel-vault theory of section 5.4.2 is due to U. FINSTERWALDER: Die Theorie der zylindrischen Schalengewölbe System Zeiss-Dywidag und ihre Anwendung auf die Grossmarkthalle in Budapest, Int. Assoc. Bridge Struct. Engg., Publ. **1** (1932), 127–150; see also his paper in Ing.-Arch **4**, cited under chapter 3, section 3.1. The simplified barrel-vault theory of section 5.4.3 has been created by H. SCHORER: Line load action in thin cylindrical shells, Proc. Am. Soc. Civ. Eng. **61** (1935), 281–316. DONNELL's equation is used by N. J. HOFF, J. KEMPNER, F. V. POHLE: Line load applied along generators of thin-walled circular cylindrical shells of finite length, Qu. Appl. Math. **11** (1954), 411–425; R. RABICH: Die Statik der Schalenträger, Bauplanung u. Bautechn. **9** (1955), 115–125, 162–167. In the paper by W. T. MARSHALL: A method of determining the secondary stresses in cylindrical shell roofs, J. Inst. Civ. Eng. **33** (1949), 126–140, it is assumed that N_z depends linearly on the vertical coordinate, and an approximate shell theory is built on this assumption.

Because of the practical importance of the barrel-vault problem, a number of books have appeared which are essentially or exclusively devoted to this subject, sometimes including the design side. We mention: L. ISSENMANN PILARSKI: Calcul des voiles minces en béton armé, Paris 1935; R. S. JENKINS: Theory and Design of Cylindrical Shell Structures (Modern Building Techniques, Bull. 1), London 1947; H. LUNDGREN: Cylindrical Shells, vol. 1: Cylindrical Roofs, Copenhagen 1951; A. R. SPAMPINATO: Teoria y cálculo de las bóvedas cáscaras cilíndricas, Buenos Aires 1953; J. E. GIBSON, D. W. COOPER: The Design of Cylindrical Shell Roofs, New York 1954; and a book: Design of Cylindrical Concrete Shell Roofs, New York 1952, edited by the American Society of Civil Engineers. LUNDGREN's book also contains a new approach to the problem. Since the actual stresses in the barrel vault differ so thoroughly from the membrane stresses, LUNDGREN considers the shell inclusive the edge members as a simple beam and calculates σ_z from the straight-line law. With this stress he goes succesively through the shell equations and obtains at last a correction to σ_z. The iteration cycle may be repeated as often as needed. The condition for its convergence is established.

The theory of cylindrical shells has been applied to arch dams by A. GALLI: Sul calcolo delle dighe a semplice curvatura, Energ. Elettr. **19** (1942), 405–412. Cylindrical vaults of non-circular section have been treated by A. A. JAKOBSEN: Zylinderschalen mit veränderlichem Krümmungshalbmesser und veränderlicher Schalenstärke, Bauing. **18** (1937), 418–422, 436–444.

Section 5.5. The theory of the cylindrical tank of constant wall thickness may be found in many text books on mechanics of materials, usually under the heading "beam on elastic foundation". Linear variation of t was first considered by H. REISS-NER: Über die Spannungsverteilung in zylindrischen Behälterwänden, Beton u. Eisen **7** (1908), 150–155. THOMSON functions were used by E. MEISSNER: Beanspruchung und Formänderung zylindrischer Gefäße mit linear veränderlicher

Wandstärke, Vj.-Schr. Naturf. Ges. Zürich **62** (1917), 153–168; see also the author's earlier book, 2nd ed., p. 169, and M. HETÉNYI: Beams on Elastic Foundation, Ann. Arbor 1946, pp. 114–119. For t proportional $\sqrt{x}$ a power series solution has been given by A. CATTIN: Serbatoio cilindrico a sezione meridiana di spessore variabile, Ric. Ing. **7** (1939), 80–87. For t proportional x^2 a closed-form solution is possible, see K. FEDERHOFER: Berechnung der kreiszylindrischen Flüssigkeits-behälter mit quadratisch veränderlicher Wandstärke, Österr. Ing.-Arch. **6** (1951), 43–64.

Shells slightly deviating from the cylindrical form have been studied by K. FEDERHOFER: Spannungen in schwach ausgebauchten Behältern, Österr. Bau-Z. **6** (1951), 149–153; and by R. A. CLARK, E. REISSNER: On axially symmetric bending of nearly cylindrical shells of revolution, J. Appl. Mech. **21** (1956), 59–67. Nonlinear bending has been studied by B. R. BAKER: A large-deformation bending theory for thin cylindrical shells, Diss. Stanford 1959.

The theory of plasticity, in particular the idea of limit analysis, has been applied to cylindrical shells in the following papers: D. C. DRUCKER: Limit analysis of cylindrical shells under axially-symmetric loading, Proc. 1st Midwest. Conf. Solid Mech., Urbana 1953, pp. 158–163; E. T. ONAT: The plastic collapse of cylindrical shells under axially symmetric loading, Qu. Appl. Math. **13** (1955), 63–72; P. G. HODGE: The rigid-plastic analysis of symmetrically loaded cylindrical shells, J. Appl. Mech. **21** (1954), 336–342; E. T. ONAT, W. PRAGER: Limits of economy of material in shells, De Ingenieur **67**, O (1955), 46–49; W. FREIBERGER: Minimum weight design of cylindrical shells, J. Appl. Mech. **23** (1956) 576–580; P. G. HODGE, S. V. NARDO: Carrying capacity of an elastic-plastic cylindrical shell with linear strain hardening, J. Appl. Mech. **25** (1958), 79–85. A comprehensive presentation of the subject may be found in P. G. HODGE: Plastic Analysis of Structures, New York 1959, pp. 270–309.

Section 5.6. The differential equations for a very general case of anisotropy were established by W. FLÜGGE: Ing.-Arch. **3** (see section 5.1). They have been applied to barrel vaults by F. DISCHINGER, Beton u. Eisen **34** (see section 5.4). Tubular shells (airplane fuselages) have been studied by H. WAGNER, H. SIMON, Luftf.-Forschg. **13** (see section 5.2); W. SCHNELL: Krafteinleitung in versteifte Kreiszylinderschalen, Z. Flugwiss. **3** (1955), 385–399.

Section 5.7. More details about bending stresses in folded structures may be found in the books by J. BORN: Faltwerke, Stuttgart 1954; and D. YITZHAKI: The Design of Prismatic and Cylindrical Shell Roofs, Haifa 1958; and in the following papers: G. GRÜNING: Die Nebenspannungen in prismatischen Falt-werken, Ing.-Arch. **3** (1932), 319–337; E. GRUBER: Berechnung prismatischer Scheibenwerke, Int. Assoc. Bridge Struct. Engg., Publ. **1** (1932), 225–246; E. GRU-BER: Die Berechnung äusserlich statisch unbestimmter prismatischer Scheibenwerke, Int. Assoc. Bridge Struct. Engg., Publ. **3** (1935), 134–158; R. OHLIG: Beitrag zur Theorie der prismatischen Faltwerke, Ing.-Arch. **6** (1935), 346–354; R. OHLIG: Mehrfache prismatische Faltwerke, Ing.-Arch. **12** (1941), 254–258.

Chapter 6. Bending Stresses in Shells of Revolution

Section 6.1. The derivation of the differential equations may be found at many places. Most of the earlier papers on bending stresses in shells of revolution are based on the equations in LOVE's Treatise. In recent years the discussion of the basic equations has attracted many authors, see e. g. E. REISSNER: A new deriva-tion of the equations for the deformation of elastic shells, Am. J. Math. **63** (1941),

177–184; and: Note on the expressions for the strains in a bent, thin shell, Am. J. Math. **64** (1942), 768–772. The following papers use tensor notation and establish sets of basic equations for shells of arbitrary shape: J. L. SYNGE, W. Z. CHIEN: The intrinsic theory of elastic shells and plates, Kármán Anniv. Vol., Pasadena 1941, pp. 103–120; W. Z. CHIEN: The intrinsic theory of thin shells and plates, Qu. Appl. Math. **1** (1944), 297–327, **2** (1944), 43–59, 120–135; H. NEUBER: Allgemeine Schalentheorie, Z. angew. Math. Mech. **29** (1949), 97–108, 142–146; W. ZERNA: Beitrag zur allgemeinen Schalenbiegetheorie, Ing.-Arch. **17** (1949), 149–164; H. PARKUS: Die Grundgleichungen der Schalentheorie in allgemeinen Koordinaten, Österr. Ing.-Arch. **4** (1950), 160–174. The papers of this group have the merit that they provide an extremely general formulation of the theory. However, it is a long way from such general formulations to a solution applicable to a concrete engineering problem. Another very general study of basic relations is found in the paper by E. WEIBEL: The strains and the energy in thin elastic shells of arbitrary shape for arbitrary deformation, Diss. Zürich 1955. A critical study of the basic equations has been made by F. B. HILDEBRAND, E. REISSNER, G. B. THOMAS: Notes on the foundation of the theory of small displacements of orthotropic shells, NACA, Techn. Note 1833 (1949). In this paper it is assumed that the shell material is anisotropic such that the elastic properties for the third normal stress and for the transverse shears are different from those for the stresses parallel to the middle surface. This makes it possible to isolate the influence of these stresses on the deformation. It also leads to a discussion of the KIRCHHOFF forces as defined by eqs. (V–32) and on pp. 396 and 404. A further study of the shear deformation is found in a paper by P. M. NAGHDI: The effect of transverse shear deformation on the bending of elastic shells of revolution, Qu. Appl. Mech. **15** (1957), 41–52. For sandwich shells see E. REISSNER: Small bending and stretching of sandwich-type shells, NACA, Techn. Note 1832 (1949). A nonlinear theory of large deformations is given by E. REISSNER: On the theory of thin elastic shells, REISSNER Anniv. Vol., Ann Arbor 1949, pp. 231–247; and: On axisymmetrical deformation of thin shells of revolution, Proc. Sympos. Appl. Math. **3** (1950), 27–52. For a survey of linear and nonlinear edge disturbance effects see E. REISSNER: Rotationally symmetric problems in the theory of thin elastic shells, Proc. 3rd US Nat. Congr. Appl. Mech., Providence, R. I., 1958, pp. 51–69.

Section 6.2. The use of Q_ϕ and χ as unknowns, which has opened the way to workable solutions, has been introduced by H. REISSNER, MÜLLER-BRESLAU-Festschr. (see chapter 2, section 2.4). The work has been continued by E. MEISSNER: Das Elastizitätsproblem für dünne Schalen von Ringflächen-, Kugel- und Kegelform, Phys. Z. **14** (1913), 343–349; Über Elastizität und Festigkeit dünner Schalen, Vj.-Schr. Naturf. Ges. Zürich **60** (1915), 23–47; Zur Elastizität dünner Schalen, Atti Congr. Int. Mat., Bologna 1928, vol. 5. pp. 155–158. For a further study of the problem see H. MÜNZ: Ein Integrationsverfahren für die Berechnung der Biegespannungen achsensymmetrischer Schalen unter achsensymmetrischer Belastung, Ing.-Arch. **19** (1951), 103–117, 255–270. A complex form of MEISSNER's equations was given by P. M. NAGHDI, C. N. DESILVA: On the deformation of elastic shells of revolution, Qu. Appl. Math. **12** (1955), 369–374.

The solution for the sphere has been worked out in detail by L. BOLLE: Festigkeitsberechnung von Kugelschalen, Diss. Zürich 1916. Further progress was made in two papers by F. TÖLKE: Zur Integration der Differentialgleichungen der drehsymmetrisch belasteten Rotationsschale bei beliebiger Wandstärke, Ing.-Arch. **9** (1938), 282–288; Die geschlossene Integration der Differentialgleichungen der drehsymmetrisch belasteten Kugelschale durch Zylinderfunktionen, Forschg.-H. Geb. Stahlb. **6** (1943), 166–182. The idea of asymptotic integration was introduced by O. BLUMENTHAL: Über die asymptotische Integration von Differentialgleichungen mit Anwendung auf die Berechnung von Spannungen in Kugelschalen, Z. Math. Phys. **62** (1914), 343–358; see also F. B. HILDEBRAND: On asymptotic integration in shell theory, Proc. Sympos. Appl. Math. **3** (1950), 53–66. The mathematical theory is found in the book by A. ERDÉLYI: Asymptotic Expansions, New York 1956; see also R. E. LANGER: On the asymptotic solution of ordinary differential equations, Trans. Am. Math. Soc. **33** (1931), 23–64; and: On the asymptotic solution of ordinary differential equations with reference to STOKES' phenomenon about a singular point. Trans. Am. Math. Soc. **37** (1935), 397–416. The highly simplified solution of section 6.2.1.4 is due to J. W. GECKELER: Über die Festigkeit achsensymmetrischer Schalen, Forschg.-Arb. Ingwes., vol. 276, Berlin 1926. In another paper: Zur Theorie der Elastizität flacher rotationssymmetrischer Schalen, Ing.-Arch **1** (1930), 255–270, the same author gave the solution presented in sections 6.2.1.5 and 6.2.1.6. The practical importance of the problem has caused a great number of special investigations. Tank bottom problems are treated by J.-E. EKSTRÖM: Studien über dünne Schalen von rotationssymmetrischer Form und Belastung mit konstanter oder veränderlicher Wandstärke, Ing. Vetensk. Akad.-Handl. **121** (1933); J.-E. EKSTRÖM: Die allgemeine Lösung des achsensymmetrischen Schalenproblems mit besonderer Rücksicht auf elliptische Dampfkesselböden, Ing. Vetensk. Akad., Handl. **140** (1936). In a book by M. ESSLINGER: Statische Berechnung von Kesselböden, Berlin 1952, boiler ends are treated in much detail. In a paper by R. A. CLARK, E. REISSNER: On stresses and deformations of ellipsoidal shells subject to internal pressure, J. Mech. Phys. Solids **6** (1957), 63–70, it is shown that the high curvature near the equator of an oblate ellipsoid of revolution causes large bending stresses in its vicinity. The following papers are reports on strain gage measurements and other tests on boiler end shells: E. SIEBEL, F. KÖRBER: Versuche über die Anstrengung und die Formänderungen gewölbter Kesselböden mit und ohne Mannloch bei der Beanspruchung durch inneren Druck, Mitt. KWI Eisenforschg. **8** (1926), 1–51; F. KÖRBER. E. SIEBEL: Modellversuche an Kesselböden mit Bohrungen und Mannlöchern, Mitt. KWI Eisenforschg. **9** (1927), 13–32; E. SIEBEL, S. SCHWAIGERER: Neue Untersuchungen an Dampfkesselteilen und Behältern, Ver. Deutsch. Ing., Forschg.-H. **400** (1940), 1–18. Applications to vaulted dams are found in the book by F. TÖLKE: Talsperren (vol. 2.1 of Wasserkraftanlagen, edited by F. LUDIN), Berlin 1938, pp. 493–503.

A number of papers are more or less exclusively devoted to toroidal shells H. WISSLER: Festigkeitsberechnung von Ringflächenschalen, Diss. Zürich 1916 applied MEISSNER's general theory to toroids of circular meridian. K. STANGE: Der Spannungszustand einer Kreisringschale, Ing.-Arch. **2** (1931), 47–91, has carried this work further. In the following papers toroids with elliptic or similar meridians are also considered: C. B. BIEZENO, J. J. KOCH: On the elastic behaviour of the so-called Bourdon pressure gauge, Akad. Wetensk. Amsterdam, Proc. **44** (1941), 779–786, 914–920; E. REISSNER: On bending of curved thin-walled tubes, Proc. Nat. Acad. Sci: **35** (1949), 204–208; R. A. CLARK: On the theory of thin elastic toroidal shells, J. Math. Phys. **29** (1950), 146–178; R. A. CLARK, E. REISSNER: Bending of curved tubes, Advances Appl. Mech. **2** (1951), 93–122; R. A. CLARK, T. I. GILROY, E. REISSNER: Stresses and deformations of toroidal shells of elliptical cross section, J. Appl. Mech. **19** (1952), 37–48; R. A. CLARK: Asymptotic solutions of toroidal shell problems, Qu. Appl. Math. **16** (1958), 47–60.

The mathematical difficulties of the problem may be reduced by restricting the discussion to shells which are almost plane plates. Earlier papers on such shallow shells were mainly concerned with a nonlinear buckling phenomenon, and some of them are listed with chapter 7. A general theory of shallow shells has been formulated by K. MARGUERRE: Zur Theorie der gekrümmten Platte grosser Formänderung, Proc. 5th Int. Congr. Appl. Mech., Cambridge, Mass. 1939, pp. 93–101. A shallow spherical cap has been studied in detail by E. REISSNER: Stresses and small displacements of shallow spherical shells, J. Math. Phys. 25 (1946–47), 80–85, 279–300, 27 (1948), 240. In another paper by the same author: Symmetric bending of shallow shells of revolution, J. Math. Mech. 7 (1958), 121–140, linear and nonlinear edge disturbance effects are studied. A few papers on shallow shells go beyond surfaces of revolution: Shallow helicoids: E. REISSNER: On finite twisting and bending of circular ring sector plates and shallow helicoidal shells, Qu. Appl. Math. 11 (1954), 473–483; Small rotationally symmetric deformations of shallow helicoidal shells, J. Appl. Mech. 22 (1955), 31–34; Shallow spheres, cylinders, and hyperboloids: D. A. CONRAD: Singular solutions in the theory of shallow shells, Diss. Stanford 1956; W. FLÜGGE, D. A. CONRAD: A note on the calculation of shallow shells, J. Appl. Mech. 26 (1959), 683–685.

Thermal stresses have been mentioned by many authors, see e. g. H. PARKUS: Wärmespannungen in Rotationsschalen mit drehsymmetrischer Temperaturverteilung, S.-B. Österr. Akad. Wiss., Math.-Nat. Kl., IIa, 160 (1951), 1–13; and the thesis of D. A. CONRAD just cited. G. E. STRICKLAND: Temperature stresses in shells caused by local heating, Diss. Stanford 1959, goes beyond the axisymmetric cases.

Plastic deformations have been studied theoretically by F.K.G. ODQVIST: Plasticity applied to the theory of thin shells and pressure vessels, REISSNER Anniv. Vol., Ann Arbor 1948, pp. 449–460; E. T. ONAT, W. PRAGER: Limit analysis of shells of revolution, Akad. Wetensch. Amsterdam, Proc. B, 57 (1954), 534–548; R. T. SHIELD, D. C. DRUCKER: Limit strength of thin walled pressure vessels with ASME standard torispherical head, Proc. 3dr US Nat. Congr. Appl. Mech., Providence, R. I., 1958, pp. 665–672; D. C. DRUCKER, R. T. SHIELD: Limit analysis of symmetrically loaded thin shells of revolution, J. Appl. Mech. 26 (1959), 61–68; see also the book by P. G. HODGE, mentioned under chapter 5, section 5.5. Tests on boiler ends stressed beyond the elastic limit were made by E. SIEBEL, A. POMP: Der Zusammenhang zwischen der Spannungsverteilung und der Fließlinienbildung an Kesselböden mit und ohne Mannloch bei der Beanspruchung durch inneren Überdruck, Mitt. KWI Eisenforschg. 8 (1926), 63–77.

The bending theory of the conical shell begins with F. DUBOIS: Über die Festigkeit der Kegelschale, Diss. Zürich 1917; and E. HONEGGER: Festigkeitsberechnung von Kegelschalen mit linear veränderlicher Wandstärke, Diss. Zürich 1919. Further contributions are: F. KANN: Kegelförmige Behälterböden, Dächer und Silotrichter, Forschg.-Arb. Eisenbet. 29, Berlin 1921; E. LICHTENSTERN: Die biegungsfeste Kegelschale mit linear veränderlicher Wandstärke, Z. angew. Math. Mech. 12 (1932), 347–350; C. TOLOTTI: Sul calcolo delle molle Belleville o discoidali, Rend. Mat. Univ. Roma, V, 1 (1940), 65–83; J. L. MERIAM: Stresses and displacements in a rotating conical shell, Yale Univ., School of Engg. Publ. no. 73 (1942); J. H. HUTH: Thermal stresses in conical shells, J. Aeron. Sci. 20 (1953), 613–616.

Section 6.3. The general solution for spherical shells was first established by A. HAVERS: Asymptotische Biegetheorie der unbelasteten Kugelschale, Ing.-Arch. 6 (1935), 282–312. A further contribution was made by A. A. JAKOBSEN: Beitrag zur Theorie der Kugelschale auf Einzelstützen, Ing.-Arch. 8 (1937), 275–294. For conical shells see H. NOLLAU: Der Spannungszustand der biegungssteifen Kegelschale mit linear veränderlicher Wandstärke unter beliebiger Belastung, Z. angew. Math. Mech. 24 (1944), 10–34; N. J. HOFF: Thin circular conical shells under arbitrary loads, J. Appl. Mech. 22 (1955), 557–562. Shells of arbitrary meridian have been treated by F. A. LECKIE: Bending theory for shells of revolution subjected to nonsymmetric edge loads, Diss. Stanford 1957. The toroid has been studied by C. R. STEELE: Toroidal shells with nonsymmetric loading. Diss., Stanford 1959.

Chapter 7. Buckling of Shells

Section 7.1. The thoughts presented in this section may be found in many books on the theory of elasticity.

Section 7.2. For uniform axial load P, uniform radial load p and combinations of both, the differential equations have been formulated and solutions have been found in the following papers:

Axial load only: R. LORENZ: Die nichtachsensymmetrische Knickung dünnwandiger Hohlzylinder, Phys. Z. **12** (1911), 241–260; E. CHWALLA: Das rotationssymmetrische Ausbeulen axial gedrückter freier Flanschenrohre, Z. angew. Math. Mech. **10** (1930), 72–77; Axial load on rectangular panels: K. MARGUERRE: Der Einfluss der Lagerungsbedingungen und der Formgenauigkeit auf die kritische Druckkraft des gekrümmten Plattenstreifens, Jb. deutsch. Luftf.-Forschg. 1940, vol. 1, pp. 867–872; Radial load only: R. V. SOUTHWELL: On the collapse of tubes by external pressure, Phil. Mag. **25** (1913), 687–698, **26** (1913), 502–511, **29** (1915), 67–77; R. v. MISES: Der kritische Außendruck zylindrischer Rohre, Z. Ver. Deutsch. Ing. **58** (1914), 750–755; Axial and radial loading, $P = p\,a/2$: R. v. MISES: Der kritische Außendruck für allseits belastete zylindrische Rohre, STODOLA-Festschr., Zürich 1929, pp. 418–430; Arbitrary combination of P and p: K. v. SANDEN, F. TÖLKE: Über Stabilitätsprobleme dünner, kreiszylindrischer Schalen, Ing.-Arch. **3** (1932), 24–66; W. FLÜGGE: Die Stabilität der Kreiszylinderschale, Ing.-Arch. **3** (1932), 463–506. The presentation in this book largely follows the latter paper. The simplified plot in fig. 13 is due to A. KROMM, see the last of his four papers mentioned below.

Cylinders under constant pressure p but with a wall thickness varying in steps, have been studied by C. B. BIEZENO, J. J. KOCH: The buckling of a cylindrical tank of variable thickness under external pressure, Proc. 5th Int. Congr. Appl. Mech., Cambridge, Mass., 1939, pp. 34–39; Strengthening cylindrical tanks of variable thickness under external pressure by circular stiffening rings, J. Appl. Mech. **7** (1940), 106–108; H. EBNER: Theoretische und experimentelle Untersuchung über das Einbeulen zylindrischer Tanks durch Unterdruck, Stahlbau **21** (1952), 153–159. For axial load and continuously variable thickness, but axisymmetric deflections only, see K. FEDERHOFER: Stabilität der Kreiszylinderschale mit veränderlicher Wandstärke, Österr. Ing.-Arch. **6** (1952), 277–288. A few attempts to penetrate the problem of plastic buckling are found in these papers: J. W. GECKELER: Plastisches Knicken der Wandung von Hohlzylindern und einige andere Faltungserscheinungen an Schalen und Blechen, Z. angew. Math. Mech. 8 (1928), 341–352; W. KAUFMANN: Plastisches Knicken dünnwandiger Hohlzylinder infolge axialer Belastung, Ing.-Arch. **6** (1935), 334–337.

Cylinders in pure shear (torsion) were studied by E. SCHWERIN: Die Torsionsstabilität des dünnwandigen Rohres, Z. angew. Math. Mech. **5** (1925), 235–243, also Proc. 1st Int. Congr. Appl. Mech., Delft 1924, pp. 255–265; L. H. DONNELL: NACA Rep. 479 (see chapter 5, section 5.1). In this latter paper the drastic simplifications were introduced which led to eqs. (V–15) of the stress problem and which may also be used with the stability problem, subject to the same advantages and objections. This work has been carried on by S. B. BATDORF: A simplified method of elastic-stability analysis for thin cylindrical shells, NACA Rep. 874 (1947).

Several cases of combined loads including shear have been treated in the following papers by A. KROMM: Die Stabilitätsgrenze eines gekrümmten Plattenstreifens bei Beanspruchung durch Schub- und Längskräfte, Luftf.-Forschg. 15 (1938), 517–526; Knickfestigkeit gekrümmter Plattenstreifen unter Schub- und Druckkräften, Jb. deutsch. Luftf.-Forschg. 1940, vol. 1, pp. 832–840; Beulfestigkeit von versteiften Zylinderschalen mit Schub und Innendruck, Jb. deutsch. Luftf.-Forschg. 1942, vol. 1, pp. 596–601; Die Stabilitätsgrenze der Kreiszylinderschale bei Beanspruchung durch Schub- und Längskräfte, Jb. deutsch. Luftf.-Forschg. 1942, vol. 1, pp. 602–616.

Nonuniform axial compression has been treated by W. FLÜGGE, Ing.-Arch. 3 (see above). When a shell is subjected to axisymmetric bending stresses, it may still have a buckling problem. This has been worked out for a special case by H. V. HAHNE: A stability problem of a cylindrical shell subject to direct and bending stresses, Diss. Stanford 1954.

Beam-column type problems of circular cylinders have been investigated by L. FÖPPL: Achsensymmetrisches Ausknicken zylindrischer Schalen, S.-B. Bayr. Akad. Wiss. 1926, 27–40; and in the paper by J. W. GECKELER (Z. angew. Math. Mech. 6), mentioned before. Both authors restricted the theory to axisymmetric cases. The linear theory of general imperfections was given by W. FLÜGGE, Ing.-Arch. 3.

Nonlinear theories of instability have been presented by L. H. DONNELL: A new theory for the buckling of thin cylinders under axial compression and bending, Trans. Am. Soc. Mech. Eng. 56 (1934), 795–806; H. S. TSIEN: The buckling of thin cylindrical shells under axial compression, J. Aeron. Sci. 9 (1942), 373–384, L. H. DONNELL, C. C. WAN: Effect of imperfections on buckling of thin cylinders and columns under axial compression, J. Appl. Mech. 17 (1950), 73–83; T. T. LOO: Effects of large deflections and imperfections on the elastic buckling of cylinders under torsion and axial compression, Proc. 2nd US Nat. Congr. Appl. Mech., Ann Arbor 1954, pp. 345–357; W. A. NASH: Buckling of initially imperfect clamped-end shells subject to torsion, Engg. Ind. Exp. Sta., Univ. of Florida, 1956. A review of the state of this problem and much new material are presented by W. THIELEMANN: New developments in the non-linear theories of buckling of thin cylindrical shells, Proc. Durand Centennial Conference, Stanford 1959.

Another nonlinear problem occurs when a long, thinwalled tube is subjected to pure bending. The circular cross section assumes an oval shape, and the bending moment increases less than proportional to the deflection, reaches a maximum, and then decreases. The maximum defines the collapse load of the shell. This problem was studied first by L. G. BRAZIER: On the flexure of thin cylindrical shells and other thin sections, Proc. Roy. Soc. London A 116 (1927), 104–114. Further papers on this "Brazier effect": E. CHWALLA: Elastostatische Probleme schlanker dünnwandiger Rohre mit gerader Achse, S.-B. Akad. Wiss. Wien, IIa, 140 (1931), 163–198; E. CHWALLA: Reine Biegung schlanker, dünnwandiger Rohre, Z. angew. Math., Mech. 13 (1933), 48–53; O. BELLUZZI: Un caso di instabilità per ovalizzazione nei tubi sollecitati a flessione, Ric. Ing. 1 (1933), 70–87. The idea has been applied to barrel vaults in two other papers by O. BELLUZZI: La stabilità dell'equilibrio delle volte a botte inflesse secondo le generatrici, Ric. Ing. 1 (1933), 124–129; Sulla stabilità dell'equilibrio delle volte Zeiss e Dywidag, Ric. Ing. 3 (1935), 35–40.

Rectangular cylindrical panels surrounded by stiffeners do not collapse as readily as an EULER column when the buckling load has been reached. As with flat plates, there exists a problem of post-buckling behavior. The following papers may be consulted about it: T. E. SCHUNCK: Der zylindrische Schalenstreifen ober-

halb der Beulgrenze, Ing.-Arch. 16 (1948), 403–432; D. A. M. LEGGETT, R. P. N. JO-
NES: The behaviour of a cylindrical shell under axial compression when the buckling load has been exceeded, Aeron. Res. Comm., Rep. Mem. 2190 (1942); H. L. LANG-
HAAR, A. P. BORESI: Buckling and post-buckling behavior of a cylindrical shell
subjected to external pressure, Univ. of Illinois, 1956; H. F. MICHIELSEN: The
behavior of thin cylindrical shells after buckling under axial compression, J. Aeron.
Sci. 15 (1948), 738–744; J. KEMPNER: Postbuckling behavior of axially compressed
circular cylindrical shells, J. Aeron. Sci. 21 (1954), 329–335.

Section 7.8. The stability of the spherical shell under external pressure has
been studied by R. ZOELLY: Über ein Knickungsproblem an der Kugelschale, Diss.
Zürich 1915; E. SCHWERIN: Zur Stabilität der dünnwandigen Hohlkugel unter
gleichmäßigem Außendruck, Z. angew. Math. Mech. 2 (1922), 81–91. Both authors
consider only axisymmetric deformations. The complete solution of the problem
was given by A. V. D. NEUT: The elastic stability of the thin-walled sphere (in
Dutch), Diss. Delft 1932. Further papers on the subject: L. S. LEIBENSON: Die
Anwendung der Methode der harmonischen Funktionen von W. THOMSON bei der
Frage der Stabilität der gepressten sphärischen Hüllen, Rec. math. Moscou 40
(1933), 429–442; K. O. FRIEDRICHS: On the minimum buckling load for spherical
shells, Kármán Anniv. Vol., Pasadena 1941, pp. 258–272.

A shallow spherical cap with a load acting in its convex side presents a nonlinear stability problem. The following papers may be consulted about it: C. B. BIE-
ZENO: Über die Bestimmung der Durchschlagskraft einer schwach gekrümmten
kreisförmigen Platte, Z. angew. Math. Mech. 15 (1935), 10–22; T. v. KÁRMÁN,
H. S. TSIEN: The buckling of spherical shells by external pressure, J. Aeron. Sci. 7
(1939), 43–50; E. L. REISS, H. J. GREENBERG, H. B. KELLER: Nonlinear deflections of shallow spherical shells, J. Aeron. Sci. 24 (1957), 533–543.

The stability of a hyperbolic paraboloid (fig. IV-9) under its own weight has
been studied by A. RALSTON: On the problem of buckling of a hyperbolic parab-
oloidal shell loaded by its own weight, J. Math. Phys. 35 (1956), 53–59.

INDEX

The manufacturer's authorised representative in the EU is Springer
Nature Customer Service Centre GmbH, Europaplatz 3, 69115 Heidelberg,
Germany. If you have any concerns regarding our products, please
contact ProductSafety@springernature.com

Printed and bound by CPI Group (UK) Ltd, Croydon, CR0 4YY
28/04/2026
02098508-0004